Smart Innovation, Systems and Technologies

Volume 364

Series Editors

Robert J. Howlett, KES International Research, Shoreham-by-Sea, UK

Lakhmi C. Jain, KES International, Shoreham-by-Sea, UK

The Smart Innovation, Systems and Technologies book series encompasses the topics of knowledge, intelligence, innovation and sustainability. The aim of the series is to make available a platform for the publication of books on all aspects of single and multi-disciplinary research on these themes in order to make the latest results available in a readily-accessible form. Volumes on interdisciplinary research combining two or more of these areas is particularly sought.

The series covers systems and paradigms that employ knowledge and intelligence in a broad sense. Its scope is systems having embedded knowledge and intelligence, which may be applied to the solution of world problems in industry, the environment and the community. It also focusses on the knowledge-transfer methodologies and innovation strategies employed to make this happen effectively. The combination of intelligent systems tools and a broad range of applications introduces a need for a synergy of disciplines from science, technology, business and the humanities. The series will include conference proceedings, edited collections, monographs, handbooks, reference books, and other relevant types of book in areas of science and technology where smart systems and technologies can offer innovative solutions.

High quality content is an essential feature for all book proposals accepted for the series. It is expected that editors of all accepted volumes will ensure that contributions are subjected to an appropriate level of reviewing process and adhere to KES quality principles.

Indexed by SCOPUS, EI Compendex, INSPEC, WTI Frankfurt eG, zbMATH, Japanese Science and Technology Agency (JST), SCImago, DBLP.

All books published in the series are submitted for consideration in Web of Science.

Pradeep Kumar Jha · Brijesh Tripathi ·
Elango Natarajan · Harish Sharma
Editors

Proceedings of Congress on Control, Robotics, and Mechatronics

CRM 2023

Editors
Pradeep Kumar Jha
Department of Mechanical and Industrial
Engineering
Indian Institute of Technology Roorkee
Roorkee, India

Elango Natarajan
Department of Mechanical
and Mechatronics Engineering
UCSI University
Kuala Lumpur, Malaysia

Brijesh Tripathi
Department of Mechanical Engineering
Rajasthan Technical University
Kota, Rajasthan, India

Harish Sharma
Department of Computer Engineering
Rajasthan Technical University
Kota, Rajasthan, India

ISSN 2190-3018　　　　　　　ISSN 2190-3026　(electronic)
Smart Innovation, Systems and Technologies
ISBN 978-981-99-5520-6　　　ISBN 978-981-99-5180-2　(eBook)
https://doi.org/10.1007/978-981-99-5180-2

© The Editor(s) (if applicable) and The Author(s), under exclusive license to Springer Nature Singapore Pte Ltd. 2024

This work is subject to copyright. All rights are solely and exclusively licensed by the Publisher, whether the whole or part of the material is concerned, specifically the rights of translation, reprinting, reuse of illustrations, recitation, broadcasting, reproduction on microfilms or in any other physical way, and transmission or information storage and retrieval, electronic adaptation, computer software, or by similar or dissimilar methodology now known or hereafter developed.
The use of general descriptive names, registered names, trademarks, service marks, etc. in this publication does not imply, even in the absence of a specific statement, that such names are exempt from the relevant protective laws and regulations and therefore free for general use.
The publisher, the authors, and the editors are safe to assume that the advice and information in this book are believed to be true and accurate at the date of publication. Neither the publisher nor the authors or the editors give a warranty, expressed or implied, with respect to the material contained herein or for any errors or omissions that may have been made. The publisher remains neutral with regard to jurisdictional claims in published maps and institutional affiliations.

This Springer imprint is published by the registered company Springer Nature Singapore Pte Ltd.
The registered company address is: 152 Beach Road, #21-01/04 Gateway East, Singapore 189721, Singapore

Paper in this product is recyclable.

Preface

This book contains outstanding research papers as the proceedings of the Congress on Control, Robotics, and Mechatronics (CRM2023). CRM 2023 has been organized by the Modi Institute of Technology, Kota, during March 25–26, 2023. CRM 2023 was technically sponsored by Soft Computing Research Society, India. The conference was conceived as a platform for disseminating and exchanging ideas, concepts, and results of the researchers from academia and industry to develop a comprehensive understanding of the challenges of the advancements on Control, Robotics, and Mechatronics. This book will help in strengthening amiable networking between academia and industry. The conference focused on Control, Robotics, Mechatronics and their applications.

We have tried our best to enrich the quality of the CRM 2023 through a stringent and careful peer-review process. CRM 2023 received many technical contributed articles from distinguished participants from home and abroad. After a very stringent peer-reviewing process, only 53 high-quality papers were finally accepted for presentation and the final proceedings. This book presents novel contributions to Control, Robotics, and Mechatronics and serves as reference material for advanced research.

Roorkee, India	Pradeep Kumar Jha
Kota, India	Brijesh Tripathi
Kuala Lumpur, Malaysia	Elango Natarajan
Kota, India	Harish Sharma

Contents

1 **Image-Based Weld Joint Type Classification Using Bag of Visual Words** 1
Satish Sonwane, Shital Chiddarwar, Mohsin Dalvi, and M. R. Rahul

2 **Speech Recognition-Based Prediction for Mental Health and Depression: A Review** 13
Priti Gaikwad and Mithra Venkatesan

3 **A Strategic Technique for Optimum Placement and Sizing of Distributed Generator in Power System Networks Employing Genetic Algorithm** 25
Prashant, Nirmal Kumar Agarwal, Sanjiba Kumar Bisoyi, Arun Kumar Rawat, and M. P. Kishore

4 **Model-Based Neural Network for Predicting Strain-Rate Dependence of Tensile Ductility of High-Performance Fibre-Reinforced Cementitious Composite** 43
Diu-Huong Nguyen and Ngoc-Thanh Tran

5 **Performance Analysis of Various Machine Learning Classifiers on Diverse Datasets** 53
Y. Jahnavi, V. Lokeswara Reddy, P. Nagendra Kumar, N. Sri Sishvik, and M. Srinivasa Prasad

6 **Three-Finger Robotic Gripper for Irregular-Shaped Objects** 63
Shripad Bhatlawande, Mahi Ambekar, Siddhi Amilkanthwar, and Swati Shilaskar

7 **Prediction of Energy Absorption Capacity of High-Performance Fiber-Reinforced Cementitious Composite** 77
Ngoc-Minh-Phuong To and Ngoc-Thanh Tran

8   Data Analysis on Determining False Movie Ratings 85
Ridhika Sahni and Karmel Arockiasamy

9   Power Consumption Saving of Air Conditioning System Using
Semi-indirect Evaporative Cooling 95
Manish Singh Bharti, Alok Singh, T. Ravi Kiran,
and K. Viswanath Allamraju

10   Recrudesce: IoT-Based Embedded Memories Algorithms
and Self-healing Mechanism 113
Vinita Mathur, Aditya Kumar Pundir, Raj Kumar Gupta,
and Sanjay Kumar Singh

11   Design and Performance Analysis of InSb/InGaAs/InAlAs
High Electron Mobility Transistor for High-Frequency
Applications 121
Prajjwal Rohela, Sandeep Singh Gill, and Balwinder Raj

12   Ant Lion Optimizer with Deep Transfer Learning Model
for Diabetic Retinopathy Grading on Retinal Fundus Images 133
R. Presilla and Jagadish S. Kallimani

13   Finite Element Simulation on Ballistic Impact of Bullet
on Metal Plate 147
Moreshwar Khodke, Milind Rane, Abhishek Suryawanshi,
Omkar Sonone, Bhushan Shelavale, Pranav Shinde,
and Aman Sheikh

14   Performance Analysis of Bionic Swarm Optimization
Techniques for PV Systems Under Continuous Fluctuation
of Irradiation Conditions 159
Shaik Rafi Kiran, CH Hussaian Basha, M. Vivek,
S. K. Kartik, N. L. Darshan, A. Darshan Kumar, V. Prashanth,
and Madhumati Narule

15   Study of Voltage-Controlled Oscillator for the Applications
in K-Band and the Proposal of a Tunable VCO 175
Rajni Prashar and Garima Kapur

16   Detailed Performance Study of Data Balancing Techniques
for Skew Dataset Classification 187
Vaibhavi Patel and Hetal Bhavsar

17   Compact Dual-Band Printed Folded Dipole for WLAN
Applications 203
Abhishek Javali

18   Performance Analysis of Various Feature Extraction Methods
for Classification of Pox Virus Images 211
K. P. Haripriya and H. Hannah Inbarani

19	**An Elucidative Review on the Current Status and Prospects of Eye Tracking in Spectroscopy** .... V. Muneeswaran, P. Nagaraj, L. Anuradha, V. Lekhana, G. Vandana, and K. Sushmitha	225
20	**Speech Recognition and Its Application to Robotic Arms** .... V. P. Prarthana, G. Sahana, A. Sheetal Prasad, and M. S. Thrupthi	237
21	**Machine Learning Robustness in Predictive Maintenance Under Adversarial Attacks** .... Nikolaos Dionisopoulos, Eleni Vrochidou, and George A. Papakostas	245
22	**Modelling and Grasping Analysis of an Underactuated Four-Fingered Robotic Hand** .... Deepak Ranjan Biswal, Alok Ranjan Biswal, Rasmi Ranjan Senapati, Abinash Bibek Dash, Shibabrata Mohapatra, and Poonam Prusty	255
23	**Design and Development of Six-Axis Robotic Arm for Industrial Applications** .... D. Teja Priyanka, G. Narasimha Swamy, V. Naga Prudhvi Raj, E. Naga Lakshmi, M. Maha Tej, and M. Purna Jayanthi	273
24	**A Solution to Collinear Problem in Lyapunov-Based Control Scheme** .... Kaylash Chaudhary, Avinesh Prasad, Vishal Chand, Ahmed Shariff, and Avinesh Lal	285
25	**A Detailed Review of Ant Colony Optimization for Improved Edge Detection** .... Anshu Mehta and Deepika Mehta	297
26	**Machine Learning-Based Sentiment Analysis of Twitter COVID-19 Vaccination Responses** .... Vishal Shrivastava and Satish Chandra Sudhanshu	311
27	**Exploring Sentiment in Tweets: An Ordinal Regression Analysis** .... Vishal Shrivastava and Dolly	331
28	**Automated Classification of Alzheimer's Disease Stages Using T1-Weighted sMRI Images and Machine Learning** .... Nand Kishore and Neelam Goel	345
29	**Employing Tuned VMD-Based Long Short-Term Memory Neural Network for Household Power Consumption Forecast** ..... Sandra Petrovic, Vule Mizdrakovic, Maja Kljajic, Luka Jovanovic, Miodrag Zivkovic, and Nebojsa Bacanin	357

30	**A 4-element Dual-Band MIMO Antenna for 5G Smartphone** Preeti Mishra and Kirti Vyas	373
31	**Optimization of Controller Parameters for Load Frequency Control Problem of Two-Area Deregulated Power System Using Soft Computing Techniques** Dharmendra Jain, M. K. Bhaskar, and Manish Parihar	385
32	**Quantization Effects on a Convolutional Layer of a Deep Neural Network** Swati, Dheeraj Verma, Jigna Prajapati, and Pinalkumar Engineer	403
33	**Non-linear Fractional Order Fuzzy PD Plus I Controller for Trajectory Optimization of 6-DOF Modified Puma-560 Robotic Arm** Himanshu Varshney, Jyoti Yadav, and Himanshu Chhabra	415
34	**Solving the Capacitated Vehicle Routing Problem (CVRP) Using Clustering and Meta-heuristic Algorithm** Mohit Kumar Kakkar, Gourav Gupta, Neha Garg, and Jajji Singla	433
35	**Drone Watch: A Novel Dataset for Violent Action Recognition from Aerial Videos** Nitish Mahajan, Amita Chauhan, Harish Kumar, Sakshi Kaushal, and Sarbjeet Singh	445
36	**Performance Analysis of Different Controller Schemes of Interval Type-2 Fuzzy Logic in Controlling of Mean Arterial Pressure During Infusion of Sodium Nitroprusside in Patients** Ayushi Mallick, Jyoti Yadav, Himanshu Chhabra, and Shivangi Agarwal	461
37	**Early Detection of Alzheimer's Disease Using Advanced Machine Learning Techniques: A Comprehensive Review** Subhag Sharma, Tushar Taggar, and Manoj Kumar Gupta	477
38	**Navigation of a Compartmentalized Robot Fixed in Globally Rigid Formation** Riteshni Devi	487
39	**Motion Planning and Navigation of a Dual-Arm Mobile Manipulator in an Obstacle-Ridden Workspace** Prithvi Narayan, Yuyu Huang, and Ogunmokun Olufeni	501
40	**A Real-Time Fall Detection System Using Sensor Fusion** Moape Kaloumaira, Geffory Scott, Asesela Sivo, Mansour Assaf, Shiu Kumar, Rahul Ranjeev Kumar, and Bibhya Sharma	513

Contents

41 A Vision-Based Feature Extraction Techniques for Recognizing Human Gait: A Review 529
Babita D. Sonare and Deepika Saxena

42 Right Ventricle Volumetric Measurement Techniques for Cardiac MR Images 539
Anjali Abhijit Yadav and Sanjay R. Ganorkar

43 Statistical Evaluation of Classification Models for Various Data Repositories 551
V. Lokeswara Reddy, B. Yamini, P. Nagendra Kumar, M. Srinivasa Prasad, and Y. Jahnavi

44 Hierarchical Clustering-Based Synthetic Minority Data Generation for Handling Imbalanced Dataset 561
Abhisar Sharma, Anuradha Purohit, and Himani Mishra

45 Regenerative Braking in an EV Using Buck Boost Converter and Hill Climb Algorithm 577
Vandana Kumari Prajapati, Arya Jha, C. R. Amrutha Varshini, and P. V. Manitha

46 An Enhanced Classification Model for Depression Detection Based on Machine Learning with Feature Selection Technique 589
Praveen Kumar Mannepalli, Pravin Kulurkar, Vaishali Jangade, Ayesha Khan, and Pardeep Singh

47 Design & Analysis of Grey Wolf Optimization Algorithm Based Optimal Tuning of PID Structured TCSC Controller 603
Geetanjali Meghwal, Shruti Bhadviya, and Abhishek Sharma

48 Design of an Adaptive Neural Controller Applied to Pressure Control in Industrial Processes 621
Lucas Vera, Adela Benítez, Enrique Fernández Mareco, and Diego Pinto Roa

49 A Comparative Analysis of Real-Time Sign Language Recognition Methods for Training Surgical Robots 641
Jaya Rubi, R. J. Hemalatha, I. Infant Francis Geo, T. Marutha Santhosh, and A. Josephin Arockia Dhivya

50 Design and Development of Rough Terrain Vehicle Using Rocker-Bogie Mechanism 649
Vankayala Sri Naveen, Veerapalli Kushin, Kudimi Lohith Kousthubam, Kudimi Lokesh Nandakam, R. S. Nakandhrakumar, and Ramkumar Venkatasamy

51	**Development of Swarm Robotics System Based on AI-Based Algorithms** 661
	Aniket Nargundkar, Shreyansh Pathak, Anurodh Acharya, Arya Das, and Deepak Dharrao
52	**Application of Evolutionary Algorithms for Optimizing Wire and Arc Additive Manufacturing Process** 671
	Vikas Gulia and Aniket Nargundkar
53	**Healthcare System Based on Body Sensor Network for Patient Emergency Response with Monitoring and Motion Detection** 685
	Maaz Ahmed, Diptesh Saha, Aditya Pratap Singh, Gunjan Gond, and S. Divya

About the Editors

Dr. Pradeep Kumar Jha is presently working as Professor at the Department of Mechanical and Industrial Engineering. After receiving bachelor's and master's degrees from M.I.T. Muzaffarpur, Bihar, and National Institute of Foundry and Forge Technology, Ranchi, Jharkhand, in the year 1995 and 1999, respectively, Prof. Jha obtained his Ph.D. degree from IIT Kharagpur in the year 2004. Prior to joining as Assistant Professor in IIT Roorkee in the year 2007, he also served as Sr. Lecturer and Assistant Professor at IIT Guwahati and IIT (ISM) Dhanbad, respectively. Prof. Jha specializes in the areas of manufacturing engineering, with special emphasis on casting operations, modeling and simulation of continuous casting operations, and metal matrix composites. He has his credit in more than 100 publications in international and national journals and conferences and two book chapters. He earns immense respect among the researchers working in the domain of process modeling studies in continuous casting, owing to novel and impactful research contributions in that domain, the most recent being use of magneto-hydrodynamic applications in continuous casting. In his broad area of expertise, he has supervised so far 10 doctoral theses, and nine students are already on roll. He has also supervised about 40 M.Tech. theses and 25 B.Tech. theses. He has dealt with many sponsored research projects from funding agencies like DST, CSIR, SERB (Government of India). Professor Jha has been involved with the teaching of subjects with Manufacturing and Thermal Streams.

Dr. Brijesh Tripathi received his Ph.D. in Mechanical Engineering in 2008 from Indian Institute of Technology Kharagpur, India. He was a post-doctorate fellow at the Michigan State University, USA, for two and half years and worked on a US government-sponsored project on Computational Fluid Dynamics. Presently, he holds the associate professor positions at Rajasthan Technical University Kota. Dr. Tripathi received 06 research grants for various sponsored projects from GNIDA, NAL, BRNS, MHRD, and AICTE. He has guided 08 Ph.D.s and 26 M.Tech. scholars. He has published more than 130 research articles in peer-reviewed international journals and conferences. Dr. Tripathi's primary interest is working on practical problems using fundamental concepts of Computational Fluid Dynamics especially in diesel engine, heating, ventilation and air[1]conditioning, turbulent flows, unconventional energy systems, etc.

Elango Natarajan has obtained B.E. and M.E. from University of Madras in 1997 and 1999, respectively. He has received Ph.D. in the areas of Soft Robotics in February 2010. He then completed his post-doctoral research in Universiti Teknologi Malaysia in the areas of soft actuators. He has received Chartered Engineer (CEng.) from Engineering Council, UK, in 2014. He has more than 20+ years of teaching experience. He is currently attached to Department of Mechanical and Mechatronic Engineering, UCSI University, Kuala Lumpur, Malaysia as Associate Professor. He has completed two Malaysian Government-funded projects and two internal funded projects worth of about MYR 594,000. He is currently working on four Malaysian Government-funded projects worth of MYR 457,800 and three internal grant projects worth of MYR 120,000. He has published 90 research articles in SCOPUS/WoS Journals. His current h-index is 17 in SCOPUS. He has edited three conference proceedings and a book. He is now editing a special issue in Materials (MDPI), two books (Elsevier and T&F). He is Reviewer of many top WoS journals, and he has reviewed more than 200 journal articles for refereed journals. He is involved in many professional societies including IEEE, IET, and ASM international. He was Executive Committee Member of IEEE Robotics and Automation Society Malaysia from 2018 to 2020. He is now Secretary of IEEE Robotics and Automation Society Malaysia, from 2021. He is Professional Review Interviewer appointed by Institution of Engineering and Technology, UK.

Harish Sharma is Associate Professor at Rajasthan Technical University, Kota, in Department of Computer Science and Engineering. He has worked at Vardhaman Mahaveer Open University Kota and Government Engineering College Jhalawar. He received his B.Tech. and M.Tech. degrees in Computer Engineering from Government Engineering College, Kota, and Rajasthan Technical University, Kota, in 2003 and 2009, respectively. He obtained his Ph.D. from ABV-Indian Institute of Information Technology and Management, Gwalior, India. He is Secretary and one of the founder members of Soft Computing Research Society of India. He is Lifetime Member of Cryptology Research Society of India, ISI, Kolkata. He is Associate Editor of *International Journal of Swarm Intelligence (IJSI)* published by Inderscience. He has also edited special issues of the many reputed journals like *Memetic Computing, Journal of Experimental and Theoretical Artificial Intelligence, Evolutionary Intelligence*, etc. His primary area of interest is nature-inspired optimization techniques. He has contributed in more than 125 papers published in various international journals and conferences.

Chapter 1
Image-Based Weld Joint Type Classification Using Bag of Visual Words

Satish Sonwane, Shital Chiddarwar, Mohsin Dalvi, and M. R. Rahul

Abstract Increased shortage of skilled workers and increased demand for goods tend to strain manufacturing activity. This issue, along with a poor, hazardous working environment, furthers the need to robotize the activity. Welding, one of the major manufacturing processes, has witnessed automation in the last two decades. Welding using robots is mainly accomplished in 'teach and playback' mode. It necessitates reconfiguration every time the robot engages in a new task. Knowing the weld joint beforehand allows the programmer to set relevant parameters in advance. Hence, this study aims to solve the issue by proposing an alternate way to automatically recognize weld joint types. This paper suggests an effective way to classify the weld joint type using the image processing and feature extraction technique. The method works in two stages: features extraction and bag of visual words (BoVW) model building. First, image processing algorithms are used to condition the greyscale image. Image conditioning involves noise removal using a contrast-limited adaptive histogram equalization (CLAHE) and enhancement to improve the image's contrast. Then SURF features of processed images are extracted and input into a support vector machine (SVM)-based bag of visual words classifier for classification. The method is capable of recognizing five types of weld joints. The bag of features strategy combined with SVM yields 97% accuracy.

Keywords Weld joint type recognition · SURF features · CLAHE

S. Sonwane (✉) · S. Chiddarwar · M. Dalvi · M. R. Rahul
Mechanical Engineering Department, Visvesvaraya National Institute of Technology, Nagpur, MH, India
e-mail: satish.sonwane@gmail.com

© The Author(s), under exclusive license to Springer Nature Singapore Pte Ltd. 2024
P. K. Jha et al. (eds.), *Proceedings of Congress on Control, Robotics, and Mechatronics*, Smart Innovation, Systems and Technologies 364,
https://doi.org/10.1007/978-981-99-5180-2_1

1.1 Introduction

A critical development in modernizing the traditional welding processes for ongoing reductions in manufacturing costs has been the development of industrial robots with welding capabilities. Welding robots have always been essential to advanced manufacturing techniques. However, welding personnel are still required to oversee and monitor the operation. Currently, none of the two most common programming modes of welding robots, namely teach and playback mode and off-line programming mode (OLP), depend on sensors input during the welding operation. Kim and Croft [1] stated that these two modes are indispensable for the static and stringently constrained environment. However, the real-world welding scenario might not be so standard, and the current methods do not exhibit the required resilience and briskness to handle the complexities and dynamism of the welding environment. Alternately welding robots can be programmed using external sensors, which help them to perceive the environment around them and make sense of the situation. The works of Rout et al. [2] and Wang et al. [3] stated that arc sound, arc spectra, ultrasonic, and electromagnetic sensing have been used in intelligent welding system models. However, Xiao et al. [4] pointed out that visual sensors are preferred nowadays for their non-contact operation, high precision, faster operation and high adaptability. Various authors [5–11] showed that visual sensors generally consist of a camera, a LASER and filters. Their applications typically include seam tracking, seam extraction, path planning and weld quality control.

Parts lend themselves to the process in specific ways based on the shape. Therefore, various applications need particular types of welds. Different types of welding joints are made to fulfil the needs of every individual application. Fan et al. [12] segregated practical ways to do it in one of the five labels: butt, corner, lap, fillet and vee joint. Identifying the type of weld joint is crucial as the related parameters governing the process change depending on the type. Currently, the parameters are set manually as per the weld joint. This step, however, has the potential for automation, making it independent, efficient and time-saving. Automatic recognition of weld joints and subsequent autonomous setting of parameters will result in enhanced productivity. Many researchers have recently presented conforming single weld joint recognition algorithms for detecting the weld joint's position. Fan et al. [13] Researched recognition of narrow butt weld joints, Zou et al. [14] studied feature recognition of lap joints using image processing and convolution operator, and Fang et al. [15] considered image-based recognition of fillet joints. Due to the single type of weld joint scope, these methods were meant to identify just one type. Many welding parameters, like welding current and voltage, vary depending on the weld joint. As a result, in a realistic welding setting, the capability to distinguish multiple types of weld junctions is essential.

Some researchers have recently explored multiple types of weld joint detection. For weld joint classification, the spatial component relationship of the vertices elements and the intersections of weld joints was utilized by Tian et al. [16]. This approach is viable but has trouble distinguishing between distinct weld joints with identical compositional properties. Wang et al. [17] proposed weld joint type recognition using ensemble learning. The method suffers because it requires costly hardware and is computationally and mathematically intensive. Fan et al. [12] developed a technique for building an SVM classifier by forming the feature map from the weld joint extremities towards the bottom of the weld; it outperforms other approaches regarding detection performance and computing cost but suffers from the fact that its application is cumbersome. Li et al. [18] used Hausdorff distance as a match for measuring laser stripes from the standard template. This technique, however, suffers from computation cost issues and is not flexible. He et al. [19] achieved multiple weld joint type identification utilizing a visual attention model and extraction of the feature points of the weld seam profile. This work lacks adaptability. All the mentioned approaches above use one or the other kind of visual sensor, which includes a laser. The cost of such sensors may hinder the market-wide adoption of these methods; therefore, developing a weld joint recognition system with high recognition accuracy, low computational complexity and cost suitable for multiple types of weld joints is necessary.

This paper uses the bag of features utilizing SURF features for scale invariance from input images to recognize a weld join type from the image. The algorithm then arranges the texture elements or textons in histograms, essentially reducing the dimensionality of the input. These histograms are then passed to the SVM-based model to achieve welding seam type recognition. The technique gathers SURF attributes from all pictures in all classes and then builds the visual vocabulary by lowering the number of features through feature space reduction using K-means clustering. The optimum SVM model for weld joint type detection is then developed based on the retrieved characteristics. It can then be used to predict the type of welding edge. The advantage of this method over the previous one enumerated by Fan et al. [12] is that the model, instead of trying to encode features directly, learns them from inputs and desired outputs, considerably saving the mathematical chicanery and expensive imaging and scanning equipment to get images from which features could be extracted. The results demonstrate that the method correctly recognizes welding joint classes and that calculation and equipment costs may be lowered compared to earlier techniques.

The critical contributions of the paper are as follows:

(1) A method for classifying the kind of weld joints from images is suggested that has the potential to increase the level of robotic welding automation.
(2) Show that non-handcrafted image features are superior to custom features that require a lot of arithmetic and processing.

This paper is organized into four sections. First, in Sect. 1.2, the methods used are explained. Next, experimental results are given in Sect. 1.3. Conclusions are then given in Sect. 1.4.

1.2 Related Theories

1.2.1 Histogram Equalisation

The conventional contrast enhancement histogram equalisation (HE) increases visual contrast by emphasizing highly recurring luminance numbers in the image. This method, however, suffers from the disadvantage of noise augmentation in similar regions. To overcome this CLAHE algorithm was proposed. CLAHE functions on small tiles in the image, simultaneously computing several histograms corresponding to the tile under consideration. The algorithm then redistributes the image's intensity values, avoiding over-amplification. The adjacent tiles are then joined using bi-linear interpolation. It is an image processing technique used to enhance an image's contrast. The main advantage of CLAHE over traditional histogram equalization is that it limits the contrast enhancement in the areas where the contrast is already high. This prevents over-amplification of noise in these areas, which can lead to an unnatural image appearance. CLAHE is particularly useful for images with a wide range of brightness levels or containing areas with low contrast. It can be used in a variety of applications, such as welding.

Zimmerman et al. [20] initially applied CLAHE to improve the contrast of medical images. The CLAHE resorts to a clipping limit to overcome the problem of noise augmentation. Its contrast limitation distinguishes it from typical AHE. The CLAHE constrains the augmentation by clipping the histogram at a predetermined value. In CLAHE, the fundamental parameters controlling image quality are block size (β) and clipping limit ((Ć)). The image gets brighter when Ć is increased. The spectrum of the picture widens, and the contrast improves as the β increases. The image quality significantly improves when these attributes are determined at the greatest entropy point. Clipping is applied to the initial histogram; the pixels are reordered, corresponding to individual grey intensity. The dispersed histogram differs from the regular histogram in that each pixel intensity is restricted to a predetermined upper limit.

The CLAHE method to boost the contrast of the image consists of the following steps:

1. Taking the original luminance picture and dividing it into noncongruent contextual sections. M × N reflects the sum quantity of picture tiles.
2. Determine the histogram of each contextual field based on the pixel intensities provided.
3. Calculate the CLAH of the contextual region by Ç value as,

$$N_{\text{avg}} = (N_s I \times N_s J)/N_{\text{grey}} \quad (1.1)$$

Here,
 N_{avg} is the arithmetic mean of the number of pixels,
 N_{grey} is the amount of pixel intensities in the immediate neighbourhood,

1 Image-Based Weld Joint Type Classification Using Bag of Visual Words

$N_s I$ and $N_s J$ are the numbers of pixels in the relevant region's X and Y directions.

The real value of Ç is written as

$$N\acute{Ç} = N_{cl} \times N_{avg} \qquad (1.2)$$

where $NN\acute{Ç}$ is the actual value of Ç, and N_{cl} is the regularized Ç and lies within the range of [0, 1]. $N_{\Sigma cl}$ denotes the total number of trimmed picture elements, and then, the mean of the remaining pixels to allocate to each pixel intensity is

$$N_{avggrey} = N_{\Sigma cl}/N_{grey} \qquad (1.3)$$

The trimming logic is given by.

If

$$H_{reg}(i) > N\acute{Ç} \text{ then } H_{reg\_cl}(i) = N\acute{Ç} \qquad (1.4)$$

Else if

$$(H_{reg}(i) + N_{avggrey}) > N\acute{Ç} \text{ then}$$
$$H_{reg_{cl}}(i) = N\acute{Ç} \qquad (1.5)$$

Else,

$$H_{reg\_cl}(i) = H_{reg}(i) + N\acute{Ç} \qquad (1.6)$$

Here, $H_{reg}(i)$ and $H_{reg\_cl}(i)$ are unique histograms and trimmed histograms of the individual region at the i-th pixel intensity.

4. Reassign the leftover pixels until they are evenly distributed. The step of pixel reassignment is provided by

$$Step = N_{gr}/N_{rem} \qquad (1.7)$$

where N_{rem} is the residual number of pixels that are trimmed.

In the preceding stage, the algorithm searches from the minimum to the maximum pixel intensity values. If the pixel value in the pixel intensity map is lesser than the threshold, the algorithm adds one-pixel intensity. If the pixels are not evenly distributed after the search concludes, the algorithm will calculate the new step according to (1.7) and begin a fresh round of search until the remaining pixels are evenly distributed.

5. Using the Rayleigh transform to boost intensity values in each local zone. The trimmed histogram is transformed into pooled probability, $P_{in}(i)$, which is then utilized to build a transfer function. Rayleigh transform is defined as

$$y(i) = y_{\min} + \sqrt{2\alpha^2 \ln \frac{1}{1 - P_{\text{in}}(i)}} \qquad (1.8)$$

Here, $y_{\min}$ is the pixel value's lower limit, and α is a Rayleigh distribution adjustment factor defined for each input picture. Each intensity value's output probability distribution may be stated as follows:

$$p(y(i)) = \frac{(y(i) - y_{\min})^1}{\alpha^2} \cdot \exp \frac{(y(i) - y_{\min})^2}{2\alpha^2} \qquad (1.9)$$

for $y(i) \geq y_{\min}$

6. Linear contrast stretch is used to dynamically resize the output of the transfer function in (1.9). The linear contrast stretch may be expressed as follows:

$$y(i) = (x(i) - x_{\min})/(x_{\max} - x_{\min}) \qquad (1.10)$$

$x(i)$ is the transfer function's input value, and $x_{\min}$ and $x_{\max}$ are the transfer function's min and max values.

7. Interpolation using bi-linear to calculate new grey pixel levels and avoid region boundaries' visibility.

CLAHE triumphs over other methods because it has a low computational cost, is easy to use and works as intended on most images.

1.2.2 Speeded-Up Robust Features (SURF)

SURF is a local rotation-invariant interest point (feature) detector. It is designed to be invariant to scale and rotation, i.e. regardless of changes in the scale or orientation of an object, SURF should be able to detect and match it accurately. And this is a requirement in the welding scenario as the captured weld joint image could be taken from any distance and may be randomly oriented.

The algorithm is divided into four sections: (1) integral picture production, (2) feature point identification, (3) descriptor alignment assignment and (4) descriptor generation.

SURF is partly inspired by SIFT, which is faster. SURF uses the integer approximation of the Hessian Blob detector to detect interest points, as it performs well in terms of computing time and accuracy. If we have a point $P = (i, j)$ in an image 'im', the Hessian matrix $ℏ(P, \sigma)$, where σ is the scale, is defined as

$$ℏ(P, \sigma) = \begin{matrix} L_{ii}(P, \sigma) & L_{ij}(P, \sigma) \\ L_{ij}(P, \sigma) & L_{jj}(P, \sigma) \end{matrix} \qquad (1.11)$$

1 Image-Based Weld Joint Type Classification Using Bag of Visual Words

where $L_{ii}(P, \sigma)$ is the Gaussian second-order derivative's convolution with the image at P, and similar descriptions can be given for $L_{ij}(P, \sigma)$ and $L_{jj}(P, \sigma)$.

Its descriptor is based on HAAR wavelet response around the point of interest. First, 'interest spots' are chosen in various areas within the picture. These areas can be corners, blobs and junctions. The area around each location of interest is then expressed as a feature vector. The pair is matched if it is closer than 0.7 times the distance of the second nearest neighbour. Finally, these feature vectors are matched based on the Euclidean distance between distinct photos to be compared. The SURF approach's key attraction is its quick computing of operators using box screens, which enables real-world applications such as tagging, object detection and recognition.

1.2.3 BoVW

The BoVW is the machine vision adaptation of the natural language processing algorithm termed 'Bag of Words' as discussed by Schmid [21]. It is a significant improvement over existing models in object detection and recognition. The BoVW model forms groups of similar input image features and internally assigns them a visual word. To achieve this, the process usually follows the following three steps, namely

(1) Detection of features
(2) Description of features
(3) Generation of the codebook.

For an image with 'm' features, the model distributes the characteristics among 'k' clusters. This determines the vocabulary size. The last step is to create a codebook using the K-means clustering. A minimum Euclidean distance from the centre puts each feature into the group. This maps the feature to the visual word.

Further vocabulary histogram is built. It has bins that are the same size as the dictionary. Each SURF-computed feature is allocated to the best cluster and shown in the histogram. Once all features have been classified, the task is a multi-label problem. While using SURF, the Gridstep of [8 8] and Blockwidth of [32 64 92 128] was used.

1.3 Experiments

All experiments in this study were conducted on a Workstation with AMD Ryzen Threadripper 2950X CPU @ 3.50 GHz, 128 GB of DDR4 RAM and NVIDIA Quadro RTX 6000 24 GB GPU. The camera used is 'Neo' by 'Soliton Technologies Pvt. Ltd.'.

Table 1.1 Data distribution of various weld joint types

Butt	Corner	Fillet	Lap	Vee
300	300	300	300	300

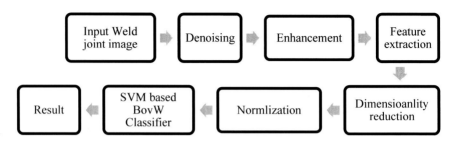

Fig. 1.1 Pipeline of the proposed method

1.3.1 Dataset

The dataset used in this study combines the Kaggle dataset [22] and a custom dataset generated by the authors. Table 1.1 shows dataset distribution as per various weld joints.

The experiment was conducted in the steps shown in Fig. 1.1. Input weld image is first denoised and enhanced using CLAHE. These images are then fed to the feature extractor. The extracted features' dimensionality is reduced in the next step to keep the most relevant features and discard the others. These are then normalized and fed to the SVM-based BoVW classifier to get the required output.

1.3.2 Noise Reduction and Enhancement

Obtained images may sometimes be noisy or blurry. We have to denoise and enhance images to pre-process them to feed to the feature extractor. We used CLAHE to perform the task. Figures 1.2 and 1.3 show images of the butt and fillet joint before and after pre-processing. As can be seen, the image's contrast has improved considerably. The gap in the butt joint becomes noticeably clear after the operation.

1.3.3 Image Feature Extraction

The objective of feature vector extraction is to obtain a vector $\{x1, x2, x3... xk\}$, of length k, which can thoroughly represent its parent image, and so. A feature vector is an image representation that is used to characterize and mathematically

1 Image-Based Weld Joint Type Classification Using Bag of Visual Words

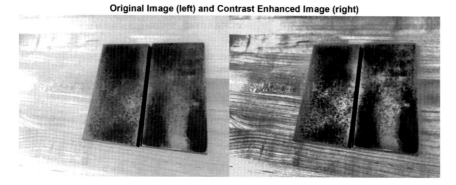

Fig. 1.2 Before and after processing an image of the butt joint

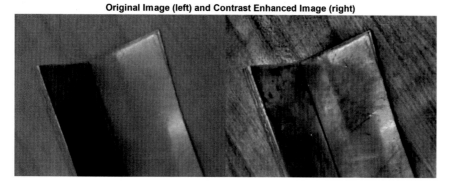

Fig. 1.3 Before and after processing an image of the fillet joint

evaluate a picture's contents. Therefore, feature extraction directly affects the quality of classification results and recognition rate.

Figure 1.4 shows one example output of the SURF feature extractor.

1.3.4 SVM Classifier

SVM is a supervised machine learning (ML) technique in classification tasks. SVM is a powerful and versatile algorithm with several advantages over other machine learning algorithms in specific applications. Its ability to perform well in high-dimensional spaces, be robust to outliers, efficiently perform nonlinear classification, avoid overfitting and provide a global optimization solution makes it popular among data scientists and machine learning practitioners. Each data item in the SVM method is represented as a point in the n-dimensional plane (where n represents the total of characteristics), with the value of each feature becoming a value of a specific

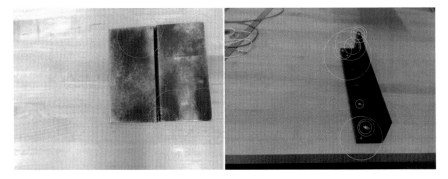

Fig. 1.4 SURF output of butt joint (left) and T joint (right)

coordinate. Then, we accomplish classification by locating the hyper-plane that suitably discriminates the number of classes. It was initially a binary classifier. However, multiple SVM can be used in a program in the one-versus-one form to achieve multi-classification. For example, we have used the error-correcting output codecs (ECOC) classifier in this study [23]. It uses the Gaussian kernel function.

1.3.5 Experimental Results

300 images of each of all five types of joints were clicked. Of them, 70% were utilized for training and the rest for validation. The program calculated 2,160,000 features, of which the most vital 80% of features from each category were kept. The features were segregated into 500 clusters. The model converged in 24 iterations, each taking about 19 s. This bag of visual words was then passed to the SVM classifier. The trial of the trained classifier on an entirely new image returned the result in 152 ms.

1.4 Conclusion and Future Scope

In this work, BoVW was used along with image processing for weld joint classification. The model provides adequate accuracy using the Gaussian RBF kernel. Table 1.2 shows the presented technique compared to some related research work, and the experimental findings demonstrate that the proposed technique achieved competitive results in classification performance. It may be noted that the joint recognition accuracy obtained in Zeng et al. [24] work is on the higher side. However, the mentioned study uses complex and costly hardware and could only improve the result by 1%.

Table 1.2 Comparative image recognition results

References	Accuracy (%)
Reference [24]	98.4
This paper	97
Reference [12]	89.2

Furthermore, experiments with new input images proved that the method has practical application in automated robotic welding due to its ease of use and requirement of modest computational power and that equipment costs are lower than earlier techniques. The results also show that non-handcrafted image features are superior to custom features that require a lot of arithmetic and processing.

This paper proposes a weld image feature extraction algorithm that demonstrates matching or superior recognition accuracy and lower computational cost than the feature extraction method presented in Refs. [12, 24]. Furthermore, the method presented in Ref. [12] misidentifies some types of weld joints, such as lap, splice and fillet weld joints, due to the relatively small size of their grooves. In contrast, the proposed method is simple and economical, but with sufficient training data sample size and tweaking of SVM parameters, it can achieve a higher accuracy rate. This approach is thus preferable to the method explained in the other two references.

For further extension of the study, the model's performance under the influence of splash noise will be tested and improved.

References

1. Kim, J., Croft, E.A.: Online near time-optimal trajectory planning for industrial robots. Robot. Comput. Integr. Manuf. **58**(February), 158–171 (2019). https://doi.org/10.1016/j.rcim.2019.02.009
2. Rout, A., Deepak, B.B.V.L., Biswal, B.B.: Advances in weld seam tracking techniques for robotic welding: a review. Robot. Comput. Integr. Manuf. **56**(2017), 12–37 (2019). https://doi.org/10.1016/j.rcim.2018.08.003
3. Wang, B., Hu, S.J., Sun, L., Freiheit, T.: Intelligent welding system technologies: state-of-the-art review and perspectives. J. Manuf. Syst. **56**(June), 373–391 (2020). https://doi.org/10.1016/j.jmsy.2020.06.020
4. Xiao, R., Xu, Y., Hou, Z., Chen, C., Chen, S.: "An adaptive feature extraction algorithm for multiple typical seam tracking based on vision sensor in robotic arc welding. Sens. Actuators, A Phys. **297**, 111533 (2019). https://doi.org/10.1016/j.sna.2019.111533
5. Hong, B., Jia, A., Hong, Y., Li, X., Gao, J., Qu, Y.: Online extraction of pose information of 3d zigzag-line welding seams for welding seam tracking. Sens. (Switzerland) **21**(2), 1–19 (2021). https://doi.org/10.3390/s21020375
6. Villán, A.F., et al.: Low cost system for weld tracking based on artificial vision. In: Conference Record—IAS Annual Meeting (IEEE Industry Applications Society, 2009. https://doi.org/10.1109/IAS.2009.5324960
7. Xu, Y., Zhong, J., Ding, M., Chen, H., Chen, S.: The acquisition and processing of real-time information for height tracking of robotic GTAW process by arc sensor. Int. J. Adv. Manuf. Technol. **65**(5–8), 1031–1043 (2013). https://doi.org/10.1007/s00170-012-4237-6

8. Rout, A., Deepak, B.B.V.L., Biswal, B.B., Mahanta, G.B.: Weld seam detection, finding, and setting of process parameters for varying weld gap by the utilization of laser and vision sensor in robotic arc welding. IEEE Trans. Ind. Electron. **69**(1), 622–632 (2021). https://doi.org/10.1109/TIE.2021.3050368
9. Patil, V., Patil, I., Kalaichelvi, V., Karthikeyan, R.: Extraction of weld seam in 3D point clouds for real time welding using 5 DOF robotic arm. In: 2019 5th International Conference on Control, Automation and Robotics, ICCAR 2019, 2019, pp. 727–733. https://doi.org/10.1109/ICCAR.2019.8813703
10. De, A., Jantre, J., Ghosh, P.K.: Prediction of weld quality in pulsed current GMAW process using artificial neural network. Sci. Technol. Weld. Join. **9**(3), 253–259 (2004). https://doi.org/10.1179/136217104225012328
11. Rodríguez-Gonzálvez, P., Rodríguez-Martín, M.: Weld bead detection based on 3D geometric features and machine learning approaches. IEEE Access **7**, 14714–14727 (2019). https://doi.org/10.1109/ACCESS.2019.2891367
12. Fan, J., Jing, F., Fang, Z., Tan, M.: Automatic recognition system of welding seam type based on SVM method. Int. J. Adv. Manuf. Technol. **92**(1–4), 989–999 (2017). https://doi.org/10.1007/s00170-017-0202-8
13. Fan, J., Jing, F., Yang, L., Long, T., Tan, M.: A precise seam tracking method for narrow butt seams based on structured light vision sensor. Opt. Laser Technol. **109**, 616–626 (2019). https://doi.org/10.1016/j.optlastec.2018.08.047
14. Zou, Y., Chen, T.: Laser vision seam tracking system based on image processing and continuous convolution operator tracker. Opt. Lasers Eng. **105**(January), 141–149 (2018). https://doi.org/10.1016/j.optlaseng.2018.01.008
15. Fang, Z., Xu, D.: Image-based visual seam tracking system for fillet joint. In: 2009 IEEE International Conference on Robotics and Biomimetics, ROBIO 2009, pp. 1230–1235. https://doi.org/10.1109/ROBIO.2009.5420852
16. Tian, Y., et al.: Automatic identification of multi-type weld seam based on vision sensor with Silhouette-mapping. IEEE Sens. J. **21**(4), 5402–5412 (2021). https://doi.org/10.1109/JSEN.2020.3034382
17. Wang, Z., Jing, F., Fan, J.: Weld seam type recognition system based on structured light vision and ensemble learning. In: Proceedings of the 2018 IEEE International Conference on Mechatronics and Automation ICMA 2018, no. 61573358, pp. 866–871. https://doi.org/10.1109/ICMA.2018.8484570
18. Li, Y., Xu, D., Tan, M.: Welding joints recognition based on Hausdorff distance. Gaojishu Tongxin/Chin. High Technol. Lett. **16**(11), 1129–1133 (2006)
19. He, Y., Xu, Y., Chen, Y., Chen, H., Chen, S.: "Weld seam profile detection and feature point extraction for multi-pass route planning based on visual attention model. Robot. Comput. Integr. Manuf. **37**, 251–261 (2016). https://doi.org/10.1016/j.rcim.2015.04.005
20. Zimmerman, J.B., Pizer, S.M., Staab, E.V., Perry, J.R., Mccartney, W., Brenton, B.C.: An evaluation of the effectiveness of adaptive histogram equalization for contrast enhancement. IEEE Trans. Med. Imaging **7**(4), 304–312 (1988). https://doi.org/10.1109/42.14513
21. Schmid, C.: Bag-of-features models for category classification category recognition
22. "Weld-Joint-Segments | Kaggle." https://www.kaggle.com/datasets/derikmunoz/weld-joint-segments. Accessed 01 Apr 2022
23. MathWorks: ClassificationECOC. [Online]. Available: https://in.mathworks.com/help/stats/classificationecoc.html
24. Zeng, J., Cao, G.Z., Peng, Y.P., Huang, S.D.: A weld joint type identification method for visual sensor based on image features and SVM. Sens. (Switzerland) **20**(2) (2020). https://doi.org/10.3390/s20020471

Chapter 2
Speech Recognition-Based Prediction for Mental Health and Depression: A Review

Priti Gaikwad and Mithra Venkatesan

Abstract A person with a mental disorder exhibits a significant disturbance in his or her behavior. Generally, mental disorders are associated with distress or impairment of normal functioning. Lack of adequate resources and facilities, as well as a lack of awareness of the symptoms of mental illness, prevent people from getting the help they need. The ability to assess depression through speech is a critical factor in improving the diagnosis and treatment of depression. The spoken language is said to provide access to the mind, and a wide range of speech capture and processing technologies can be used to analyze mental health. Speech processing is about recognizing spoken words. The automatic recognition and extraction of information from speech enables the determination of some physiological characteristics that make a speaker unique to identify their mental health status. In this paper, we describe how mental health-related problems can be predicted by speech processing. This paper identifies the gaps in the literature review that lead to the proposed methodology.

Keywords Mental health · Speech processing · Natural language processing

2.1 Introduction

Mental health refers to the state of being aware of one's abilities, coping with daily stresses, working, and making a positive contribution to society. According to the World Health Organization (WHO), mental health is the absence of mental disorders. Our capacity to think, feel, interact with one another, and carry out daily tasks depends on both the mental health of each individual and the state of our society as a whole.

P. Gaikwad (✉) · M. Venkatesan
Dr. D. Y. Patil Institute of Technology, Pimpri Pune, Maharashtra, India
e-mail: ppgaikwad.scoe@sinhgad.edu

M. Venkatesan
e-mail: mithra.v@dypvp.edu.in

Due to promotion, protection, and restoration, mental health has become a central concern of communities and societies around the world [1].

In humans, speech production is a result of physiological processes that are naturally affected by physical stress. There are major changes in the fundamental frequency level, the speaking rate, the pause pattern, and the breathiness of speech. A speech is one of the most natural and common ways in which we communicate with each other on a daily basis, and it contains a profound array of information that goes far beyond the verbal message it conveys. Listening to the speaker's utterances can reveal the speaker's gender, age, dialectal background, emotional status, and personality. A speaker's physiological and health condition is part of the paralinguistic information in his or her speech. Through signal processing techniques and statistical modeling, it is possible to capture natural changes in the human body by analyzing the sound and linguistic content of speech signals. Speech sounds in patients with depression tend to have a lower pitch, in the form of acoustic signals [2]. It is possible to detect and quantify disorders, diseases, monotonous speech, lower sound intensity, and slower speech rates, as well as more hesitations, stutters, and whispers. Speech has several advantages: it is difficult to hide symptoms, it directs emotion and thought through its language content, and it is an inexpensive medium. Due to similar vocal anatomy, it may generalize across languages, which is particularly useful when natural language processing technology is not available for low-resource languages. Since most clinical interviews are already recorded, it is easy to obtain using Smartphones, tablets, and computers rather than more costly wearable or invasive neuroimaging methods.

970 million people worldwide, or 1 in 8, experience mental disorders, primarily anxiety and depression. Due to COVID-19 pandemic, the number of people experienced anxiety and depression. According to preliminary projections, the prevalence of anxiety and major depressive disorders will rise by 26% and 28%, respectively, in 2020 [3].

There could be a "mental health epidemic" in India, according to President Ram Nath Kovind, who noted that 10% of the country's 1.3 billion people suffer from mental illness. The WHO estimates that about 15% of the world face the issue related to mental health in India. A meta-analysis of community surveys found that 33 out of 1,000 people experience depression or anxiety [4].

Physical health conditions like cancer, diabetes, and chronic pain can have underlying, life-altering effects on mental health conditions like stress, depression, and anxiety. So, given the aforementioned issue, depression must be automatically detected.

Thinking, feeling, and behavior are all impacted by depression. Depression makes day-to-day living more challenging and interferes with relationships, work, and study. If a person feels down, sad, or miserable most of the time for longer than two weeks, has lost interest in or pleasure from most of their usual activities, and exhibits multiple symptoms from at least three of the categories listed below, they may be depressed [5]. It is important to remember that everyone occasionally experiences some of these symptoms, and they may not signify depression per se. Likewise, not everyone who is depressed will exhibit all of these symptoms.

2.2 Literature Review

Researchers have developed many new methods proposed for speech patterns that indicate mental health. By analyzing depression detection, it can be seen that to evaluate patient health from an electronic health record is to map speech signals to depression features. Researchers proposed several models for detecting depression which leads to mental illness which are discussed below.

2.2.1 Related Work

In this study, Nanath et al. [6] have shown how social media data can be used to predict the mental health characteristics of people using text features and natural language processing. According to Alghowinem et al. [7], speech patterns, eye movements, and head posture were each analyzed for statistical features. A support vector machine (SVM) was used in emotion classification tasks. The Reddit database was used by Rssola et al. [8], who were able to identify trends in the writing style, emotional expression, and online behavior of the users in question by visualizing and analyzing some probabilistic features. Sarkara et al. [9] used emotions.csv dataset from the Kaggle Web site and used different machine learning (ML) and deep learning (DL) methods, multi-layer perceptron, convolution neural network, recurrent neural network with long short-term memory, SVM, and linear regression which are used as classifiers, to solve real-world problem; among them the RNN model has the highest accuracy 97.50% in the training set and in the test set 96.50%. Liu et al. [10] the goal of the NetHealth study was to predict people's mental health status by using network methods and DMF (a method from RS). Smartphone data, data from wearable sensors (Fitbit), and people's trait data from surveys were collected.

Due to behavioral interference from interviewers and problems in matching audio transcripts, Dong et. al. [11] only considered depression detection from non-interaction databases like DAIC-WOZ. In the future, it may be possible to use interaction databases to confirm the generalizability of the model. According to research by Ye et al. [12], patients with mild and minor depression have a higher recognition error rate than average individuals and patients with major depression. Di Matteo et al. [13] developed an Android app that collects regular audio recordings of participants' surroundings and recognizes English words with automatic speech recognition. Amanat et al. [14] obtained a large imbalanced dataset of tweets from the Kaggle Web site, implemented the one-hot coding method and principal component analysis (PCA), LSTM, and RNN for further improvement, and proposed a hybrid recurrent neural network for a large database. In the case of the real-time datasets, better results are obtained than other classification algorithms, which is in agreement with Gupta et al. [15]. However, the accuracy of the proposed algorithm can be improved for real-time recorded files by recording the speech in a professional environment and making an appropriate selection of neurons and values for drop-out layers.

Rejaibi et al. [16] worked with the dataset DAIC-WOZ, Ryerson Audio-Visual Database of Emotional Speech and Song (RAVDESS). An LSTM was chosen for high-level audio feature extraction, suggesting that textual features may accurately represent depression. Rutowski et al. [17] use two corpora of American English speech collected by Ellipsis Health. They work with a deep learning model classifier based on deep NLP and transfer learning showed excellent transferability across age, gender, and ethnicity. Schultebraucks et al. [18] suggested that to fully incorporate the vast clinical knowledge of clinicians, larger samples are required to confirm the findings and test for interactions between verbal and facial modalities. Another drawback is the dependence of the feature extraction process on pre-trained models. Although the author used standard techniques, it is pointed out that facial expression recognition is known to have shortcomings and is subject to bias. Aloshban et al. [19] conducted the experiments using 59 interviews recorded in three psychiatric centers in Italy. Feature extraction is done by a bidirectional long-term memory network (BLSTM). The primary drawback of the study is that the interviews were manually transcribed, particularly in depressed patients, the results of the Beck Depression Inventory-II (BDI-II) are unreliable. El Shazly et al. [20] studied 48 Egyptian EFL learners, there is no control group, the sample is small, and the data are descriptive. Although the sample size was small, the use of a mixed methods design provides a better understanding. According to Garoufis et al. [21], a speech analysis system that acknowledges (anomalous) pre- and relapse states in persons with psychotic disorders utilizing unsupervised learning with convolutional autoencoders. Daus et al. [22] use the Linguistic Inquiry and Word Count (LIWC) method to analyze verbal information that has been automatically translated in terms of the number of words of emotional categories. According to Sharma et al. [23], the diagnosis of mental illnesses is based on standardized interviews with a deterministic set of questions and scales. The machine learning model created using the imbalance dataset results in predictions that are biased toward the majority class; as a result, the model will consistently forecast that depression is absent, even when it is. There are no agreed-upon and accepted standards for biomarker scores in various nations and ethnic groups, so the XGBoost model created under this study cannot be adapted to other nations and racial groups. Machine learning should be used to accurately study the different types of depression. According to Wang et al. [24], research should focus on extracting additional speech signal features to describe them more accurately and cross-language learning to increase the reuse rate of models. Villatoro-Tello et al. [25] worked on the (DAIC-WOZ) Alzheimer's dementia dataset. Bag-of-words (BoW), (LIWC), and Third BERT techniques were used. He suggested that the LA method can be fused with raw waveform based CNN to increase the performance.

According to Araño [26], it is challenging to infer happiness from speech characteristics. Future research can therefore concentrate on developing new descriptors that accurately depict this emotional state. Mou et al. [27] used CNN and LSTM and suggested expanding the sample size and participant age range to boost the model's generalizability. Unsupervised learning can identify driver stress-related features from unlabeled data.

Current research in the automatic detection of depression has a number of limitations. First, some methods rely heavily on manually selected questions, which require the involvement of psychologists with relevant expertise. In addition, the interview must cover all predetermined questions; otherwise, the analysis may be flawed. The question of how to enhance detection performance without pre-programmed questions remains a challenge. In addition, due to ethical concerns, there are not many publicly available depression datasets.

2.2.2 Gaps Identified

Despite the fact that speech depression recognition (SDR) has advanced using current datasets, the following are the dataset problems from the literature survey which impede its further advancement.

- Database annotation objectivity: Data annotation serves as the foundation for future work, but the performance of the developed model will be impacted by the distribution of depression values because annotators' perceptions are not always accurate.
- Small in scope and unavailability: Due to ethical issues and the sensitivity of depression speech, most institutions were unable to obtain sufficient samples. AVEC2013, AVEC2014, DAIC-WOZ, and BD are the only public depression databases currently available, and they are not suitable for scientific research. It is critical to address ethical concerns in publishing datasets.
- Non-universality: At the moment, interactive clinical interviews, where questions are carefully crafted such that there is no noise or interference. As a result, these data cannot represent depressed patients' daily lives accurately. Additionally, the issue of linguistic and cross-cultural communication has not yet been considered.
- Model generalization: The models are challenging to generalize to other datasets or data from different languages because the majority of studies only use one or a few small datasets. It is also necessary to improve the model's reliability and predictive validity across corpora, societies, languages, and crowded environments.
- Types of depression disorder: For instance, compared to the most prevalent major depressive disorder, the pathogenesis and behavior of bipolar disorder are different. Few studies have been conducted on how speech signals can differentiate between these two.
- Multi-modality fusion mechanism: Since various modalities can successfully complement one another, future research trends in depression analysis cannot be avoided, including the combining of multiple modalities. However, the success of multimodal research depends on an effective and appropriate mechanism.

According to the identified gaps in the literature, the following objectives have been formulated in the paper.

2.2.3 Objectives

- To study existing literature where speech and language processing is applied for mental well-being
- To build, collect, and process datasets based on speech for the diagnostic model
- To propose a diagnostic model capable of finding mental illness based on speech processing.

2.3 Proposed Methodology

The goal is to develop a safe speech diagnostic model for people with depression. Speech depression datasets are typically recorded by in-person, telephone, or virtual interviewers as clinical clinicians talk with depressed patients. Other modalities, including information from the depression scale, facial expressions, physiological dynamics, etc., are also sometimes recorded during data collection for supplemental analysis.

The proposed methodology is depicted in Fig. 2.1. The different levels involved in the model are database, preprocessing of speech data, feature extraction from speech data, and validation of model and classification as depressed or normal.

2.3.1 Dataset

The following datasets are public records.

The audio-visual depression language corpus for AVEC2013 includes the AVEC2013 and AVEC2014 datasets. AVEC2014 is a set of AVEC2013 consisting of 300 German videos with shorter video clips than in AVEC2013. One component of the Distress Analysis Interview Corpus used for AVEC2016 and AVEC2017 is the Distress Analysis Interview Corpus—Wizard of Oz (DAIC-WOZ).

The traditional method of diagnosing depression uses clinical interviews to screen potential patients for depression. However, these assessments rely heavily on physician questions, patient verbal reports, actions reported by family or friends, and mental status tests such as the Beck Depression Inventory, the Hamilton Rating Scale for Depression, and the Scale for the Assessment of Negative Symptoms. The PHQ-9 is the Health Status Questionnaire for measuring depression scale in that The DSM-nine IV diagnostic criteria for MDD. The PHQ-9 can be applied as a screening tool, a diagnostic tool, and a tool for symptom assessment. It can be used to track changes in particular symptoms over time as well as the overall extent of a patient's depression. Based on the depression score: 0–4 none, 5–9 mild, 10–14 moderate, 15–19 moderately severe, and 20–27 severe. These are all based on subjective assessments, and since there are no reliable, quantitative measures, the results often vary depending on the situation.

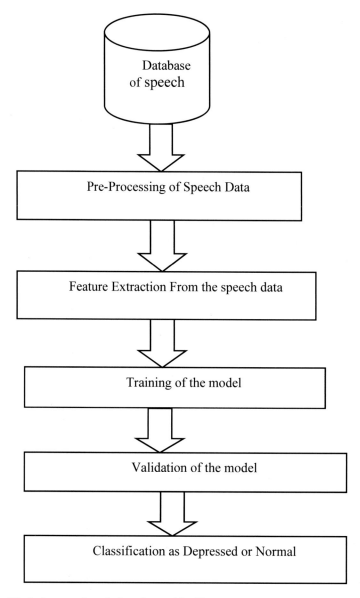

Fig. 2.1 Block diagram of prediction of mental health

The Distress Analysis Interview Corpus (DAIC) comprises an English database. It includes clinical interviews intended to assist in the identification of mental illnesses like anxiety, depression, or post-traumatic stress disorder. The "Wizard of Oz" interviews, conducted by a virtual interviewer, make up the depression section of the corpus. The PHQ-8 Depression Inventory, which is different from the databases used in Germany and Turkey, was used to determine patients' depression scores. In total, there are 189 records of 189 patients.

Data Availability To download one of the databases users must complete the registration process and submit the form in order. Due to consent restrictions, only academics and other non-profit scholars are allowed to access datasets. Thus, when requesting, the user needs only provide their academic email address in order to download.

The DAIC-WOZ Database & Extended DAIC Database data available at https://dcapswoz.ict.usc.edu/

Extended DAIC Database The DAIC-WOZ database for the evaluation of PTSD and depression was created by ICT, and this is its expanded version. Additionally available upon request, this information was used for the AVEC 2019 Challenge.

We are going to use this E-DIAC English dataset for our proposed work, as it is secure and available as per request, it has large no of files, and annotation of the data is also done.

The E-DAIC is the next version of the DAIC-WOZ, collected from semi-clinical interviews to help with the treatment of mental disease like anxiety and depression. Participant information is labeled with age, gender, and PHQ-8 score, the dataset includes 163 developmental patterns, 56 samples for training, and 56 samples for testing. For AVEC2019, and this database will be used.

2.3.2 Preprocessing of Speech Data

Speech recognition faces many difficulties. First, there are not enough datasets in the field of speech, as the creation of a high-quality speech emotion database requires a lot of time and effort. Second, the different data in the database have different speakers, each with different gender, age, language, culture, and so on. Finally, sentences rather than specific words are often the basis for the emotions expressed in speech. Therefore, a challenge in current research is to increase the accuracy of emotion recognition by using low-level descriptors (LLDs) and sentence-level features. There are typically three methods in conventional techniques for speech emotion recognition. Data preprocessing, which includes data normalization, speech segmentation, and other operations, is the first step.

The original speech data must be enhanced by changing the speech playing speed, and the problem of an unbalanced distribution of speech data must be resolved using the balancing datasets weight method. We can use data enhancement and speech segmentation to increase the number of training samples to address the issue of the limited number of training samples.

2.3.3 Speech Feature Extraction

Features Speech features like Mel-frequency cepstral coefficients (MFCC), pitch, jitter, shimmer, energy, the zero-crossing rate (ZCR), the harmonic-to-noise ratio, fundamental frequency (F0), formant, low speech volume, monotone intonation, reduced articulation and the harmonic distribution, as well as perceptual linear prediction (PLP) coefficients have performed better in classifying an individual as a depressed or a healthy one.

The spectral features: related to spectral centroid; the cepstral features: related to the cepstrum analysis (an anagram to the spectrum signal) like the Mel-frequency cepstral coefficients (MFCCs); prosodic features: fundamental frequency F0 (the first signal harmonic) and the loudness the voice quality: like the formants (the spectrum maxima), the jitter (the signal fluctuation), and the shimmer (the peaks variation). Source feature: voice quality feature; deep audio feature: raw audio input for acoustic feature.

Techniques Recently, deep learning models have used convolution neural networks (CNNs) in particular in conjunction with automatic feature extraction, either explicitly from time-domain samples or using a frequency-domain representation of the signal, like the discrete Fourier transform (DFT) or spectrogram.

Successful approaches combine spectral features and their time derivatives with machine learning algorithms like hidden Markov models (HMMs), Gaussian mixture models (GMMs), or hybrid GMM-PLP coefficients. Convolution neural systems (CNNs) [4, 23] in particular have recently been used in deep learning models, utilizing either a frequency-domain representation of the signal or automatically extracting features from time-domain samples.

For spectral feature MFCC feature extraction MFCC-CNN, MFCC-RNN so we are proposed to use **MFCC Multichannel CNN-BLSTM** by fusion of magnitude and phase spectral feature.

We can combine spectral features with prosodic features, and an autocorrelation technique can extract pitch prediction like fundamental frequency (f0) raw. In prosodic features, we can extract probability of voicing (POV), F0 intensity, loudness, voice quality, and F0 envelope. The jitter and shimmer algorithm can be used for voice quality.

We can also use speech features like amplitude envelope, zero-crossing rate, and spectral flux.

2.3.4 Classification for Mental Health Data

To support the diagnosis of depression, it is therefore necessary to develop depression classification methods. Several techniques have recently been developed for assisting clinicians during the diagnosis and monitoring of clinical depression, the recent development of machine learning and artificial neural networks. Utilizing sufficiently sizable speech corpora will allow for the development of the necessary models for mental health information prediction. Machine learning algorithms like SVM, decision trees, logistic regression, and KNN classifiers were used in the investigation. The model's accuracy was increased using the SMOTE method. On the other hand, deep learning models like CNN, ANN [4], and LSTM outperformed conventional machine learning techniques.

Deep learning models commonly used include CNN, LSTM, and BERT; however, BiLSTM [11] can offer higher accuracy. MT-CNN multitasks CNN [4] to perform better at predicting depression.

We can implement hierarchical depression models so that we can combine classification and regression for better performance parameters. In the hierarchical model in the first layer, multiple classifiers can be used as ensemble models and in the second layer for each recording regression algorithm can be used.

2.4 Discussion

The following points are suggested based on the thorough literature review and methodical meta-analysis. It details the approach taken, the benefits and difficulties encountered while using the datasets for depression. In comparison with volume-based features, SVM, multivariate regression, performs better depression prediction. A variety of mental illnesses can be quickly identified with the aid of RF, NB, and SVM. Spatiotemporal data can be extracted using the CNN and RNN combination. In order to detect depression more effectively and accurately, hybrid models like CNN with LSTM are used. Algorithms for machine learning and boosting are useful in identifying the sociodemographic and psychological factors that contribute to depression. It has been noted that the voice change study may aid in the early detection of depression.

2.5 Conclusion

Our lives depend on having a healthy mind. Serious issues brought on by mental instability are challenging to diagnose and treat. A serious mental health condition with high societal costs is depression. One of the objective indicators for the early detection of depression can be speech signal characteristics. To address the issue of the representation of speech signals by conventional feature extraction techniques is difficult; so in this study we proposed MFCC multichannel CNN-BLSTM spectral feature. We can add different speech features such as spectral feature, prosodic feature, and voice quality feature for extraction of the speech so that exact level of mental health can be identified. The potential of AI algorithms to address mental health questions in mental health care will give the best results.

References

1. Liu, S., Vahedian, F., Hachen, D., Lizardo, O., Poellabauer, C., Striegel, A., Milenković, T.: Heterogeneous network approach to predict individuals' mental health. ACM Trans. Knowl. Discov. Data **15**(2), Article 25
2. Stasak, B., Huang, Z., Joachim, D., Epps, J.: Automatic elicitation compliance for short-duration speech based depression detection. In: ICASSP 2021—2021 IEEE International Conference on Acoustics, Speech and Signal Processing (ICASSP), 978-1-7281-7605-5/20/$31.00 ©2021 IEEE. https://doi.org/10.1109/ICASSP39728.2021.9414366
3. https://www.who.int/news-room/factsheets/detail/mental-disorders
4. https://www.ideasforindia.in/topics/human-development/understanding-india-s-mental-healthcrisis.html#:~:text=In%202017%2C%20the%20President%20of,49%20million%20from%20anxiety%20disorders
5. Priya, A., Garga, S., Tigga, N.P.: Predicting anxiety, depression and stress in modern life using machine learning algorithm. In: International Conference on Computational Intelligence and Data Science (ICCIDS 2019). Procedia Comput. Sci. **167**, 1258–1267 (2020)
6. Nanath, K., Balasubramanian, S., Shukla, V., Islam, N., Kaitheri, S.: Developing a mental health index using a machine learning approach: assessing the impact of mobility and lockdown during the COVID-19 pandemic. Technol. Forecast. Soc. Change **178**, 121560 (2022)
7. Alghowinem, S., Goecke, R., Wagner, M., Epps, J., Hyett, M., Parker, G., Breakspear, M.: Multimodal depression detection: fusion analysis of paralinguistic, head pose and eye gaze behaviors. IEEE Trans. Affect. Comput. **9**(4) (2018)
8. Ríssola, E.A. Aliannejadi, M., Crestani, F.: Mental disorders on online social media through the lens of language and behaviour: analysis and visualization. Inf. Process. Manag. **59**, 102890 (2022)
9. Sarkara, A., Singh, A., Chakraborty, R.: A deep learning-based comparative study to track mental depression from EEG data. Neurosci. Inform. **2**, 772–5286 100039 (2022)
10. Liu, S., Vahedian, F., Hachen, D., Lizardo, O., Poellabauer, C., Striegel, A., Milenković, T.: Heterogeneous network approach to predict individuals' mental health. ACM Trans. Knowl. Discov. Data **15**(2), Article 25. Publication date: April 2021
11. Dong, Y., Yang, X.: A hierarchical depression detection model based on vocal and emotional cues. Neurocomputing **441**, 279–290 (2021)
12. Ye, J., Yu, Y., Wang, Q., Li, W., Liang, H., Zheng, Y., Fu, G.: Multi-modal depression detection based on emotional audio and evaluation text. J. Affect. Disord. **295**, 904–913 (2021)
13. Di Matteo, D., Fotinos, K., Lokuge, S., Mason, G., Sternat, T., Katzman, M.A., Rose, J.: Automated screening for social anxiety, generalized anxiety, and depression from objective smartphone-collected data: cross-sectional study. J. Med. Internet Res. **23**(8), e28918 (2021)
14. Amanat, A., Rizwan, M., Javed, A.R., Abdelhaq, M., Alsaqour, R., Pandya, S., Uddin, M.: Deep learning for depression detection from textual data. Electronics **11**, 676 (2022). https://doi.org/10.3390/electronics11050676
15. Gupta, M., Vaikole, S.: Audio signal based stress recognition system using AI and machine learning. J Algebraic Stat. 13(2), 1731–1740 (2022)
16. Rejaibi, E., Komaty, A., Meriaudeau, F., Agrebi, S., Othmani, A.: MFCC-based recurrent neural network for automatic clinical depression recognition and assessment from speech. Biomed. Signal Process. Control **71**, 103107 (2022)
17. Rutowski, T., Shriberg, E., Harati, A., Lu, Y., Oliveira, R., Chlebek, P.: Cross-demographic portability of deep NLP-based. depression models. In: 2021 IEEE Spoken Language Technology Workshop (SLT), 978-1-7281-7066-4/20/$31.00 ©2021 IEEE. https://doi.org/10.1109/SLT48900.2021.9383609
18. Schultebraucks, K., Yadav, V., Shalev, A.Y. Bonanno, G.A., Galatzer-Levy, I.R.: Deep learning-based classification of posttraumatic stress disorder and depression following trauma utilizing visual and auditory markers of arousal and mood. PsychologicalMedicine 1–11. https://doi.org/10.1017/S0033291720002718

19. Aloshban, N., Esposito, A., Vinciarelli, A., What you say or how you say it? Depression detection through joint modeling of linguistic and acoustic aspects of speech. Cognitive Comput. https://doi.org/10.1007/s12559-020-09808-3
20. El Shazly, R.: Effects of artificial intelligence on English speaking anxiety and speaking performance: a case study. Expert Syst. **38**, e12667 (2021)
21. Garoufis, C., Zlatintsi, A., Filntisis, P.P., Efthymiou, N., Kalisperakis, E., Garyfalli. V., Karantinos, T., Mantonakis, L., Smyrnis N., Maragos, P.: An unsupervised learning approach for detecting relapses from spontaneous speech in patients with psychosis. In: Proceedings 2021 IEEE EMBS International Conference on Biomedical and Health Informatics (BHI), Athens, Greece, July 2021
22. Daus, H., Backenstrass, M. Feasibility and acceptability of a mobile-based emotion recognition approach for bipolar disorder. Int. J. Interact. Multim. Artif. Intell. **7**(2)
23. Sharma, A., Verbeke, W.J.M.I.: Improving diagnosis of depression with XGBOOST machine learning model and a large biomarkers Dutch Dataset ($n = 11,081$). Front. Big Data. **3**, Article 15 (2020). www.frontiersin.org
24. Wang, H., Liu, Y., Zhen, X., Tu. X.: Depression speech recognition with a three-dimensional convolutional network. Front. Hum. Neurosci. **15**, Article 713823 (2021). www.frontiersin.org
25. Villatoro-Tello, E., Pavankumar Dubagunta, S., Fritsch, J., Ramírez-de-la-Rosa, G., Motlicek, P., Magimai-Doss, M.: Late Fusion of the available lexicon and raw waveform-based acoustic modeling for depression and dementia recognition
26. Araño, K.A., Gloor, P., Orsenigo, C., Vercellis, C.: When old meets new: emotion recognition from speech signals. Cognitive Comput. **13**, 771–783 (2021). https://doi.org/10.1007/s12559-021-09865-2
27. Mou, L., Zhou, C., Zhao, P., Nakisa, B., Rastgoo, M.N., Jain, R., Gao, W.: Driver stress detection via multimodal fusion using attention-based CNN-LSTM. Expert Syst. Appl. **173**, 114693 (2021)

Chapter 3
A Strategic Technique for Optimum Placement and Sizing of Distributed Generator in Power System Networks Employing Genetic Algorithm

Prashant, Nirmal Kumar Agarwal, Sanjiba Kumar Bisoyi, Arun Kumar Rawat, and M. P. Kishore

Abstract Reducing power loss and meeting escalating load requirements are one of the most important goals in the distribution system. This paper discusses ways to optimize placement and sizing of distributed generators (DGs) to reduce power loss and meet increasing load requirements in the most efficient way, thereby supplying clean energy. The best size of distributed generating unit is found through a genetic algorithm inclusive of minimizing the losses in real power, and location is determined based on minimum real power loss and improving the stability index. The two goals of power loss reduction and stability are in conflict with each other. So, an optimal solution is achieved which meets both objectives. The genetic algorithm is employed to search out a group of optimal solutions satisfying those two objectives. To solve the extremely nonlinear issue of computing the total power loss under operational equality and inequality requirements, the genetic algorithm (GA) is applied. A simulation-based analysis is performed on the IEEE-14 bus system to verify the simulation results using Simulink.

Prashant (✉) · N. K. Agarwal · S. K. Bisoyi · A. K. Rawat
Department of Electrical Engineering, JSS Academy of Technical Education, Noida 201301, India
e-mail: prashant27@jssaten.ac.in

N. K. Agarwal
e-mail: nirmalkragarwal.eed@jssaten.ac.in

S. K. Bisoyi
e-mail: sanjibabisoyi.eed@jssaten.ac.in

A. K. Rawat
e-mail: arunrawat@jssaten.ac.in

M. P. Kishore
Department of Electrical and Electronics Engineering, Vidya Vikas Institute of Engineering and Technology, Mysuru 570028, India
e-mail: kishoremp3853@gmail.com

© The Authors(s), under exclusive license to Springer Nature Singapore Pte Ltd. 2024
P. K. Jha et al. (eds.), *Proceedings of Congress on Control, Robotics, and Mechatronics*, Smart Innovation, Systems and Technologies 364,
https://doi.org/10.1007/978-981-99-5180-2_3

Keywords Distributed generator · Stability index · Genetic algorithm · Power loss reduction

Nomenclature

Symbols	Description
DG	Distributed generator
S_{ij}	Complex power between 2 buses i and j
P_k	Active power
Q_k	Reactive power
k	Bus number
MW	Mega Watts
NR	Newton Raphson method
δ_i	Load angle
V_i	Voltage at ith bus
θ_i	Impedance angle at "i" bus
PDG	Active power of DG
QDG	Reactive power of DG
Y_k	Admittance at bus k
Z_{ij}	Impedance between bus i and j
I_k	Current at bus k
JR	Jacobian matrix operator

3.1 Introduction

System operators in distribution systems have implemented a number of operational measures to guarantee a steady and dependable supply of electricity to consumers while taking economic efficiency into account. The distribution systems in India are mostly radial because they are easier to work with and operate. The radial distribution systems are fed from a single point, and therefore, the power travels within it is unidirectional [1]. Distribution lines have a high ratio of R/X which then results in large bus voltage deviations, higher loss in power flow, and low stability of the system [2]. Loss of power line communication and improper voltage distribution are major problems for utilities [3]. Several researchers have tried to boost voltage profiles and increase efficiency of distribution systems [4, 5]. There are many different strategies proposed within the literature that to embed in small amounts of electricity with the distribution network in order to achieve the objectives mentioned. DG is a power supply near small generators or the load supplied. Modern advances in renewable technologies are promoting DG as a secure solution to these problems [6, 7]. The decentralized generation of electricity is becoming increasingly popular in the new technological era [8]. In such a distribution system incorporating DGs, the impact of minimizing voltage disturbances was assessed in [9]. By taking into account mending fault durations, the positioning and size of DGs in a loop arrangement were improved in [10]. Installing distributed generating units at a location other than the optimal one could increase line losses and may additionally bring down the voltage profile of the system [11]. As a result, it is crucial that perhaps the DG units' installation and size in the distribution network be toward the ideal and suitable location to optimize their advantages for both utilities and customers [12]. There are numerous different ways and approaches projected inside the literature for best placement of DGs to the distribution system [13–16]. In order to achieve proportionate reactive power distribution across DGs while preserving relatively low power loss, a mixed-integer linear programming (MILP) issue suitable for a droop based microgrid has indeed been suggested [17]. First, the ideal positioning and then sizing of DGs are examined. The water cycle algorithm is suggested for the ideal positioning and size of DGs and banks of capacitors. The advantages of the suggested technique include technological, financial, and ecological [18]. In [19] authors proposed a stochastic distributed power generation model that accounts for the concurrent deployment of capacitors and distributed energy. Authors in [20] offer a fresh method for solving the issue of distributed generation planning optimization utilizing a new heuristic known as the artificial hummingbird approach. The optimization problem is constructed in [21, 22] with numerous objectives, including power loss reduction, voltage stability margin, voltage deviation reduction, and annual economic savings, while taking into account different operational restrictions. The use of distributed generation in the power system makes the entire system more efficient, reliable, and secure. The DG should be used where it will have the most impact. It should be the size that is most effective. This paper focuses on finding the best location of DG based on minimal losses and also finding its optimal sizing using genetic algorithms for optimized operations of the power system.

3.2 Standard 14 Bus System

The network shown in Fig. 3.1 will be utilized for determining the optimal location and sizing of distributed generators as per load requirements.

3.3 Analysis of Optimal Flow Through Newton Raphson Method

Optimal power flow is a key for dependable and efficient operation of power grids. OPF deals with minimizing both the power losses in the distribution system and the cost of electricity that is charged by the utilities, without negatively impacting

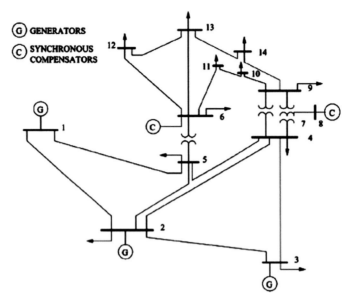

Fig. 3.1 Standard IEEE 14 bus model

the voltage profiles. OPF methods are used to find the best state of any system under system constraints, such as minimizing loss, meeting reactive power limits, and optimizing thermal conditions. By incorporating renewable energy sources into the power flow of the system, we were able to optimize the power flow under different constraints. This paper focuses about the Newton Raphson method, which is employed to enhance the operations of power systems. This method provides a faster convergence of the solution for load flow analysis with a technically optimized and economically stable system as compared to any other method. Newton Raphson method works with one initial condition and works more efficiently for systems with high loads as compared to other methods. The results expected for load current are phase angle, voltage magnitude, and real and reactive power. This paper simulates Newton Raphson's method for optimal power flow analysis [23] with IEEE 14 bus topology. Let Pk and Qk be the net active and reactive power injected into the network at the bus k.

$$I_k = Y_{k1}V_1 + Y_{k2}V_2 + \cdots + Y_{kn}V_N = \sum_{n=1}^{N} Y_{kn}V_n \tag{3.1}$$

$$P_k - jQ_k = V_k I_k^* \tag{3.2}$$

$$P_k - jQ_k = V_k^* \sum_{n=1}^{N} Y_{kn}V_n \tag{3.3}$$

On equating imaginary parts and real parts of Eq. (3.3),

$$P_k = \sum_{n=1}^{N} |Y_{kn}||V_k||V_n| \cos\cos(\theta_{kn} + \delta_n - \delta_k) \quad (3.4)$$

$$Q_k = \sum_{n=1}^{N} |Y_{kn}||V_k||V_n|(\theta_{kn} + \delta_n - \delta_k) \quad (3.5)$$

3.4 Genetic Algorithm

The genetic algorithm is an optimization approach drawn from Darwin's theory that is survival of the fittest. It is a type of search heuristic algorithm that employs natural biological evolution concepts like mutations and crossovers. It makes use of principle of natural selection to find solutions for finding the optimal DG locations in a distributed system. For solving most practical problems, genetic algorithms make use of three operators: reconstruction, crossover, and mutation.

- Genetic algorithms use encoding of a group of parameters in place of encoding the parameters themselves.
- Genetic algorithms help in developing use for objective function values in place of making use of traditional or additional knowledge present.
- Genetic algorithms at all instances do population search in place of an individual point search.
- Genetic algorithms always employ rules of probability instead of rules of determination. The entire procedure is shown in Fig. 3.2.

3.5 Problem Formulation

3.5.1 Minimize the Real Power Loss, Where

$$\alpha_{ij} = \frac{\left(r_{ij}^* \text{Cos}(\delta_i - \delta_j)\right)}{V_i V_j} \text{ and } \beta_{ij} = \frac{\left(r_{ij}^* \text{Sin}(\delta_i - \delta_j)\right)}{V_i V_j}$$

V_i = Voltage at ith bus and δ_i = load angle at ith bus

$$r_{ij} + x_{ij} = Z_{ij} \text{ is the impedance of line}$$

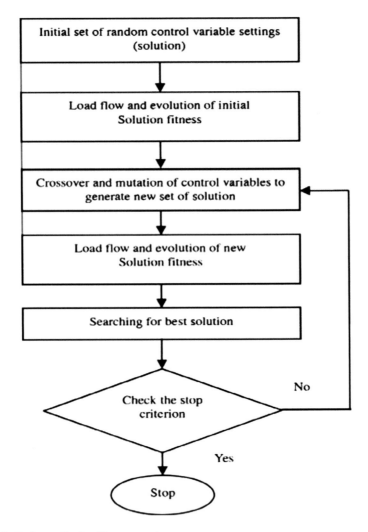

Fig. 3.2 Basic genetic algorithm approach

P_i and Q_i is the real and reactive power injected into ith bus

$$P_i = P_{DGi} - P_{Di} \tag{3.6}$$

$$Q_i = Q_{DGi} - Q_{Di} = aP_{DGi} - Q_{Di} \tag{3.7}$$

where $a = \dfrac{Q}{P}$

After substitution of these two equations in the exact loss formula, we can obtain the new equation as:

$$P_L = \sum_{i=1}^{N}\sum_{j=1}^{N}[\alpha_{ij}\{(P_{DGi} - P_{Di})P_j + (aP_{DGi} - Q_{Di})Q_j\} \\ + \beta_{ij}\{(P_{DGi} - P_{Di})P_j + (P_{DGi} - Q_{Di})Q_j\}] \quad (3.8)$$

On differentiating following equation with respect to P_{DGi},

$$\frac{\partial P_L}{\partial P_{DGi}} = 2 \times \sum_{j=1}^{N}[\alpha_{ij}\{(P_j + aQ_j)+\} + \beta_{ij}\{(aP_j - P_{Di})\}] = 0 \quad (3.9)$$

Finally, after substitution and simplification, we get Eq. (3.10)

$$P_{DGi} = \frac{\{\alpha_{ij}(P_{Di} + aQ_{Di}) + \beta_{ij}(aP_{Di} - Q_{Di}) - X_i - aY_i\}}{(a^2\alpha_{ii} + \alpha_{ii})} \quad (3.10)$$

where $X_i = \sum_{j=1}^{n}(\alpha_{ij}P_j - \beta_{ij}Q_j)$ and $Y_i = \sum_{j=1}^{n}(\alpha_{ij}Q_j - \beta_{ij}P_j)$ and $j \neq i$

By using the above formula, we can estimate the losses. For improving the stability index of power system, the formula for calculation of stability index is given as:

$$J_R = L - (\text{inverse of } H \times N) \quad (3.11)$$

jacobian matrix $= [HNJL]$ and $[E_1D_1N_1] =$ Eigen values of JR

The stability index is created to offer the signal for the stability of the buses and will also integrate the influence of DG in order to identify the weakest bus in the systems.

$$\text{stability index} = 4r_{ij}\frac{(P_{Di} - P_{DG})}{\{V_i Cos(\theta - \delta)\}^2} \quad (3.12)$$

The value of stability index should be less than 1 after placement of DG.
The constraints in the operation of the system are:
Voltage calculated at each bus of the power system should lie in the range given by:

$$V_i(\min) \leq v_i \leq V_i(\max) \quad (3.13)$$

The DG size should be restricted between:

$$P_{DGi}(\min) \leq P_{DGii} \leq P_{DGi}(\max) \qquad (3.14)$$

3.6 Genetic Algorithm Used for Finding Size and Location of DG

3.6.1 Minimization of Power Loss and Improving Stability Index

Genetic algorithms are a class of heuristic algorithms that are primarily based on the evolutionary concept of genetics and biological selection. The simple ideas of genetic algorithms encompass designing machines to simulate the techniques of natural systems essential for evolution, in particular for those the first principle of Charles Darwin "Survival of the fittest" is observed. Simple genetic algorithms (SGAs) had been originally defined by John Holland, who explained them as a notion of genetic evolution and supplied a conceptual structure of mathematics for adaptation. Genetic algorithms continuously modify a population of individual solutions. These models encompass three simple factors, which are "fitness" measures to adjust an individual's capacity to influence. The process of reproducing and selecting to generate children for the subsequent generation is used in this algorithm, and different genetic operators decide the biological composition of the offspring.

Algorithm

1. Analyze distribution system data—bus data and line data
2. Various parameters are initialized as:

   Population Size = 30
   No. of iteration = 100
   Rate of mutation = 0.3
   Rate of crossover = 0.7
   Size of string = 8.

   Inverse type mutation is used, and whole arithmetic recombination is used for crossover.
3. The initial population is randomly generated.
4. The iteration count is set to 1.
5. The distributed power line is run for the initial parent, and the fitness value is calculated. The voltage limits are checked for violation, and next parents are chosen.
6. Parents for crossbreeding are selected based on their characteristics.
7. Using crossover and mutation, a new seed is generated.

8. Aptitude for the new outcome is assessed (child). Parent and child solutions are combined (n), and the best solution array is selected (2n) dependent on the matching.
9. The termination conditions were analyzed. If the count of iteration is more than max, iteration is terminated; otherwise the process moves to the next step.

3.7 Result and Discussion

Tables 3.1, 3.2, 3.3 and 3.4 shows the real power loss in the network under investigation. The possible location for placement of DG for various loading conditions using Newton Raphson method based on minimum line losses are shown in Table 3.5 which is evident from Tables 3.2, 3.3 and 3.4. These locations are with minimum line losses. Figures 3.3, 3.4 and 3.5 shows the voltage profile enhancement after placement of DG for different loading conditions.

Tables 3.6, 3.7 and 3.8 shows the total real power loss reduction and stability index values after placement of DG showing the efficiency of the implemented algorithm.

Table 3.1 Real power line losses for existing loading conditions

S. No.	From bus	To bus	Real power loss (MW)
1	1	2	3.865
2	1	5	2.602
3	2	3	2.267
4	2	4	1.648
5	2	5	0.885
6	3	4	0.359
7	4	5	0.493
8	4	7	0
9	4	9	0
10	5	6	0
11	6	11	0.085
12	6	12	0.07
13	6	13	0.222
14	7	8	0
15	7	9	0
16	9	10	0.006
17	9	14	0.096
18	10	11	0.043
19	12	13	0.011
20	13	14	0.1

Table 3.2 Real power line loss for 10% increase in loading using NR Method

S. No.	From bus	To bus	Real power loss by using NR method (MW)
1	1	2	3.916
2	1	5	2.212
3	2	3	2.745
4	2	4	1.795
5	2	5	1.058
6	3	4	0.442
7	4	5	0.6
8	4	7	0.012
9	4	9	0.013
10	5	6	0.012
11	6	11	0.108
12	6	12	0.087
13	6	13	0.275
14	7	8	0.014
15	7	9	0.013
16	9	10	0.007
17	9	14	0.116
18	10	11	0.052
19	12	13	0.014
20	13	14	0.12

Table 3.9 shows the optimal sizing of DG obtained corresponding to optimal locations meeting the desired loading conditions. Tables 3.10, 3.11 and 3.12 shows the improved line losses connecting different buses for various loading conditions after placement of DG using genetic algorithm. It is observed that with the improvisation of stability index after placement of DG line losses reduce greatly in all the loading conditions.

3.8 Conclusions

This work shows an effective method for selecting the most optimized size and position of the distributed generation (DG) so as to meet the increasing load requirements in the most optimized manner including loss minimization and improvement in stability index for various loading conditions. The results conclusively present the fact that DG integration can be a very effective tool for reduction of various losses of the distribution system in addition to meeting the load demands. The studies have proven that the benefits of DG can only be attained efficiently if proper planning of

Table 3.3 Real power line loss for 30% increase in loading using NR Method

S. No.	From bus	To bus	Real power loss by using NR method (MW)
1	1	2	4.278
2	1	5	2.712
3	2	3	2.945
4	2	4	1.995
5	2	5	1.658
6	3	4	0.942
7	4	5	0.8
8	4	7	0.013
9	4	9	0.014
10	5	6	0.015
11	6	11	0.408
12	6	12	0.887
13	6	13	0.775
14	7	8	0.014
15	7	9	0.015
16	9	10	0.007
17	9	14	0.416
18	10	11	0.082
19	12	13	0.018
20	13	14	0.22

DG is made in terms of its placement and sizing. The optimal DG model will differ for every system, largely dependent on the system configuration, load requirements to be connected and its practical applications in the sustainable manner for clean energy.

Table 3.4 Real power line loss for 50% increase in loading using NR Technique

S. No.	From bus	To bus	Real power loss by using NR method (MW)
1	1	2	8.416
2	1	5	5.284
3	2	3	4.761
4	2	4	3.244
5	2	5	1.88
6	3	4	0.887
7	4	5	1.124
8	4	7	0.014
9	4	9	0.013
10	5	6	0.014
11	6	11	0.224
12	6	12	0.173
13	6	13	0.546
14	7	8	0.014
15	7	9	0.015
16	9	10	0.013
17	9	14	0.213
18	10	11	0.098
19	12	13	0.026
20	13	14	0.222

Table 3.5 Possible location for placement of DG based on minimum real power loss

Increased loading by (%)	Possible location for placement of DG based on minimum line loss
10	5.8
20	6.8
50	7.9

Availability of data and material: All data generated or analyzed during this study are included in this research article and any relevant information related to the current study are available from the corresponding author on reasonable request.

Competing interests: The authors have no conflicts of interest to declare.

Funding: No funding or any financial aid has been received for this particular research work.

3 A Strategic Technique for Optimum Placement and Sizing of Distributed …

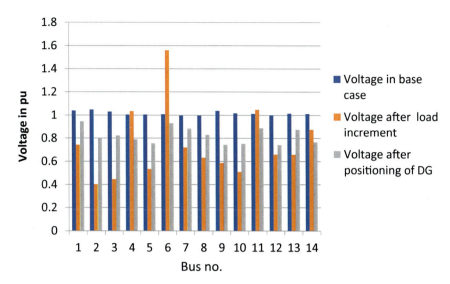

Fig. 3.3 Voltage profile for 10% increase in loading

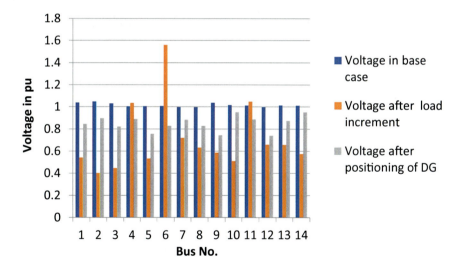

Fig. 3.4 Voltage profile for 30% increase in loading

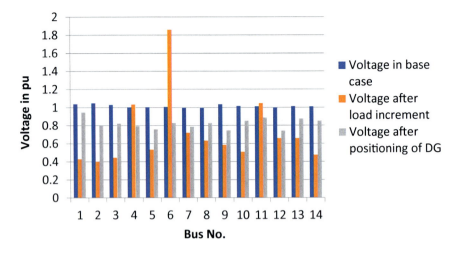

Fig. 3.5 Voltage profile for 50% increase in loading

Table 3.6 Total losses and stability index for 10% increase in loading

10% Increased loading	Total losses (MW)	Stability index
Without DG	13.54	2.0473
With DG	2.49	0.9573

Table 3.7 Total losses and stability index for 30% increase in loading

30% Increased loading	Total losses (MW)	Stability index
Without DG	18.14	2.0473
With DG	3.49	0.9573

Table 3.8 Total losses and stability index for 50% increase in loading

30% Increased loading	Total losses (MW)	Stability index
Without DG	27.11	2.0473
With DG	8.13	0.9573

3 A Strategic Technique for Optimum Placement and Sizing of Distributed ...

Table 3.9 Optimal size and optimal place of DG for different loading condition

Increased loading by (%)	Optimal location	Optimal size (in p.u.)
10	2	0.1367 s
	8	1.2625
	5	0.0109
	3	0.5765
30	3	0.0806
	6	0.0084
	8	0.0123
	4	0.1118
50	3	0.4396
	9	0.1552
	5	0.1157
	7	0.0202

Table 3.10 Real power line loss for 10% increase in loading after DG placement

S. No.	From bus	To bus	Real power losses before DG placement (MW)	Real power losses after DG placement (MW)
1	1	2	3.916	0.844
2	1	5	2.212	0.451
3	2	3	2.745	0.596
4	2	4	1.795	0.043
5	2	5	1.058	0.028
6	3	4	0.442	0.107
7	4	5	0.6	0.016
8	4	7	0.012	0
9	4	9	0.013	0
10	5	6	0.012	0
11	6	11	0.108	0.022
12	6	12	0.087	0.146
13	6	13	0.275	0.078
14	7	8	0.014	0
15	7	9	0.013	0
16	9	10	0.007	0.002
17	9	14	0.116	0.099
18	10	11	0.052	0.041
19	12	13	0.014	0.009
20	13	14	0.12	0.017

Table 3.11 Real power line loss for 30% increase in loading after DG placement

S. No.	From bus	To bus	Real power losses before DG placement (MW)	Real power losses after DG placement (MW)
1	1	2	4.278	1.046
2	1	5	2.712	0.705
3	2	3	2.945	0.429
4	2	4	1.995	0.383
5	2	5	1.658	0.209
6	3	4	0.942	0.203
7	4	5	0.8	0.133
8	4	7	0.013	0
9	4	9	0.014	0
10	5	6	0.015	0
11	6	11	0.408	0.008
12	6	12	0.887	0.078
13	6	13	0.775	0.102
14	7	8	0.014	0
15	7	9	0.015	0
16	9	10	0.007	0.009
17	9	14	0.416	0.124
18	10	11	0.082	0.049
19	12	13	0.018	0.011
20	13	14	0.22	0.007

Table 3.12 Real power line loss for 50% increase in loading after DG placement

S. No.	From bus	To bus	Real power losses before DG placement (MW)	Real power losses after DG placement (MW)
1	1	2	8.416	2.812
2	1	5	5.284	1.255
3	2	3	4.761	1.083
4	2	4	3.244	0.713
5	2	5	1.88	0.461
6	3	4	0.887	0.294
7	4	5	1.124	0.524
8	4	7	0.014	0
9	4	9	0.013	0
10	5	6	0.014	0
11	6	11	0.224	0.105
12	6	12	0.173	0.234
13	6	13	0.546	0.335
14	7	8	0.014	0
15	7	9	0.015	0
16	9	10	0.013	0.036
17	9	14	0.213	0.047
18	10	11	0.098	0.051
19	12	13	0.026	0.111
20	13	14	0.222	0.077

References

1. Kaneko, A., Hayashi, Y., Anegawa, T., Hokazono, H., Kuwashita, Y.: Evaluation of an optimal radial-loop configuration for a distribution network with PV systems to minimize power loss. IEEE Access **8**, 220408–220421 (2020). https://doi.org/10.1109/ACCESS.2020.3043055
2. Xu, Y., Dong, Z.Y., Wong, K.P., Liu, E., Yue, B.: Optimal capacitor placement to distribution transformers for power loss reduction in radial distribution systems. IEEE Trans. Power Syst. **28**(4), 4072–4079 (2013). https://doi.org/10.1109/TPWRS.2013.2273502
3. Wang, W., Jazebi, S., de León, F., Li, Z.: Looping radial distribution systems using superconducting fault current limiters: feasibility and economic analysis. IEEE Trans. Power Syst. **33**(3), 2486–2495 (2018). https://doi.org/10.1109/TPWRS.2017.2749144
4. Rajaram, R., Sathish Kumar, K., Rajasekar, N.: Power system reconfiguration in a radial distribution network for reducing losses and to improve voltage profile using modified plant growth simulation algorithm with distributed generation (DG). Energy Rep. **1**, 116–122 (2015)
5. Rao, R.S., Ravindra, K., Satish, K., Narasimham, S.V.L.: Power loss minimization in distribution system using network reconfiguration in the presence of distributed generation. IEEE Trans. Power Syst. **28**(1), 317–325 (2013)
6. Fathi, V., Seyedi, H., Ivatloo, B.M.: Reconfiguration of distribution systems in the presence of distributed generation considering protective constraints and uncertainties. Int. Trans. Electr. Energy Syst. **30**(5), 1–25 (2020)

7. Almabsout, E.A., El-Sehiemy, R.A., An, O.N.U., Bayat, O.: A hybrid local search-genetic algorithm for simultaneous placement of DG units and shunt capacitors in radial distribution systems. IEEE Access **8**, 54465–54481 (2020). https://doi.org/10.1109/ACCESS.2020.2981406
8. Amanulla, B., Chakrabarti, S., Singh, S.N.: Reconfiguration of power distribution systems considering reliability and power loss. IEEE Trans. Power Del. **27**(2), 918–926 (2012)
9. Xiao, J., Wang, Y., Luo, F., Bai, L., Gang, F., Huang, R., Jiang, X., Zhang, X.: Flexible distribution network: Definition, configuration, operation, and pilot project. IET Gener., Transmiss. Distrib. **12**(20), 4492–4498 (2018)
10. Hamad, A.-E., Hoballah, A., Azmy, A.M.: Defining optimal DG penetration for minimizing energy losses concerning repairing fault periods. In: Proceedings of the 18th International Middle East Power Systems Conference (MEPCON), 2016, pp. 1–6.
11. Yang, Z., Li, Y., Xiang, J.: Coordination control strategy for power management of active distribution networks. IEEE Trans. Smart Grid **10**(5), 5524–5535 (2019)
12. Viral, R, Khatod, D.K.: Optimal planning of distributed generation systems in distribution system: a review. Renew. Sustain. Energy Rev. **7**(16), 5146–5165 (2012)
13. Gandomkar, M., Vakilian, M., Ehsan, M.: Optimal distributed generation allocation in distribution network using Hereford Ranch algorithm. In: 2005 International Conference on Electrical Machines and Systems, 2005, vol. 2, pp. 916–918. https://doi.org/10.1109/ICEMS.2005.202678
14. Selim, A., Kamel, S., Alghamdi, A.S., Jurado, F.: Optimal placement of DGs in distribution system using an improved Harris Hawks optimizer based on single- and multi-objective approaches. IEEE Access **8**, 52815–52829 (2020) https://doi.org/10.1109/ACCESS.2020.2980245
15. Kizito, R., Li, X., Sun, K., Li, S.: Optimal distributed generator placement in utility-based microgrids during a large-scale grid disturbance. IEEE Access **8**, 21333–21344 (2020). https://doi.org/10.1109/ACCESS.2020.2968871
16. Bilal, M., Rizwan, M., Alsaidan, I., Almasoudi, F.M.: AI-Based Approach for Optimal Placement of EVCS and DG With Reliability Analysis. IEEE Access **9**, 154204–154224 (2021). https://doi.org/10.1109/ACCESS.2021.3125135
17. Gupta, Y., Doolla, S., Chatterjee, K., Pal, B.C.: Optimal DG allocation and volt–var dispatch for a droop-based microgrid. IEEE Trans. Smart Grid **12**(1), 169–181 (2021). https://doi.org/10.1109/TSG.2020.3017952
18. El-Ela, A.A.A., El-Sehiemy, R.A., Abbas, A.S.: Optimal placement and sizing of distributed generation and capacitor banks in distribution systems using water cycle algorithm. IEEE Syst. J. **12**(4), 3629–3636, Dec. 2018, doi: https://doi.org/10.1109/JSYST.2018.2796847
19. Pereira, B.R., da Costa, G.R.M., Contreras, J., Mantovani, J.R.S.: Optimal distributed generation and reactive power allocation in electrical distribution systems. IEEE Trans. Sustain. Energy 7(3), 975–984, July 2016. https://doi.org/10.1109/TSTE.2015.2512819
20. Shadman Abid, M., Apon, H.J., Morshed, K.A., Ahmed, A.: Optimal planning of multiple renewable energy-integrated distribution system with uncertainties using artificial hummingbird algorithm. IEEE Access **10**, 40716–40730 (2022). https://doi.org/10.1109/ACCESS.2022.3167395.
21. Ameli, A., Bahrami, S., Khazaeli, F., Haghifam, M.-R.: A multiobjective particle swarm optimization for sizing and placement of DGs from DG Owner's and Distribution Company's viewpoints. IEEE Trans. Power Deliv. **29**(4), 1831–1840 (2014). https://doi.org/10.1109/TPWRD.2014.2300845
22. Li, X., Wang, L., Yan, N., Ma, R.: Cooperative dispatch of distributed energy storage in istribution network with PV generation systems. IEEE Trans. Appl. Superconductivity **31**(8), 1–4 (2021), Art no. 0604304. https://doi.org/10.1109/TASC.2021.3117750.
23. Chen, J.-F., Chen, S.-D.: Multiobjective power dispatch with line flow constraints using the fast Newton-Raphson method. IEEE Trans. Energy Convers. **12**(1), 86–93 (1997). https://doi.org/10.1109/60.577285

Chapter 4
Model-Based Neural Network for Predicting Strain-Rate Dependence of Tensile Ductility of High-Performance Fibre-Reinforced Cementitious Composite

Diu-Huong Nguyen and Ngoc-Thanh Tran

Abstract High-performance fibre-reinforced cementitious composite (HPFRCC) has been demonstrated to provide superior tensile ductility and fracture energy compared to normal concrete at both quasi-static and dynamic strain rates. For this reason, this material becomes potential material for application to structures subjected to dynamic loading. However, there is still a lack of accuracy model for estimating strain-rate dependence of tensile ductility of HPFRCCs since most current empirical regression models have been proposed based on individual limited test data. In this study, a model-based neural network has been trained to estimate the strain-rate dependence of tensile ductility of HPFRCCs using 150 tensile test results. There are six input variables: matrix strength, fibre type, fibre length, fibre diameter, and fibre volume content, while strain-rate dependence of tensile ductility is output parameter. The results of prediction showed that the machine learning-based model was an efficient method to estimate strain-rate sensitivity in tensile ductility of HPFRCCs with high accuracy. By performing sensitivity analysis, the relative importance of all influencing factors was determined.

Keywords HPFRCC · Tensile ductility · Machine learning model · Prediction

D.-H. Nguyen · N.-T. Tran (✉)
Institute of Civil Engineering, Ho Chi Minh City University of Transport, Ho Chi Minh City, Vietnam
e-mail: ngocthanh.tran@ut.edu.vn

4.1 Introduction

High-performance fibre-reinforced cementitious composite (HPFRCC) has been demonstrated to provide superior tensile ductility and fracture energy compared to normal concrete at both quasi-static and dynamic strain rates by exhibiting strain-hardening behaviour with multiple fine cracks under tension [1, 2]. After reaching first cracking strength, fibres provide crack bridging capacity, and the tensile load-carrying capacity continues to increase up to the post cracking strength, resulting in strain-hardening behaviour accompanied by multiple micro-cracks. At the same time, the tensile ductility as well as energy absorption capacity increases significantly. Thus, tensile ductility is considered as the most important property which proves the significant difference between HPFRCC and normal concrete under both quasi-static and dynamic strain rates loading [3–5].

In order to evaluate the behaviour of HPFRCC and further apply it to structures subjected to dynamic loading, the strain-rate dependence of tensile ductility is very important. Several researches have demonstrated the strain-rate dependence of tensile ductility of HPFRCCs, and they investigated that the strain-rate sensitivity of tensile ductility of HPFRCCs was dependent on many factors [6–11]. Kim et al. [6] found that the strain-rate sensitivity in tensile ductility of HPFRCCs under low rates loading corresponding to seismic load was strongly influenced by strain rate, fibre type, volume fraction, and matrix strength. In the same manner, Wille et al. [7] proved that the strain-rate sensitivity in tensile ductility of HPFRCCs under low rates loading was also affected by the strain rate, fibre type, and volume content. Tran et al. [8] found that the strain-rate sensitivity in tensile ductility of HPFRCCs under high rates loading (impact loading) was influenced by the fibre length and specimen size. Similarly, fibre type, fibre volume content, fibre diameter, and specimen shape were investigated to effect on the strain-rate dependence of tensile ductility of HPFRCCs under high rates loading [9, 10]. On the other hand, Park et al. [11] investigated that the strain-rate dependence of tensile ductility of HPFRCCs exhibited relative improvements with increasing the strain rates from static to high rate of 170 s^{-1}.

Since strain-rate dependence of tensile ductility of HPFRCCs has been affected by a lot of factors including strain rate, fibre type, length, diameter, volume fraction, and matrix strength, this is a huge challenge to predict it exactly. Although many empirical regression models [11, 12] have been employed to estimate the strain-rate dependence in tensile ductility of HPFRCCs, these models showed low accuracy because they were developed based on fixing of individual limited test data, and they considered only the effect of strain rate, while other factors were not mentioned. Thus, it is necessary to develop more effective models for estimating tensile strength of HPFRCCs and the strain-rate dependence of tensile ductility of HPFRCCs with high accuracy and with the consideration of many factors at the same time. Recently, machine learning-based models have been demonstrated as an effective tool for estimating mechanical properties of HPFRCC with high accuracy and reliability [13, 14]. Unfortunately, machine learning-based estimation study for strain-rate dependence of tensile ductility of HPFRCCs is rarely mentioned.

The purpose of this study is to estimate the strain-rate dependence of tensile ductility of HPFRCCs using machine learning techniques. The main objectives are (1) to propose a model-based neural network to predict strain-rate dependence of tensile ductility of HPFRCCs; and (2) to find out the contribution of each variable factor.

4.2 Machine Learning-Based Model

4.2.1 Artificial Neural Network Approach

An approach-based neural network has been proposed to evaluate strain-rate dependence of tensile ductility of HPFRCCs. Normally, an artificial neural network (ANN) layout consists of three layers which are an input layer, hidden layers, and an output layer, as indicated in Fig. 4.1. In the operation of ANN method, the input layer receives the raw information and then transfers to the network. Next, hidden layers treat the information through the use of powerful nonlinear equations. Hidden layers play an important role in solving complex problems and obtaining outstanding results. Finally, the information is passed to output layer, and the prediction results are generated.

In the training process of neural network, each input neuron is received by the first hidden layer, and its output is computed, as performed in Eq. 4.1. Then, the output neurons in first hidden layer will become the input neurons in the next hidden layer until the final hidden layer.

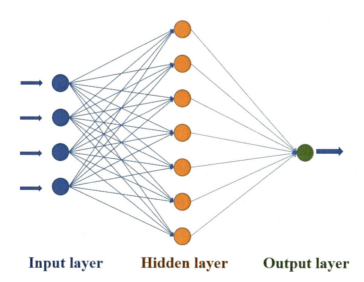

Fig. 4.1 Artificial neural network layout

Table 4.1 Tuning parameters of the ANN model

Parameters	Description
Input layer	1 layer
Hidden layer	1 layer
Output layer	1 layer
Input node	6 nodes
Hidden node	9 nodes
Output node	1 nodes
Activation function	Sigmoid
Optimizer	Gradient descent method
Loss function	Mean squared error
Learning rate	0.001

$$y = f\left(\sum_{i=1}^{n} x_i \times w_i + b\right) \quad (4.1)$$

where f is an activation function, x_i defines the input value, w_i is the weight, b defines the bias, and n defines number of node in input layer.

In this research, the model-based neural network includes three layers: an input, a hidden, and an output layers. The input layer consists six neurons, while the output layer includes one neuron. The reasonable size of hidden layers and number of neurons per hidden layer will be determined by experiments regarding the balance between the result quality and the running time. After trials with different scenarios, this study accepts one hidden layer with 9 nodes. During the trial process, the sigmoid activation function is used to shift the values between nodes of layers, and the gradient descent method is used in the back-propagation learning algorithm. The learning rate in the gradient descent method is maintained as 0.001. Table 4.1 provides all the tuning parameters of the ANN model.

4.2.2 *Experimental Data Collection*

To develop ANN model, 150 test results have been collected from 15 published studies [5–11, 15–22]. Table 4.2 provides the range of input and output parameters from direct tensile tests. Total six input parameters are evaluated, first input parameter is fibre type with three types of twisted, hooked and smooth, five other parameters are compressive strength of matrix, fibre volume fraction, fibre diameter, fibre length, and strain rate with the range of 28–230 MPa, 1–3%, 0.1–0.4 mm, 13–30 mm, and 0.0001–300 s^{-1}, respectively, while one output parameter is strain-rate sensitivity of tensile ductility with range of 0.09–10.5. Because of the wide range of the input variables, the data need to be preprocessed before the training of ANN. In order to obtain high quality data for the learning algorithm, each input variable is normalized

Table 4.2 Range of input and output parameters from experimental tests

Variable	Unit	Lower bound value	Upper bound value	Type
Fibre type		Twisted, hooked and smooth		Income
Matrix strength	MPa	28	230	Income
Fibre volume fraction	%	1	3	Income
Fibre diameter	mm	0.1	0.4	Income
Fibre length	mm	13	30	Income
Strain rate	s^{-1}	0.0001	300	Income
Strain-rate sensitivity of tensile ductility		0.09	10.5	Outcome

using its standard deviation and the mean. Among 150 test data, the training and testing samples are 80% and 20% of the total data of each case, respectively. The training process is iterated to minimize the mean squared error (MSE) between the predicting outcomes and the actual values. The maximum epoch is set to 1,000,000. Initial weights with a mean of zero are chosen randomly within the range (− 1, 1).

4.2.3 Statistical Measures

The accuracy of model-based neural network was investigated using three following measures: correlation coefficient (R), three statistical metrics, root mean squared error (RMSE), and mean bias error (MBE). These measures are formulated as follows.

$$R = \frac{s \sum_{i=1}^{n} \hat{t}_i t_i - \left(\sum_{i=1}^{n} \hat{t}_i\right)\left(\sum_{i=1}^{n} t_i\right)}{\sqrt{s \sum_{i=1}^{n} \left(\sum_{i=1}^{n} \hat{t}_i^2\right) - \left(\sum_{i=1}^{n} \hat{t}_i\right)^2} \sqrt{s \sum_{i=1}^{n} \left(\sum_{i=1}^{n} t_i^2\right) - \left(\sum_{i=1}^{n} t_i\right)^2}} \quad (4.2)$$

$$\text{RMSE} = \sqrt{\frac{\sum_{i=1}^{n}\left(\hat{t}_i - t_i\right)^2}{s}} \quad (4.3)$$

$$\text{MBE} = \frac{\sum_{i=1}^{n}\left(\hat{t}_i - t_i\right)}{s} \quad (4.4)$$

where s defines number of experimental data, $\hat{t}_i$ defines the outcome data, and t_i defines the actual data.

4.3 Results and Discussion

4.3.1 Performance of the Model-Based Neural Network

Figure 4.2 illustrates the accuracy of the model-based neural network for estimating strain-rate dependence of tensile ductility in the training set. The correlation between estimated data and experimental test data in the training dataset was found to perform well with R value of 0.967. Moreover, the RMSE and MBE values in the training set were 0.389 and 0.004, respectively, performing good accuracy of the model-based neural network in the estimation of the strain-rate sensitivity of tensile ductility.

The accuracy of the model-based neural network for estimating strain-rate dependence of tensile ductility in the testing set is shown in Fig. 4.3. Similarly, the correlation between estimated data and experimental test data in the testing set also exhibited very good with the R value of 0.941. Additionally, the RMSE and MBE values in the testing set were 0.792 and $-$ 0.065, respectively. Thus, the proposed model-based neural network could estimate strain-rate dependence of tensile ductility with good accuracy and reliability in both training and testing sets.

4.3.2 Sensitivity Analysis

In order to investigate the relative importance of all input variables, the sensitivity analysis was performed. In this study, an approach mentioned by God [23] was carried out to obtain the contribution of each input variable. Figure 4.4 illustrates the results of sensitivity analysis for all input variables. The prediction results indicated that strain rate was the most important variable influencing the strain-rate dependence of tensile ductility. On the contrary, fibre length was the less important variable. In

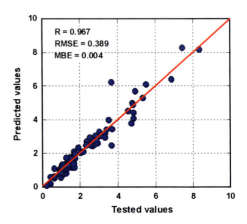

Fig. 4.2 Performance of the model-based neural network in training set

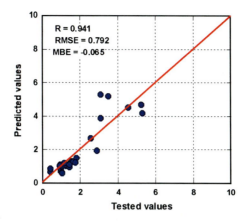

Fig. 4.3 Performance of the model-based neural network in testing set

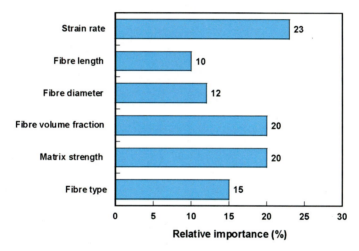

Fig. 4.4 Contribution of all input variables

addition, the performance ranking was found to be as follows: Strain rate > Fibre volume fraction ~ Matrix strength > Fibre type > Fibre diameter > Fibre length.

4.4 Conclusion

Total 150 test results have been collected from 15 published studies to construct a model-based neural network for predicting strain-rate dependence of tensile ductility of HPFRCCs. The proposed model considered the effects of six input parameters with complex relationships. From the results of the accuracy of model and the sensitivity analysis, the following conclusions can be figured out as follows:

- ANN model can be an effective tool for predicting strain-rate dependence of tensile ductility of HPFRCCs by considering influence of several input parameters at the same time.
- The proposed model-based neural network could estimate strain-rate dependence of tensile ductility with good accuracy and reliability in both training and testing sets.
- The strain rate was the most important variable influencing the strain-rate dependence of tensile ductility. The performance ranking was found to be as follows: Strain rate > Fibre volume fraction ~ Matrix strength > Fibre type > Fibre diameter > Fibre length.

References

1. Tran, T.K., Nguyen, T.K., Tran, N.T., Kim, D.J.: Improving the tensile resistance at high strain rates of high-performance fiber-reinforced cementitious composite with twisted fibers by modification of twist ratio. Structures **39**, 237–248 (2022)
2. Ngo, T.T., Tran, N.T., Kim, D.J., Pham, T.C.: Effects of corrosion level and inhibitor on pullout behavior of deformed steel fiber embedded in high performance concrete. Constr. Build. Mater. **280**(3), 122449 (2021)
3. Tran, N.T., Nguyen, D.L., Kim, D.J., Ngo, T.T.: Sensitivity of various fiber features on shear capacities of ultra-high-performance fiber-reinforced concrete. Mag. Concr. Res. **74**(4), 190–206 (2021)
4. Tran, N.T., Nguyen, D.L., Vu, Q.A., Kim, D.J., Ngo, T.T.: Dynamic shear response of ultra-high-performance fiber-reinforced concretes under impact loading. Structures **41**, 724–736 (2022)
5. Tran, T.K., Tran, N.T., Kim, D.J.: Enhancing impact resistance of hybrid ultra-high-performance fiber-reinforced concretes through strategic use of polyamide fibers. Constr. Build. Mater. **271**, 121562 (2021)
6. Kim, D.J., El-Tawil, S., Naaman, A.E.: Rate-dependent tensile behavior of high performance fiber reinforced cementitious composites. Mater Struct **42**, 399–414 (2009)
7. Wille, K., Xu, M., El-Tawil, S., Naaman, A.E.: Dynamic impact factors of strain hardening UHP-FRC under direct tensile loading at low strain rates. Mater. Struct. **49**, 1351–1365 (2016)
8. Tran, T.N., Tran, T.K., Kim, D.J.: High rate response of ultra-high-performance fiber-reinforced concretes under direct tension. Cem. Concr. Res. **69**, 72–87 (2015)
9. Tran, N.T., Tran, T.K., Jeon, J.K., Park, J.K., Kim, D.J.: Fracture energy of ultra-high-performance fiber-reinforced concretes at high strain rates. Cem. Concr. Res. **79**, 169–184 (2016)
10. Tran, N.T., Kim, D.J.: Synergistic response of blending fibers in ultra-high-performance concrete under high rate tensile loads. Cem. Concr. Compos. **78**, 132–145 (2017)
11. Park, S.H., Kim, D.J., Kim, S.W.: Investigating the impact resistance of ultra-high-performance fiber-reinforced concrete using an improved strain energy impact test machine. Constr. Build. Mater. **125**, 145–159 (2016)
12. Thomas, R.J., Sorensen, A.D.: Review of strain rate effects for UHPC in tension. Constr. Build. Mater. **153**, 846–856 (2017)

13. Ngo, T.T., Le, Q.H., Nguyen, D.L., Kim, D.J., Tran, N.T.: Experiments and prediction of direct tensile resistance of strain-hardening steel-fiber-reinforced concrete. Magazine of Concrete Research, Ahead of Print (2023). https://doi.org/10.1680/jmacr.22.00060
14. Tran NT, Nguyen TK, Nguyen DL, Le QH: Assessment of fracture energy of strain-hardening fiber-reinforced cementitious composite using experiment and machine learning technique. Structural Concrete, Early View (2022). https://doi.org/10.1002/suco.202200332
15. Tran, T.K., Tran, N.T., Nguyen, D.L., Kim, D.J., Park, J.K., Ngo, T.T.: Dynamic fracture toughness of ultra-high-performance fiber-reinforced concrete under impact tensile loading. Struct. Concr. **22**, 1845–1860 (2021)
16. Tran, T.K., Kim, D.J.: Investigating direct tensile behavior of high performance fiber reinforced cementitious composites at high strain rates. Cem. Concr. Res. **50**, 62–73 (2013)
17. Tran, T.K., Kim, D.J.: High strain rate effects on direct tensile behavior of high performance fiber reinforced cementitious composites. Cement Concr. Compos. **45**, 186–200 (2014)
18. Pyo, S., Wille, K., El-Tawil, S., Naaman, A.E.: Strain rate dependent properties of ultra high performance fiber reinforced concrete (UHP-FRC) under tensions. Cement Concr. Compos. **56**, 15–24 (2015)
19. Pyo, S., El-Tawil, S., Naaman, A.E.: Direct tensile behavior of ultra high performance fiber reinforced concrete (UHP-FRC) at high strain rates. Cem. Concr. Res. **88**, 144–156 (2016)
20. Fujikake, K., Senga, T., Ueda, N., Ohno, T., Katagiri, M.: Effects of strain rate on tensile behavior of reactive powder concrete. J. Adv. Concr. Technol. **4**, 79–84 (2006). https://doi.org/10.3151/jact.4.79
21. Cadoni, E., Meda, A., Plizzari, G.A.: Tensile behaviour of FRC under high strain-rate. Mater. Struct. **42**, 1283–1294 (2009)
22. Caverzan, A., Cadoni, E., Di Prisco, M.: Tensile behaviour of high performance fibre reinforced cementitious composites at high strain rates. Int. J. Impact Eng. **45**, 28–38 (2012)
23. God, A.T.C.: Back-propagation neural networks for modeling complex systems. Artif. Intell. Eng. **9**, 143–151 (1995)

Chapter 5
Performance Analysis of Various Machine Learning Classifiers on Diverse Datasets

Y. Jahnavi, V. Lokeswara Reddy, P. Nagendra Kumar, N. Sri Sishvik, and M. Srinivasa Prasad

Abstract Machine learning is used to analyze data from different perspectives, summarize it into useful information, and use that information to predict the likelihood of future events. Classification is one of the main problems in the field of machine learning. The aim here is to study various classification algorithms in machine learning applied on different kinds of datasets. The algorithms used for this analysis are J48, Naive Bayes, multilayer perceptron, and ZeroR. The performance is analyzed using various metrics such as true positive rate, false positive rate, and error rates such as root mean squared error and mean absolute error. The performance of J48 algorithm is better than other algorithms for large datasets. The proposed algorithm still increases the performance in terms of error rates for large datasets. The contemplated algorithm is eventuated by mutating the splitting paradigm in the tree-based algorithms. The experimental analysis demonstrates that the proposed algorithm has reduced error rate as compared with the traditional J48 algorithm.

Y. Jahnavi (✉)
Department of Computer Science, Dr. V S Krishna Govt Degree and PG College (Autonomous), Andhra University TDR-HUB, Visakhapatnam, Andhra Pradesh, India
e-mail: yjahnavi.2011@gmail.com

V. Lokeswara Reddy
Department of Computer Science and Engineering, K.S.R.M College of Engineering (Autonomous), Kadapa, Y. S. R (Dt), Andhra Pradesh, India

P. Nagendra Kumar
Department of Computer Science and Engineering, Geethanjali Institute of Science and Technology, Nellore, Andhra Pradesh, India

N. Sri Sishvik
Department of Computer Science and Engineering, Vellore Institute of Technology, Kelambakkam-Vandalur Road, Chennai, Tamil Nadu, India

M. Srinivasa Prasad
Department of Library Science, Dr. V S Krishna Govt Degree and PG College (Autonomous), Visakhapatnam, Andhra Pradesh, India

© The Author(s), under exclusive license to Springer Nature Singapore Pte Ltd. 2024
P. K. Jha et al. (eds.), *Proceedings of Congress on Control, Robotics, and Mechatronics*, Smart Innovation, Systems and Technologies 364,
https://doi.org/10.1007/978-981-99-5180-2_5

Keywords Machine learning · Classification · Tree based classifier · Splitting criteria

5.1 Introduction and Preliminaries

During the past several years, investigation has been concentrated on diverse groups based on the machine learning algorithms due to the extreme require of accurate prophecies. Machine learning is not about giving tight rules by analyzing the datasets rather it is used to predict the likelihood of future events with some certainty. Classification is a machine learning approach used to fore tell cluster association for documents illustration and is a widely used technique in various fields [1].

Machine learning, pervasive computing, statistical analysis, data analytics, etc., are the applications of artificial intelligence (AI), whereas machine learning allows training and strengthens from practice to estimate the eventualities [2–5, 5–8].

A well-known test sample label is correlated with a separate result from the model. The extent of precision of the proportion of instances of the test set is grouped consequently by the framework. If precision is tolerable, then this model is used to separate tuples of data class labels, which are unknown [9, 10].

5.2 Literature Work and Methodologies

Classification has been considered as a seminal issue in the area of machine learning [11]. All the time, there has been absolutely a number of enormous surveys on classification algorithms [12, 13], performance evaluation [14–16], collations, and assessment of various classification algorithms [2, 17] beside their uses in figuring out real-life problems in the applications of business [9, 18–21], engineering, medicine [1, 22, 23], etc.

Amudha and Abdul Rauf [24, 25] applied data mining techniques as an approach for intrusion detection to identify whether the deviation from normal usage patterns can be flagged as intrusions and performed a correlative investigation of various classification algorithms.

Voznika and Viana [26, 27], described different approximation algorithms such as statistical algorithms, genetic programming, neural networks and concluded that the best model can be found by trial and error trying different algorithms in order to obtain the best results possible.

Kesavaraj, Sukumaran [13] performed investigation on multifold categorization methods to furnish an exhaustive analysis of machine learning algorithms.

Chintan Shah and Anjali Jeevani [17] compared decision tree, K-nearest neighbor, Naive Bayesian using parameters like correctly classified illustrations, time taken, relative absolute error, kappa statistic, and root relative absolute error on breast cancer dataset.

Dogan and Tanrikulu [18], performed a study that collate and contrast the precision of the classification algorithms. The application of certain classification models on multiple datasets is done in three stages. The research addressed the reliability of the classifiers, studied by demonstration on various datasets.

Rutvija and Pandya [28] performed the extensive analysis on various categorization techniques.

Keerthana [29] focused on image classification approach in order to identify better algorithm for medical image classification.

Classification is an approach of grouping or allocating class labels to a pattern set under the direction of an instructor. Classification is also termed as supervised learning. The patterns are primarily segregated into training and test sets. Training set is used to prepare the classifier, and the test set is prone to estimate the precision of a classifier. The classifiers are categorized into tree-based, rule-based, Bayes, functions, etc. The algorithms that have been chosen for this predictive data mining task include J48 from trees, multilayer perceptron from functions, ZeroR from rules, and Naive Bayes from Bayes.

The most popular supervised classifier which can work well on noisy data is decision tree classifier. There are various other types of classifiers such as Bayesian classification, neural network-based classifier, and support vector machine. A great deal of research has been done for developing efficient methods in the field of machine learning.

The J48 algorithm uses information gain and gain ratios to construct the decision tree for a given dataset. It works by recursively dividing the data on a single attribute, according to the information gain calculated. Each split in the tree represents a node where a decision must be taken, and you go to the following node and the next till you reach the leaf that expresses you the predicted output.

The steps in the J48 algorithm are as follows:

(i) If the requirements are the identical group, the tree illustrates the leaf so that the leaf is substituted by designating in the identical class.
(ii) The feasible information is intended for every characteristic, determined by a check on the attribute. Then, the gain in information is premeditated that would outcome from a examination on the characteristic (attribute).
(iii) Then the best characteristic (attribute) is identified on the foundation of the current selection criterion and that attribute is adopted for ramification.

5.3 Information Gain and Gain Ratio

The information gain is based on the entropy after a dataset is split on an attribute, where entropy is used to estimate the similarity of a sample. If the instance is completely identical, the entropy is zero, and if the instance is evenly separated, it has entropy of one.

The entropy is calculated using the following formula.

$$\text{entropy}(p1, p2, \ldots, pn) = -p1 \log p1 - p2 \log p2 - \cdots - pn \log pn \quad (5.1)$$

$$\text{entropy}(p1, p2, \ldots, pn) = -\sum pi \log pi \quad (5.2)$$

Entropy on the other hand is an estimate of *impurity*. It is characterized for a binary class with estimates a and b as:

$$\text{entropy} = -p(a) * \log(P(a)) - p(b) * \log(p(b)) \quad (5.3)$$

Using the above formula, we calculate two entropy values, namely entropy before and entropy after. The entropy before value is calculated before splitting, and entropy after is computed after considering the split. Now by assimilating the entropy before and after the split, we derive an estimate of information gain as denoted below:

$$\text{Information gain} = \text{entropy before} - \text{entropy after} \quad (5.4)$$

At each node of the tree, this computation is carried out for every feature, and the feature with the largest information gain is chosen for the split in a greedy manner. This process is applied recursively from the root-node down and stops when a leaf node contains instances all having the same class, i.e., it stops when the node cannot be divided further. Constructing a decision tree is all about finding attribute that has the highest information gain.

Gain ratio is a modification of the information gain that reduces its bias. It takes into account the number and size of branches while choosing an attribute. There are chances of getting negative values in the existing information gain and gain ratio algorithms.

The idea of the proposed algorithm is to eliminate the negative values. The accuracy of the algorithm can be improved by eliminating the negative values. The proposed algorithm checks if the entropy before value is less than entropy after value and return 0; otherwise, it returns the unknown rate calculated.

Because of the outliers pruning is a significant step to the result. Some instances are present in all datasets which are not well defined and differ from the other instances on its neighborhood.

The classification is performed on the instances of the training set, and tree is formed. There exist various algorithms for performing classification, extracting salient features, opinion mining, processing of scalable web log data using map reduce framework etc. [30–37]. The pruning is performed for decreasing classification errors which are being produced by specialization in the training set. Pruning is performed for the generalization of the tree.

5.4 Results and Discussion

Evaluation of the datasets has been done by using the proposed classification algorithm. Various classification algorithms analyzed are evaluated by the evaluation criteria such as true positive rate, false positive rate, mean absolute error, and root mean square error. Heart disease, mushrooms, and birds are the datasets used for the analysis. These two datasets are taken from UCI machine learning repository. The classification algorithms are applied on the data using tenfold cross validation technique, and the results are then recorded. A sample description of datasets has been represented in Table 5.1.

The considered sample heart disease dataset has 14 attributes and 303 instances that are categorized into 5 classes. Mushrooms dataset has 23 attributes and 8124 instances that are categorized into 2 classes. Experimentation has been done on each dataset.

5.4.1 Data Set 1 (Heart Disease Dataset)

The considered sample heart disease dataset has 14 attributes and 303 instances that have only 5 classes. This dataset has taken from UCI machine learning repository.

True positive rate, false positive rate, root mean square error, and mean absolute error are calculated for J48, Naive Bayesian, multilayer perceptron, ZeroR, and the proposed algorithm, which are represented in Table 5.2.

True positive rate of J48, Naive Bayesian, multilayer perceptron, ZeroR, and the proposed algorithm are 0.558, 0.559, 0.574, 0.541, and 0.558, respectively. It shows that the proposed modified algorithm is able to show the same performance as J48 algorithm.

Table 5.1 Sample description of datasets

Dataset	Attributes	Instances	Classes
Heart disease	14	303	5
Mushrooms	23	8124	2

Table 5.2 Results of the classification algorithms on heart disease dataset

	TP	FP	Mean absolute error	Root mean squared error
J48	0.558	0.238	0.2	0.3867
NB	0.559	0.273	0.1838	0.3368
MLP	0.574	0.189	0.2768	0.378
ZERO R	0.541	0.541	0.2591	0.3592
Modified algorithm	0.558	0.238	0.2	0.3367

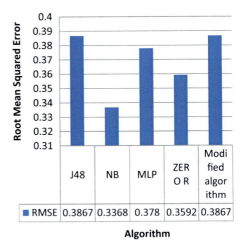

Fig. 5.1 Comparison of root mean squared error of classifiers on heart disease dataset

False positive rate of J48, Naive Bayesian, Multilayer Perceptron, ZeroR, and the proposed algorithm are 0.238, 0.273, 0.189, 0.541, and 0.238, respectively. It shows that the proposed modified algorithm is able to show the same performance as J48 algorithm.

Mean absolute error of J48, Naive Bayesian, multilayer perceptron, ZeroR, and the proposed algorithm are 0.2, 0.1838, 0.2768, 0.2591, and 0.2, respectively. It shows that J48 and the proposed modified algorithm are better in terms of mean absolute error, compared with ZeroR and multilayer perceptron. But for this dataset Naive Bayesian performs better by exhibiting low mean absolute error, i.e., 0.1838.

Root mean squared error of J48, Naive Bayesian, multilayer perceptron, ZeroR, and the proposed algorithm are 0.3867, 0.3368, 0.378, 0.3592, and 0.3367, respectively. It shows that the proposed modified algorithm is better in terms of root mean squared error, compared with the other algorithms, i.e., 0.3367.

For the considered dataset, the proposed modified algorithm shows better performance than other algorithms in terms of the root mean square error. The root mean square error values of various algorithms are pictorially represented in Fig. 5.1 on heart disease dataset.

5.4.2 Data Set 2 (Birds Dataset)

Birds dataset used here is an images dataset. It contains 600 images of 34 samples of 6 types of birds. We applied some filters before classifying the data. This dataset has taken from a Ponce research group repository. True positive rate, false positive rate, root mean square error, and mean absolute error are calculated for J48, Naive Bayesian, multilayer perceptron, ZeroR, and the proposed algorithm, which are represented in Table 5.3.

5 Performance Analysis of Various Machine Learning Classifiers ...

Table 5.3 Results of the classification algorithms on a birds dataset

	TP	FP	Mean absolute error	Root mean squared error
J48	0.372	0.126	0.213	0.2536
NB	0.325	0.095	0.2726	0.3421
MLP	0.390	0.102	0.2774	0.3784
ZERO R	0.165	0.169	0.2778	0.3727
Modified algorithm	0.372	0.026	0.213	0.2436

True positive rate of J48, Naive Bayesian, multilayer perceptron, ZeroR, and the proposed algorithm are 0.372, 0.325, 0.390, 0.165, and 0.372, respectively. It shows that the proposed modified algorithm is able to show the same performance as J48 algorithm.

False positive rate of J48, Naive Bayesian, multilayer perceptron, ZeroR, and the proposed algorithm are 0.126, 0.095, 0.102, 0.169, and 0.026, respectively. It shows that the proposed modified algorithm is able to show better performance compared with all the considered algorithms.

Mean absolute error of J48, Naive Bayesian, multilayer perceptron, ZeroR, and the proposed algorithm are 0.213, 0.2726, 0.2774, 0.2778, and 0.213, respectively. It shows that J48 and the proposed modified algorithm is better for the considered dataset in terms of mean absolute error, compared with Naive Bayesian, ZeroR, and multilayer perceptron.

Root mean squared error of J48, Naive Bayesian, multilayer perceptron, ZeroR, and the proposed algorithm are 0.2536, 0.3421, 0.3784, 0.3727, and 0.2436, respectively. It shows that the proposed modified algorithm is better in terms of root mean squared error, compared with the other algorithms, i.e., 0.2436.

For the considered dataset, the proposed modified algorithm shows better performance than other algorithms in terms of the root mean square error. The root mean square error values of various algorithms are pictorially represented in Fig. 5.2 on birds dataset.

The experimentation has shown that the proposed algorithm outperforms other existing algorithms on various considered datasets.

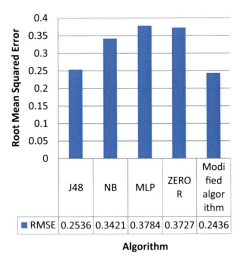

Fig. 5.2 Comparison of root mean squared error of classifiers on a bird's dataset

5.5 Conclusion

This study focuses on finding the right algorithm for classification of diverse datasets. The datasets that used are heart disease and mushrooms. For mushrooms dataset, the decision tree algorithm J48 and the proposed modified algorithm gave better results in terms of root mean squared error (RMSE) and for heart diseases dataset; the proposed modified algorithm gave reduced error rates than the traditional J48 algorithm, Naïve Bayes, multilayer perceptron, and ZeroR. However, it is noticed that the performance of a classifier depends on the dataset used. Although there are many algorithms available, the best one is often found by trial and error. For better results, one must compare or even combine the available algorithms. The performance of the existing algorithms can even be improved with minor modifications.

References

1. Vandehei, B., et al.: Leveraging the defects life cycle to label affected versions and defective classes. ACM Trans. Softw. Eng. Methodol. **30**, 2 (1–35) (2021), Online publication date: 1-Mar-2021
2. Jahnavi, Y., Radhika, Y.: A cogitate study on text mining. Int. J. Eng. Adv. Technol. **1**(6), 189–196 (2012)
3. Jahnavi, Y., Radhika, Y.: Hot topic extraction based on frequency, position, scattering and topical weight for time sliced news documents. In: 15th International Conference on Advanced Computing Technologies, ICACT 2013
4. Jahnavi, Y.: Statistical data mining technique for salient feature extraction. Int. J. Intell. Syst. Technol. Appl. (Inderscience Publishers), **18**(4) (2019)
5. Jahnavi, Y., et al.: A novel ensemble stacking classification of genetic variations using machine learning algorithms. Int. J. Image Graph. 2350015(2021)

6. Jahnavi, Y., Radhika, Y.: FPST: a new term weighting algorithm for long running and short lived events'. Int. J. Data Anal. Tech. Strategies **7**(4), 366–383 (2015)
7. Jahnavi, Y.: A new algorithm for time series prediction using machine learning models. Evol. Intell. (Springer), Accepted, 2022
8. Jahnavi, Y.: Analysis of weather data using various regression algorithms. Int. J. Data Sci. (Inderscience Publishers), **4**(2) (2019)
9. Parneetkaur, et al.: Classification and prediction based data mining algorithms to predict slow learns in education sector. In: 3rd International Conference on Recent Trends in Computing, 2015
10. Shiqun, et al.: Research and implementation of classification algorithm on web text mining. In: IEEE International Conference on Semantics Knowledge and Grid, 2007, pp. 446–449
11. Singh, P., et al.: Machine learning: a comprehensive survey on existing algorithms (2021)
12. Gulati, H.: Predictive analytics using data mining technique. In: 2015 2nd International Conference on Computing for Sustainable Global Development (INDIACom), 2015, pp. 713–716
13. Binjubeir, M., et al.: Comprehensive survey on Big Data privacy protection. Access IEEE **8**, 20067–20079 (2020)
14. Kumar, et al.: Performance evaluation of decision tree versus artificial neural network based classifiers in diversity of datasets. In: 2011 World Congress on Information and Communication Technologies, 2011, pp. 798–803. https://doi.org/10.1109/WICT.2011.6141349
15. Rjeily, B., Andres, E., et al.: Medical data mining for heart diseases and the future of sequential mining in medical field. In: Tsihrintzis, G., Sotiropoulos, D., Jain, L. (eds.) Machine Learning Paradigms. Intelligent Systems Reference Library, vol 149. Springer, Cham (2019). https://doi.org/10.1007/978-3-319-94030-4_4.
16. Sharma, S., et al.: Machine learning techniques for data mining: a survey. In: IEEE National Conference on Computational Intelligence and Computing Research (ICCIC), 2013
17. Panda, M., et al.: International proceedings on advances in soft computing. Intell. Syst. Appl. **628**, 363 (2018)
18. Dogan, W., Tanrikulu, Z.: A comparitive analysis of classification algorithms in data mining for accuracy, speed & robustness. Springer **14**(2), 105–124 (2013)
19. Panchal, G., et al.: Performance analysis of classification techniques using different parameters. In: Second International Conference, ICDEM, 2010
20. Tamizharasi, K., UmaRani, Performance analysis of various data mining algorithms. Int. J. Comput. Commun. Inf. Syst. (IJCCIS) **6**(3) (2014)
21. Korting, T.S.: "C4.5 algorithm and multivariate decision trees", image processing division. In: National Institute for Space Research--INPE
22. Ian, et al.: Data Mining Practical Machine Learning Tools and Techniques, 3rd edn. Elsevier (2011)
23. Zuveda, Comparison of the performance of several data mining methods for bad debt recovery in the health care industry. J. Appl. Bus. Res. **21** (2005)
24. Gayatri, N., et al.: Performance analysis of data mining algorithms for software quality prediction. In: International Conference on Advances in Recent Technologies in Communication and Computing, ARTCom'09, India, October 2009
25. Amudha, P., Abdul Rauf, H.: Performance analysis of data mining approaches in intrusion detection. In: PACC, International Conference, 2011
26. Voznika, F., Viana, L.: Data Mining Classification. Springer (2001)
27. Peter, T.J., Somasundaram, K.: An empirical study on prediction of heart disease using classification data mining techniques. In: IEEE-International Conference on Advances in Engineering Service and Management, 2012
28. Pandya, R., Pandya, J.: C5.0 Algorithm to improved decision tree with feature selection and reduced error pruning. Int. J. Comput. Appl. **117**(16) (2015), (0975-8887)
29. Keerthana, P., et al.: Performance analysis of data mining algorithms for medical image classification. Int. J. Comput. Sci. Mob. Comput. **5**(3) (2016)
30. Jahnavi, Y.: A New Term Weighting Algorithm for Identifying Salient Events. LAP LAMBERT Academic Publishing (2018)

31. Tiwari, V., et al.: Applications of the Internet of Things in healthcare: a review. Turk. J. Comput. Math. Educ. **12**(12), 2883–2890 (2021)
32. Haripriya, et al.: Using social media to promote E-commerce business. Int. J. Recent Res. Aspects **5**(1), 211–214 (2018)
33. Sukanya, et al.: Country location classification on Tweets. Indian J. Public Health Res. Dev. **10**(5) (2019)
34. Vijaya, et al.: Community-based health service for Lexis Gap in online health seekers
35. Bhargav, et al.: An extensive study for the development of web pages. Indian J. Public Health Res. Dev. **10**(5) (2019)
36. Srivani, et al.: An approach for opinion mining by acumening the data through exerting the insights
37. Jahnavi, Y., et al.: A novel processing of scalable web log data using map reduce framework. In: Proceedings of CVR 2022, Computer Vision and Robotics, ISBN: 978-981-19-7891-3

Chapter 6
Three-Finger Robotic Gripper for Irregular-Shaped Objects

Shripad Bhatlawande, Mahi Ambekar, Siddhi Amilkanthwar, and Swati Shilaskar

Abstract The development of industrial and service robots in recent years has attracted a lot of attention to robotics research. Commercially available robotic grippers are sometimes costly and difficult to customize for individual applications. Therefore, an open-source, low-cost three-finger robotic gripper has been introduced in this paper. The main focus was to create a 3D-printed model. 3D printing technology solves the issues of cost and weight for the implementation of designs. Linear-bearing LM8UU hard-chrome smooth rods were used in the design. Flexible couplings were used to provide motion to the fingers of the gripper. DC servomotor was used as an actuator in the model. The proposed system used Raspberry Pi as the processor and L298N as the motor driver. The gripper was designed to provide the ability to perform grasping a variety of household objects. The gripper was tested on 7 household objects which included a square box, cylindrical object, broom, cup mug, bag, and bottle. The gripper performed well while grasping different objects. The gripper was successful in grabbing objects of various shapes and sizes to the weight of 1000 gm.

Keywords Grasping · Gripper · 3D Printing · Raspberry Pi · Motor driver

---

S. Bhatlawande · M. Ambekar · S. Amilkanthwar · S. Shilaskar (✉)
Vishwakarma Institute of Technology, Pune, India
e-mail: swati.shilaskar@vit.edu

S. Bhatlawande
e-mail: Shirpad.bhatlawande@vit.edu

M. Ambekar
e-mail: mahi.ambekar19@vit.edu

S. Amilkanthwar
e-mail: siddhi.amilkanthwar19@vit.edu

6.1 Introduction

In today's world, robots are evolving and transforming at a high pace. Handling equipment has enabled industrial robots to compete with humans for occupations that were formerly performed by humans. There are robots for nursing in hospitals, service robots for administering, search and rescue robots in many areas, underwater oil transfer lines robots to inspect underwater, and medical robots for surgery. Robots have a variety of applications. Simple, dependable, adaptable, and low-cost robotic grippers are in high demand. Assembly jobs, machine tending, handling sensitive hardware in laboratories, and picking and placing different objects are the generalized uses for grippers. As a result, grippers are a good asset for reducing physical labor.

Many existing robotic grippers have been studied and analyzed in the following section.

The study [1] describes the creation of a three-fingered soft robotic gripper based on pneumatic networks. The gripper is capable of working in both positive and negative pressure room. Designing and building a flexible gripper with three fingers [2] based on hand gestures was proposed in the study. 3D printing was used to fabricate the fingers of the proposed gripper. An adaptive gripper [3] has been used to carry out gripping and holding operations, using a data hand glove for command input. Additionally, the paper describes how to control the adaptive robotic gripper with finger gestures and how to create a graphic user interface using an MFC C++ application. A rigid and flexible coupling for a three-finger gripper [4] has been proposed for picking fruits. A motor was used to control the soft finger consistently in order to simplify the control system. The soft gripper system used the proportional integral derivative (PID) control method. The research proposed to build and develop a seven-degree-of-freedom underactuated three-finger gripper [5]. The goal was to use the robot operating system framework to mimic a flexible gripper that can grasp objects of various shapes. The paper goes over the various design and simulation steps involved, including specification creation, CAD modeling, and torque computation for motor selection.

A study [6] proposes a motor-driven, three-finger robotic gripper to grasp size-variable and fragile objects with proper grip. In this study, the proposed finger design is produced using 3D printing and thermoplastic elastomers (TPE) with a 50% infill percentage. The developed gripper is put on an industrial robot to demonstrate its capacity to handle a variety of fragile objects of various sizes. In order to design and build a three-finger gripper [7], a study used prototype materials along with the insertion of force sensing resistors (FSRs) on each finger. This project intended to build and implement an electric gripper based on conventional three-finger. To demonstrate its use in real-world circumstances, the constructed gripper was connected to an industrial robotic arm. An accessible, non-expensive three-finger robotic gripper [8] was a 3D underactuated gripper built for scientific and learning applications. SolidWorks CAD software was used to generate the 3D model of the gripper. A prosthetic gripper with three opposed fingers functioned by a hydraulic actuator [9] was built without any electronic components. Pushing a lever on an operating interface located on the affected side of the user's upper arm controlled all three fingers at the same time. Users could open the fingers with 16.6 N of force due to the hydraulic actuator. A paper demonstrated the design and performance of an underactuated three-fingered robotic hand [10]. The Yale OpenHand team used methods from the Yale prototype to create a hand. The sole drawback is that the maximum distance between adjacent fingers for a given prismatic joint diminishes as the number of fingers increases. A teleoperated three-finger robot [11] that moves by dragging objects is presented in a study. The components used here are an analog joystick, Arduino, servomotors, and flexible force sensors. Its demerit is that it could not grab small objects due to its large physical size. Researchers have created a three-finger eight-degree-of-freedom (DoF) hand that exerts a 100-N grabbing force with force-magnification drives. Another study developed an eight-degree hand with three fingers [12]. The trials showed that the hand can grab objects with a force of 100 N and that all joints can move at speeds above 400/s. It is a more complex hand than a normal hand. A three-finger robotic hand with seven degrees of freedom [13] was designed with the help of SolidWorks. Aluminum and perspective plastic was used for the design. An underwater three-finger gripper with cable drive [14] instead of focusing on the use of commercial components limited the number of sealed parts. A three-finger multi-Degree of freedom [15] was an easily-controllable robotic gripper based on a four-bar linkage. Recent theoretical advances in modeling and analyzing rigid structure frictional constraints to a three-finger gripper [16], drawing on implications for gripper design were analyzed in a study. A robotic gripper capable of grabbing cylindrical objects [17] is built. The key characteristic of this design is its dependability and efficiency while remaining simple. However, its size was a major disadvantage.

Steps for deploying a multi-link mechanism-based handling gripper [18] were studied. The step that constrains the driven portion will be contained when the driving control bar is released. This prototype used an embedded computer and a data acquisition system (DAQ) for signal processing.

An innovative adaptable critic-based NN controller [19] for commanding the fingers to maintain a trajectory was described in detail. A critical signal and an outer PD force control loop make up the gripper controller. The processes were shown with a three-finger gripper in this work, and the methods were then translated into constructing an appropriate hybrid position force controller. Using modular three-finger grippers, a new method for "peg-in-hole" assembly [20] was proposed. Using finger position data, a method was developed for computing the peg's center point and changing the center point to determine the hole's direction. Using finger position information, this approach determined the height of the peg's center and predicted the direction of the hole when the peg was moved with the fingers. A grip controller that deals strongly with uncertainty by leveraging inputs from multiple contact-based sensors [21] created a strong sensor-based grip primitive that requires minimal information in order to complete its duty and can rectify and adapt to variances and imperfections. Oz et al. [22] presented a 3D-printed soft robotic gripper that is capable of handling objects with varying shapes and sizes. Singh et al. proposed a deep learning-based approach to enhance the performance of robotic grippers when grasping irregular objects [23]. Wagle et al. developed a soft robotic gripper equipped with sensors to improve its ability to grasp irregular objects. The studies collectively emphasize the importance of developing adaptable robotic grippers for better handling of irregularly shaped objects, with potential applications in various industries such as manufacturing and health care [24].

6.2 Methodology

To execute a variety of jobs that are performed by a human hand, a robotic gripper with three fingers is an effective solution. So, taking into account the goal of the proposed system, which is grasping various household objects, a gripper with three fingers was executed. The overall block diagram of the proposed system is shown in Fig. 6.1. A processor-based system provides commands to the motor driver which then provides inputs to the motor. The motor connected to the gripper changes the motion of all three fingers of the gripper. The gripper moves in forward and backward directions and grasps objects.

6.2.1 Design

The proposed system helps in grasping various objects. SolidWorks software was used to design the gripper. There is one phalange in each of the three fingers. The three fingers are designed at a 120-degree angle to each other. This is due to the fact that fingers at this angle have the potential to grasp items more effectively. The implemented design of the gripper along with circuits is shown in Fig. 6.2.

6.2.2 Mechanism of Finger

The joint determines the motion of the fingers. The actuation mechanism in the three-finger gripper application allows the fingers to flex and perform basic tasks like

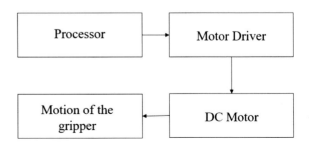

Fig. 6.1 Block diagram of the system

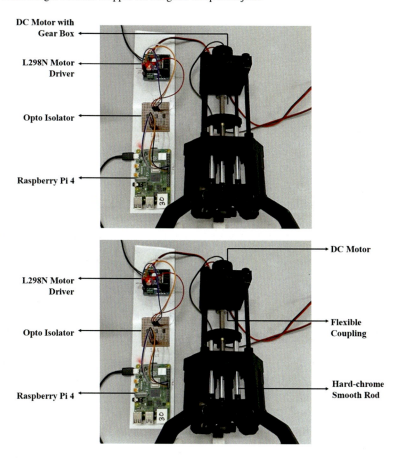

Fig. 6.2 Implemented design of the model

gripping. The flexible coupling mechanism has been used in the model to provide motion and limit the number of actuators. DC motor has been used as an actuator in the model because of its low cost, tremendous torque, and ease of control. Due to its tiny size and lightweight nature, DC motors can be inserted in joints or robotic hands. As a result, the size, cost, and control mechanism of the gripper are reduced. It has a large number of controlled degrees of freedom, making it ideal for grasping objects. It rotates to an angle of 360°. The gripper has the ability to firmly grab a variety of things by combining the grasping states of three finger.

Fig. 6.3 CAD design of the gripper

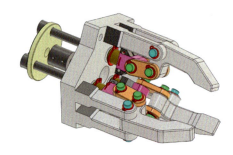

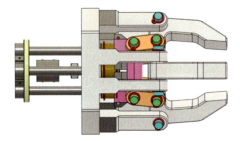

6.2.3 3D Printing

The three-finger gripper model is made with ABS material and 3D printers. ABS is a robust, heat-resistant, and long-lasting thermoplastic. The three fingers and the handle are 3D printed in the gripper. To establish revolute joints between each segment of the finger, joint nut bolts are put into each phalanx. The complete design of a 3D printed gripper is shown in Fig. 6.3 [25].

6.2.4 DC Motor-Based Actuations

The actuator used in this system is a 12v DC motor. One DC servomotor is used to control the fingers. Three hard-chrome smooth rods of 8mm diameters are used as a tendon at each joint to complete the device. The rods are used to give support to the handle and gripper. A flange screw is used for flexible coupling which has a 1.25 mm pitch size. In addition to hard-chrome smooth rods, linear bearing LM8UU was used to handle the load of the object which has 15 mm as outer diameter and 8mm as inner diameter. Torque is generated through a screw mechanism and linear bearing. The grasping operation occurred when the motor received a power supply to flex the fingers and complete the task. The proposed prototype weighs 1.1 kg, and its lifting capacity is 1 kg.

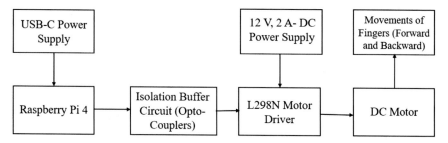

Fig. 6.4 Block diagram of interfacing of gripper and processor

6.2.5 Control

One of the most significant aspects of the grasping operation is the interaction between the gripper and the object. To perform a motion with the proper force, the screw mechanism, flexible coupling, and control systems have been used. A DC motor controlled by the Raspberry Pi 4 processor and driven by an L298N motor driver was used in the model. The block diagram for the same is presented in Fig. 6.4. A direct current (DC) motor converted electrical energy into mechanical energy which then actuated the fingers of the gripper. The Raspberry Pi processor is the main processing unit of the system through which the gripper is controlled using various commands. The power supply to the Raspberry Pi processor was provided by a USB Type-C cable. The motions and actions of the fingers of the gripper controlled by the processor are run (forward and backward), stop, only forward, and only backward. The Raspberry Pi processor cannot be directly connected to the motor or the motor driver because the latter has a very high operating voltage. If there is any malfunctioning in the motor driver, an inductive load will be generated which could damage the entire Raspberry Pi. Algorithm 1 depicts the algorithm for Raspberry Pi.

Therefore, a buffer circuit is created between the Raspberry Pi and the motor driver. The isolation buffer circuit has opto isolators. The opto isolators in the circuit act as a safeguard, preventing harmful electrical currents from flowing across the device. The opto isolator works by converting an electrical signal into a light signal using a light-emitting diode that operates in the near-infrared range.

Algorithm 1 Algorithm for Raspberry Pi interfacing

Input: DC power - 5v Raspberry Pi
Output: Gripping objects feedback
1. **Pin mode** (input–output)
2. Input1 (pin for motor)
3.   Input2 (pin for motor)
4. *Loop*
5.   Set raw input()
6.   *If* x=='f' (string)
7.   Display "run"
8.   Set input1 high
9.   Set input2 low
10.   *Display "forward"*
11.   *Else if* x=='b'
12.   Set input1 low
13.   Set input2 high
14.   *Display "backward"*
15.   *Else if* x=='s'
16.   Set input1 low
17.   Set input2 low
18.   Display "stop"
19.   *Else if* x=='e'
20.   Display "GPIO clean up"
21.   Cleanup GPIO pins
22.   *Else*
23.   Display "wrong data"
24. *End loop*

This algorithm describes a system that reads raw input from an external source and performs motor control operations on a gripping mechanism based on the input received. The input is interpreted as follows: "f" for forward, "b" for backward, "s" for stop, and "e" for GPIO cleanup. The corresponding motor control signals are sent to Input1 and Input2 pins, and the system provides feedback through display messages. The GPIO pins are cleaned up at the end of the process.

The output signals of the buffer circuit are then fed as input signals to the L298N motor driver. The L298N motor driver is also connected to a separate 12 V, 2 amperes power supply containing DC power. Motor speeds and directions are controlled by the L298N dual H-bridge driver. The motor driver is then connected to the motor which moves in a forward and backward direction according to the commands provided by the user.

Flexiforce A101 force sensors have been used to measure the force. They are placed at the tip of the fingers of the gripper. It then measures the force between the fingers of the gripper and the object. It is connected to Raspberry Pi which sends the signal to the gripper to stop when an object has been totally gripped.

The connections of Raspberry Pi with buffer circuit and buffer circuit with L298N motor driver are shown in Fig. 6.5.

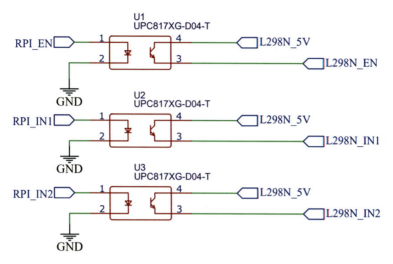

Fig. 6.5 Opto isolator circuit for motor drivers

6.3 Results

This gripper is tested to grasp several household objects of various shapes and sizes. To determine the maximum weight that the gripper can hold, the testing object's weight was gradually increased from 110 to 1000 gm. To calculate the average grabbing weight, this experiment was performed seven times on seven different items. The goal of these experiments was to demonstrate the gripping and holding capacity of the gripper.

Figure 6.6. shows the objects that have been tested in the experiment. The first object tested is a square metal box which is shown in Fig. 6.4a. Its dimensions are 12 cm × 12 cm. A cylindrical object whose diameter is 10 cm and length is 6 cm has been grasped in the experiment, which is shown in Fig. 6.4b. In Fig. 6.4c, the cylindrical object has been changed by a bigger size broom whose diameter is 7 cm and length is 80 cm. The force and the friction make the broom stable in the robotic gripper. A cup and a mug whose diameters are 8 cm and 14 cm, respectively, have been tested, and the results are shown in Fig. 6.4d, e, respectively. The gripper can not only grasp things but also can hook and grasp, thus lifting things up. Figure 6.4f shows a bag that was lifted up by the gripper which demonstrates the ability of the gripper to mimic human hand capability. The last object which was tested was a bottle as shown in Fig. 6.4g.

Table 6.1 sums up all the results showing the name of the object tested, the number of spins, and the weight of the objects in grams. Spins represent the number of times the screw has been rotated in the coupling of the gripper.

Fig. 6.6 Examples of grasps: **a** square box; **b** cylindrical object; **c** broom; **d** cup; **e** mug; **f** bag; **g** bottle

Table 6.1 Experimental results of the gripper

Object	Distance between the fingers (in cm)	Number of spins	Weight (in gm)
Square box	4	16	110
Cylindrical object	7	19	293
Broom	0.5	30	306
Cup	3.4	15	289
Mug	6	13	327
Bag	0	33	722
Bottle	**2.9**	**26**	**1000**

Bold values denote parameter for heaviest payload of 1000 gms.

6.4 Conclusion

This work outlines the design of a 3D-printed robotic gripper. The gripper prototype's low cost of production in comparison with similar commercially available robotic end effectors is made possible by the use of 3D printing technology. The experimental prototype and gripper design model are introduced and thoroughly explained in this study. It is demonstrated that the proposed robotic gripper with a single actuator satisfies the design criteria in terms of (a) a simple mechanical structure of the gripper due to flexible finger coupling and a single actuation; (b) low cost as a result of the use of a single actuator and 3D printing technologies; and (c) a relatively high payload. Because it is manufactured utilizing 3D printing technology, the gripper is lightweight, low-cost, portable, and simple to operate.

This system has a basic mechanism for gripping tasks. A DC servomotor has been used for driving the fingers of the gripper. Using three fingers, this gripper can hold items weighing up to 1000 g. Gripping operations on various household items have been evaluated. It is observed that the gripper successfully grasps all the objects within 1 kg of weight. It is observed that when the object is of less diameter, the grip will be less for holding the objects. In the future, this design will be implemented for soft grippers. The soft grippers can then be used in many applications, including food, beverage, manufacturing, and packaging industries due to their ability to grasp a wide range of irregular shapes and delicate items.

The gripper has performed adequately in all of the experiments conducted. But there were a few challenges associated with using the three-finger robotic gripper. Irregular-shaped objects may be difficult to position correctly with the gripper, which may result in an unstable grip or the object falling out of the gripper. Additionally, the limited range of motion of the gripper can make it hard to achieve a secure grip. To address these challenges, grippers with more fingers or degrees of freedom can be used, or adaptive grippers which adjust their shape to match the object being grasped can be used as desired. By doing so, the range of motion of the gripper can be increased, allowing for better grasping and manipulation of irregularly shaped objects. Another limitation that can be highlighted is that the gripper needs continuous Internet and power supply as it currently operates by code and Raspberry Pi. Finding alternatives would be a focus of future research to reduce these limitations.

Acknowledgements The authors express deep gratitude to the doctors, supporting staff, and participants in this study. The authors sincerely thank the AICTE, Government of India, New Delhi, and VIT Pune for providing financial support (File No. 8-53/FDC/RPS (POLICY-I)/2019-20) to carry out this research work.

References

1. Ariyanto, M., Munadi, M., Setiawan, J.D., Mulyanto, D., Nugroho, T.: Three-fingered soft robotic gripper based on pneumatic network actuator. In: 2019 6th International Conference on Information Technology, Computer and Electrical Engineering (ICITACEE), pp. 1–5. IEEE, 2019
2. Sadeghian, R., Sedigh, P., Azizinezhad, P., Shahin, S., Masouleh, M.T.: Design, development and control of a three flexible-fingers gripper based on hand gesture. In: 2018 6th RSI International Conference on Robotics and Mechatronics (IcRoM), pp. 359–363. IEEE, 2018
3. Tabassum, M., Ray, D.D.: Intuitive control of three fingers robotic gripper with a data hand glove. In: 2013 International Conference on Control, Automation, Robotics and Embedded Systems (CARE), pp. 1–6. IEEE, 2013
4. Li, L, Tian, W., OuYang, Z., Yu, S., Sun, W.: A rigid-flexible coupling three-finger soft gripper for fruit picking. In: 2021 40th Chinese Control Conference (CCC), pp. 4068–4072. IEEE, 2021
5. Mohan, A., Soman, G., Srichitra, S., Cletus, J.: Design and simulation of 3 fingered underactuated gripper. In 2020 IEEE Recent Advances in Intelligent Computational Systems (RAICS), pp. 86–90. IEEE, 2020
6. Liu, C.-H., Chung, F.-M., Chen, Y., Chiu, C.-H., Chen, T.-L.: Optimal design of a motor-driven three-finger soft robotic gripper. IEEE/ASME Trans. Mechatron. **25**(4), 1830–1840 (2020)
7. Matos, A., Caballa, S., Zegarra, D., Guzman, M.A.A., Lizano, D., Fernando Gonzales Encinas, D., Oscanoa, H., Arce, D.: Three-fingered gripper for multiform object grasping with force feedback sensing control. In: 2020 IEEE ANDESCON, pp. 1–5. IEEE, 2020
8. Telegenov, K., Tlegenov, Y., Shintemirov, A.: A low-cost open-source 3-D-printed three-finger gripper platform for research and educational purposes. IEEE access **3**, 638–647 (2015)
9. Yamanaka, Y., Yoshikawa, M.: A prosthetic gripper with three opposing fingers driven by a hydraulic actuator. In: 2020 42nd Annual International Conference of the IEEE Engineering in Medicine & Biology Society (EMBC), pp. 4947–4950. IEEE, 2020
10. Backus, S.B., Dollar, A.M.: An adaptive three-fingered prismatic gripper with passive rotational joints. IEEE Robot. Autom. Lett. **1**(2), 668–675 (2016)
11. Kalid, K.S, Sebastian, P., Saman, A.B.S.: TriBot: dragging locomotion three-finger robot. In: 2011 IEEE Symposium on Industrial Electronics and Applications, pp. 387–392. IEEE, 2011
12. Takayama, T., Yamana, T., Omata, T.: Three-fingered eight-DOF hand that exerts 100-N grasping force with force-magnification drive. IEEE/ASME Trans. Mechatron. **17**(2), 218–227 (2010)
13. Shauri, R.L.A., Remeli, N.H., Jani, S.A.M., Jaafar, J.: Development of 7-DOF three-fingered robotic hand for industrial work. In: 2014 IEEE International Conference on Control System, Computing and Engineering (ICCSCE 2014), pp. 75–79. IEEE, 2014
14. Bemfica, J.R., Melchiorri, C., Moriello, L., Palli, G., Scarcia, U.: A three-fingered cable-driven gripper for underwater applications. In: 2014 IEEE International Conference on Robotics and Automation (ICRA), pp. 2469–2474. IEEE, 2014
15. Li, G., Fu, C., Zhang, F., Wang, S.: A reconfigurable three-finger robotic gripper. In: 2015 IEEE International Conference on Information and Automation, pp. 1556–1561. IEEE, 2015
16. Kerr, D.R., Sanger, D.J.: Grasping using a three-fingered gripper
17. Samavati, F.C., Feizollahi, A., Sabetian, P., Ali, S., Moosavian, A.: Design, fabrication and control of a three-finger robotic gripper. In: 2011 First International Conference on Robot, Vision and Signal Processing, pp. 280–283. IEEE, 2011
18. Park, K.T., Kim, D.H.: Robotic handling gripper using three fingers. In: 2012 9th International Conference on Ubiquitous Robots and Ambient Intelligence (URAI), pp. 588–592. IEEE, 2012
19. Galan, G., Jagannathan, S.: Adaptive critic neural network-based object grasping control using a three-finger gripper. In: Proceedings of the 40th IEEE Conference on Decision and Control (Cat. No. 01CH37228), vol. 4, pp. 3140–3145. IEEE, 2001

20. Choi, M.-S., Lee, D.-H., Park, J.-H., Bae, J.-H.: Kinesthetic sensing of hole position by 3-finger gripper. In: 2020 17th International Conference on Ubiquitous Robots (UR), pp. 520–525. IEEE, 2020
21. Felip, J., Morales, A.: Robust sensor-based grasp primitive for a three-finger robot hand. In: 2009 IEEE/RSJ International Conference on Intelligent Robots and Systems, pp. 1811–1816. IEEE, 2009
22. Oz, M.T., Turgut, A.E., Temeltas, H.: Design of a 3D printed soft robotic gripper for handling irregular objects. Sensors **21**(8), 2814 (2021)
23. Singh, S.K., Rao, A.S., Behera, A.K.: A deep learning-based approach for grasping irregular objects using robotic grippers. In: 2021 IEEE 13th International Conference on Advanced Computational Intelligence (ICACI) (pp. 171–176). IEEE (2021)
24. Wagle, B.N., Lee, C.H., Yoon, J.W.: Development of a soft robotic gripper with embedded sensors for grasping irregular objects. In: 2021 IEEE/ASME International Conference on Advanced Intelligent Mechatronics (AIM) (pp. 651–656). IEEE (2021)
25. https://grabcad.com/library/3-jaw-gripper-2

Chapter 7
Prediction of Energy Absorption Capacity of High-Performance Fiber-Reinforced Cementitious Composite

Ngoc-Minh-Phuong To and Ngoc-Thanh Tran

Abstract High-performance fiber-reinforced cementitious composite (HPFRCC), a new class of concrete technology, exhibits outstanding mechanical resistance, especially in terms of superior post cracking strength, strain capacity, and energy absorption capacity. Among mechanical properties, energy absorption capacity of HPFRCCs has become one of the most popular properties that received much attention from researchers to discover and model. However, a more accurate model for prediction of energy absorption capacity is still discouraged to develop since current empirical regression models based on limited data have shown their limitations. In this research, the energy absorption capacity of HPFRCCs is predicted through a proposed machine learning-based model using 103 tensile test results. The input variables include matrix strength, fiber type, fiber length, fiber diameter, and fiber volume content, while the output variable consists of energy absorption capacity. From the prediction results, the energy absorption capacity could be predicted well using machine learning based models. From the results of sensitivity analysis, the contribution of each input variable to the energy absorption capacity of HPFRCCs was figured out.

Keywords HPFRCC · Energy absorption capacity · Machine learning model · Prediction

N.-M.-P. To · N.-T. Tran (✉)
Institute of Civil Engineering, Ho Chi Minh City University of Transport, Ho Chi Minh City, Vietnam
e-mail: ngocthanh.tran@ut.edu.vn

7.1 Introduction

High-performance fiber-reinforced cementitious composite (HPFRCC), a new class of concrete technology, exhibits outstanding mechanical resistance, especially in terms of high post cracking strength, strain capacity, and energy absorption capacity [1, 2]. Under tension load, the first cracking primarily appears, then the fiber crossing cracked section will provide a bridging capacity along the crack interface of composite, leading to the continuous increase of tensile strength with increasing tensile strain. At that time, the composite with fibers shows strain-hardening behavior with the formation of multiple micro-cracks, resulting in the significant improvements of the composite post cracking strength, strain capacity, and energy absorption capacity in comparison with normal concrete. For this reason, energy absorption capacity (EAC) of HPFRCCs has become one of the most popular properties that received much attention from researchers to discover and model [3–5].

Several researches have investigated the EAC of HPFRCCs, and they found that the EAC of HPFRCCs was dependent on many factors [6–10]. Xu et al. [6] and Smarzewski et al. [7] concluded that the EAC of HPFRCCs was dependent on fiber type. In the same manner, Nguyen et al. [8] proved that the EAC of HPFRCCs was also affected by the fiber type, fiber size, and aspect ratio. Yoo et al. [9] found that the EAC of HPFRCCs was affected by on the fiber volume fraction. Similarly, matrix strength, fiber volume content, and fiber strength were investigated to affect energy absorption capacity of HPFRCCs by Sahin et al. [10].

From the above literature, many factors such as fiber type, length, diameter, volume content, and compressive strength of matrix have been demonstrated to have an influence on the energy absorption capacity of HPFRCCs, and thus, the accurate prediction of it has become a huge challenge. Although the energy absorption capacity of HPFRCCs has been estimated by many empirical regression models [6, 11], these models exhibited low accuracy because they were proposed based on fixing of individual limited test data as well as considering of the effects of limited factors. Thus, a more effective model for estimating the energy absorption capacity of HPFRCCs with high accuracy and with the consideration of many factors at the same time should be developed. Recently, machine learning-based models have proved their capacity in estimating mechanical properties of HPFRCC with high accuracy and reliability [1, 12]. Unfortunately, very little studies focus on the prediction of the energy absorption capacity of HPFRCCs using machine learning-based models.

To overcome the above knowledge gaps, this research aims to estimate the energy absorption capacity of HPFRCCs using machine learning-based models. The main objectives are (1) to build a neural network model for predicting energy absorption capacity of HPFRCCs; and (2) to evaluate the contribution of each input factor to the energy absorption capacity of HPFRCCs.

7.2 Machine Learning-Based Model

7.2.1 Neural Network Model

An artificial neural network-based model has been used to estimate energy absorption capacity of HPFRCCs. Normally, an artificial neural network (ANN) layout composes of one input layer, single or multiple hidden layers, and one output layer, as illustrated in Fig. 7.1. The input layer receives the raw information, while the hidden layer will treat it as well as possible and the output layer produces the prediction results.

In the ANN system, the first hidden layer receives the input data from the input layer and then calculates its outcome as formulated in Eq. 7.1. After that, the outcome in first hidden layer will become the input data for next hidden layer.

$$y = f\left(\sum_{i=1}^{n} x_i \times w_i + b\right) \tag{7.1}$$

where x_i defines the input variables, w_i is the weight, b defines the bias, n is number of node in input layer, and f is an activation function.

In this work, the neural network model concludes an input layer, a hidden layer, and an output layer. The input layer consists of five neurons while the output layer composes of one node. The reasonable quantity of hidden layers and number of their neurons will be determined by experiments regarding the balance between the result quality and the running time. After trials with different scenarios, this study accepts one hidden layer with 30 nodes. During the trial process, the hyperbolic tangent activation function is used to shift the values between nodes of layers, and the Adam method is used in the back-propagation learning algorithm. The learning rate in the

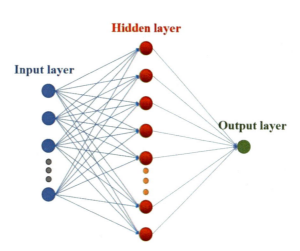

Fig. 7.1 Artificial neural network layout

Table 7.1 Tuning parameters within ANN model

Parameters	Description
No. of input layer	1
No. of hidden layer	1
No. of output layer	1
No. of input neurons	5
No. of hidden neurons	30
No. of output neurons	1
Activation function	Hyperbolic tangent
Optimizer	Adam method
Loss function	Mean squared error (MSE)
Learning rate	0.002

gradient descent method is maintained as 0.002. Table 7.1 presents all the tuning parameters within ANN model.

7.2.2 Database

To build the ANN model, 103 test results have been collected from 11 published studies [8, 11, 13–21]. Table 7.2 provides the range of input and output parameters from direct tensile tests. Total five input parameters are evaluated, first input parameter is fiber type with three types of twisted, hooked, and smooth. Four other parameters are matrix compressive strength, fiber volume fraction, fiber diameter, and fiber length with the range of 28–230 MPa, 0.6–3%, 0.2–0.775 mm, and 13–62 mm, respectively. Meanwhile, one output parameter is energy absorption capacity with range of 0.6–120.5 kJ/m^3. Because of the wide range of the input variables, the data need to be re-processed before the training of ANN. In order to obtain high-quality data for the learning algorithm, each input variables is normalized using its standard deviation and the mean. Among 103 test data, the training and testing samples are random about 80% and 20% of the total data of each case, respectively. The training process is iterated to minimize the MSE between the predicting outcomes and the experimental test data. The maximum epoch is set to 1,000,000. Initial weights with a mean of zero are chosen randomly within the range ($-$ 1, 1).

7.2.3 Statistical Measures

Three following measures are used to evaluate the accuracy of neural network model: correlation coefficient (R), three statistical metrics, root mean squared error (RMSE), and mean absolute error (MAE). These measures are formulated as following equations.

Table 7.2 Range of input and output parameters from direct tensile tests

Variable	Unit	Minimum value	Maximum value	Type
Type of fiber		Twisted, hooked, and smooth		Input
Compressive strength of matrix	MPa	28	230	Input
Fiber volume fraction	%	0.6	3	Input
Fiber diameter	mm	0.2	0.775	Input
Fiber length	mm	13	62	Input
Energy absorption capacity	kJ/m^3	0.6	120.5	Output

$$R = \frac{n\sum_{i=1}^{n} \hat{t}_i t_i - \left(\sum_{i=1}^{n} \hat{t}_i\right)\left(\sum_{i=1}^{n} t_i\right)}{\sqrt{n\sum_{i=1}^{n}\left(\sum_{i=1}^{n} \hat{t}_i^2\right) - \left(\sum_{i=1}^{n} \hat{t}_i\right)^2} \sqrt{n\sum_{i=1}^{n}\left(\sum_{i=1}^{n} t_i^2\right) - \left(\sum_{i=1}^{n} t_i\right)^2}} \quad (7.2)$$

$$\text{RMSE} = \sqrt{\frac{\sum_{i=1}^{n}\left(\hat{t}_i - t_i\right)^2}{n}} \quad (7.3)$$

$$\text{MAE} = \frac{\sum_{i=1}^{n}\left|\hat{t}_i - t_i\right|}{n} \quad (7.4)$$

where n defines number of experimental test results, $\hat{t}_i$ defines the predicted data, and t_i is the tested data from experiments.

7.3 Results and Discussion

7.3.1 Performance of the Neural Network Model

Figure 7.2 performs the correlation of the neural network model for predicting energy absorption capacity in the training set. The correlation between estimated data and experimental test data in the training data set was found to perform well with an R value of 0.95. Moreover, the RMSE and MAE values in the training set were 7.12 and 4.66, respectively, indicating high accuracy of machine learning-based models in the estimation of the energy absorption capacity.

The prediction result of the neural network model for predicting energy absorption capacity in the testing set is shown in Fig. 7.3. Similarly, the correlation between estimated data and test data in the testing set also exhibited good performance with the R value of 0.94. Moreover, the RMSE and MAE values in the testing set were 5.60 and 4.55, respectively. Thus, the proposed machine learning-based model could estimate energy absorption capacity with high accuracy and reliability in both training and testing sets.

Fig. 7.2 Prediction result of the neural network model in training set

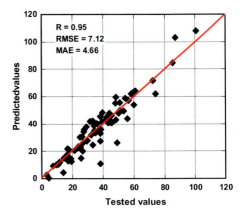

Fig. 7.3 Prediction result of the neural network model in testing set

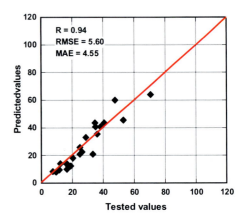

7.3.2 Sensitivity Analysis

The contribution of each input parameter was found out by calculating their relative importance. In this study, an approach mentioned by God [22] was carried out to obtain contribution of each input parameter. Figure 7.4 shows the results of sensitivity analysis for each input parameter. From the estimation results, matrix compressive strength became the most important variable effecting the energy absorption capacity. On the contrary, fiber diameter was the less important variable. In addition, the performance ranking was found to be as follows: Matrix strength > Fiber length > Fiber type > Fiber volume fraction ~ Fiber diameter.

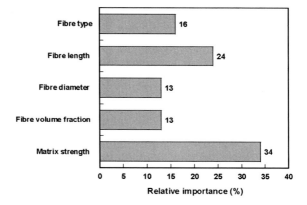

Fig. 7.4 Sensitivity analysis of each input parameter

7.4 Conclusion

The energy absorption capacity of HPFRCCs was estimated by using a machine learning-based model developed based on total 103 test results. Five input parameters with complex relationships were considered. From the results of the prediction and the sensitivity analysis, the following observations and findings drawn in this research include:

- The energy absorption capacity could be predicted well using a machine learning-based model considering the effects of several input parameters at the same time.
- The proposed neural network model could estimate the energy absorption capacity with high accuracy and reliability in both training and testing sets.
- The compressive strength of matrix became the most influential parameter affecting the energy absorption capacity. The performance ranking was found to be as follows: Matrix strength > Fiber length > Fiber type > Fiber volume fraction ~ Fiber diameter.

References

1. Ngo, T.T., Le, Q.H., Nguyen, D.L., Kim, D.J., Tran, N.T.: Experiments and prediction of direct tensile resistance of strain-hardening steel-fiber-reinforced concrete. Mag. Concr. Res. Ahead Print (2023). https://doi.org/10.1680/jmacr.22.00060
2. Ngo, T.T., Tran, N.T., Kim, D.J., Pham, T.C.: Effects of corrosion level and inhibitor on pullout behavior of deformed steel fiber embedded in high performance concrete. Constr. Build. Mater. **280**(3), 122449 (2021)
3. Tran, N.T., Nguyen, D.L., Kim, D.J., Ngo, T.T.: Sensitivity of various fiber features on shear capacities of ultra-high-performance fiber-reinforced concrete. Mag. Concr. Res. **74**(4), 190–206 (2021)
4. Tran, N.T., Nguyen, D.L., Vu, Q.A., Kim, D.J., Ngo, T.T.: Dynamic shear response of ultra-high-performance fiber-reinforced concretes under impact loading. Structures **41**, 724–736 (2022)

5. Tran, T.K., Nguyen, T.K., Tran, N.T., Kim, D.J.: Improving the tensile resistance at high strain rates of high-performance fiber-reinforced cementitious composite with twisted fibers by modification of twist ratio. Structures **39**, 237–248 (2022)
6. Xu, M., Wille, K.: Fracture energy of UHP-FRC under direct tensile loading applied at low strain rates. Compos. B Eng. **80**, 116–125 (2015). https://doi.org/10.1016/j.compositesb.2015.05.031
7. Smarzewski, P.: Comparative fracture properties of four fibre reinforced high performance cementitious composites. Materials **13**, 2612 (2020). https://doi.org/10.3390/ma13112612
8. Nguyen, D.L., Lam, M.N.T., Kim, D.J., Song, J.: Direct tensile self-sensing and fracture energy of steel-fiber-reinforced concretes. Compos. B Eng. **183**, 107714 (2020). https://doi.org/10.1016/j.compositesb.2019.107714
9. Yoo, D.Y., Shin, H.O., Yang, J.M., Yoon, Y.S.: Material and bond properties of ultra high performance fiber reinforced concrete with micro steel fibers. Compos. B Eng. **58**, 122–133 (2014). https://doi.org/10.1016/j.compositesb.2013.10.081
10. Sahin, Y., Koksal, F.: The influences of matrix and steel energy of high-strength concrete. Constr. Build. Mater. **25**, 1801–1806 (2011). https://doi.org/10.1016/j.conbuildmat.2010.11.084
11. Wille, K., El-Tawil, S., Naaman, A.E.: Properties of strain hardening ultra high performance fiber reinforced concrete (UHP-FRC) under direct tensile loading. Cement Concr. Compos. **48**, 53–66 (2014)
12. Chaabene, W.B., Flah, M., Nehdi, M.L.: Machine learning prediction of mechanical properties of concrete: critical review. Constr. Build. Mater. **260**, 119889 (2020)
13. Yoo, D.Y., Kim, S., Kim, J.J., Chun, B.: An experimental study on pullout and tensile behavior of ultra-high-performance concrete reinforced with various steel fibers. Constr. Build. Mater. **206**, 46–61 (2019)
14. Tran, N.T., Kim, D.J.: Synergistic response of blending fibers in ultra-high-performance concrete under high rate tensile loads. Cem. Concr. Compos. **78**, 132–145 (2017)
15. Kim, D.J., Wille, K., Naaman, A.E., El-Tawil, S.: Strength dependent tensile behavior of strain hardening fiber reinforced concrete. In: Parra-Montesinos, G.J., Reinhardt, H.W., Naaman, A.E. (eds.) High Performance Fiber Reinforced Cement Composites 6. RILEM State of the Art Reports, pp. 23–10 (2012)
16. Park, S.H., Kim, D.J., Ryu, G.S., Koh, K.T.: Tensile behavior of ultra high performance hybrid fiber reinforced concrete. Cem. Concr. Compos. **34**, 172–184 (2012)
17. Wille, K., Kim, D.J., Naaman, A.E.: Strain-hardening UHP-FRC with low fiber contents. Mater. Struct. **44**, 583–598 (2011)
18. Tran, T.K., Kim, D.J.: High strain rate effects on direct tensile behavior of high performance fiber reinforced cementitious composites. Cem. Concr. Compos. **45**, 186–200 (2014)
19. Pyo, S., Wille, K., El-Tawil, S., Naaman, A.E.: Strain rate dependent properties of ultra high performance fiber reinforced concrete (UHP-FRC) under tensions. Cem. Concr. Compos. **56**, 15–24 (2015)
20. Tran, N.T., Tran, T.K., Jeon, J.K., Park, J.K., Kim, D.J.: Fracture energy of ultra-high-performance fiber-reinforced concretes at high strain rates. Cem. Concr. Res. **79**, 169–184 (2016)
21. Donnini, J., Lancioni, G., Chiappini, G., Corinaldesi, V.: Uniaxial tensile behavior of ultra-high performance fiber-reinforced concrete (uhpfrc): experiments and modeling. Compos. Struct. **258**, 113433 (2021)
22. God, A.T.C.: Back-propagation neural networks for modeling complex systems. Artif. Intell. Eng. **9**, 143–151 (1995)

Chapter 8
Data Analysis on Determining False Movie Ratings

Ridhika Sahni and Karmel Arockiasamy

Abstract Online movie ratings have evolved into a serious business. Hollywood generates around $10 billion in box office revenue in the United States each year, and online ratings aggregators may have an increasing influence over where that money goes. A single film critic can no longer make or break a film, but perhaps thousands of critics, both professional and amateur, can how well can these platforms be trusted especially if they are showing the ratings and making money by selling the tickets? Do they have a bias by rating movies higher than it should be? The objective of the analysis is to provide brief research of one such online movie-ticket selling company, Fandango. The data collected from Fandango are analyzed and compared with the data collected from sites like Rotten Tomatoes, IMDB, and Metacritic. With the help of visualization, fraud and risk can be reduced offered by such online sites.

Keywords Online movie ratings · Critics · Conflict · Fraud · Risk · Visualization · Data analysis · Comparison

8.1 Introduction

Data visualization can be used to represent and interpret the results of data analysis techniques used for false movie detection, e.g., visualizing the results of image processing and machine learning algorithms can help analysts understand the features that are being used to classify movies as real or fake, and identify any patterns or anomalies in the data. Some examples of data visualization techniques that can be used for false movie detection include:

R. Sahni · K. Arockiasamy (✉)
School of Computer Science and Engineering, Vellore Institute of Technology, Chennai, Tamil Nadu 600127, India
e-mail: Karmel.a@vit.ac.in

© The Author(s), under exclusive license to Springer Nature Singapore Pte Ltd. 2024
P. K. Jha et al. (eds.), *Proceedings of Congress on Control, Robotics, and Mechatronics*, Smart Innovation, Systems and Technologies 364,
https://doi.org/10.1007/978-981-99-5180-2_8

- Scatter plots: These can be used to plot different features of a movie, such as color histograms or edge detection, against each other to identify patterns or clusters of real and fake movies.
- Heat maps: These can be used to show the relative importance of different features in classifying movies as real or fake.

Online movie partners play a vital role in our recreation. With the advent of new tools and technologies, like Web development and Data Science, etc., this industry has modernized its approach by suggesting movies according to user preference. Most people today like to go out and watch movies with their friends or family. These online platforms have eased the process of buying tickets. Earlier, people had to wait in long queues to buy tickets. But, with the help of Internet and technology, this problem has been solved as with just one click anybody can buy tickets of their favorite movie and can begin to watch it.

This industry gives customized experience to the user. The user needs to register to their app or website. Then, select his location. After this, the user needs to select his preferred place, i.e., cinema hall. The app/website then asks the user to enter for how many people he is buying the ticket for, thus giving a real-life experience. After all the choices, the user is finally taken to the payment section. As reliable this industry looks, in reality they are not! Some movie ratings in the site are intentionally inflated to attract more customers into buying them. This fraud is committed without the knowledge of customers. Even bad movies are given a decent star which confuses the customers into buying tickets for such movies. One of such misleading company is Fandango. According to an article by FiveThirtyEight, on Fandango, no movie is rated less than three stars.

According to another article posted on April 10, 2017, Fandango's movie ratings algorithm is inaccurate. When the author of the article pulled out Fandango's data and investigated upon it, the algorithm could not justify the data at the backend. Data analysis and visualization can be useful tools for detecting false movie reviews. By analyzing patterns and trends in the data, it can be possible to identify features or characteristics that are indicative of fake reviews.

8.2 Literature Survey

The false ratings have a greater impact not only on the film industry but also on consumers. It also have impact on the potential methods that are used for detecting as well as mitigating the false ratings.

It is important to be critical when evaluating online movie ratings [1], as they may not always be reliable. Some ratings may be influenced by factors such as paid reviews or a bias toward certain films. Additionally, Fandango's summary of movie ratings may be based on a small sample size of reviews, which could lead to inaccuracies. The investigation started when the author's friend complained to him that a bad movie is given descent rating on Fandango app. After this, the author and his team found out that Fandango company intentionally changed the backend data and presented the data to the user in an inflated matter. The author also wrote that it is always a good idea to read a variety of reviews from different sources before deciding about a movie.

The article described IMDb (Internet Movie Database) [2] is a database of information related to films, television programs, and video games, including cast, production crew, personal biographies, plot summaries, trivia, fan reviews, and ratings. Rotten Tomatoes is a website that aggregates reviews from professional critics to create a "fresh" or "rotten" rating for a movie. Metacritic is another review aggregator that assigns a weighted average score to a movie based on reviews from various critics.

Fandango is an online movie ticketing platform which also provides movie ratings based on a small sample of reviews. All these sources have their own unique way of collecting and calculating the ratings and reviews. But, when the author pulled data of Fandango, he found that Fandango's algorithm is inaccurate and rounds off the stars instead of pushing them to next decimal value. The article concluded that IMDB, Rotten Tomatoes, and Metacritic are generally considered to be more reliable sources for movie ratings compared to Fandango.

This paper [3] provides an overview of the recent research on deep learning for fake review detection. The fake reviews along with its consequence on online service platforms and e-commerce are introduced. The various deep learning models applied for fake review detection, such as convolutional neural networks (CNNs), generative adversarial networks (GANs), and recurrent neural networks (RNNs) taking the inputs as text, images along with network structure are analyzed.

The paper [4] proposes a methodology called multi-granularity multi-modality attention network especially for detecting fake reviews. The information such as word level, review level, sentence level, and multiple modalities, including text and images along with network structure are exploited for detecting whether the reviews are fake or not. This model is built with three components namely a multi-granularity encoder which extract the features granularity levels, a multi-modality attention mechanism that analyzes the various modalities, and a classifier for finalizing the decisions. The evaluation was done using various datasets, and its performance are analyzed.

In paper [5], the forecasting movie ratings are discussed. The data analytics uses statistical techniques as well as ML for analyzing the large amounts of data such as details about the crew, cast, and budget of the film with the existing work. The research paper [6] involved collecting and analyzing a large dataset of online reviews from different platforms and using natural language processing techniques to classify the reviews as positive or negative. The research also investigated the prevalence of positive and negative reviews and examine the factors that influence them (e.g., product type, price, brand, etc.). The research also displayed the implications of these findings, such as how the prevalence of positive and negative reviews affects consumer behavior and the reputation of the reviewed entity. The paper could be useful for businesses and marketers, as they can use the findings to understand the drivers of online reviews and take appropriate actions to improve their products and services. The paper was also targeted toward a technical audience with some knowledge of data analysis and online movie reviews. The findings of paper could help the audience and businessmen.

This paper [7] focused on the use of large, up-to-date movie datasets to investigate how prereleased attributes (such as trailers, posters, and cast information) impact a movie's gross revenue. The research was conducted on IMDb dataset and gathered a huge dataset of 7.5 million movie titles. The research involved collecting and analyzing a large dataset of movies and their associated prereleased attributes and gross revenue. The paper [8] focuses on using big data analytics techniques, and applies alternating least squares algorithm for handling the collaborative filtering problem.

In this study [9], a method for finding the determinants of movie review rating-based big data are analyzed. The dataset is then preprocessed, and machine learning algorithms are applied to identify the key features and sentiments that influence movie review ratings. They found that factors such as the movie's cast, director, genre, and plot are strongly correlated with movie review ratings. Additionally, they found that the sentiment of the review text, such as the use of positive or negative words, also plays a significant role in determining the movie review rating. The proposed method can be used to predict movie review ratings more accurately and can be useful for movie producers, marketers, and movie-goers. The algorithm used was IBM SPSS Text Analytics for survey. But one of the ambiguities was that tea ceremony was associated with positive emotional words and had a positive effect on grading evaluation. The result itself gives limited interpretable information.

The paper [10] states that the film industry has a significant impact on the global economy. It is the world's most visible contributor to the economy. Every year, hundreds of thousands of films are released to the public in the hope that one of them will be the next blockbuster. According to movie industry statistics, six to seven films out of ten are unprofitable, and only one-third of the films are successful. The movie industry's producers, studios, investors, and sponsors are all interested in predicting the film's box office success. This paper analyzes the film genre, release date around holidays, release month of movies, languages, and country with more movies from the movie review dataset. There are attributes taken from the dataset (country, languages, genre, movie release date, budget, and revenue), and the derived attributes (release month of the movie derived from release date of movie and profit from budget and revenue) are analyzed to determine the movie performance. The analyzed data is graphed for statistical analysis of the film's success.

8.3 Methodology

8.3.1 Web Scrapping

Scraping a dataset from the real-world involves using a program or script to automatically extract data from a specific source, such as a website or an API.

Table 8.1 Correlation of stars, rating, and votes

	Stars	Rating	Votes
Stars	1.000000	0.994649	0.164218
Rating	0.994696	1.000000	0.163764
Votes	0.164218	0.163764	1.000000

Identify the source of the data: This could be a website, an API, or any other digital source that contains your information.

Inspect the source: Use a Web browser's developer tools to inspect the source code of the website or the API endpoint to understand the structure of the data and identify the specific elements for scraping.

Write the scraping script: Use a programming language, such as Python or Java Script, to write a script that can extract the data from the source. This script will typically use libraries such as Beautiful Soup or Scrapy for parsing HTML, and requests for sending HTTP requests.

8.3.2 Importing and Preprocessing the Dataset

Once we have scraped a dataset, the next step is to import it into our analysis environment and begin exploring its properties and structure. Some of the steps are:

- Load the dataset: Used a library such as pandas in Python to load the dataset into your analysis environment.
- Clean and preprocess the data: Removed any irrelevant or duplicate data and formatted the data in a way that is suitable for analysis. For preprocessing the dataset, lambda function was used which is a data preprocessing concept. It helped in tasks like splitting columns and then adding the newly created column in the dataset.
- Correlation: Calculating correlation between columns.

Table 8.1 presents that stars and rating are not perfectly correlated. This shows that there is some difference between the stars being shown to the user versus the true ratings.

8.3.3 Visualization

- Create a KDE plot: Used libraries such as Seaborn in Python to create a KDE plot of the actual ratings held by Fandango on its backend.

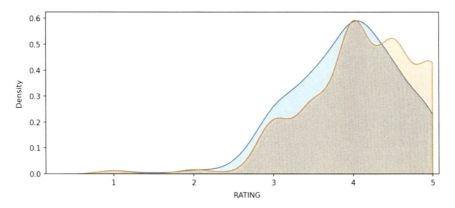

Fig. 8.1 Displays ratings (stars) versus true ratings KDE plot

- Create another KDE plot: Used Seaborn to create a KDE plot of the ratings displayed to users. This plot will show the distribution of the ratings that users see.
- Compare the plots: Compared the two plots to identify any discrepancies between the actual ratings held by Fandango and the ratings displayed to users. For example, if the plot of the actual ratings has a higher density of ratings around 4.5, while the plot of the ratings displayed to users has a higher density of ratings around 5, this could indicate that Fandango is inflating the ratings displayed to users as shown in Fig. 8.1.

8.3.4 Comparing Fandango's Data with Other Sites

- Importing the dataset collected from all the other sites. Importing is done with the help of pandas. After importing, data is preprocessed before exploration.
- Exploring: Created a scatter plot to compare critic and user scores for online movies can help to understand the relationship between the two types of ratings. These included data from sites like Rotten Tomatoes, IMDb, and Metacritic. KDE plots were formed to compare actual ratings versus true ratings to know more about the data and about how accurate the data is (Fig. 8.2).

8.3.5 Comparing Fandango's Data with Other Sites

- Importing the dataset collected from all the other sites. Importing is done with the help of pandas. After importing, data is preprocessed before exploration.
- Exploring: Created a scatter plot to compare critic and user scores for online movies can help to understand the relationship between the two types of ratings.

8 Data Analysis on Determining False Movie Ratings   91

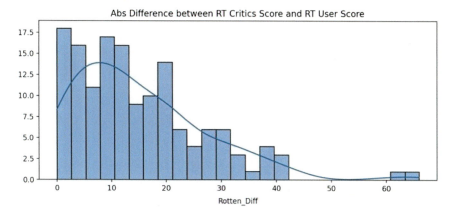

Fig. 8.2 Absolute value difference between Rotten Tomatoes critics score and Rotten Tomatoes user

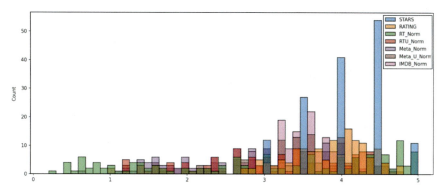

Fig. 8.3 Comparison of normalized results

These included data from sites like Rotten Tomatoes, IMDb, and Metacritic. KDE plots were formed to compare actual ratings versus true ratings to know more about the data and about how accurate the data is (Fig. 8.3).

In Fig. 8.4, the darkest color displays the harshest critic. The darkest color is maintained along the row except for Fandango's stars and ratings. This is another visual example for how much Fandango is pushing the ratings for movies.

8.3.6 Score

- Normalizing the values between 0 and 5 stars for a fair comparison can help to standardize the ratings data and make it easier to compare different ratings scales.

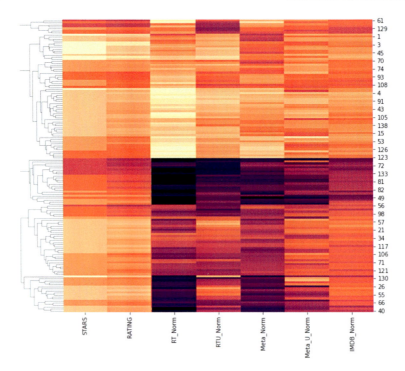

Fig. 8.4 Cluster map visualization

RT, Metacritic, and IMDB do not use a score between 0 and 5 stars like Fandango does. To do a fair comparison, these values are normalized, so they all fall between 0 and 5 stars and the relationship between reviews stays the same. In this project, all the ratings collected from other sites are converted by simple formula-

100/20 = 5.

10/2 = 5.

Figure 8.3 shows that all the site's data is normalized first and then compared with one another. Also, we can see that Rotten Tomatoes, IMDb, and Metacritic's ratings are close to one another, but Fandango's ratings are much higher than the rest. In fact, no movie on Fandango is rated less than 3 stars. This shows that even for bad movies which are not critically acclaimed by other sites, Fandango has rated them decent (Fig. 8.5).

- Comparing: After normalization, Fandango's data was merged with dataset bearing all site's data. For comparing, the project used KDE plot. KDE plot compared the distribution of RT critic ratings against the STARS displayed by Fandango.

Figure 8.6 displays the workflow of the project. Real-world data is collected and scraped, then with the use of Pandas and Seaborn, data is imported and explored,

8 Data Analysis on Determining False Movie Ratings   93

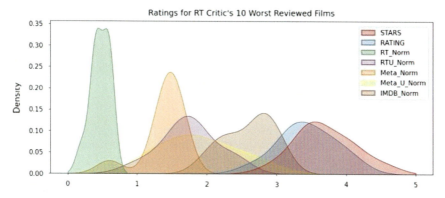

Fig. 8.5 Results displaying that no movie on Fandango is given less than 3 stars

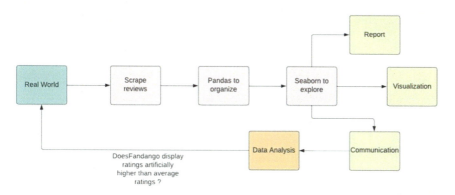

Fig. 8.6 Workflow

then with different visualization techniques, the data is analyzed, and a solution is formed for the real-world problem.

8.4 Results

It is important to be aware that online movie ratings can be manipulated or biased. Some users may give artificially high or low ratings for a variety of reasons, such as personal bias or a desire to influence others' opinions. Additionally, some websites may have their own rating systems that are not necessarily representative of the consensus. Therefore, it is always a good idea to read a variety of reviews and ratings from different sources to get a more accurate understanding of a movie's quality.

The visualization on Fandango's dataset proved that the company's algorithm was inaccurate and instead of showing the next decimal value, it rounded off the whole number. As shown in Fig. 8.6, no movie on Fandango is rated less than 3 stars. This creates a mistrust and a question mark on the company's integrity.

8.5 Conclusion

The research work revolves around how valid are Fandango's movie ratings. It aims to deliver, the one of its kind research which compared Fandango's movie dataset along with the dataset of famous sites like IMDb, Rotten tomatoes, and Metacritic. With the help of Data Analysis, exploring and comparing are much easier.

The paper aims at giving a brief about how online websites are source of risk and fraud. With Data Analysis, this fraud and risk can be mitigated and solved.

The paper also provides a brief about the platforms selling and booking online movie tickets intentionally inflating the stars/ratings to attract more customers into buying the tickets (Fig. 8.5). With different data analysis concepts like KDE, the results of website are compared, and fraud is detected.

This research paper can also be considered for prototype building for projects which detect fraud. Also, by this research paper, users will be opened to harsh reality and can escape scams.

References

1. Hickey, W.: Be suspicious of online movie ratings, especially Fandango's. Five Thirty-Eight (2015)
2. Olteanu, A.: Whose ratings should you trust? IMDB, Rotten Tomatoes, Metacritic, or Fandango? (2017)
3. Hu, L., Wei, S., Zhao, Z., Wu, B.: Deep learning for fake news detection: a comprehensive survey. AI Open (2022)
4. Zhang, T., Wang, D., Chen, H., Zeng, Z., Guo, W., Miao, C., Cui, L.: BDANN: BERT-based domain adaptation neural network for multi-modal fake news detection. In: 2020 International Joint Conference on Neural Networks (IJCNN), pp. 1–8. IEEE (2020)
5. Kharb, L., Chahal, D.: Forecasting movie rating through data analytics. In: Data Science and Analytics: 5th International Conference on Recent Developments in Science, Engineering and Technology, REDSET 2019, Gurugram, India, 15–16 Nov 2019, Revised Selected Papers, Part II 5, pp. 249–257. Springer, Singapore (2020)
6. Schoenmueller, V., Netzer, O., Stahl, F.: The polarity of online reviews: prevalence, drivers and implications. J. Mark. Res. **57**(5), 853–877 (2020)
7. Sharma, A.S., Roy, T., Rifat, S.A., Mridul, M.A.: Presenting a larger up-to- date movie dataset and investigating the effects of pre-released attributes on gross revenue. arXiv preprint arXiv:2110.07039
8. Choudhry, N., Xie, J., Xia, X.: Big data analytics of movie rating predictive system. J. Phys. Conf. Ser. **1575**(1), 012063 (2020). IOP Publishing
9. Yagi, T., Murata, S.: Determinants of movie review ratings—new method by using big data (2015)
10. Vanitha, V., Sumathi, V.P., Soundariya, V.: An exploratory data analysis of movie review dataset. Int J Rec Technol Eng **7**(4), 380–384 (2019)

Chapter 9
Power Consumption Saving of Air Conditioning System Using Semi-indirect Evaporative Cooling

Manish Singh Bharti, Alok Singh, T. Ravi Kiran, and K. Viswanath Allamraju

Abstract In the last couple of decades, efforts were made to accomplish a concordance between indoor air quality, air distribution, and energy efficiency and similarly in the center of thermal comfort in the indoor atmosphere. We have shown up at a time when air conditioners add to a significant piece of the power interest in the structure cooling sector. In India, about 70% of the power request is satisfied by thermal power plants, and from this time forward, more energy consumption means more coal consumption causing higher emissions of Earth-wide temperature boost gases. At an outside temperature of 40 °C or above, air forming system plays a basic capacity in keeping up the inside temperature. Global warming has increased the temperature of the atmosphere and that in the end has caused more conspicuous use of air-shaping systems in building cooling. It has become essential equipment to control the internal thermal comfort of business and residential buildings around the globe. The essential concern of any air conditioner is to keep up, with the indoor air quality, temperature consistently, humidity, and air velocity; all these atmospheric properties add to the state of thermal comfort. In this paper, energy saving through semi-indirect evaporative cooling of a vapor compression refrigeration system is studied.

Keywords Power consumption · SIEC · COP

M. S. Bharti (✉) · T. Ravi Kiran
Mechanical Engineering Department, Rabindranath Tagore University, Raisen, Madhya Pradesh, India
e-mail: manish.bharti23@gmail.com

A. Singh
Mechanical Engineering Department, MANIT Bhopal, Bhopal, Madhya Pradesh, India

T. Ravi Kiran
Centre for Renewable Energy, Rabindranath Tagore University, Raisen, Madhya Pradesh, India

K. Viswanath Allamraju
Mechanical Engineering Department, IARE, Hyderabad, Telangana, India

9.1 Introduction

Conditions of human comfort are determined by humidity, indoor temperature, air quality, and air motion. But some of these factors lie out of the comfort range of human, one does not believe calm as well his/her effective toward work, otherwise think to depreciate considerably along with, air conditioning assist in improving the efficiency of work as well as the whole quality and the outcome quantity of some human attempt [1–3]. But for the comfort of humans, air conditioning is as well necessary for good interior conditions on huge sizes such as ordnance factories, food processing, and coloring of the vehicle as well as a lot of developed processes along with the procedure.

Air conditioning as well recovers the life and presentation of lots of semiconductors supporting electronic strategy similar to microprocessors (magnetic recorders, supercomputers, and mainframes are good quality examples), supply units of power control, etc. [4, 5]. Air conditioning procedure might be definite as the concurrent organization of the humidity of the air, air velocity, temperature, positive pressure, and quality distinction of the hardened room, underneath predefined restrictions intended for human health and interior comfort. Classification of the air conditioning system is shown in Fig. 9.1 [6].

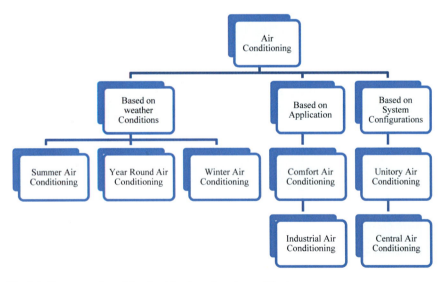

Fig. 9.1 Representation of the classification of an air conditioning

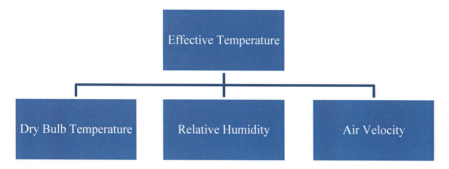

Fig. 9.2 Temperature feature's affects [6]

9.1.1 Effect of Temperature of Ambient Air

The cold or warm degree felt through the human body mainly depends on lying on the subsequent three aspects (Fig. 9.2). The mutual consequence of the over aspect can be resolute through a single name temperature effect, definite as that catalog which associates the consequence of every one of three lying on the human body comfort. Numerically, it is established equivalent to still saturated air's temperature, i.e., airspeed of 5–8 m/min, which generates the identical reaction of coldness or warmth as shaped underneath the known conditions.

The comfort plan ready through ASHRAE following the research completed lies on a variety of people subjected to an extensive variety of relative humidity, air velocity, and close temperatures, and in supplementary words, it is based lying on the thought of the temperature's effect. The plan is appropriate toward the situation of rationally motionless air wherever the occupant is doing light work or spaces at rest, as well as whose with this exterior are on the denote temperature equivalent toward dry-bulb temperature of the air.

9.1.2 Evaporative Cooling System

An evaporative cooling process is a mass and heat relocation process that utilizes the air cooling water evaporation, in which a huge quantity of heat is relocated from water to air and air to water, as well as therefore the air temperature reduces. Figure 9.3 demonstrates an evaporative cooling system universal classification intended for building cooling.

An evaporative cooler can be categorized into the:

- Indirect evaporative coolers, wherever a plate/surface split among the working fluids;
- Direct evaporative coolers, in which the operational liquids (air/ water) are within direct contact;

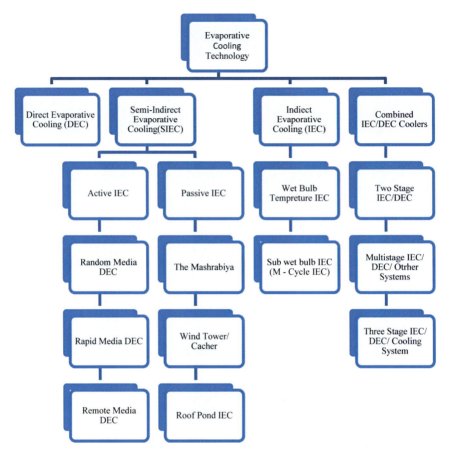

Fig. 9.3 Evaporative cooling system inside the building cooling's classification

- Mutual system of indirect and direct evaporative coolers or/and through further cooling cycles [7, 8].

The evaporative cooling procedure is the technique of decreasing the air temperature through water vapor evaporation. Evaporative cooling is a procedure of the removal of air-sensible heat as well as an equal addition of latent heat within the water vapor form. The direct evaporative cooling method is most excellent suited within dry weather (i.e., higher ambient temperature and low humidity) such as experiential in central India [9–11].

In an air conditioning system's early stage, the direct evaporative cooling processes play an important function. The DEC is the oldest, simplest, as well as most general appearance of the evaporative air conditioning method. Though, the conservative air conditioning system arrival for air conditioning has restricted the direct evaporative cooling method use. Recently, this DEC method has gained a few impetuses due to

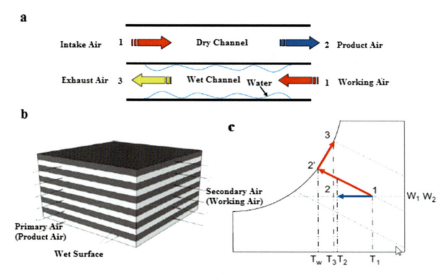

Fig. 9.4 IEC arrangement [13]

the ecological anxiety leading toward global warming and ozone layer depletion; moreover, it uses extra energy [12].

9.1.2.1 Indirect Evaporative Cooling System

In the IEC process, the derived air of the conditioned space is accepted throughout one face of the heat exchanger. These flows of air are frozen through direct evaporative cooling. The heat exchanger's walls are frozen through the derived air. The derived air as well as the main air is allowed to stream independently inside the heat exchanger without contact, wherever the main air is wisely frozen through the derived air flow exclusive of some direct contact. Water from the wet pads in the working air channel is evaporated and thus air is cooled and humidified. This cooling effect is transferred to the dry channel where product air is sensibly cooled. This process is known as indirect evaporative cooling. The IEC system is appropriate for humid and hot weather situations, i.e., coastal regions likewise—the city of Bangalore, Chennai, Mumbai, etc., within India. The IEC method is exposed in Fig. 9.4.

9.1.2.2 Wet-Bulb Temperature of IEC System

The temperature of the wet bulb in the IEC system is wrapped up in a unit of the cross-flow heat exchanger, flat-plate stack, the most universal flow pattern, and configuration, which can lower the temperature of air close to, except not below, the inlet air's wet-bulb temperature.

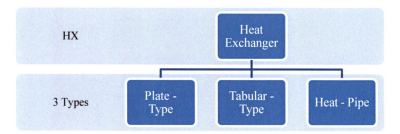

Fig. 9.5 Types of heat exchanger

Figure 9.4 demonstrates a representation of the operational principles of a classic heat exchanger pattern of a wet-bulb temperature IEC arrangement which encompasses more than a few pairs of contiguous channels: the supply (main) air dry passages and the working (derived) air wet passages [13, 14]. Heat transfer happened among the two functioning fluids throughout the heat conductive cover, and therefore, the main air is refrigerated sensibly through no extra moisture commencing keen on the refrigerated supply airflow. However, the heat transfer method among the functioning air as well as water within wet channels is through water vaporization latent heat. The wet-bulb efficiency of the IEC arrangement is inside the variety of 40% to 80%, which is inferior to the DEC systems [9]. IEC system types are survived which is categorized in Fig. 9.5, according to the heat exchanger's types.

9.2 Methodology

As referred, the structure inside an improvement diversion arrangement in a room for product air inlet. SIEC system (semi-indirect evaporative cooler) has two autonomous supplies of the air stream, firstly used for purpose of cooling, simultaneously by a subsequent, return air stream, which indirectly gets in touch with water and there is transfer of heat and mass. Water is compulsory alongside the return air stream as well as it circulates constantly.

- Analysis of thermodynamics of joint structure for dry and hot climatic circumstances was carried out.
- After that, investigational setup for a joint structure was fabricated and installed.
- Investigations of the experimental and theoretical data revealed from the setup as well as create the contrast of the two lying on the comfort airbase, economy, and consumption of power.

9.2.1 Equipment Used

The sorts of rigging used to finish the examinations are going with anemometer, flow meter, and voltmeter.

9.2.1.1 Structure for Supply

That contains scheduled a follower through a potentiometer toward screening the air streams.

9.2.1.2 AHU (Air Handling Unit)

This apparatus grants us the way to mirror the circumstances of the space given (humidity as well as temperature).

9.2.1.3 ADS (Air Distribution System)

Every one of the evaluated tools is placed there.

9.2.1.4 WDS (Water Distribution System)

A water siphon gives water out of the container toward the force sprinkle structure through plunging intended for spouts.

9.2.1.5 SIEC

Air in a semi-indirect evaporative cooler goes into the air handling unit and gets sensibly cooled and humidified. These principal air streams are known as fundamental air streams as well as these air streams are adjusted within the semi-indirect evaporative cooling that goes into space. Consequently, the ambient air from the outside space comes across the semi-indirect evaporative cooling within a cross-stream. After leaving semi-indirect evaporative cooler, the air is at higher humidity content level and lower dry bulb temperature.

9.2.1.6 Adjusted Space

These estimations are (02 m * 02 m * 2.5 m), which have a high temperature indoors toward assurance while requiring that the break is suitably adjusted.

9.2.1.7 Data Acquisition and Monitoring Structure

The PC stores as well as controls every one of the outcomes of the evaluating apparatus. The appraisals sensors employed are:

T: Assessment sensor for temperature
Category: Techno term 60
Precision: 0.1 °C.

HR: Assessment sensor for relative humidity
Category: HIH-3610 Honeywell
Precision: ± 2% RH; here, the range is 0–100% RH.

DP: Transducer for differential pressure
Category: (range is 603–2) DWYER
Precision: ± 2% range expand (70 °F).

9.3 Experimental Setup

The following exploratory plan could be prepared for testing particular cooling pads made of different materials, e.g., sugarcane, banana, coconut, honeycomb paper cooling pad, and khus fibers.

Depiction of the cooling cushions and the external spreads for the cooling cushions were made utilizing a smooth steel wire fill-in, as cuboids of estimations are 0.30 m * 0.15 m * 0.05 m.

These cuboids combined wraps were then piled up with accurately 100 g of the entirety of the cooling media material to be unequivocal (as Fig. 9.6) honeycomb paper. This was done to guarantee the same thickness and subsequently uniform credits of all cooling cushions. Here, we use honeycomb pad. A ceramic pipe was unnecessarily used for ingestion structure and channels, regardless of the way that they do not seem to have been as essential as lead pipes. Wood pipes were used for channels and confirmation systems. Here, we show in Fig. 9.7 the ceramic pipes which are used for an evaporative cooler. In this course of action, we used various limits with fixed ranges and got ready for the appraisal of our structure which is evaporative cooling as given in Table 9.1.

9 Power Consumption Saving of Air Conditioning System Using … 103

Fig. 9.6 Honeycomb pads for lowering down the power consumption of compressor

Fig. 9.7 SIES, made up of ceramic pipes for the experimental setup

Table 9.1 Range of parameters of various equipments

S. no	Equipments	Range
1	Fan power and pump	300W
2	Air mass flow rate	0.28 kg/s
3	Bypass factor of coil X	0.2
4	Effectiveness of wet bulb	1
5	Specific humidity w_0	2–6 gm/kg
6	Ambient temperature T_0 (^{0}C)	30–47 °C
7	ADP	2 °C
8	VCR system's COP	3

9.3.1 *Experimental Proposed Feature*

The depiction of the improved structure was ended through subsequent examining agreement process. A whole factorial arrangement was useful toward research torpid as well as rational heat improved inside arrangement toward getting the whole heat improved through the evaporative cooling structure. Figures 9.8, 9.9, and 9.10 show the components researched which were:

9.3.1.1 V (Air Flow)

There are three air stream stages are:

9.3.1.2 HL (Humidity Level)

The level of humidity (Fig. 9.8) shows HL1: relative humidity under 30%, HL2: relative humidity someplace within the scope of 30% as well as 60%, and HL3: relative humidity over 60%.

9.3.1.3 T (Temperature)

The level of temperatures is Fig. 9.10): T5: 40 °C, T4: 36.5 °C, T3: 33 °C, T2: 29.5 °C, as well as T1: 26 °C.

Fig. 9.8 Air flow levels

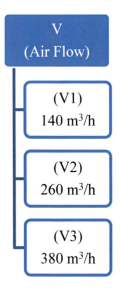

9 Power Consumption Saving of Air Conditioning System Using … 105

Fig. 9.9 Humidity levels

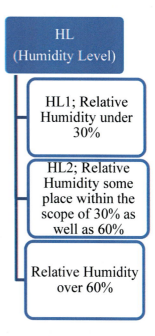

Fig. 9.10 Selection of temperature levels in a window AC

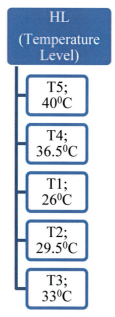

At the point when the prohibited aspects as well as the related points are chosen, a balanced arrangement grid was made. These assessments were unsystematic toward staying away from the opportunity of making a "demand sway" because it inclined through the gathering of the tests. Every one of the assessments was reproduced on different occasions. Correspondences between factors are similar while the special effects of personage aspects were explored. Observations are confirmed by calculation using analysis of variance (ANOVA) [7]. The ambiguity principles intended for the traits examined are given in Table 9.2, which addresses the ambiguity assessments for all volumetric stream portrayed [8].

The semi-indirect evaporative cooler (SEIC) has two self-sufficient air stream materials, one employed for cooling, alongside a subsequent, the return air stream, indirect contact with water toward positive discrimination of heat as well as mass transfer. There is water which is compelled alongside the comeback air stream as well as it is persistently flowing.

- Thermodynamic analysis of consolidated structure for hot and dry climatic conditions was conducted.
- Then, we will make a preliminary plan for a joint system.
- Analyses of the theoretical and test data show that the relationship of the power consumption is dependent on comfort air, power usage, and economy.

The evaporative air cooler (Fig. 9.11) is of prime essentialness in the summer season and boiling conditions. A cooler is commonly used in any place from high

Table 9.2 Procedures for $k = 2$ (Latent as well as sensible heat): Ambiguity worth for the distinctiveness examination [8]

Humidity level (HL)	Temperature	Latent heat			Sensible heat		
		U_v (1%)	U_v (2%)	U_v (3%)	U_v (1%)	U_v (2%)	U_v (3%)
HL1	40	4.4	5.6	5.6	4.3	5.7	5.6
	36.5	4.4	5	5.5	4.5	5.1	5.4
	33	4.4	5.3	5.9	4.4	5.3	6
	29.5	4.5	6.1	5.6	4.7	6.1	5.6
	26	4.4	5.6	5.6	4.3	5.7	5.6
HL2	40	4.3	5.7	6	4.4	5.7	6
	36.5	4.3	5.9	5.9	4.5	6.1	6.2
	33	4.4	6.1	6	4.5	6.1	6.3
	29.5	4.4	6	6.3	4.6	6.1	6.4
	26	4.5	6.1	5.9	4.5	6.3	6.1
HL3	40	4.3	5.7	6.4	4.3	5.8	6.5
	36.5	4.3	5.3	6.4	4.5	5.6	6.5
	33	4.4	6.2	6.4	4.5	6.3	6.6
	29.5	4.4	6.2	6.3	4.3	6.3	6.4
	26	4.5	6.1	6.4	4.6	6.4	6.6

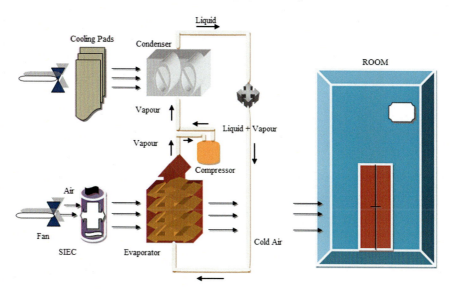

Fig. 9.11 Schematic and working of experimental set up

class to average families. The center of a modified evaporative cooling system is the ceramic material chamber where the water goes inside the chamber and as a result of the property of porosity of the chamber water comes out the outer surface of the chamber and cooperates with air passing other than the chamber and air get cooled.

9.4 Results and Discussion

Power Consumption

In this examination, it was unspecified that the certain cooling load and vapor density air influential structure would effort on a COP as of 3. Therefore, the general power utilization of the two cooler types is shown in Fig. 9.12. These are too probable that the evaporative cooling structures association would reduce top power requests from the lattice by around 163 W. The full scale of cooling units of the evaporative system's operational cost greater than the excursion hour is standing on the existing structure 0.175 kWh. Afterward, established evaporative cooling structure accumulates on common more than 70% within the cost of operation energy. Figure 9.13 demonstrates that the funds are renowned through every design as well as the equivalent COP.

This evaporative cooling structure has a constructive collision together to give frequent fresh air as well as lessening power usage along with consequent emissions of carbon. Based on the discharge aspect of the power age mix of 0.75 kg CO_2 l/

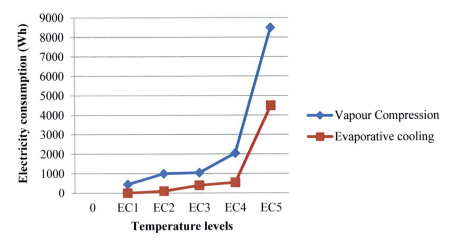

Fig. 9.12 Electricity consumption of a window AC with and without SIEC system

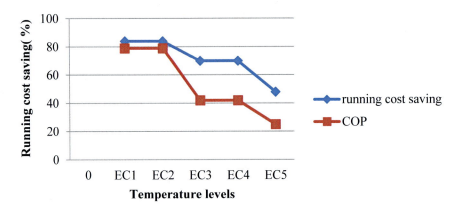

Fig. 9.13 Comparison between running cost saving and evaporative cooling unit

kWh, this total amount of CO_2 decreases while taking a gander at vapor compression systems along with an evaporative cooling structure to provide an equivalent predictable cooling load of 0.246 tons (i.e., 78% savings). Figure 9.14 illustrates the running cost saving on or after all evaporative coolers. It represents each evaporative cooler's emission savings in terms of reduction of CO_2 emission in percentile toward each evaporative cooler.

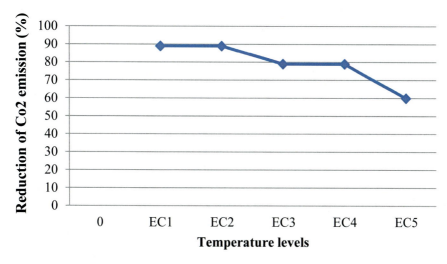

Fig. 9.14 Compression between the reduction of CO_2 emission and evaporative cooling unit

9.5 Conclusion

In favor of near-to-the-ground air supply with high temperatures and relative humidity contents, the model outcome is evaporative as of the surface of the ceramic pipe. We have to compare the air conditioning system by semi-evaporative cooling as a contrast to the conventional system in terms of the disparity of cooling coil load. Also, we have to study the air conditioning system by semi-evaporative cooling as a contrast to the usual structure in terms of disparity of air velocity with the mass flow to months. Also, compare the compilation of the proposed system cooling coil load to the temperature of the apparatus dew point of the exterior situation during summer. Also, study on cooling coil load for the cooling pads' dispersion efficiency and the temperature of apparatus dew point. Here, proposed for the air supply with relative humidifies and high temperatures, dehumidification obtains mark as well as thus reduction comes out in the pipes external surface, as well as the recovered sensible heat and torpid are included.

1. The cooling coil load of the experimental setup is reduced by 29.6% at optimum pad thickness $\delta = 0.15$ m.
2. All developed configurations have been compared based on saturation efficiency. The highest saturation efficiency of 95% was observed.
3. In this experimental study, different cooling pads are developed from five different materials. It may be conducted that from this study coconut fibers and banana fibers' cooling pads give higher saturation efficiency than conventional cooling pads.
4. Wet-bulb effectiveness is having a moderate value at 2 m/s intake air velocity.

5. Cooling capacity decreases with a decrease in relative humidity. At a relative humidity of 55%, wet-bulb effectiveness is 21% higher than dew point effectiveness.
6. Wet-bulb effectiveness increases with an increase in air inlet velocity. At 2.5 m/s inlet air velocity, the wet-bulb effectiveness increases by 26.2% at 35 °C $\underline{T_{di}}$.
7. Power consumption is reduced by 46% at evaporative cooling stage five (EC-5) as compared to the conventional vapor compression refrigeration system.
8. Latent, as well as sensible heat improvement with a variety of aspect levels similar to the flow of air, and levels of temperature relative humidity were focused on. There is considering of the possessions of three structures (temperature of moving toward the water, the velocity of the air, and the temperature of dry bulb for air) lying on effecting cooling.

All restrictions were distorted, whereas observance of every other noteworthy variable was stable, as well as data was collected by numerous parameters. These effects moreover prove that the interior evaporative cooler for supply air has superior humidity content than the evaporative cooler in a direct mode which constructs it a high-quality humidifier during quick storage wherever succulently close to diffusion is necessary. In that case, the air temperature of the dew point is close to the air temperature of the dry bulb; its way to the relative humidity will be high, while the dew point is going below the air temperature of the dry bulb through the low relative humidity.

References

1. Ma, H., Du, N., Yu, S., Lu, W., Zhang, Z., Deng, N., Li, C.: Analysis of typical public building energy consumption in northern China. Energy Buildings **136**(1), 139–150 (2017)
2. Malli, A., Seyf, H.R., Layeghi, Md., Sharifian, S., Behravesh, H.: Investigating the performance of cellulosic evaporative cooling pads. Energy Buildings **52**, 2598–2603 (2011)
3. Gómez, E.V., Martínez, F.J., Diez, F.V., Leyva, M.J., Martín, R.: Description and experimental results of a semi-indirect ceramic evaporative cooler. Int. J. Refrig. **28**, 654–662 (2005)
4. Camargo, J.R., Ebinuma, C.D., Silveira, J.L.: Experimental performance of a direct evaporative cooler operating during summer in a Brazilian city. Int. J. Refrig. **28**(7), 1124–1132 (2005)
5. Sheng, C., Nnanna, A.G.A.: Empirical correlation of cooling efficiency and transport phenomena of direct evaporative cooler. Appl. Therm. Eng. **40**, 48–55 (2012)
6. Ashrae, A.H.: Systems and equipment. American society of heating, refrigerating, and air-conditioning engineers (2008)
7. Kolokotsa, D., Santamouris, M., Synnefa, A., Karlessi, T.: Passive solar architecture. Compr. Renew. Energy **3**, 637–665 (2012)
8. Jain, D.B., Mishra, M., Sahoo, P.K.: A critical review on the application of solar energy as a renewable regeneration heat source in solid desiccant—vapor compression hybrid cooling system. J. Building Eng. **18**, 107–124 (2018). https://doi.org/10.1016/j.jobe.2018.03.012
9. Guan, L., Bennett, M., Bell, J.: Development of a climate assessment tool for hybrid air conditioner. Build. Environ. **82**, 371–380 (2014)
10. Pescod, D.: A heat exchanger for energy saving in an air-conditioning plant. ASHRAE Trans. **85**, 238–251 (1979)
11. Watt, J.R.: Evaporative air conditioning handbook. 3rd edn. Prentice-Hall (1997)

12. Gilani, N., Poshtiri, A.H.: Heat exchanger design of direct evaporative cooler based on outdoor and indoor environmental conditions. J. Therm. Sci. Eng. Appl. **6**(4), 16–41 (2014)
13. Gillan, L.: Maisotsenko cycle for cooling processes. Int. J. Energy Clean Environ. **9**, 47–64 (2008)
14. Stoitchkov, N.J., Dimitrov, G.I.: Effectiveness of cross-flow plate heat exchanger for indirect evaporative cooling. Int. J. Refrig. **21**, 463–471 (1998)

Chapter 10
Recrudesce: IoT-Based Embedded Memories Algorithms and Self-healing Mechanism

Vinita Mathur, Aditya Kumar Pundir, Raj Kumar Gupta, and Sanjay Kumar Singh

Abstract As rapid increase in IoT framework in edge computing, the data storage requirement is also increased. This data is stored in RAID 5 (Redundant array of independent drives) disks which require memory testing and a repair algorithm for a reliable design system. MBIST design system depends on the memory testing algorithm. Various fault detection approaches are introduced using March test on SRAM and DRAM. But, still, some focus is required in terms of time penalty and fault coverage. This paper introduces a novel approach for fault detection in memory with less time penalty, better fault coverage, and device utilization performance with the help of the IoT framework in edge computing. An IoT framework system is proposed for proper monitoring of memory under test for fault occurrence and number of fault repair for SRAM. Results show the optimal faults repair and with less time penalty. The simulation is conducted on MATLAB and Xilinx ISE suite. Proposed work can be used in commercial and space application where radiation hardened memories is used.

Keywords IoT device · RAID 5 · MBIST · MBISR

V. Mathur (✉) · R. K. Gupta · S. K. Singh
Amity University Rajasthan, Jaipur, India
e-mail: vinita.mathurdec@gmail.com

R. K. Gupta
e-mail: rkgupta@jpr.amity.edu

A. K. Pundir
Arya College of Engineering and IT, Jaipur, India

10.1 Introduction

In the era of smart and intelligent system for optimization and designing, Internet of Things (IoT) play an integral role. IoT is a process that allows to access number of devices like sensors and actuators to be connected through Internet. Testing of the process connectivity and result analysis of the software is pivotal process because it provides the results of specification, design used, and code generation [1, 2]. Hence, the main agenda of software testing is to reduce the faults that occur in the devices which are in process [3, 4].

With the advancement in submicron VLSI technology, the capacity and density of the memories are also drastically increased [1]. Memory is the main part of the integrated circuit as 90–92% of the embedded system is occupied by the memories than that of the logical memories. There are two types of memories: static and dynamic memories. Static memories are highly used due to their less power requirement and faster speed [2]. As the memory density increase, the number of fault occurrences in memory cell increase. Nowadays, built-in self-test (BIST) controller is used for memory testing with a hardware and software testing approach [5]. Boundary-scan register [6] architecture is used for memory testing which provides a convenient and better option than probing, but the chip used in this method operates in one test session only for more test session designers to define new instructions. Programmable built-in self-test (PBIST) [7] structures are designed for SRAM; it adopts a micro–macro code to select seven March algorithms to detect the fault, but in this method, required testing time is increased [8, 9]. They have proposed testing multiple memory cores in parallel sequence, but it can detect address decoder and stuck-at faults.

This paper is categorized into five sections. Section 10.2 describes the proposed IoT framework in edge computing using gateway. Sect. 10.3 presents the proposed work flowchart for IoT-based memory testing and repair. Section 10.4 describes March testing and repair algorithm for better fault coverage with less time penalty. Section 10.5 shows the simulation profile of the proposed work. Section 10.6 describes the conclusion and future scope of the proposed work.

10.2 IoT Devices and Framework Architecture

IoT is process in which large amount of data is transmitted/processed and being communicated across the networks. IoT architecture consists of several layers which provide the optimal solution for the system. Figure 10.1 shows the primary stages of IoT framework architecture which are sensor/actuators, IoT gateways for data acquisition, data processing layer, and application layer of smart device, i.e., Cloud data center.

1. **Sensors/Actuators**: These can be wired or wireless sensor used to transmit or process the information like GPS, RFID, etc. Some sensors require the gateway connectivity; for this, a local area network or personal area network is used.

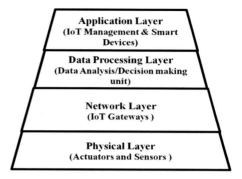

Fig. 10.1 IoT framework architecture

2. **Gateways and Data Acquisition**: A huge amount of data is produced by the sensors which require a high-speed network and gateway like LAN and WAN for data transfer.
3. **Edge IoT**: An edge computing is required for transferring the data in cloud. This process can be done through hardware and software. Advantage of edge IoT is that if the data is same as the previous one then that data is not transferred to cloud, that data is saved in its memory.
4. **Data Cloud**: It provides the management services, it process the data in term of analytics, security controls, and management of device. It also provides the data to end user.

10.3 Proposed Work Flowchart for IoT-Based Memory Testing and Repair

Figure 10.2 shows the flowchart of the proposed work. A master accelerator unit is used to update the data of memory under test (MUT). In this paper, we have considered RAID 5 (Redundant array of independent drives) for data storage which consist of following memory array $256 \times 8 \times 1$, $1K \times 8 \times 1$, $256 \times 16 \times 1$, and $1K \times 16 \times 1$ memories for testing. RAID 5 offers an optimized performance in low cost with high reliability [5].

An IoT framework module and smart gateways are used to update the information to master accelerator that the MUT is faults free or not, and if there is any faults, then that memory can be repaired using repair algorithms and available for run time.

For memory testing, an CHECKERMARC [10] algorithm is used as it provide better faults coverage 0.8% as compare to parent March C-. For repair, an MMBISR [11] (modified memory built-in self-repair) algorithm is used; it is a hybrid redundancy algorithm, it provides the fault dictionary which contains updated or fixed concurrent information of MBIST controller.

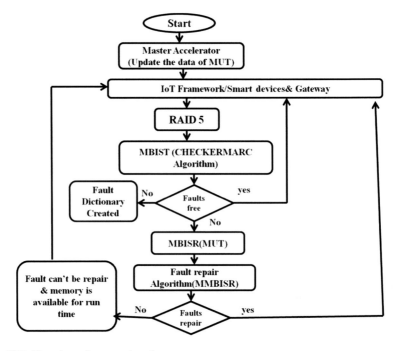

Fig. 10.2 Flowchart of proposed work

10.4 March Testing and Repair Algorithm

10.4.1 Memory Testing Algorithms

A March test consists of a sequence of March elements. These elements provide the sequence of operations to be performed on every cell of memories [12, 13]. The order of operation performed can be ascending, descending, and irrelevant [5]. Notations used in the March test are given in Table 10.1.

In this paper as given in Table 10.2, an CHECKERMARC [10] algorithm is used for MBIST (memory built-in self-test); this is an hybrid testing algorithm designed

Table 10.1 Notation used in March algorithms

S. No.	Notation	Meaning
1	↑	Ascending
2	↓	Descending
3	↕	Irrelevant
4	r0/1	Read 0/Read 1
5	w0/1	Write 0/Write 1

Table 10.2 CHECKERMARC algorithm for EVEN and ODD type [10]

Algorithm 1 CHECKERMARC
Step1: ↕ [wr0(even) &wr1(odd)]
Step2: ↑[rd0(even)& rd1(odd), (wr1(even) &wr0(odd)]
Step3: ↑[rd1(even)& rd0(odd), (wr0(even) &wr1(odd)]
Step 4:↓ [rd0(even)& rd1(odd), (wr1(even) &wr0(odd)]
Step 5:↓ [rd1(even)& rd0(odd), (wr0(even) &wr1(odd)]
Step 6: ↕ [rd0(even)& rd1(odd)]

with the help of CHECKER_BOARD and MARCH C-. This provides better results in terms of fault coverage and hardware easiness with slight area overhead issues.

An MMBISR algorithm [11] (modified memory built-in self-repair) is used for repairing the faulty memory; it is an also an hybrid redundancy analysis algorithm which is modified with the help of essential spare pivoting (ESP) and local repair most (LRM); it provides optimized set of row and column combination that is suitable for the repair process.

10.5 Simulation Results

10.5.1 Memory Testing Profile

Table 10.3 elaborates the simulation profile for diagnosis of faults in memories array $256 \times 8 \times 1$, $1k \times 8 \times 1$, $256 \times 16 \times 1$, and $1k \times 16 \times 1$ which is performed on the basis of testing time (ns) and number of fault coverage. Results show that the CHECKERMARC [10] algorithm covers more numbers of faults with less testing time. Figure 10.3 show the graphical analysis of the algorithm used with the help of Xilinx ISE suite.

10.5.2 Memory Testing Repair

Table 10.4 presents the simulation profile of memories array $256 \times 8 \times 8$ and $512 \times 8 \times 1$ to be repaired using MMBISR [11] algorithm. This hybrid algorithm is analysis with the least repair most (LRM) and essential spare pivot (ESP) [13–15] on the basis of repair rate, area overhead, and repair testing time (ms). In Fig. 10.4, we have analyzed the graphical analysis of the algorithm.

These testing and repairing processes are applied to IoT framework system which enhances the tracing time capability.

Table 10.3 Simulation profile for diagnosis of faults

S. No.	Testing algorithms	Memory size	Testing time (ns)	Fault coverage	Environment/tool used
1	CHECKERBOARD	256 × 8 × 1	35.79	94.13	VHDL/Virtex 4/ Xilinx
		1k × 8 × 1	38.72	92.17	
		256 × 16 × 1	32.45	93.25	
		1k × 16 × 1	33.47	95.17	
2	MARCH C-	256 × 8 × 1	30.5	96.72	
		1k × 8 × 1	31.93	97.12	
		256 × 16 × 1	28.82	96.72	
		1k × 16 × 1	29.38	95.35	
3	CHECHKERMARC	256 × 8 × 1	36.12	96.8	
		1k × 8 × 1	37.51	97.92	
		256 × 16 × 1	30.5	97.51	
		1k × 16 × 1	35.3	95.1	

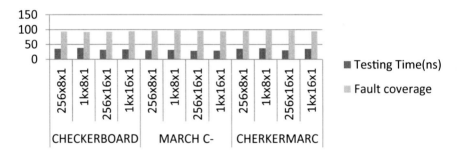

Fig. 10.3 Simulation profile for diagnosis of faults

Table 10.4 Simulation profile for repair algorithm

Algorithm used	256 × 8 × 8			512 × 8 × 8		
	Repair rate	Area overhead	Repair time (ms)	Repair rate	Area overhead	Repair time (ms)
LRM	99.5	4929	2500	99.5	5954	2500
ESP	93.82	1200	1500	93.82	2753	1500
MMBISR	95.6	2500	2100	96.12	3922	2215

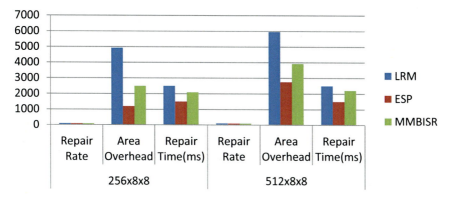

Fig. 10.4 Simulation profile for repair algorithm

10.6 Conclusion

In this paper, an IoT framework in edge computing system is proposed for monitoring the memory for testing and repairing. An RAID 5 (Redundant array of independent drives) disk is used for memories under test array $256 \times 8 \times 1$, $1K \times 8 \times 1$, $256 \times 16 \times 1$, and $1K \times 16 \times 1$ as it provides reconstruction of data in case of disk failure. RAID 5 provides the better balancing between high performance and security. RAID 5 MUT is under-processed for fault diagnosis and repaired with the help of hybrid algorithm CHECKERMARC and MMBISR. Result shows the optimal fault repair with less time penalty. The simulation is conducted on MATLAB and Xilinx ISE suite. Proposed work can be used in commercial work like in space application where radiation-hardened memories are used.

References

1. Hagar, J.D.: Software test architectures and advanced support environments for IoT. In: IEEE International Conference on Software Testing, Verification and Validation Workshops (ICSTW), pp. 252–256. Sweden (2018
2. Sahu, K., Srivastava, R.K.: Soft computing approach for prediction of software reliability. ICIC Express Lett. **12**(12), 1213–1222 (2018)
3. Lima, J.A.P., Vergilio, S.R.: Test case prioritization in continuous integration environments: a systematic mapping study. Inf. Softw. Technol. **121**, 1–16 (2020)
4. Sahu, K., Srivastava, R.K.: Revisiting software reliability in data management, analytics and innovation, vol. 801, pp. 221–235. Analytics and Innovation Springer, Singapore (2020)
5. Yuan, Z., You, X., Lv, X., Xie, P.: SS6: online short-code RAID-6 scaling by optimizing new disk location and data migration. Comput. J. **64**(10), 1600–1616 (2021). https://doi.org/10.1093/comjnl/bxab134
6. Tian, C., Li, Y., Wu, S., Chen, J., Yuan, L., Xu, Y.: Popularity-based online scaling for raid systems under general settings. IEEE Trans. Comput. Aided Des. Integr. Circuits Syst. **39**(10), 2911–2924 (2020). https://doi.org/10.1109/TCAD.2019.2930580

7. Pundir, A., Sharma, O.P.: CHECKERMARC: a modified novel memory-testing approach for bit-oriented SRAM. Int. J. Appl. Eng. Res. **12**, 3023–3028 (2017)
8. Pundir, A.: Novel modified memory built in self-repair (MMBISR) for SRAM using hybrid redundancy-analysis technique. IET Circ. Dev. Syst. (2019). https://doi.org/10.1049/iet-cds.2018.5218
9. Manikandan, P., Larsen, B.B., Aas, E.J., Areef, M.: A programmable BIST with macro and microcodes for embedded SRAMs. In: 2011 9th East-West Design & Test Symposium (EWDTS), pp. 144–150 (2011). https://doi.org/10.1109/EWDTS.2011.6116584
10. John, P.K., Rony Antony, P.: BIST architecture for multiple RAMs in SoC. Procedia Comput. Sci. **115**, 159–165 (2017). https://doi.org/10.1016/j.procs.2017.09.121
11. Harutyunyan, G., Shoukourian, S., Vardanian, V., Zorian, Y.: A new method for March test algorithm generation and its application for fault detection in RAMs. IEEE Trans. Comput. Aided Des. Integr. Circuits Syst. **31**(6), 941–949 (2012). https://doi.org/10.1109/TCAD.2012.2184107
12. Manasa, R., Verma R., Koppad, D.: Implementation of BIST technology using March-LR algorithm. In: 2019 4th International Conference on Recent Trends on Electronics, Information, Communication & Technology (RTEICT), pp. 1208–1212 (2019). https://doi.org/10.1109/RTEICT46194.2019.9016784
13. Hamdioui S., van de Goor, A.J., Rodgers, M.: March SS: a test for all static simple RAM faults. In: Proceedings of the 2002 IEEE International Workshop on Memory Technology, Design and Testing (MTDT2002), pp 95–100 (2002). https://doi.org/10.1109/MTDT.2002.1029769
14. Lakshmi, G.S., Neelima, K., Subhas, D.: A march ns algorithm for detecting all types of single bit errors in memories. Int. J. Emerg. Technol. Innov. Res. **6**(4), 289–296 (2019). ISSN: 2349-5162
15. Benso, A., Bosio, A., Di Carlo, S., Di Natale, G., Prinetto, P.: March test generation revealed. IEEE Trans. Comput. **57**(12), 1704–1713 (2008). https://doi.org/10.1109/TC.2008.105

Chapter 11
Design and Performance Analysis of InSb/InGaAs/InAlAs High Electron Mobility Transistor for High-Frequency Applications

Prajjwal Rohela, Sandeep Singh Gill, and Balwinder Raj

Abstract This paper investigates the performance of a high electron mobility transistor (HEMT) with a 0.4 μm gate size enhancement mode. The device is composed of an InGaAs/InAlAs structure grown on an InSb substrate, with heavily doped In0:6Ga0:4As source/drain (S/D) regions and dual δ(sigma)-doping linear layers. The transistor described in this paper incorporates a buried Au metal gate technique to minimize short channel effects and enhance transconductance. The device also features heavily doped In0:6Ga0:4As source/drain (S/D) regions, Si dual sigma-doping linear layers at the edges of the In0:75Ga0:25As channel area. The high electron mobility transistor (HEMT) InSb/InGaAs/InAlAs provides outstanding high-frequency performance. Silvaco TCAD simulations that use the accurate methodology at room temperature indicated that the investigated device exhibited good pinch-off performances of IDS = 222.8 A at VGS = − 0.6 V, with a high transconductance of 894.8 A/V and a threshold voltage (IDS) of 3 V.

Keywords HEMT · InAlAs · InGaAs · InSb · Silvaco TCAD · Pinch-off

11.1 Introduction

High electron mobility transistors (HEMTs) have evolved as an essential innovation in high-speed microwave applications. HEMTs frequently use Group III-V devices because of their high electron mobility and density. With differences in electron activity, the device having a higher charge density induces a movement in the device's conductance oscillations. HEMTs have emerged as an essential technology in high-speed microwave applications. Group III-V devices are often employed as HEMTs

P. Rohela (✉) · S. S. Gill · B. Raj
VLSI Design Lab, Department of Electronics and Communication Engineering, NITTTR Chandigarh, Chandigarh 160019, India
e-mail: prajjwal.ece20@nitttrchd.ac.in

© The Author(s), under exclusive license to Springer Nature Singapore Pte Ltd. 2024
P. K. Jha et al. (eds.), *Proceedings of Congress on Control, Robotics, and Mechatronics*, Smart Innovation, Systems and Technologies 364,
https://doi.org/10.1007/978-981-99-5180-2_11

due to their high electron mobility and density. With differences in electron behavior, higher charge density in the transistor induces a shift in the device's conductance oscillation [1]. Takashi Mimura invented the HEMT at Fujistu, while work into modulation-doped heterostructures by Raymond Dingle and others at Bell Lab was equally important [2]. These features enable the device to operate at a high switching speed while consuming less power. Because of the devices' fast switching speeds and low power requirements, it is ideal for integrating multiple devices onto a single chip. The device may be sized and scaled without compromising performance, which will make it more appropriate for use in contemporary electronics applications [3].

InGaAs/GaAs, AlGaN/GaN, and other Group III-V compounds were used in the construction of HEMT transistors. Traditional HEMTs have operated at frequencies more than 60 GHz and at data rates greater than 10 Gb/s [4]. The composition and architecture of the gadget should be upgraded well above the required operation velocity and data throughput. Higher electron percentages in Group III-V elements permit faster electron transport. With a smaller effective mass and larger electron intensity, and a perfect lattice constant with a tiny energy gap, InSb possesses the finest electrical characteristics of any III-V element [5].

Several applications in telecommunication and imaging are emerging at frequencies exceeding 100 GHz. To provide excellent yield while consuming little power, it is necessary to use high-performance equipment and circuits [6].

In recent years, semiconductors containing narrower band gaps have been employed for HEMT channels to attain substantially higher electron mobilities due to decreased effective mass. In order to enhance the efficiency of HEMT in higher speed, lowest noise, and lower power applications, heterostructure is based on the semiconductor with narrow bandgap, such as AlSb/InAs and AlInSb/InSb, also known as Sb-based heterojunctions, which basically emerged as an excellent choice, by increasing their potential for use in military/army and space applications with ultra-low power requirements [7].

AlSb, GaSb, InSb, and InAs binary compound semiconductors, as well as their related alloys, are candidates for high-speed, low-power electronic devices. High-speed analog and digital systems could be used in portable equipment such hand-held devices and satellites for data processing, communications, imaging, and sensing. Transistors made from Sb materials have the potential to provide a technological breakthrough for digital, mixed-signal, and low-noise high-frequency amplifiers.

According to the graph in Fig. 11.1, there is a trend toward higher frequencies and lower power consumption as the lattice constant increases. In the recent advances in HEMTs, RTDs, and HBTs (lattice constants greater than 6.0A) antimonide-arsenide materials system have been made [10].

11 Design and Performance Analysis of InSb/InGaAs/InAlAs High ...

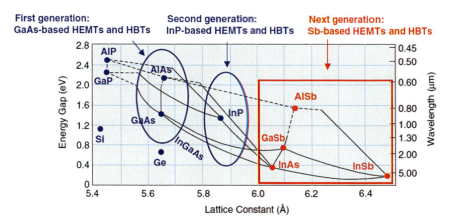

Fig. 11.1 Energy gap versus lattice constant plot demonstrates the trend of transistors toward larger lattice constants and narrower bandgaps to achieve high-frequency and low-power operation [10]

11.2 Device Physics

Over the past decade, there has been considerable progress in researching the properties and applications of antimonide-based semiconductor materials. In 2001, the Defense Advanced Research Projects Agency (DARPA) initiated the antimonide-based compound semiconductors program (ABCS program) [8], which has led to noteworthy advancements in the development of antimonide-based microstructure materials and device applications worldwide [9]. For reference, Table 11.1 presents a comparison of the physical properties of III-V compound semiconductors at room temperature.

To fully deplete the InAlAs supply/barrier layer, one can calculate the required gate voltage (VGS) by subtracting the pinch-off voltage of the n + supply/barrier area from the built-in Schottky barrier voltage, expressed as [10]

$$v_{Gs} = v_b - v_P \tag{11.1}$$

The drain current of the HEMT is significantly influenced by the width of the barrier layer. A wider barrier layer leads to an increase in sheet charge density (NS) or 2DEG density. Additionally, the mobility of electrons initially increases proportionally to the barrier thickness. Nonetheless, once the barrier thickness exceeds a specific critical threshold, the electron mobility starts to decline [11]. Selecting an appropriate thickness for the barrier layer can result in a significant increase in the charged-sheet intensity (NS) while maintaining high mobility [12].

$$ID = (qNs\mu nWVDS)/L \tag{11.2}$$

Table 11.1 Comparison of the physical properties of III-V compound semiconductors at room temperature

Properties/ Materials	Si	Ge	GaAs	GaN	InP	InAs	InSb
Bandgap (eV)	1.12	0.67	1.43	3.4	1.344	0.35	0.17
Dielectric constant (static; high frequency)	11.9	16.2	Static-12.9 HF-10.89	Static-8.9;10.4;9.5 HF-5.35;5.8	Static-12.61 HF-9.61	15.2	16.8
Electron mobility (cm²/Vs)	≤ 1400	3900	≤ 8500	≤ 1000	≤ 5400	40,000	78,000
Hole mobility (cm²/Vs)	≤ 450	1900	≤ 4000	350;200	≤ 200	500	850
Saturated electron drift velocity (cm/s)	2 × 10⁷	6	(3.5–4.2) × 10⁷	2 × 10⁷ at 130 kV/cm	1.2 times of GaAs at 150 kV/cm; at 30 kV/cm 1.34 times of GaAs	4 × 10⁷	4 × 10⁷
Electric breakdown field (V/cm)	3 × 10⁵	10⁵	4 × 10⁵	5 × 10⁶	5 × 10⁵	4 × 10⁴	10³
Thermal conductivity (W/cm °C)	1.5	0.58	0.55	1.3	0.68	0.27	180

The product of μnNS increases as the width (W) of the device increases and the distance (L) between the source and drain decreases, given a certain thickness of the barrier layer. In simpler terms, increasing the device's width and reducing the distance between the source and drain can lead to a higher μnNS product for a given barrier thickness [11]. According to Eq. (11.2), an increase in the barrier layer thickness results in an increase in current. However, there is a trade-off associated with this increase in thickness. Increasing the thickness of the barrier layer results in a larger distance between the channel electrons (2-DEG) and the gate terminal. Consequently, the electrostatic control of the gate over the channel electrons becomes insufficient, causing a decrease in the transconductance (gm) as the gap between the gate and channel increases [10]. This impact is represented by:

$$gm = (W\, Er\, E0 V_{\text{eff}})/T_{g-\text{ch}} \tag{11.3}$$

However, V_{eff} represents the effective electron velocity, while $t_{g-\text{ch}}$ represents the gate to channel separation [12]. As a result, it is obvious that increasing the barrier width up to a critical level results in a rise in drain current but at the expense of reduced transconductance [10]. As previously stated, increasing the barrier thickness up to a particular amount enhances the drain current. However, due to the greater gate-to-channel spacing, the gate end drops electrostatic command throughout the channel electrons, by decreasing transconductance (gm) [11]. To address this trade-off, the delta doping method (also known as pause doping) is employed. Rather than heavily doping the entire barrier layer, this technique involves heavily doping only a thin layer at the bottom of the barrier, while leaving the rest of the barrier undoped or mildly doped [10].

11.3 Device Structure

The structure of the proposed HEMT is depicted in Fig. 11.2, which comprises an In0.48-Al0.52As supply/barrier layer with a thickness of 30 nm, an In0.53Ga0.47-As channel layer with a thickness of 16 nm, and a 1 nm thick delta-doped layer sandwiched between them. These layers are separated by a 2 nm wide spacer region. The delta-doped layer and spacer layer are made of undoped and strongly doped materials, respectively, similar to the supply/barrier layer.

The active region of the device is formed by the channel layer, which comprises a low-bandgap material. The bandgap difference between InGaAs and InAlAs generates a 2DEG in the channel layer. A potential barrier confines the electrons to a narrow charge sheet, which is referred to as the 2DEG. A 40 nm thick InSb layer serves as the buffer layer for the proposed device, which is deposited on a 60 nm thick InSb substrate [13].

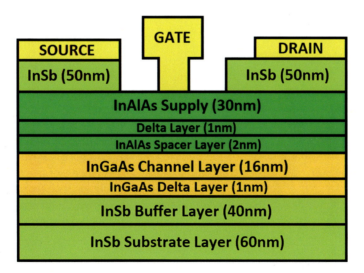

Fig. 11.2 Proposed design of an InSb-based HEMT, without any indication of scale

Table 11.2 Performance of a HEMT device based on InAlAs/InGaAs/InSb and incorporating a gate with a rectangular shape which is analyzed

Parameters	Dimensions
Gate length	0.40 μm
Source length	0.05 μm
Drain length	0.05 μm
Gate work function	4.73

The simulation using Silvaco TCAD is conducted at room temperature as it helps to analyze the parameters contributing to the drain-induced barrier lowering (DIBL) effect. Key device dimensions such as gate terminal length (L_g), spacing of side recess (L_{side}), thickness of a channel (t_{ch}), and insulator thickness between gate and channel (t_{ins}) are taken into consideration during the simulation, which significantly affects the DC and RF performance.

DIBL, or drain-induced barrier lowering, is a phenomenon where the threshold voltage of a device shifts toward a more negative value as the channel barrier decreases and the drain-to-source voltage (VDS) increases [14].

Table 11.2 displays a range of device parameters, such as using a 0.40 μm gold gate (with a work function value of 4.73 eV) and gold source and drain electrodes for the transistor. A T-gate topology is recommended in Fig. 11.3 as it simplifies the design process. This structure reduces gate resistance by offering a more prominent area while retaining a shorter foot length. To avoid plagiarism, the following sentence could be used: Adding a T-gate to the device has been found to improve its performance in both the DC and RF domains.

11.4 Results and Discussions

It should be noted that the meshing strategy used in device simulations can have a significant impact on the accuracy and computational efficiency of the results. Therefore, the authors likely spent considerable effort optimizing their meshing strategy to strike a balance between accuracy and efficiency. The accuracy of the solution is highly dependent on the mesh density. Fine grids are necessary for critical regions, whereas coarse grids are sufficient for less significant areas.

The device's doping configuration, depicted in Fig. 11.4, consists of heavily doped cap and donor layers to ensure good ohmic contact and the supply of free electrons to the channel region, which is unintentionally doped. Simulation tools were employed to investigate the device's physical phenomena and electrical parameters, aiding in determining semiconductor and device properties. We simulated the InAlAs/InGaAs/InSb HEMT device using Silvaco TCAD's Atlas and Deck Build environment.

In order to study the switching behavior of the transistor, the I-V characteristics of the device were obtained by keeping the gate voltage fixed, as depicted in Fig. 11.6. This enabled the conversion of I_{off} to Ion and vice versa, mimicking the functioning of a conventional transistor. The drain currents were measured at different gate voltage

11 Design and Performance Analysis of InSb/InGaAs/InAlAs High ... 127

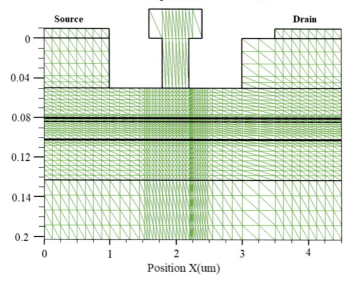

Fig. 11.3 Simulation of InAlAs/InGaAs/InSb HEMTs with T-gates conducted using meshing

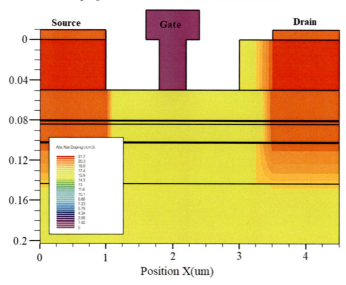

Fig. 11.4 Illustration of doping in a 2D InAlAs/InGaAs/InSb HEMT device

values and various drain voltage values. The transistor presented in Fig. 11.2 demonstrates a saturated drain current of 216.94 Amperes at a gate voltage (V_{GS}) of -0.2 V, which is the highest among the analyzed devices.

The HEMT device analyzed in this study demonstrates strong pinch-off behavior with a drain saturation current of 222.8 mA achieved at $V_{gs} = -1$ V and $V_{ds} = 0$ V. The subthreshold slope is visualized in Fig. 11.9, while Fig. 11.8 shows the transconductance (gm) plotted against gate voltage. The device achieves a maximum transconductance gm of 894.8 mS/mm at $V_{gs} = -0.6$ V.

$$\text{Gm} = \frac{\delta I_{ds}}{\delta V_{GS}} \tag{11.4}$$

The HEMT device is characterized at a drain voltage (V_{DS}) of 3 V, and the threshold voltage (V_t) is found to be -0.2 V at this bias point. The maximum drain current (I_{DS}) value of 222.8 µA is obtained at a gate voltage of -0.2 V and a drain voltage of 3 V, as shown in Fig. 11.6.

The transfer characteristics and transconductance of the device were obtained by measuring the variations in drain current (I_D) resulting from changes in gate voltage (V_{GS}), as illustrated in Figs. 11.5, 11.6, 11.7, 11.8 and 11.9. The device's transconductance reaches its maximum value of 894.8 mS/mm at a gate bias of -0.6 V, after which it starts to decrease.

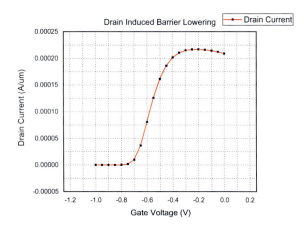

Fig. 11.5 A study was carried out to analyze the drain-induced barrier lowering (DIBL) in an HEMT that utilizes InSb as a material

Fig. 11.6 One studied the dependence of the drain current of a HEMT based on InSb on the variation of the source-to-drain voltage VDS

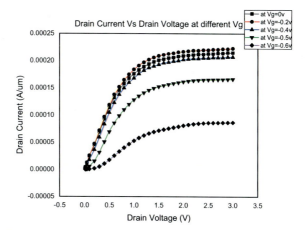

Fig. 11.7 Outcome plot displays the maximum value of ID for multiple VDS variations while changing V_{GS}

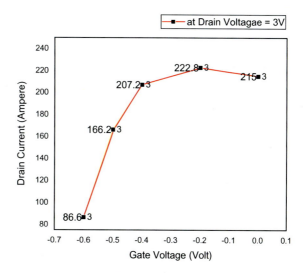

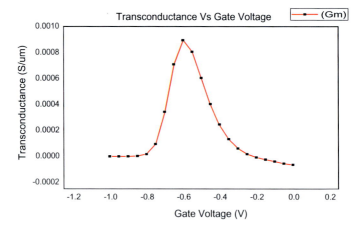

Fig. 11.8 Dependence of the transconductance Gm (S/μm) on the source to gate voltage VGS was analyzed in HEMTs based on InSb

Fig. 11.9 InSb-based HEMT on InSb's sub-threshold slope

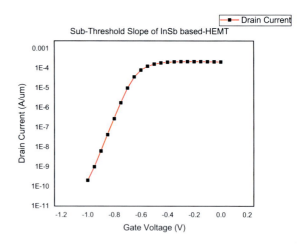

11.5 Conclusion

In this study, we have successfully demonstrated the operational capabilities of HEMTs that utilize an In0.75Ga0.25As channel and have $L_g = 0.40$ μm on InSb-based substrates. These devices have exhibited high drain current and transconductance, which can be attributed to the higher electron mobility in the 2DEG. Moreover, the lateral strain induced by the mismatch between the lattice constants of the strongly doped In0.6Ga0.4As source/drain regions and the In0.75Ga0.25As channel layer further enhances electron mobility. These results suggest that these HEMTs have enormous potential for various applications in the fields of military, space, medicine, and communication.

References

1. Kastner, M.A.: The review of high electron mobility transistor. Rev. Mod. Phys. **64**(3), 849 (1992)
2. Zhu, J., III.: Review of III–V based high electron mobility transistors. IOSR J. Eng. **5**(4), 2278–8719 (2015)
3. Yu, S., Hwang, S.W., Ahn, D.: Macromodeling of single-electron transistor for efficient circuit simulation. IEEE Trans. Nanotechnol. **46**(8), 1667–1671 (1999)
4. Takser, P.J.: High electron monility transitor. In: Morgan, D.V., Williams, R.H. (eds.) Physics and Technology of Heterojunction Devices. 1st edn. Peter Peregrinus Ltd, United Kingdom (1991)
5. Munusami, R., Prabhakar, S.: Group III–V semiconductor high electron mobility transistor on Si substrate. In: Different Types of Field-Effect Transistors-Theory and Applications. IntechOpen (2017)
6. Rodwell, M., Lee, Q., Mensa, S.D., et al.: Heterojunction bipolar transistors with greater than 1 THz extrapolated power-gain cut off frequencies. Proc. 7th IEEE THz Conf. **25** (1999)
7. Rodilla, H., González, T., Pardo, D., Mateos, J.: High-mobility heterostructures based on InAs and InSb: a Monte Carlo study. J. Appl. Phys. **105**(11), 113705 (2009)
8. Rosker, M., Shah, J.: DARPA's program on antimonide based compound semiconductors (ABCS). IEEE GaAs Digest, 293 (2003)
9. Liu, C., Li, Y., Zeng, Y.: Progress in antimonide based III-V compound semiconductors and devices. Engineering **02**(08), 617–624 (2010). https://doi.org/10.4236/eng.2010.28079s
10. Bennett, B., Magno, R., Boos, J., Kruppa, W., Ancona, M.: Antimonide-based compound semiconductors for electronic devices: a review. Solid-State Electron. **49**(12), 1875–1895 2005). https://doi.org/10.1016/j.sse.2005.09.008
11. Deen, D.A., Storm, D.F., Meyer, D.J., Bass, R., Binari, S.C., Gougousi, T., Evans, K.R.: Impact of barrier thickness on transistor performance in AlN/GaN high electron mobility transistors grown on free-standing GaN substrates. Appl. Phys. Lett. **105**(9), 093503 (2014). https://doi.org/10.1063/1.4895105
12. Robertson, I.D., Lucyszyn, S. (eds.): RFIC and MMIC Design and Technology, vol. 13. IET (2001)
13. Subash, T., Gnanasekaran, T.: Indium antimonide based HEMT for RF applications. J. Semiconductors **35**(11), 113004 (2014). https://doi.org/10.1088/1674-4926/35/11/113004/meta
14. Ajayan, J., Nirmal, D.: 22 nm In${}_{0.75}{{\rm{Ga}}}_{0.25}$As channel-based HEMTs on InP/GaAs substrates for future THz applications. J. Semiconductors **38**(4), 044001 (2017). https://www.researchgate.net/profile/Dnirmal-Phd/publication/315781969_22_nm_In_075_Ga_025_As_channel-based_HEMTs_on_InPGaAs_substrates_for_future_THz_applications/links/5a05250aaca2726b4c74a05e/22-nm-In-075-Ga-025-As-channel-based-HEMTs-on-InP-GaAs-substrates-for-future-THz-applications.pdf

Chapter 12
Ant Lion Optimizer with Deep Transfer Learning Model for Diabetic Retinopathy Grading on Retinal Fundus Images

R. Presilla and Jagadish S. Kallimani

Abstract Diabetic retinopathy (DR) becomes a sight-threatening complication because of diabetes mellitus which affects the retina. Initial identification of DR turns out to be a significant one as it might cause permanent impaired vision in the late stages. The automatic grading of DR seems to have effective benefits in solving such impediments, like rising efficiency, scalability, and coverage of analyzing process, extending applications in developed areas, and enhancing patient prevention by offering premature diagnosis and referral. In recent times, the performances of deep learning (DL) systems in the analysis of DR are close to that of expert-level diagnoses for grading fundus images. This article introduces an Ant Lion Optimizer using ALODTL-DRG technique on retinal fundus images. The presented ALODTL-DRG model performs preprocessing via interpolation image resizing, weighted Gaussian blur, and CLAHE-based contrast enhancement. For feature extraction, Inception with ResNet-v2 model is utilized in this study. At last, the ALO algorithm can be exploited as a hyper parameter tuning strategy to accomplish enhanced DR detection performance. The experimental assessment of the ALODTL-DRG method can be tested by making use of benchmark datasets. A widespread comparison study stated the enhanced performance of the ALODTL-DRG model over recent approaches.

Keywords Diabetic retinopathy · Ant Lion optimizer · Transfer learning · Retinal fundus images · Computer-aided diagnosis

---

R. Presilla
Department of Computer Science and Engineering, M S Ramaiah Institute of Technology, Bangalore, India

Visvesvaraya, Technological University, Belagavi, Karnataka, India

R. Presilla
e-mail: 1ms20pcs07@msrit.edu

J. S. Kallimani (✉)
Department of Artificial Intelligence and Machine Learning, M S Ramaiah Institute of Technology, Bangalore, India
e-mail: jagadish.k@msrit.edu

© The Author(s), under exclusive license to Springer Nature Singapore Pte Ltd. 2024
P. K. Jha et al. (eds.), *Proceedings of Congress on Control, Robotics, and Mechatronics*, Smart Innovation, Systems and Technologies 364, https://doi.org/10.1007/978-981-99-5180-2_12

12.1 Introduction

Diabetes is the most advantageous disease for using the concept of deep learning algorithms [1]. Lot of researchers have been working on the prediction of diabetes disease and complications arising from diabetes [2]. There exist variety of applications available that assist the practitioner to study the disease and complications, but those application have their own pros and cons. Diabetic retinopathy (DR) is the most important complication in human eye of diabetic patients [3]. DR is a compilation of diabetes which causes blood vessels of retina to leak and swell blood and fluids [4]. One of the commonest ways to diagnose the diabetic eye is to study the severity of the disease and examine fundus images. There are four major levels of DR; most advanced stage, abnormal blood vessel propagates on the retina surface that might result in cell loss and scarring in the retina [5].

With the development of computer vision technique, many automated systems were introduced by researcher workers for the diagnoses of DR [6]. There exist number of problems related to the enhancement in computer-aided diagnoses (CAD) systems, namely segmentation of blood vessels, detection of lesions from a retinal image, subdivision of optic disk, and so on. Even though machine learning (ML)-based system has shown effective performance in DR diagnosis, their efficiency is dependent highly on handcrafted feature that is highly complex to generalize [7]. In order to address this problem, deep learning (DL) method provides automated classification and feature extraction from fundus images. DL model has greater performance, but to train them, huge datasets and a lot of time are needed [8]. Since only a limited amount of images are available in the medical image classification task, training DL model is a challenging task. To overcome these shortcomings, we apply transfer learning (TL) method [9]. TL method is referred to as learning a new task via transfer of knowledge from previously learned related tasks [10]. Current study shows that TL approach does not require large dataset. Furthermore, the required training time is minimized because model is already pretrained.

This article introduces an Ant Lion Optimizer with deep transfer learning model for diabetic retinopathy grading (ALODTL-DRG) technique on retinal fundus images. The presented ALODTL-DRG model performs preprocessing via interpolation image resizing, weighted Gaussian blur, and CLAHE-based contrast enhancement. For feature extraction, Inception with ResNet-v2 model is utilized in this study. Next, deep belief network (DBN) model is applied for classification and rating DR. At last, the ALO algorithm can be exploited as a hyper parameter tuning strategy to accomplish enhanced DR detection performance. The experimental assessment of the ALODTL-DRG model is tested using benchmark dataset.

12.2 Related Works

In [11], the model is implemented with preprocessing and augmenting approaches which are helpful in improving the precision of feature extraction. Shankar et al. [12] introduce a new DNN together with moth search optimization (DNN-MSO) method related classification and detection method for DR images. The proposed DNN-MSO method adds various processes like preprocessing, classification, segmentation, and feature extraction. Primarily, in DR images, the contrast level was improvised by utilizing contrast limited adaptive histogram equalizing technique. Then, the images which are preprocessed were segmented by making use of histogram method. Next, Inception-ResNet-v2 method can be implemented for feature extraction. At last, feature vectors which are extracted were provided to the DNN-MSO related classification method for classifying the distinct phases of DR.

They make an effort in discovering an automated way for classifying a provided fundus image set [13, 14]. And bringing forth CNNs power to DR detection has three major difficulties; they are detection, classification, and segmentation. Associating with TL and hyper parameter tuning, adopts GoogleNet, AlexNet, ResNet, and VggNet and examines how well such methods perform with the DR image classifications.

Sungheetha and Sharma [15] work derives the features by integrating deep networks via CNN. The micro-aneurysm is seeming in initial phases of mild patients. The primary detection of diabetic condition was attained via the hard executes (HE) presented in the blood vessel of an eye by employing devised CNN structure. It is also utilized for detecting a diabetic condition of person. Bilal et al. [16] introduced a new and hybrid technique for prior DR classification and detection. It merged different methods for achieving less error-prone and robust DR identification processes whenever determining the classifications related to the majority voting technique. The projected work will follow preprocessing feature extracting and classifying steps. The preprocessing phase fosters abnormality presence along with segmentation; the extracting step gains only appropriate features; and the classifier step employs methods like binary trees (BT), SVM, and KNN.

12.3 The DR Classification and Rating Model

In this study, a novel ALODTL-DRG technique is modeled for DR classification on fundus images. The presented ALODTL-DRG technique encompasses several subprocesses, namely image preprocessing, Inception with ResNet-v2 feature extraction, DBN classification, and ALO hyper parameter tuning. Figure 12.1 depicts the overall process of ALODTL-DRG approach.

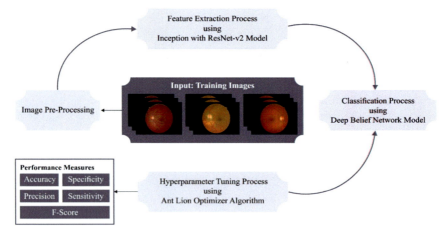

Fig. 12.1 Overall process of ALODTL-DRG approach

12.3.1 Image Preprocessing

This study adapted different preprocessing stages to normalize the fundus image. Firstly, resize the fundus image into a uniform size through bi-cubic interpolation over 4×4 neighborhood pixels. The σ standard deviation and Gaussian function in 2D (x, y) are mathematically expressed in Eq. (12.1).

$$G(x, y) = \frac{1}{2\pi\sigma^2} \varepsilon^{\frac{x^2+y^2}{2\sigma^2}} \qquad (12.1)$$

12.3.2 Feature Extraction

Google Company launched an Inception-ResNet-V2 (IRV2) that severs as a state-of-the-art and categorizes the images. It progressed from concatenation of ResNet and GoogleNet (Inception). Generally, Inception is associated with the respective layer as exploited in GoogleNet. Largescale convolutional kernel will improve the matrix parameter, where the small-scale convolutional kernel is exchanged for limiting the functional parameter for receptive field [17]. After that, the small-scale convolutional kernel is employed to extract image features efficiently and minimize parameters of the model. At last, it is accurate and extensive in comparison with the present network with Inception. Lately, Inception v1–v4 was a traditional method of GoogleNet. The primary goal of ResNet was to enclose a direct connection, which was named Highway Network. As well, it allows new input dataset that should be directly transmitted to the successive layer. Concurrently, ResNet safeguards data privacy through directly transmitting data to output. The variations among input and output are essential and study the advantages and disadvantages.

The Inception module is employed as it has low processing difficulty in comparison with real Inception for Residual-Inception network. Inception-ResNet-A, Inception-ResNet-B, and Inception-ResNet-C are the layer of Inception-ResNet. The reduction layers of IRV2 are Reduction-A and Reduction-B. According to the current research, IRV2 has been deployed from IRV1 through matching real expenses of Inception v4. Figure 12.2 showcases the layers in Inception-ResNet-v2 technique. Eventually, minimal variations among non-residual and residual Inception are, for Inception-ResNet, batch normalization (BN) which is employed. It was obvious that employing good activation size intakes maximal Inception module and high GPU memory, and it is appended by eliminating BN layer afterward the completion of activation function. In addition, once the filter count exceeds 1000, residual network becomes inconsistent, and early death exists in network training process.

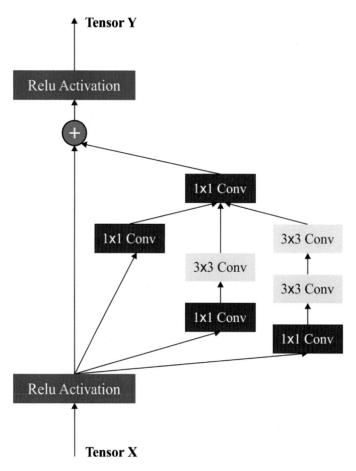

Fig. 12.2 Layers in inception-ResNet-v2

12.3.3 Image Classification

At this phase, the features extracted were fed into the DBN method for classification purposes. The key component is restricted Boltzmann machine (RBM) for constructing DBN [18]. A visible and hidden unit of RBN establishes a two-layer architecture, and it is shown in the following

$$\text{Enr}(VI, HI, \theta) = -\sum_{i=1}^{D} bV_i v_i - \sum_{j=1}^{F} a_{Hj} h_j - \sum_{i=1}^{D} \sum_{j=1}^{F} w_j VI_i HI_j$$
$$=> -b^T VI - a^T HI - VI^T VVHI \quad (12.2)$$

where $\theta = \{bV_i, aH_j, we_{ij}, we_{ij}\}$ indicates the weight among i and j visible and hidden components, bV_i and aH_j denote the bias condition of hidden and visible units correspondingly, and it is shown below

$$\Pr(VI, HI, \theta) = \frac{1}{NC(\theta)} \exp(-\text{Enr}(VI, HI, \theta)) \quad (12.3)$$

$$NC(\theta) = \sum_{VI} \sum_{HI} \text{Enr}(VI, HI, \theta) \quad (12.4)$$

From the expression, $NC(\theta)$ indicates a regularization constant. The energy function is applied as a likelihood distribution; the trained vector is attuned. For extracting the feature from dataset, the single hidden layer of RBN is applied. The output of initial layer is applied as the input of the following layer, and output of following layer is the input of third layer of RBN. This hierarchal layer-wise architecture of RBN designs the DBN, and the deeper feature extraction from the input data is efficient with a hierarchal method of DBN.

12.3.4 Parameter Tuning

The ALO algorithm is used as hyper parameter optimizer to optimize the detection efficiency. The ALO is an original meta-heuristic approach [19]. In random walk of ants, the way of ants was demonstrated based on the subsequent Eq. (12.5):

$$X(t)$$
$$= [0, \text{cumsum}(2r(t_1) - 1); \text{cumsum}(2r(t_2) - 1); \ldots; \text{cumsum}(2r(t_T) - 1);] \quad (12.5)$$

whereas cumsum is equivalent to cumulative sum; t implies the present step; T signifies the maximal count of rounds. $r(t)$ demonstrates the stochastic function. It can be determined as Eq. (12.6).

$$r(t_i) = \begin{cases} 1 & \text{if rand\_num} > 0.5 \\ 0 & \text{if rand\_num} \leq 0.5 \end{cases} \tag{12.6}$$

In which rand_num represents the arbitrary number that is created with uniform distribution from the interval of zero and one, and $1 \leq i \leq T$. T arbitrary values were created in all the iterations. All the ants utilize in Eq. (12.7) for normalizing their place for preventing ants in going out of searching space. The formula is written as:

$$X_i^t = \frac{(X_i^t - d_i)(b_i^t - a_i^t)}{c_i - d_i} + b_i^t \tag{12.7}$$

In which, d_i^t denotes the upper restraint of RW of ith variable, d_i signifies the lesser restraint of RW from ith variable, and a_i^t represents the minimal value of ith iteration to ith dimensional. c_j signifies the maximal value to the ith dimensional. The explanations of b_i^t and a_i^t are written as:

$$a_i^t = \text{Antlion}_i^t + a^t \tag{12.8}$$

$$b_i^t = \text{Antlion}_i^t - b^t \tag{12.9}$$

whereas a^t signifies the minimal value of every variable at ith iteration; b^t signifies the maximal value of every variable at ith iteration. All the ants are only preyed on by one AL using Roulette approach. The AL with a superior fitness value was highly possibly for capturing the ant [20]. The subsequent Eqs. (12.10) and (12.11) simulate this capture procedure.

$$a^t = \frac{c^t}{I} \tag{12.10}$$

$$b^t = \frac{d^t}{I} \tag{12.11}$$

whereas I signifies the ratio factor that is determined as:

$$I = \begin{cases} 1 & \text{if } g \leq 0.1G \\ 10^\omega * \frac{g}{G} & \text{if } g > 0.1G \end{cases} \tag{12.12}$$

In which, g denotes the present round and G stands for the maximal count of iterations. ω is fetched using the subsequent in Eq. (12.13).

$$\omega = \begin{cases} 1, & \text{if } 0 < g \leq 0.1G \\ 2, & if\ 0.1G < g \leq 0.5G \\ 3, & \text{if } 0.5G < g \leq 0.75G \\ 4, & \text{if } 0.75G < g \leq 0.9G \\ 5, & \text{if } 0.9G < g \leq 0.95G \\ 6 & \text{if } 0.95G < g \leq G \end{cases} \qquad (12.13)$$

whereas ω has utilized for controlling explorations.

The ALO method makes a derivation of a fitness function (FF) for attaining improvised classifier outcomes. It sets a positive numeral for denoting a superior outcome of the candidate solutions. In this article, the reduction of the classifier error rate can be regarded as the FF, as presented in Eq. (12.14).

$$\begin{aligned} \text{fitness}(x_i) &= \text{Classifier Error Rate}(x_i) \\ &= \frac{\text{number of misclassified samples}}{\text{Total number of samples}} * 100 \end{aligned} \qquad (12.14)$$

12.4 Experimental Evaluation

The performance validation of the ALODTL-DRG method is tested against the MESSIDOR dataset [21]. It contains 1200 fundus images with four class labels (Table 12.1) and images as in Fig. 12.3.

Table 12.2 provides a brief set of DR classification results offered by the ALODTL-DRG model. The results implied that the ALODTL-DRG model has attained enhanced classifier output under all classes. For instance, on entire dataset, the ALODTL-DRG model has offered average $accu_y$ of 99.50%, $prec_n$ of 98.82%, $sens_y$ of 98.81%, $spec_y$ of 99.67%, and F_{score} of 98.81%. Eventually, on 70% of TR data, the ALODTL-DRG technique has provided average $accu_y$ of 99.52%, $prec_n$ of 98.92%, $sens_y$ of 98.63%, $spec_y$ of 99.68%, and F_{score} of 98.77%. Likewise, on 30% of TS data, the ALODTL-DRG approach has rendered average $accu_y$ of 99.44%, $prec_n$ of 98.66%, $sens_y$ of 99.25%, $spec_y$ of 99.65%, and F_{score} of 98.95%.

Table 12.1 Considered dataset

Class	No. of instances
Normal	548
Stage 1	152
Stage 2	246
Stage 3	254
Total number of instances	1200

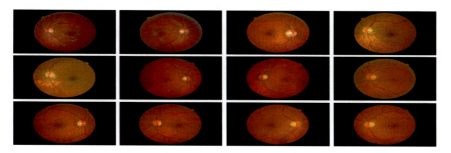

Fig. 12.3 Sample images

Table 12.2 Result analysis of ALODTL-DRG approach with different measures

Class labels	Accuracy	Precision	Sensitivity	Specificity	F-score
Entire dataset					
Normal	99.50	99.82	99.09	99.85	99.45
Stage 1	99.58	99.33	97.37	99.90	98.34
Stage 2	99.67	99.19	99.19	99.79	99.19
Stage 3	99.25	96.93	99.61	99.15	98.25
Average	99.50	98.82	98.81	99.67	98.81
Training set (70%)					
Normal	99.64	99.74	99.48	99.78	99.61
Stage 1	99.40	99.02	96.19	99.86	97.58
Stage 2	99.88	100.00	99.38	100.00	99.69
Stage 3	99.17	96.92	99.47	99.08	98.18
Average	99.52	98.92	98.63	99.68	98.77
Testing set (30%)					
Normal	99.17	100.00	98.17	100.00	99.08
Stage 1	100.00	100.00	100.00	100.00	100.00
Stage 2	99.17	97.67	98.82	99.27	98.25
Stage 3	99.44	96.97	100.00	99.32	98.46
Average	99.44	98.66	99.25	99.65	98.95

The training and validation accuracies through SOSDCNN-HAR technique are shown in Fig. 12.4. Similarly, training and validation losses through SOSDCNN-HAR approach are shown in Fig. 12.5.

A detailed comparison study is presented in Table 12.3 [22]. Figure 12.6 highlights the comparative $accu_y$ inspection of other models. The attained values represented that the ALODTL-DRG method has outperformed the other DR classification models.

Fig. 12.4 Training and validation accuracy analysis

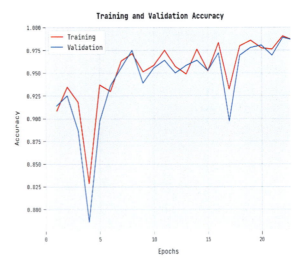

Fig. 12.5 Training and validation loss analysis

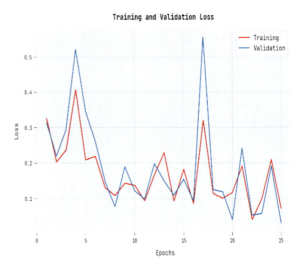

Figure 12.7 signifies the comparative $sens_y$ analysis of other models. The attained values denoted that the ALODTL-DRG algorithm has outperformed the other DR classification models.

Figure 12.8 illustrates the comparative $spec_y$ examination of other models. The attained values indicated that the ALODTL-DRG approach has outperformed the other DR classification models. These values assured that the ALODTL-DRG model has gained maximum DR classification performance.

Table 12.3 Comparative analysis of ALODTL-DRG approach with recent algorithms

Methods	Accuracy	Sensitivity	Specificity
ALODTL-DRG	99.44	99.25	99.65
AlexNet model	89.11	91.46	87.81
VGG-16 model	95.50	96.63	94.49
DenseNet model	94.26	93.29	92.87
MobileNet model	91.87	92.38	90.43
Xception model	91.56	93.14	88.88
ResNet-101 model	94.35	94.97	94.40
ResNet-50 model	95.50	95.06	95.28
Inception V3 model	97.79	98.49	96.64

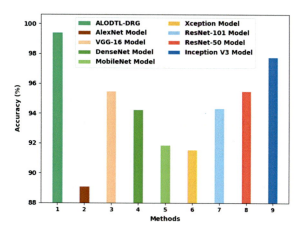

Fig. 12.6 $Accu_y$ analysis of ALODTL-DRG approach with recent methodologies

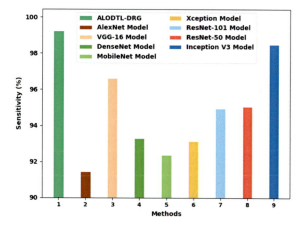

Fig. 12.7 $Sens_y$ analysis of ALODTL-DRG approach with recent methodologies

Fig. 12.8 Spec$_y$ analysis of ALODTL-DRG approach with recent methodologies

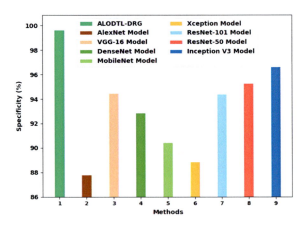

12.5 Conclusion

In this article, a novel ALODTL-DRG technique has been projected for DR classification on fundus images. The proposed ALODTL-DRG technique encompasses several sub-processes, namely image preprocessing, Inception with ResNet-v2 feature extraction, DBN classification, and ALO hyper parameter tuning. The ALO algorithm is exploited as a hyper parameter tuning strategy to accomplish enhanced DR detection performance. The experimental assessment of the ALODTL-DRG model is tested using benchmark dataset. A widespread comparison study stated the enhanced performance of the ALODTL-DRG model over recent approaches.

References

1. Hemanth, D.J., Deperlioglu, O., Kose, U.: An enhanced diabetic retinopathy detection and classification approach using deep convolutional neural network. Neural Comput. Appl. **32**(3), 707–721 (2020)
2. Tsiknakis, N., Theodoropoulos, D., Manikis, G., Ktistakis, E., Boutsora, O., Berto, A., Scarpa, F., Scarpa, A., Fotiadis, D.I., Marias, K.: Deep learning for diabetic retinopathy detection and classification based on fundus images: a review. Comput. Biol. Med. **135**, 104599 (2021)
3. Kavitha, T., Mathai, P.P., Karthikeyan, C., et al.: Deep learning based capsule neural network model for breast cancer diagnosis using mammogram images. Interdiscip Sci. Comput. Life Sci. (2021). https://doi.org/10.1007/s12539-021-00467-y
4. Hasan, D.A., Zeebaree, S.R., Sadeeq, M.A., Shukur, H.M., Zebari, R.R., Alkhayyat, A.H.: Machine learning-based diabetic retinopathy early detection and classification systems-a survey. In: 2021 1st Babylon International Conference on Information Technology and Science (BICITS), pp. 16–21. IEEE (2021)
5. Kalyani, G., Janakiramaiah, B., Karuna, A. and Prasad, L.V.: Diabetic retinopathy detection and classification using capsule networks. Complex Intell. Syst. 1–14 (2021)
6. Alyoubi, W.L., Shalash, W.M., Abulkhair, M.F.: Diabetic retinopathy detection through deep learning techniques: a review. Inf. Med. Unlocked **20**, 100377 (2020)

7. Amin, J., Sharif, M., Rehman, A., Raza, M., Mufti, M.R.: Diabetic retinopathy detection and classification using hybrid feature set. Microsc. Res. Tech. **81**(9), 990–996 (2018)
8. Abdelsalam, M.M.: Effective blood vessels reconstruction methodology for early detection and classification of diabetic retinopathy using OCTA images by artificial neural network. Inf. Med. Unlocked **20**, 100390 (2020)
9. Chen, W., Yang, B., Li, J., Wang, J.: An approach to detecting diabetic retinopathy based on integrated shallow convolutional neural networks. IEEE Access **8**, 178552–178562 (2020)
10. Jayakumari, C., Lavanya, V. and Sumesh, E.P.: Automated diabetic retinopathy detection and classification using imagenet convolution neural network using fundus images. In: 2020 International Conference on Smart Electronics and Communication (ICOSEC), pp. 577–582. IEEE (2020)
11. Patel, S.: Diabetic retinopathy detection and classification using pre-trained convolutional neural networks. Int. J. Emerg. Technol. **11**(3), 1082–1087 (2020)
12. Shanthi, T., Sabeenian, R.S.: Modified Alexnet architecture for classification of diabetic retinopathy images. Comput. Electr. Eng. **76**, 56–64 (2019)
13. Wan, S., Liang, Y., Zhang, Y.: Deep convolutional neural networks for diabetic retinopathy detection by image classification. Comput. Electr. Eng. **72**, 274–282 (2018)
14. Sungheetha, A., Sharma, R.: Design an early detection and classification for diabetic retinopathy by deep feature extraction based convolution neural network. J. Trends Comput. Sci. Smart Technol. (TCSST) **3**(02), 81–94 (2021)
15. Bilal, A., Sun, G., Li, Y., Mazhar, S., Khan, A.Q.: Diabetic retinopathy detection and classification using mixed models for a disease grading database. IEEE Access **9**, 23544–23553 (2021)
16. Jianjie, S., Weijun, Z.: Violence detection based on three-dimensional convolutional neural network with Inception-ResNet. In: 2020 IEEE Conference on Telecommunications, Optics and Computer Science (TOCS), pp. 145–150. IEEE (2020)
17. Zhang, C., He, Y., Yuan, L., Xiang, S.: Analog circuit incipient fault diagnosis method using DBN based features extraction. IEEE Access **6**, 23053–23064 (2018)
18. Mirjalili, S.: The ant lion optimizer. Adv. Eng. Softw. **83**, 80–98 (2015)
19. Roy, K., Mandal, K.K., Mandal, A.C.: Ant-Lion Optimizer algorithm and recurrent neural network for energy management of micro grid connected system. Energy **167**, 402–416 (2019)
20. https://www.adcis.net/en/third-party/messidor/
21. Nneji, G.U., Cai, J., Deng, J., Monday, H.N., Hossin, M.A., Nahar, S.: Identification of diabetic retinopathy using weighted fusion deep learning based on dual-channel fundus scans. Diagnostics **12**(2), 540 (2022)
22. Rene Beulah, J., Prathiba, L., Murthy, G.L.N., Fantin Irudaya Raj, E., Arulkumar, N.: Blockchain with deep learning-enabled secure healthcare data transmission and diagnostic model. Int. J. Model. Simul. Sci. Comput. https://doi.org/10.1142/S1793962322410069

Chapter 13
Finite Element Simulation on Ballistic Impact of Bullet on Metal Plate

Moreshwar Khodke, Milind Rane, Abhishek Suryawanshi, Omkar Sonone, Bhushan Shelavale, Pranav Shinde, and Aman Sheikh

Abstract For both military and civilian vehicles, ballistic safety systems frequently use high-strength Titanium and Aluminum plates. The choice of alloy is therefore based on its intended usage, ballistic performance, and safety for them. In this study, the effect of a bullet on Titanium and Aluminum alloy plates with fixed edges on both sides is examined. The effects of bullet thickness and impingement angle on Titanium and Aluminum alloy plates were examined using simulations. These simulations were carried out in Ansys Workbench using the Finite Element Method. Simulations using both material models also revealed a distinct variation in the plate's deformation. The targeted plate was impacted at a 45° oblique angle with velocity of 830 m/s in every test. The findings demonstrated a crucial slant angle of 45° where the piercing operation transitions to ricochet. The other simulation was run to ascertain the plate thickness where the piercing operation transitions to embedment.

Keywords Impact · Ballistic · Plate · Bullet

13.1 Introduction

With the rise in terrorism, incidents, and public violence, the public security and property security became difficult [1]. Due to the fierce rivalry among peoples and nations in the modern period, security is a crucial consideration. Fighters are faced with a lot of operating conditions and injury risks. The study of army rifles and small weapons safety is vital from both a civilian and military standpoint. The majority studies on ballistic focus on perpendicular collision, where the inclination with the projectile's velocity vector and the perpendicular vector of the focus plane at 0 [2]. However, mostly affect at an angle of impact of bullet on an Aluminum alloy and Titanium plate under various situations is investigated in this study. Mostly, the material has their own properties, so it is critical to understand that not all Titanium

M. Khodke · M. Rane (✉) · A. Suryawanshi · O. Sonone · B. Shelavale · P. Shinde · A. Sheikh
Vishwakarma Institute of Technology, Pune 411037, India
e-mail: milind.rane@vit.edu

characteristics are the same. While certain Titanium alloys are bulletproof, pure Titanium is not. Because each grade of Titanium has different advantages and primary applications, it is crucial to perform some study or consult an authority.

Aluminum is widely recognized for being a lightweight material with strong energy involved properties. To survive in any battle situation, troops require armor with these attributes that allows them to move more quickly. Despite having a larger capacity for energy absorption, steel is heavy than Aluminum. Consequently, Aluminum is a superior material to employ for armor. Contributions of paper are as follows:

- Two simulations were performed to study the impact of a 7.62 × 17.3 mm NATO Ball bullet with a speed of 830 m/s on Aluminum alloy and Titanium plates.
- The plates had a cross-section of 300 mm × 300 mm and thicknesses ranging from 6 to 10 mm and 15 mm to 25 mm were tested in the first simulation.
- In the second simulation, the target was hit at a 45° obliquity, and the critical angle was established.
- It was found that bullets were able to penetrate plates with a thickness of less than 15 mm at a 45° angle.
- However, the bullets did not penetrate through the plates with a thickness greater than 15 mm at the same angle.
- Consequently, it may be inferred that a plate's thickness has a significant influence on its capacity to withstand ballistic impact at oblique angles.

13.2 Literature Review

Manes et al. [3] have investigated the result of least velocity collision on sandwich panels experimentally and numerically. It is demonstrated that such features are extremely important in least velocity collision. The same author investigated the collision of tiny bullets on Aluminum plate. Borvik et al. [4] investigated the impact of NATO bullet and APM2 on a 20 mm thick Aluminum plate. The research was carried out both analytically and physically. The collision velocity was also 830 m/s. Laser-based optical devices were used to evaluate the initial and residual velocities. Perforation was shown to be converted to set for stalling angles smaller than 60°. Iqbal et al. [5] mild steel has been defined at various stress and strain rates; the material characteristics of 12 and 16 mm thickness targets were utilized to run numerical ballistic simulations against API projectiles. The same researcher investigated the effects of trajectory nose form, collision velocity, and required thickness on several kinds of plates [6, 7]. Insulation with many layer upon layer was investigated by White et al. [8] on excessive velocity bullet impact Iqbal et al.

Alwan et al. [9] present a comprehensive literature review of previous studies that have used finite element methods for ballistic impact analysis. The authors highlight the challenges associated with experimental testing of materials under ballistic impact conditions and propose using dynamic finite element simulations as an alternative. They discuss the advantages of using dynamic simulations over static simulations and

Fig. 13.1 CAD model of bullet

present the results of their own simulations, which show good agreement with experimental data. The study demonstrates the effectiveness of dynamic finite element simulations in predicting the damage caused by high velocity bullets on Aluminum and magnesium alloys and provides useful insights for future research in this field.

The article by Zahrin et al. [10] discusses the use of the finite element method for numerical simulation of oblique impact on a structure. They begin by emphasizing how crucial it is to comprehend how structure behaves when hit from an angle, particularly for development and defense purposes. The authors then present a thorough literature assessment of earlier research that included creation of meshes, material modeling, and boundary conditions in finite element approaches for impact studies. In order to get correct findings, the authors also go over how crucial it is to simulate the impact parameters precisely, such as motion, incidences angle, and impact position. Furthermore, the authors present the results of their own simulations, which show good agreement with experimental data. They demonstrate the effectiveness of the finite element method in predicting the behavior of structures under oblique impact and suggest future research directions, such as studying the effect of different material properties and impact conditions. Overall, the study provides valuable insights into the use of finite element methods for impact analysis of structures under oblique impact. Figure 13.1 shows the 3D model which represents CAD model of bullet, and Fig. 13.2 shows metal plate having meshing parameter by finite element analysis. Results of impact are presented in Table 13.1.

13.3 Material Properties

The material used for experimentation is Aluminum and Titanium which are mentioned below.

13.3.1 Aluminum

Table 13.2 lists the Aluminum's properties.

Fig. 13.2 Metal plate (300 × 300 mm)

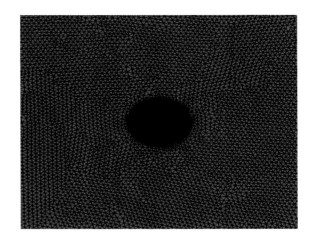

Table 13.1 Mesh for plate and bullet geometries

Geometry	Elements	Nodes
Plate	123,183	22,428
Bullet	55,168	19,376
Total	178,351	41,804

Table 13.2 Properties of aluminum

Property	Aluminum alloy	Unit
Yield strength	280	MPa
Young's modulus	71,000	MPa
Bulk modulus	69,608	Mpa
Density	2770	Kgm^{-3}
Tensile ultimate Strength	310	MPa
Poisson's ratio	0.33	MPa
Shear modulus	26,692	MPa

13.3.2 Titanium

Table 13.3 lists the Titanium's properties.

Table 13.3 Properties of titanium

Property	Titanium alloy	Unit
Yield strength	880	MPa
Young's modulus	113,800	MPa
Bulk modulus	96.8	Mpa
Density	443	Kgm^{-3}
Tensile ultimate Strength	950	MPa
Poisson's ratio	0.342	MPa
Shear modulus	44	MPa

13.4 Ballistic Impact

Ballistics is the name of the field of study that deals with trajectory. It is focused on rocket releasing, flight behavior, and striking consequences, particularly for long-range weapon explosive devices like bullets, unguided grenades, missiles, and other similar weapons. A ballistic object, in this case a bullet, is a moving entity with momentum that may be affected by factors such as gravitational forces and air drag while in aircraft. It can also be affected by pressured gases from a rifle barrel or a propulsive jet. Ballistics may be categorized into three groups:

i. Interior ballistics: are the events that take place between the hit of the firing pin or striker and the discharge of the bullet charge from the muzzle end of the barrel, which lasts for around two milli sec.
ii. Outer ballistics: The trajectory of a fire or bullet after it leaves the barrel.
iii. Terminal ballistics: Injury caused when a bullet pierces a target.

13.5 Simulation of Impact and Modeling

13.5.1 Method Selection

The finite element method (FEM) is a numerical technique that is widely used in engineering and physics for solving problems. Its applications include simulating the impact of bullets, which is particularly useful in analyzing the stress and strain distribution in a material subjected to high-velocity impact.

The ability of FEM to accurately simulate complex geometries is one of the main advantages of utilizing it for ballistic impact modeling. It can be difficult to analyze complicated deformation patterns caused by a bullet's impact on a surface using conventional analytical techniques. The geometry can be broken down into small pieces using FEM, enabling a thorough investigation of the local stress and strain fields.

Large deformations that could happen during a ballistic impact can also be simulated using FEM. This is crucial when simulating the behavior of soft materials like ballistic gel that can undergo significant deformation during impact. FEM can optimize the design of protective materials by accurately anticipating the deformation pattern, hence lowering the risk of harm. FEM is also extremely accurate since it takes into account the problem's boundary conditions, geometry, and material properties. In designing protective materials like body armor or helmets, where even a minor simulation error might have negative effects, this leads to a more accurate forecast of the impact reaction.

Its ability to model complex geometries, predict large deformations, and reduce the need for experimental testing makes it an important tool in designing protective materials.

13.5.2 Modeling Methodology

Ansys v22 R1 was used to construct the plate. Ansys Modeler was used to design the geometries of the bullet, and the plate before a separate simulation of the projectile was performed. An analysis of the entire system was then run. The problem has been solved using Ansys Explicit Dynamics with AUTODYN, and modeling output has been produced. A model's mesh and contact settings between components have been built up after each part's materials have been modeled into Ansys engineering data. Finally, the solver's attributes have been defined, including the beginning conditions, system statics, dynamic properties, and intended output.

13.6 Impact Simulation of a 7.62 mm NATO BALL Bullet at Various Plate Thicknesses

The analysis of bullet collision on various thickness plates is achieved through FEA on 300 mm × 300 mm Aluminum alloy and Titanium plates. The bullet used is a NATO BALL 7.62 mm round. Bullet hitting plates with thicknesses of 6 mm, 10 mm, 15 mm, and 25 mm were simulated. The plate boundary is constrained by a set restriction. Bullet impact on flat plate with thicknesses is ranging from 6 to 25 mm in this simulation. The bullet's velocity in the x-direction is 830 m/s. The bullet perforates the plate at thicknesses of 15 mm or less. A bullet cannot pierce a plate with a thickness more than 15 mm. Finite element simulations for bullet impacting on Aluminum plate with different thicknesses are shown in Fig. 13.3 and for Titanium plate with different thicknesses are shown in Fig. 13.4.

For various scenarios, the maximum equivalent stress in the plate was evaluated. The maximum equivalent stress in a plate is proportional to different thickness of plate. Experimentation is performed on Aluminum plate is shown in Fig. 13.5. This shows the maximum equivalent stress on the Aluminum plate as a function of its thickness. The maximum equivalence stress increases as the plate thickness increases,

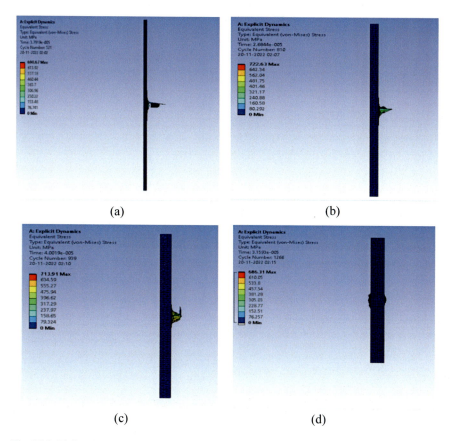

Fig. 13.3 Finite element simulations for bullet impacting Aluminum plate with different thicknesses. **a** 6 mm, **b** 10 mm, **c** 15 mm, and **d** 25 mm

and at a thickness of 15 mm, the maximum equivalence stress reaches its maximum value. Beyond that point, the maximum equivalence stress decreases as the thickness of the plate increases.

Secondly, experimentation is performed on Titanium plate which is shown in Fig. 13.6. This shows the maximum equivalent stress on the Titanium plate as a function of its thickness. The maximum equivalence stress increases as the plate thickness increases, and at a thickness of 15 mm, it is maximum. Beyond that point the maximum equivalence stress decreases as the thickness of the plate increases.

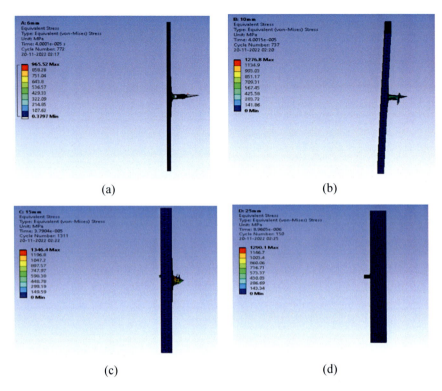

Fig. 13.4 Finite element simulations for bullet impacting Titanium plate with different thicknesses. **a** 6 mm, **b** 10 mm, **c** 15 mm, and **d** 25 mm

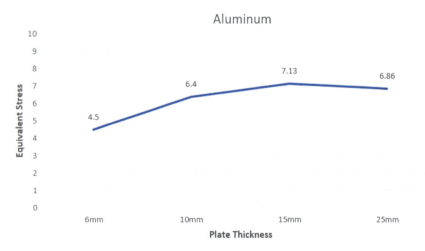

Fig. 13.5 Variation of equivalent stress in aluminum plate with change in plate thickness

13 Finite Element Simulation on Ballistic Impact of Bullet on Metal Plate   155

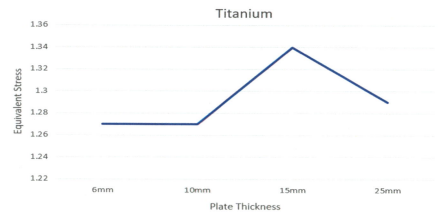

Fig. 13.6 Variation of equivalent stress in titanium plate with change in plate thickness

13.7 Impact Simulation of Bullet Impact on Plate at 45° Angle

A 7.62 mm NATO BALL bullet with an ascribed velocity of 830 m/s strikes Aluminum and Titanium 15 mm in this simulation. The AUTODYN material library contains the material model used in bullet and plate construction. The plates edges have a fixed boundary condition applied to them. The bullet is assigned a starting speed of 830 m/s in the z-direction. Results for Aluminum plate at a 45° angle is shown in Fig. 13.7, and result for Titanium plate is shown in Fig. 13.8.

Fig. 13.7 Simulation of aluminum plate at 45°

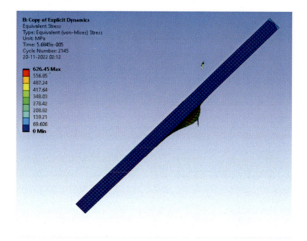

Fig. 13.8 Simulation of titanium plate at 45°

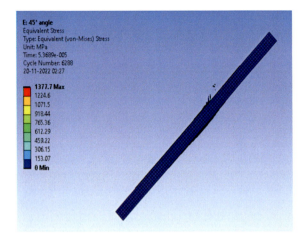

13.8 Results

When a bullet strikes on Aluminum plate with a 6 mm thickness, the plate is perforated, and the bullet passes through the plate quite easily as shown in Fig. 13.3a. For a plate with 10 mm thickness, the bullet is significantly passed through a plate as shown in Fig. 13.3b. When a plate is 15 mm thick, the bullet barely penetrates the plate as shown in Fig. 13.3c. The bullet is trapped inside a plate for 20 mm thick plates as shown in Fig. 13.3d.

The Titanium 6 mm and 10 mm thick plates are significantly penetrated by bullets as shown in Figs. 13.4a, b. Bullet scarcely penetrates through 15 mm thick plate bullets as shown in Fig. 13.4c. The bullet is trapped inside a plate for 20 mm thick plates' bullets as shown in Fig. 13.4d.

Various methods may be used to compare simulated ballistic test results. The design might provide a decent indication of the size and form of the plate's damage state. The simulation model may also forecast the bullet and target's overall and direction deformations. Figures 13.3c and 13.4c show the final pictures from a simulation of plate deform and show that was successful in stopping the 15 mm thick of the plate.

13.9 Conclusion

Bullet striking metal plate simulations were performed during this work. The finite element analysis was conducted out using the FEM, which was done in Ansys. Two types of simulations were carried out. In first scenario, Aluminum and Titanium the plate thickness vary, but the bullet strikes perpendicular to surface of plate. According to the findings of this investigation, a bullet cannot penetrate a plate thicker than

15 mm. The inclination of collision of the bullet on the plate was adjusted in the second case of experiment. It was discovered that after 45°, the bullet is unable to enter the plate since the tension created has decreased.

Research work on ballistic impact can yet be expanded upon. The targeted plate in this simulation is a on different materials, different thicknesses, and angle plate. A curve targeting plate might be used in further studies. The projectile's form may surely be changed to ensure that the results are precise and that they can endure the ballistic load of various projectiles.

References

1. Jena, D.P., Jena, D.K., Kumar, S.: Simulation of bullet penetration using finite element method. In: 2019 International Conference on Range Technology (ICORT). IEEE (2019)
2. Bhuarya, M.K., Rajput, M.S., Gupta, A.: Finite element simulation of impact on metal plate. Procedia Eng. **173**, 259–263 (2017)
3. Manes, A., Gilioli, A., Sbarufatti, C., Giglio, M.: Experimental and numerical investigations of low velocity impact on sandwich panels. Compos. Struct. **99**, 8–18 (2013)
4. Børvik, T., Olovsson, L., Dey, S., Langseth, M.: Normal and oblique impact of small arms bullets on AA6082-T4 aluminium protective plates. Int. J. Impact Eng. **38**(7), 577–589 (2011)
5. Iqbal, M.A., Senthil, K., Bhargava, P., Gupta, N.K.: The characterization and ballistic evaluation of mild steel. Int. J. Impact Eng. **78**, 98–113 (2015)
6. Gupta, N.K., Iqbal, M.A., Sekhon, G.S.: Experimental and numerical studies on the behavior of thin aluminum plates subjected to impact by blunt-and hemispherical-nosed projectiles. Int. J. Impact Eng. **32**(12), 1921–1944 (2006)
7. Gupta, N.K., Iqbal, M.A., Sekhon, G.S.: Effect of projectile nose shape, impact velocity and target thickness on the deformation behavior of layered plates. Int. J. Impact Eng. **35**(1), 37–60 (2008)
8. White, D.M., Wicklein, M., Clegg, R.A., Nahme, H.: Multi-layer insulation material models suitable for hypervelocity impact simulations. Int. J. Impact Eng. **35**(12), 1853–1860 (2008)
9. Alwan, F.H.A., et al.: Assessment of ballistic impact damage on aluminum and magnesium alloys against high velocity bullets by dynamic FE simulations. J. Mech. Behav. Mater. **31**(1), 595–616 (2022)
10. Zahrin, M.F., Kamarudin, K.A., Abdullah, A.S.: Numerical simulation of oblique impact on structure using finite element method. In: AIP Conference Proceedings, vol. 2644, no. 1. AIP Publishing LLC (2022)

Chapter 14
Performance Analysis of Bionic Swarm Optimization Techniques for PV Systems Under Continuous Fluctuation of Irradiation Conditions

Shaik Rafi Kiran, CH Hussaian Basha, M. Vivek, S. K. Kartik, N. L. Darshan, A. Darshan Kumar, V. Prashanth, and Madhumati Narule

Abstract The nonrenewable energy sources give continuous more electrical energy when compared to the renewable energy systems. But the availability nonrenewable energy sources are very less. Also, the nonrenewable energy sources are not safe for the human life. Now, most of the electricity generation companies are working on renewable power supply. The most commonly utilized renewable source is solar. The features of solar are free of cost availability and less effect on human life. But, it gives nonlinear power curves. So, the obtaining of peak power and voltage from the solar system is quite difficult. Here, the Perturb & Observe (P&O) along with Particle Swarm Optimization (P&O-PSO) method is interfaced in the photovoltaic (PV) system for finding the actual working point of the PV module. The proposed topology is studied by using a MATLAB/Simulink window.

Keywords Converter voltage oscillations · Duty value · Fast convergence speed · Dynamic response · High accuracy of MPP

S. R. Kiran
Sri Venkateswara College of Engineering (Autonomous), Tirupati 517502, India

CH Hussaian Basha (✉) · M. Vivek · S. K. Kartik · N. L. Darshan · A. Darshan Kumar · V. Prashanth
NITTE Meenakshi Institute of Technology, Bangalore 560064, India
e-mail: sbasha238@gmail.com

V. Prashanth
e-mail: prasanth.v@nmit.ac.in

M. Narule
SCRC, Nanasaheb Mahadik College of Engineering, Walwa 415407, India

14.1 Introduction

As of the current scenario, the electricity generation of nonrenewable energy sources is reducing extensively because of its low level of availability, more operational cost, extensive atmospheric pollution, and less efficient [1]. Also, the nonrenewable energy sources required high catchment area [2]. So, to limit the disadvantages of conventional power sources, the current power supplying companies are working on renewable power for giving the energy to the local customers at the time of shortage of grid dependent power. From the literature survey, the renewable sources are illustrated as solar, tidal, geothermal, plus bio energy [3]. From the all-renewable sources, PV is most predominantly used power source as of its attractive features which are high accessibility on earth, very low maintenance cost, less harmful emissions, and high security with less dependence on other power supply systems. In addition to that, it does not require more space [4].

The solar electricity generation required high installation price and less working efficiency at non-uniform solar insolation conditions which are limited by using the various advanced semiconductor manufacturing methodologies [5]. The solar power cells are broadly divided into three types which are illustrated as crystalline silicon, thin-film cells, poly crystalline, and monocrystalline solar cell [6]. From the three types of solar cells, the monocrystalline cell is popularly used for constructing the solar panels because of its features which are high working efficiency and high voltage generation capability [7]. The operation of PV cell is similar to the PN-diode nature. Each PV cell generates 0.7 V which is not sufficient for peak load customers. The PV cells are interlinked in sequential manner with one and another to enhance the voltage supply of the customers [8]. Same way, the PV cells stay in interlinked manner.

Basically, the solar systems are implemented either by utilizing one-single diode solar cell or two-diode solar cell. In article [9], the authors used the one diode solar cell for analyzing the various power point tracing methods. The features of one diode solar cell are less price, easy understanding, and implementation. Also, this one diode cell needed less space for constructing the solar panel. But it consists of less accuracy in generating the nonlinear I-V characteristics of PV module [10]. In this work, the two-diode configuration is applied to design the PV panel. The merits of dual diode solar cell are good accuracy in producing the solar output characteristics and optimum heat conduction losses.

Basically, the power versus voltage curves changes with respect to the continuous variations of environmental conditions [11]. As a result, the working point of PV on I–V curve changes. So, the power supply to the customers may vary which is not desired. To make the working step of PV which is constant under different sun insolation conditions, the MPPT block is link up near to the PV panel [12]. The MPPT technologies are sort out as normal, artificial neural network, swarm intelligence, plus soft computing optimization methods [13]. The normal MPPT methods are organized as Perturb & Observe (P&O), incremental resistance of converter, dp/di, plus hill climb.

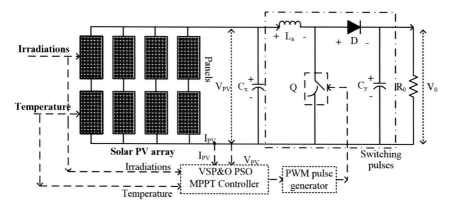

Fig. 14.1 Proposed PV standalone system with adaptive VSP&O with PSO controller

The P&O MPPT is the generally utilized controller for tracing the working point of the solar panel [14]. Here, the PV voltage variable is selected as reference parameter for observing the output power of PV at various sunlight temperature conditions. The step variation of power on P–V curve gives positive value then it starts moving in similar direction. If not, it relocates in overturn way. The qualities of this method are easy implementation and less hardware and software design complexity [15, 16]. The drawbacks of P&O are more oscillations at MPP plus drifting nature.

In the dp/di power point tracking method, the variation of current parameter and sudden changes of power variables are fed to the pulse generator block for controlling the duty step value of the power conversion circuit in order to meet the future consumer power demand [17]. The features of dp/di method are fast converter duty cycle control action, high convergence speed, and less operational cost [18]. But it has a drawback of less accuracy in power point tracking. Basically, the power converters generate distorted current and voltage waveforms [19]. The distorted ripple method is called as ripple correlation MPPT method. The advantages of ripple correlation controller are less heading and power conduction losses. However, the above methods give oscillations of MPP at instantaneous changes of environmental conditions. In this work, the adaptive variable step hybrid MPPT controller is proposed for partial shading condition of PV. The proposed MPPT controller is explained in Fig. 14.1.

14.2 Implementation of Two-Diode PV Cell

The solar panel is implemented either by applying one-diode or two-diode type PV cell. The one-diode cell design plus understanding is easy when equated to the other models of solar PV cells [20]. For every type of PV cell design, there are few parameters which are needed which are evaluated by using the various swarm optimization methods. The variables applied for implementation of one-diode type cell are maximum peak current, voltage, series and parallel resistances [21]. In this dual-diode model, there are few more parameters which are required for implementing the PV cell which are neglected.

So, here, the dual-diode cell is utilized for implementing the solar module. The block diagram of the dual diode solar cell is shown in Fig. 14.2. Figure 14.2 gives the terms I_{PV}, and I_L are the photo current plus PV cell supply current. Similarly, the terms R_h, V_L, and R_s are the shunt resistance, PV cell load voltage, plus series resistance offered by the PV panel. From the current distribution rule, the PV module generated current is obtained as follows:

$$I_{PV} = I_L \, N_{ss} + P - Q - \frac{V_{PV} + I_{PV} \, R_s (N_{ss}/N_{pp})}{R_h (N_{ss}/N_{pp})}, \tag{14.1}$$

$$P = I_{01} \, N_{pp} * \left(\exp\left(\frac{V_{PV} + I_{PV} \, R_s \left(N_{ss}/N_{pp} \right)}{A \, V_t \, N_{ss}} \right) - 1 \right), \tag{14.2}$$

$$N = I_{02} \, N_{pp} \left(\exp\left(\frac{V + I \, R_s (N_{ss}/N_{pp})}{A \, V_t \, N_{ss}} \right) - 1 \right), \tag{14.3}$$

$$I_{PV} = (I_{PVSTC} + K_i \, \Delta T) \frac{G}{G_{STC}}, \tag{14.4}$$

$$I_{02} = I_{01} = I_0 = \frac{I_{SC\_STC} + K_i \, \Delta T}{e((V_{ocSTC} + K_i * \Delta T)/(A_1 + A_2/p) \, V_t) - 1}, \tag{14.5}$$

$$V_{t2} = V_{t1} = V_t = \frac{N_s \, KT}{q}. \tag{14.6}$$

Based on Eq. (14.1), the solar cell-generated current is changing based on the sun insolation conditions, plus its working atmospheric values. Also, the PV cell-generated voltage increases when incident irradiations are enhanced. From Eq. (14.3), the series-connected cells enhance the supply voltage. The current supply of a PV is improved by making the inter connection of solar cells. Based on Eq. (14.5), the I_0 is related to the property of materials used for implementing of it. From Eq. (14.6), the

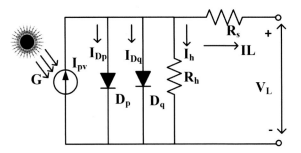

Fig. 14.2 Mathematical representation of dual-diode type PV cell

thermal potential of each cell is identical to the two shunt-connected semiconductor devices. The diode ideality factors are selected as 0.8 and 0.95.

14.2.1 Shading Behavior of PV

The solar standalone systems are mounted on the top of the hills and buildings. However, the tree falling shades, fogginess, and darkness of light are the major concerns for the partial shading behavior of the PV panel. Due to this shading nature, the power versus voltage curves consist of multiple peak power points, so that the extraction of peak output power of PV is highly difficult. In addition to this, the solar panel is having possibility of settling in any local MPP of nonlinear curve. Under shading condition, the PV behavior plus its associated features are given in Fig. 14.3 plus Fig. 14.4.

From Fig. 14.4a plus b, it is observed that the voltage characteristics of solar module are distributed nonlinear fashion at various environmental conditions. As a result, there are one global MPP and many unwanted peak power points which are not desirable for the operation of the solar cell. From the literature study, each and every PV module consists of one parallel diode across to it in order to bypass the short-circuited or faulty currents of PV under various sun insolation patterns' conditions. Here, the shaded panel current is absorbed by normal PV panel. So, the overall system losses are improved.

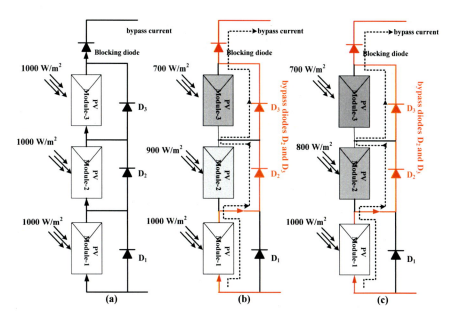

Fig. 14.3 Solar panel, **a** uniform behavior, **b** shading behavior 1, plus, **c** shading behavior 2

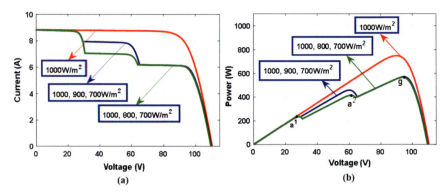

Fig. 14.4 PV panel, **a** P–V curve, plus, **b** I–V characteristics

14.3 Bionic Swarm Optimization-Based MPPT Techniques

As of from the literature study, the MPPT methodologies are playing very important role in the sun-based power plant. The conventional power point finding methods are applicable for constant solar insolation conditions because of its drawbacks which are low efficient for partial shaded condition, and more oscillations near the MPP. Due to that the soft computing, and bionic swarm optimization methods are utilized for extracting the more voltage of the solar panel. Here, the adaptive VSP&O-PSO controller is utilized plus it is associated along by the conventional PSO, and VSCSO power point tracking controllers.

14.3.1 Conventional PSO-Based MPPT Controller

The swarm intelligence is introduced by Eberhart from the hunting of birds [22]. Here, each and every bird is represented as particle. The particle's searching information is utilized for finding the optimum duty cycle of the converter. In the first iteration, all the particles search the particular region of I–V curve for evaluating the P_{best} constrain. The obtained P_{best} is stored in a particular position. After that the searching of particles is started for determining the required MPP position of PV panel. There are plenty number of iterations which are used for tracing the G_{best} position. The searching of PSO is given in Fig. 14.5.

From Fig. 14.5, it is identified that the random duty values of power converter are initiated to all the particles. The features of PSO are required very less parameters for tuning and give best solution when the search space is very low. The particle velocity plus duty cycle length are upgraded by using Eqs. (14.7) plus (14.8).

$$l^{x+1}_l = d^{x}_l + V^{x+1}_l, \qquad (14.7)$$

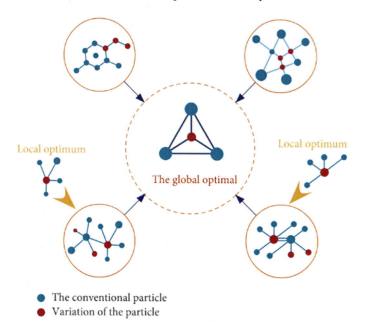

Fig. 14.5 Searching behavior of PSO-based powerpoint finding controller

$$l^{x+1} = W\, V_l^{x+1} + C_m\, r_m(P_{best} - d_l^x) + C_n\, r_n(G_{best} - d_l^x), \qquad (14.8)$$

where the terms x, l, plus V are the total iterations, length of the particle, and velocity of each agent. In addition to that the term 'W' is defined as weight of each agent. The particles' acceleration factors are 'C_m' plus 'C_n'. Finally, the variables 'r_m' plus 'r_n' are the random selected parameters.

14.3.2 Convectional CSO MPPT Controller

The basic cuckoo hunt technique is implemented from the behavior of birds. In this optimization method, the cuckoo eggs are removed with the help of host birds. The host birds create the new nests [23]. In this method, there are three rules which are involved for searching the target position of MPP. The basic searching concept of cuckoo algorithm is illustrated in Fig. 14.6.

In the first rule, all the eggs of cuckoos are consisted of good quality, then only the algorithm is shifted to second iteration. Otherwise, the first iteration is continued. In the second rule, the host nests are consisting of fixed value. Finally, the levy flights are helpful for achieving the new eggs of cuckoos. The levy flight of each cuckoo is obtained by using Eq. (14.9).

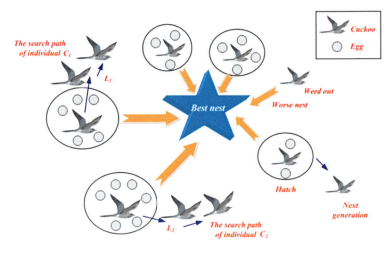

Fig. 14.6 Searching concept of cuckoo exploration controller

$$\text{levy\_flight}(\beta) = L^{-\beta}. \tag{14.9}$$

From Eq. (14.9), the term 'L' is defined as levy length which is having a value of 1.2–1.8. The cuckoo learning constant is 'β' which is selected as 1.5 for effective finding of MPP. The fitness value of cuckoo algorithm is selected randomly based on the operating duty value of converter. The duty of power converter is optimized by using Eq. (14.10).

$$D^{s+1} = D_m^n + \alpha \oplus \text{Levy}(\beta), \tag{14.10}$$

$$d_m^{s+1} = d_m^{s+1} + c_{\text{levy}}\left(\frac{N}{w^{1/\beta}}\right)(D_{\text{best}}^s - D_m^s), \tag{14.11}$$

where the term β and c_{levy} are selected as 2.0, plus multiplication factor which is identified as 2.5. The liner functions of 'A' and 'B' are identified as,

$$A = Y(0, \sigma_A^2),\ B = Y(0, \sigma_B^2), \tag{14.12}$$

$$\sigma_A = \left(\frac{r(1+\beta)\sin(\pi\beta/2)}{r((1+\beta)/2)\beta\, 2^{(\beta-1)/2}}\right)^{1/\beta};\ \sigma_B = 1. \tag{14.13}$$

14.3.3 Proposed Adaptive P&O-PSO-Based MPPT Controller

In this work, the hybrid bionic swarm controller is developed from the advantages of P&O and adaptive PSO controllers. In this proposed controller, at initial, the conventional P&O is applied for moving the working point of the PV close to the actual MPP location [24]. So, the controller convergence speed plus tracking speed are enhanced. The working point of the PV reaches the near to the true MPP position, then the swarm optimization controller is applied for compensating the oscillations of working power point of the solar PV, so that the accuracy of hybrid controller is improved. The proposed hybrid controller flowchart is shown in Fig. 14.7.

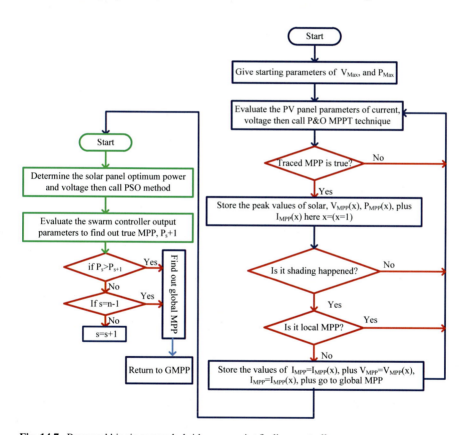

Fig. 14.7 Proposed bionic swarm hybrid power point finding controller

14.4 Discussion of Simulative Performance Results

Here, three types of solar PV modules are sequentially connected in series to improve the supply voltage of the PV. The solar module design has been done by utilizing the dual-diode type PV cell. The utilized variables for the implementation of PV cell are short-circuit current of cell I_{SC} is 8.85A, peak current of cell I_{MPP} is 8.84A, series resistor R_S is 0.45 Ω, shunt resistance R_h is 312 Ω, and peak voltage V_{MPP} is 35.55 V. Here, each PV module supply voltage rating is 250 W. The overall three module systems supply power, current, plus voltages are 750 W, 7.978 A, plus 93.99 V, respectively. The solar module supply voltage is very low which is step-up by utilizing the high voltage gain power converter. The converter is interfaced with the PV as given in Fig. 14.1.

The selected parameter values of converter are $C_x = 18$ μF, $L_a = 1.4$ mH, plus $C_y = 10$ μF. Finally, the load resistor of power converter is $R = 50$ Ω. The input side capacitor is placed in a shunt manner across the PV module to stabilize the solar voltage at various atmospheric temperature conditions. The ripple of PV power is filtered by utilizing the inductor which is placed at middle of the supply and resistor. The load capacitor is placed near to the consumer supply for enhancing the supply PV power. The direct connection of PV with converter is not helpful for high power demand application. So, the working point of PV is placed near to the power converter along with the pulse width controller. Here, the current sensor and power sensors are applied for sensing the PV module output parameters. The sensing PV values are directly supplied to the MPPT block.

14.4.1 Initial PV Shading Behavior at 1000 W/m², 900 W/m², Plus 700 W/m²

As of now, the shading occurs on PV modules due to the shadows plus cloud conditions. So, the PV generates output nonlinear curves which are consisted of more than one maximum power point. So, finding the peak power point is quite complex. In addition, the power supply of converter is depending on the duty cycle control. Here, there are three bionic swarm intelligence controllers which are analyzed at 1000, 900, 700W/m² for adjusting the duty of converter. At this shading condition, the maximum available power plus voltages of PV are 578W plus 92.7 V. The solar panel nonlinear curves at this shading behavior are illustrated in Fig. 14.8a, b, plus c. The performance values of proposed topology are shown in Table 14.1. The solar-generated power, voltage, plus current values are illustrated in Fig. 14.8d, e, plus f.

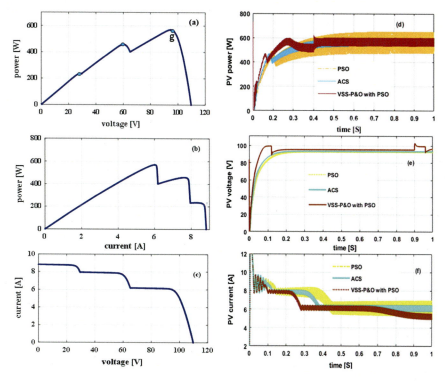

Fig. 14.8 **a** Power versus voltage, **b** power versus current, **c** current versus voltage, **d** solar module power, **e** solar module voltage, and **f** solar panel current at the initial shading condition

14.4.2 Second PV Shading Behavior at 1000 W/m², 800 W/m², Plus 700 W/m²

In this second shading behavior of solar modules, the shading effect on solar PV modules is excess. Also, the nonlinear power curves of solar module consist of multiple operating power points. As a result, the finding of accurate MPP place in the entire P–V curve region is quite difficult. The available power plus voltage of solar modules under this shading condition are 569.09 W plus 91.789 V. The overall generated current of PV is 6.186 A which indicates that the overall system power dissipation losses plus heating effects are increased. So, the power system life span will reduce. The noted parameters of solar PV modules under this shading behavior are given in Table 14.1. The related PV supply parameters are given in Fig. 14.9a, b, plus c. The PSO plus VSSP&O-PSO power point finding controllers generated power parameters are given in Fig. 14.9 d, e, plus f.

Table 14.1 Comparative performance results of various bionic swarm MPPT controllers at various environmental conditions

Variables	PSO	Adaptive CS	VSSP&O-PSO
At first PSC of PV			
Voltage of capacitor (V)	93.832	94.11	103.21
Power of PV panel (W)	567.77	568.88	579.77
Current of PV panel (A)	6.021	5.912	5.501
Actual PV power (W)	584	584	584
Efficiency of MPPT (%)	97.313	97.585	99.615
Iterations required	None	None	11.00
Time taken for settling (s)	0.4511	0.50	0.400
Waveform distortions	Moderate	Moderate	Less
Speed of MPPT (s)	0.200	0.280	0.1502
Converter duty	0.7	0.39	0.58
At second PSC of PV			
Voltage of capacitor (V)	86.534	89.213	100.122
Power of PV panel (W)	561.00	568.99	580.002
Current of PV panel (A)	6.439	6.361	5.7123
Actual PV power (W)	582.00	582.00	582.00
Efficiency of MPPT (%)	96.386	97.921	99.565
Iterations required	None	None	9.00
Time taken for settling (s)	0.34	0.450	0.300
Waveform distortions	Moderate	Moderate	Less
Speed of MPPT (s)	0.400	0.31	0.1241
Converter duty	0.5	0.7	0.4

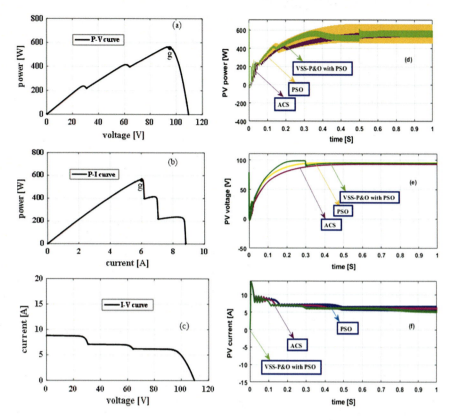

Fig. 14.9 **a** Power versus voltage, **b** power versus current, **c** current versus Voltage, **d** solar panel power, **e** solar panel voltage, and **f** solar panel current at the second shading condition

14.5 Conclusion

The adaptive VSSP&O-based bionic swarm optimization-related maximum power point finding controller is executed by applying a MATLAB tool. The proposed MPPT gives high convergence speed, low distortions of solar module voltage, requires few numbers of iterations, high output voltage, plus more accuracy of MPP finding when equated with the other bionic swarm optimization controllers. From the comparative investigation of power point finding controllers, it has been identified that the hybrid controller gives superior performance at various sun irradiation patterns conditions.

References

1. Murali, M., Basha, C.H., Kiran, S.R., Amaresh, K.: Design and analysis of neural network-based MPPT technique for solar power-based electric vehicle application. In: Proceedings of Fourth International Conference on Inventive Material Science Applications: ICIMA 2021, pp. 529–541. Springer Singapore (2022)
2. Kiran, S.R., Murali, M., Hussaian Basha, C.H., Fathima, F.: Design of artificial intelligence-based hybrid MPPT controllers for partially shaded solar PV system with non-isolated boost converter. In: Computer Vision and Robotics: Proceedings of CVR 2021, pp 353–363. Springer, Singapore (2022)
3. Rahman, A., Farrok, O., Haque, M.M.: Environmental impact of renewable energy source based electrical power plants: Solar, wind, hydroelectric, biomass, geothermal, tidal, ocean, and osmotic. Renew. Sustain. Energy Rev. **161**, 112279 (2022)
4. Osman, A.I., Chen, L., Yang, M., Msigwa, G., Farghali, M., Fawzy, S., Yap, P.S.: Cost, environmental impact, and resilience of renewable energy under a changing climate: a review. Environ. Chem. Lett. 1–24 (2022)
5. Hussaian Basha, C.H., Akram, P., Murali, M., Mariprasath, T., Naresh, T.: Design of an adaptive fuzzy logic controller for solar PV application with high step-up DC–DC converter. In: Proceedings of Fourth International Conference on Inventive Material Science Applications: ICIMA 2021, pp. 349–360. Springer Singapore (2022)
6. Kumari, P.A., Geethanjali, P.: Adaptive genetic algorithm based multi-objective optimization for photovoltaic cell design parameter extraction. Energy Procedia **117**, 432–441 (2017)
7. Kiran, S.R., Basha, C.H., Kumbhar, A., Patil, N.: A new design of single switch DC-DC converter for PEM fuel cell based EV system with variable step size RBFN controller. Sādhanā **47**(3), 128 (2022)
8. Murali, M., Hussaian Basha, C. H., Kiran, S. R., Akram, P., Naresh, T.: Performance analysis of different types of solar photovoltaic cell techniques using MATLAB/simulink. In: Proceedings of Fourth International Conference on Inventive Material Science Applications: ICIMA 2021, pp. 203–215. Springer Singapore (2022)
9. Basha, C.H., Mariprasath, T., Murali, M., Rafikiran, S.: Simulative design and performance analysis of hybrid optimization technique for PEM fuel cell stack based EV application. Mater. Today: Proc. **52**, 290–295 (2022)
10. Kiran, S.R., Mariprasath, T., Basha, C.H., Murali, M., Reddy, M.B.: Thermal degrade analysis of solid insulating materials immersed in natural ester oil and mineral oil by DGA. Mater. Today: Proc. **52**, 315–320 (2022)
11. Udhay Sankar, V., Hussaian Basha, C.H., Mathew, D., Rani, C., Busawon, K.: Application of WDO for decision-making in combined economic and emission dispatch problem. In: Soft Computing for Problem Solving: SocProS 2018, Vol. 1, pp. 907–923. Springer Singapore (2020)
12. Mellit, A., Rezzouk, H., Messai, A., Medjahed, B.: FPGA-based real time implementation of MPPT-controller for photovoltaic systems. Renew. Energy **36**(5), 1652–1661 (2011)
13. Chen, Z.Y.: Integration learning of neural network training with swarm intelligence and meta-heuristic algorithms for spot gold price forecast. Appl. Artif. Intell. **36**(1), 1994217 (2022)
14. Femia, N., Petrone, G., Spagnuolo, G., Vitelli, M.: A technique for improving P&O MPPT performances of double-stage grid-connected photovoltaic systems. IEEE Trans. Industr. Electron. **56**(11), 4473–4482 (2009)
15. Sher, H.A., Murtaza, A.F., Noman, A., Addoweesh, K.E., Al-Haddad, K., Chiaberge, M.: A new sensorless hybrid MPPT algorithm based on fractional short-circuit current measurement and P&O MPPT. IEEE Trans. Sustain. Energy **6**(4), 1426–1434 (2015)
16. Basha, C.H., Rani, C.: Different conventional and soft computing MPPT techniques for solar PV systems with high step-up boost converters: a comprehensive analysis. Energies **13**(2), 371 (2020)

17. Hussaian Basha, C.H., Bansal, V., Rani, C., Brisilla, R.M., Odofin, S.: Development of cuckoo search MPPT algorithm for partially shaded solar PV SEPIC converter. In: Soft Computing for Problem Solving: SocProS 2018, vol. 1, pp. 727–736. Springer Singapore (2020)
18. Hussaian Basha, C.H., Rani, C.: Performance analysis of MPPT techniques for dynamic irradiation condition of solar PV. Int. J. Fuzzy Syst. **22**(8), 2577–2598 (2020)
19. Basha, C.H., Rani, C., Odofin, S.: A review on non-isolated inductor coupled DC-DC converter for photovoltaic grid-connected applications. Int. J. Renew. Energy Res. (IJRER) **7**(4), 1570–1585 (2017)
20. Basha, C.H., Murali, M.: A new design of transformerless, non-isolated, high step-up DC-DC converter with hybrid fuzzy logic MPPT controller. Int. J. Circuit Theory Appl. **50**(1), 272–297 (2022)
21. Vodapally, S.N., Ali, M.H.: A comprehensive review of solar photovoltaic (PV) technologies, architecture, and its applications to improved efficiency. Energies **16**(1), 319 (2023)
22. Basha, C.H., Rani, C.: A New single switch DC-DC converter for PEM fuel cell-based electric vehicle system with an improved beta-fuzzy logic MPPT controller. Soft. Comput. **26**(13), 6021–6040 (2022)
23. Hussaian Basha, C.H., Rani, C., Brisilla, R.M., Odofin, S.: Simulation of metaheuristic intelligence MPPT techniques for solar PV under partial shading condition. In: Soft Computing for Problem Solving: SocProS 2018, vol. 1, pp. 773–785. Springer Singapore (2020)
24. Basha, C.H., Rani, C.: Design and analysis of transformerless, high step-up, boost DC-DC converter with an improved VSS-RBFA based MPPT controller. Int. Trans. Electr. Energy Syst. **30**(12), e12633 (2020)

Chapter 15
Study of Voltage-Controlled Oscillator for the Applications in K-Band and the Proposal of a Tunable VCO

Rajni Prashar and Garima Kapur

Abstract The progress in wireless technology has simplified and streamlined the transfer or sharing of data, thereby maximizing its impact on societies worldwide. However, with these advancements, more memory space is needed to store the vast amount of information being transferred. To achieve this, the size of devices must be reduced, necessitating the scaling of MOS transistors to deep submicron levels. Transceiver is being one of the crucial components which is responsible for transmitting or receiving information from the wireless device. Within the wireless transceiver, the frequency synthesizer produces an stable output frequency and further mixed with the received signal down to lower frequencies and vice versa. To operate at high frequencies between 12 and 40 GHz, where operations are carried out at high speeds and coverage is done with multiple beams, circuits must be compatible with high speed. In this paper, we study the VCO component, list the design parameters, and propose a model in which the inductor is replaced with a gyrator-C active inductor to minimize the overall area and use the frequency of oscillations as per the requirement. The design is simulated at 90 nm technology on ADS design tool.

Keywords RFIC · VCO · Oscillators

R. Prashar (✉) · G. Kapur
Department of Electronics and Communication Engineering, JIIT, Sector 62, Noida, India
e-mail: 2401.rajni@gmail.com

G. Kapur
e-mail: garima.kapur@jiit.ac.in

© The Author(s), under exclusive license to Springer Nature Singapore Pte Ltd. 2024
P. K. Jha et al. (eds.), *Proceedings of Congress on Control, Robotics, and Mechatronics*, Smart Innovation, Systems and Technologies 364,
https://doi.org/10.1007/978-981-99-5180-2_15

15.1 Introduction

Due to the advancements in wireless technology and the increasing number of users with limited bandwidth, there is a growing demand for improved performance, such as high data rates with multimedia applications. As a result, designers are developing and fabricating wireless components. With the technological advancements and automation algorithms in the last three decades, it is now possible to fabricate all the components of a transceiver in any wireless communication system on a single IC.

The demand for increased bandwidth in communication systems, particularly in wireless communication, has resulted in the development of VLSI technology and the automation industry. Frequencies ranging from 12 to 40 GHz are of particular interest, where carrier signals of very high frequency range are used as shown by Lee et al. [1] applications which are mainly focused on military arm forces, communication used in aircraft and satellite, radio and radar communication. The range from 27 to 40 GHz is used in high-throughput satellite applications and is widely available.

15.2 Voltage-Controlled Oscillator Design

In any communication system whether it is a transmitter or a receiver, low noise amplifier (LNA) power amplifier, oscillator (VCO), and PLL are the main part of the system. In every communication system, voltage-controlled oscillator holds a crucial role. The high-frequency signals which are used as a carrier signal, are obtained with the help of voltage-controlled oscillator circuits. These days with the advancement in CMOS technology where inductors can be realized using MOSFETs so by using active inductors oscillators are designed which can be used to generate the signals up to the range of GHz as presented by Fahs et al. [2] and Banu [3]. The traditional method used to design VCOs is either by using CMOS ring oscillator or by using Harley and Collpit's oscillator which uses LC as a tank circuit which are described by Akashe et al. in [4]. The evolution of wireless technology has led to the simplification and optimization of information transfer or sharing, which in turn has increased its impact on global society. Due to these advances, more memory space is required to store such a large transfer of information. This can only be done by reducing the device size which means scaling of MOS transistor to deep submicron levels. Its role is to transmit or receive the information to (or from) the wireless device. The purpose of the frequency synthesizer is to produce a steady output frequency that can be utilized to mix the incoming signal to lower frequencies, and vice versa. The generation of this consistent output frequency is achieved by implementing a Phase-Locked Loop (PLL). While working at high frequency at the range of 12–40 GHz, where the operation is carried out at a very high speed and the coverage is done with the multiple beams, the circuits used at high frequency should be compatible with high speed. So, in this paper, one very basic component which is the heart of communication system, i.e., VCO is studied and the design parameters have been listed in this paper with reference to Berroth [5] and Naseh [6]. A tunable VCO can generate a wide range of frequencies, making it useful in many applications. In a wireless communication system, for example, the VCO is used to generate a carrier signal whose frequency can be varied to transmit different data rates. The VCO frequency can be controlled by adjusting the input control voltage, which is usually derived from a phase-locked loop (PLL) or frequency synthesizer. A tunable VCO can also be used in radar systems, where the frequency of the transmitted signal needs to be varied to detect objects at different distances. In this case, the VCO frequency can be controlled by adjusting the input voltage to the oscillator circuit.

15.2.1 VCO Based on Inductor Capacitor Pair

A very basic voltage-controlled oscillator with an inductor and capacitor is shown in Fig. 15.1. The circuit contains inductor L and capacitor C which are parallel to each other. In the circuits, parasitic components are shown as RL and RC for the inductor and capacitor, respectively. To overcome the energy loss associated with these parasitic components, MOSFETs or CMOS can be utilized to have the negative resistance. The energy which is lost in the tank circuit is given by (15.1)

$$P_{loss} = 4\pi^2 RC^2 fo^2 V_{peak}^2 = \frac{R}{4\pi^2 L^2 fo^2} V_{peak}^2. \qquad (15.1)$$

The power loss experienced by the tank circuit is dependent on the value of inductance and operating frequency, with an inverse relationship between the two. By decreasing the series resistance R, which is directly proportional to the loss, the power loss in the tank circuit is reduced linearly. Additionally, an increase in the tank inductance results in a quadratic decrease in power loss. To counter the energy loss caused by the parasitic components, a MOSFET can be introduced to the circuit. To remunerate the stray resistance, a negative resistance-R can be introduced to the circuit by using active devices so that both the unwanted stray elements can be canceled out. This is done by using the transistors in cross-coupling topology in the tank circuit, and this is shown in Fig. 15.2.

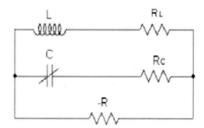

Fig. 15.1 VCO based on inductor capacitor pair

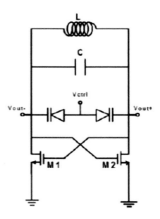

Fig. 15.2 Cross-coupled oscillator

The idea behind using the cross-coupled transistors as shown by Niaboli-Guilani [7] is to have the same value of conductance (gm) as provided by the negative resistance from the oscillator.

15.2.2 Ring VCO

A ring oscillator as shown in Fig. 15.3 can be designed by using a number of buffer stages, where all the buffers are connected serially and the output of nth stage is connected back to the first stage. The criteria for the oscillation are that the circuit must give a phase shift of 2π or 0 and voltage gain should be greater that equal to 1 depicted by Prajapati [8].

In ring oscillator, each stage produces a finite delay and the phase shift of π is generated by the active element, and the another required phase shift is provided by a dc inversion which is described by Razavi [9].

In this study paper, the previous section displays the frequently employed CMOS ring oscillator and LC tank oscillator VCOs. Specifically, the section highlights the LC oscillator circuits that can be used as they have the advantage that they have better noise characteristics, but with the drawback that this approach may have the large dimensions, so cannot be used where the phase shift is needed. On the other side, the cascaded voltage-controlled oscillator has the better performance parameters as compared to LC oscillator, but they also have the demerit that these circuits have less power requirements and minimum area on the chip, so they can be prone to noise. So,

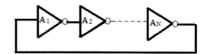

Fig. 15.3 Ring oscillator

because of these clear advantages of ring oscillators, ring oscillators are preferred over the LC oscillators which are shown by Bormontov et al. [10] and Gupta [11].

15.3 Design Topologies

Its basic function of VCO is to generate a constant RF frequency in wireless transceivers. Besides having a simple design architecture, it is the most challenging block to design because of its operation at high RF frequencies in which the phase noise becomes significant and its parameters get deviated from desired values. When voltage at the input of VCO changes, then frequency at the output is varied. The VCO can be realized either as a ring oscillator or as resonant oscillators.

As shown in Fig. 15.4a, an inductor L and capacitor C are parallel to each other. The resonance frequency of this LC circuit is given by (15.2)

$$w_{\text{res}} = \sqrt{LC}. \tag{15.2}$$

At the given resonant frequency, the impedance of the inductor and capacitor which are written as jLw_{res} and $1/(jCw_{\text{res}})$, respectively, is opposite and equal to each other, thus resulting an infinite impedance. But in actual practical circuits these passive components have stray impedance indicated as resistive components, as shown in Fig. 15.4b and the quality factor for inductor is given as, Q as given in (15.3).

$$Q = L_\omega / R_S. \tag{15.3}$$

The losses attributed to RC are deemed negligible because the value of the capacitor's quality factor (Q) significantly exceeds that of the inductor's Q. The series model of the circuit is presented in the form of Fig. 15.4c.

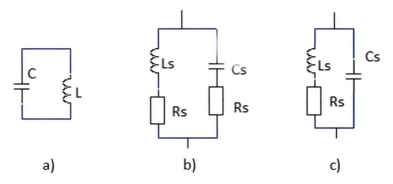

Fig. 15.4 a LC parallel circuit, b resonance circuit, c series model of the circuit

Fig. 15.5 **a** Inductor L with R_s as parasitic resistance, **b** series resistance conversion into parallel resistance

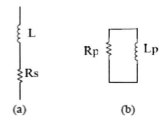

Fig. 15.6 RLC gain stage used in the circuit

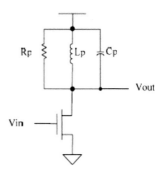

Now, converting the resistor (R_s) which in series with inductor (L) in Fig. 15.5a into the parallel form in Fig. 15.5b, we get

$$L \approx L_P, \tag{15.4}$$

$$R \approx Q^2 R_S. \tag{15.5}$$

It can be concluded from (15.5) that the quality factor of the inductor plays an important role in determining the amount of energy lost in the tank. Figure 15.6 shows a simple gain stage based on an LC tank.

15.4 Design Parameters

Almost every transreceiver which is designed for wireless applications requires a tunable reference frequency. Thus, an ideal VCO is required that will generate an output which is linearity proportional to the applied input voltage. In addition to the linearity parameter, several other parameters significantly impact oscillator design. The following are some of the key parameters discussed below.

Fig. 15.7 Ideal VCO tuning linearity

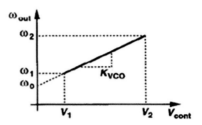

15.4.1 Linearity

Linearity and the tuning of a VCO have a trade-off. Ideally, linear tuning is required, but in actual implementation, the nonlinear behavior of VCO is observed as components used are also nonlinear. Linearity is the required to have the VCO gain (KVCO) constant as given in Fig. 15.7.

15.4.2 Range of Tuning

The range of tuning of oscillator is based on the following two specifications:

(i) The center frequency of the tuning range must be remained constant with the frequency of oscillation.
(ii) Frequency deviation due to even slight variations in temperature results nonlinearities in VCO characteristics. To minimize the effect of these variations, so, a wide tuning range is selected. But, these nonlinearities can be minimized by narrowing down the tuning range. Therefore, a trade-off exists between nonlinearities and tuning range.

15.4.3 Power Consumption

In a PLL, most power is dissipated by the VCO as compared to other components. In this paper, VCO is studied for RF transreceiver; therefore, priority is given to the tuning range and phase noise as compared to power.

15.4.4 Phase Noise

The sidebands present around the central frequency in frequency domain system are called the phase noise and the same sidebands in time-domain system are called jitter. Linear time-invariant model of phase noise given by Leeson is represented in (15.6)

$$L(f_m) = 10\log\left[\frac{2FkT}{Ps}\left(\frac{f_o}{2Q_l f_m}\right)^2\left(1+\frac{f_k}{f_m}\right) + \frac{|K_{VCO}|^2}{2f_m^2}S_{VCNT}\right.$$
$$\left. + \frac{|K_{VDD}|^2}{2f_m^2}S_{VDD} + \frac{|K_{IB}|^2}{2f_m^2}S_{IB}\right]. \tag{15.6}$$

Q_L = loaded quality factor.
fo = oscillation frequency.
Ps = signal power of oscillation.
fm = offset frequency.
F = noise factor of active devices.
k = Boltzmann's constant.
T = temperature (Kelvin).
f_k = flicker noise corner frequency in the phase noise.

$$\frac{|K_{VCO}|^2}{2f_m^2}S_{VCNT}, \frac{|K_{VDD}|^2}{2f_m^2}S_{VDD}, \text{ and } \frac{|K_{IB}|^2}{2f_m^2}S_{IB}$$

are the sensitivity of the VCO to the control voltage, supply, and bias current, respectively. We observe that among various factors that can reduce phase noise, the circuit designer can control only three factors, namely, loaded quality factor, noise factor of active device, and signal power of oscillations.

(i) The quality factor of the circuit is calculated by the amount of series resistance present in the LC tank.
(ii) Resonators with a higher quality factor inherently exhibit lower phase noise.
(iii) The inverse relationship between phase noise and output power (Ps). However, as the output power must be minimized, a trade-off between power and phase noise arises.

As the noise factor (F) is proportional to phase noise. Lowering the noise factor is directly linked to the active components used in the VCO. Therefore, devices with lower flicker noise are better for this application.

15.5 Various VCO Designs and Proposed Circuit

Various VCO structures have been studied, and their performance parameters have been compared. The comparison of performance parameters has been done for the designs which are designed for the frequency range between 24 and 40 GHz, i.e., suitable for the application of 5G circuits. The circuit presented by Allstot et al. [12] is used for the generation of a 24 GHz oscillatory signal. This signal has been generated using a 12 GHz voltage-controlled oscillator indirectly cascaded with passive mixer. The circuit has been implemented in 0.18um CMOS technology, the advantage of using the passive mixer reduces the power consumption and also increases the tuning range of the device. For the application of VCOs in Radar, it is used to find the highest frequency range of the system. In VCOs, varactor diodes are used to tune the frequency of the device. A complementary cross-coupled LC-VCO presented by Hou et al. [13] uses the MOSFET in accumulation mode for the tuning of the device. The LC tank resonator is used in the circuit which contains on-chip differential inductor and a pair of MOSFET that is used in accumulation mode for a good linearity. The circuit can be used for wireless high-quality video streaming. The circuit also uses the cross-coupled VCO used with varactor capacitance. The circuit provides the wide tuning range, less power consumption as well as low phase noise. Next approach that is indicated by Gao et al. [14] uses the concept of a varactor and MOM capacitor combination method for the tunability. The circuit indicates the high value of Q-factor and low phase noise VCO indicted by Ryu et al. [15] which provides high transconductance gm. In the given circuit, two N-type MOS are connected in parallel with the conventional VCO circuit. Capacitive division technique is used to increase the voltage swing and to lower the phase noise value. The design depicted by Sethi et al. [16] utilizes a low phase noise voltage-controlled oscillator (VCO) that has a higher transconductance (gm). This is achieved by incorporating parallel MOSFETs in the design, where two n-channel transistors are connected in parallel with the cross-coupled n-channel transistors of a typical VCO. The purpose of parallel MOSFETs increases the total negative conductance available to the circuit, which helps to cancel out the stray or small signal resistance of the LC tank. The VCO is implemented using CMOS technology, which provides low-power consumption and high integration density. The oscillator frequency can be controlled by adjusting the bias voltage, which changes the capacitance of the varactors in the circuit. The output signal from the VCO is then mixed with a frequency reference signal to generate the desired frequency output. Many designs use a MoM capacitor bank which is a high-power pulsed power supply for medical applications. This design employs a bank of MoM capacitors connected in parallel to store electrical energy and release it as a high-power pulse. The capacitors are charged using a high-voltage power supply, and a switch is used to discharge the energy into a load.

In this paper, we are proposing a basic circuit of the VCO with a minor change in the circuit which results not only reducing the area size but also provides indirect tuning for the circuit. The design of cross-coupled VCO is presented in Fig. 15.8 and has been simulated at 90 nm technology using BSIM4 MOSFETs, and the operation is carried out by the charging and discharging of inductor and capacitor. There will be the die down wave in the frequency of oscillations because of the loss of energy. This loss of energy can by represented by adding the Rp resistance in parallel to L and C.

The circuit has been designed on ADS design tool, and the simulation is done at 90 nm technology. The output has been plotted differentially between Vop and Vom nodes. To avoid the degradation of VCO tuning range, the capacitively loading on the output nodes should be avoided. The device size is decided by considering the loss in the circuit. When the net currents are balanced, net voltage across the LC tank circuit is zero, and at that point, the noise will affect the circuit performance. The

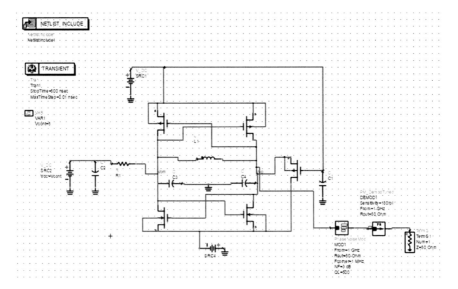

Fig. 15.8 Cross-coupled voltage-controlled oscillator

tuning of the VCO can be done by using the tunable active inductor. The oscillation produced is shown in Fig. 15.9, and the frequency of oscillations can be calculated by taking the Fourier transform of the (Vop-Vom). The tuning of the circuit is proposed by replacing the inductor L by gyrator-C-based inductor in which the direct tuning can be done by the biasing applied at the feedback MOSFET as shown in Fig. 15.10.

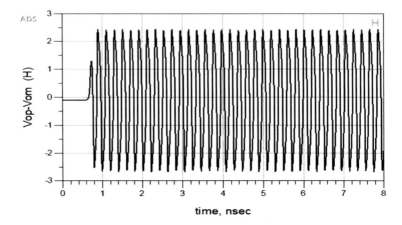

Fig. 15.9 Frequency of oscillations of cross-coupled VCO

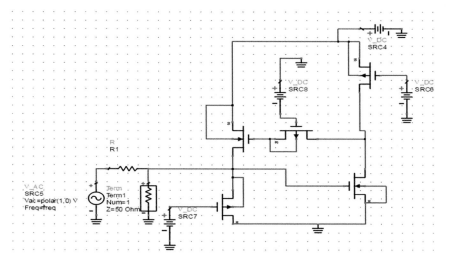

Fig. 15.10 Tunable active inductor design

15.6 Conclusion

The study of voltage-controlled oscillators in the realm of wireless communication and other fields of application has been conducted, and the paper has presented the important design parameters to consider when designing a VCO. The paper has also explored the use of VCOs in applications such as radar and other wireless communication systems. Some designs have utilized the tunable active inductor approach, which could be enhanced by modifying the design technology. The paper proposes a simple approach of replacing the inductor L with a single-ended inductor. Additionally, the modification of using varactor diodes for charge storage in the tank circuit is also suggested.

References

1. Hajimiri, A., Lee, T.H.: Design issues in CMOS differential LC oscillators. IEEE J. Solid-State Circ. **34**(5), (1999)
2. Hyvertl, J., Cordeaul, D., Paillotl, J.-M., Philippe, P., Fahs, B.: A new class-C very low phase-noise Ku-band VCO in 250 nm SiGe: C BiCMOS technology. (2015) 978-1-4799-8275-2/15/ $31.00 ©2015 IEEE
3. Banu, M.: MOS oscillators with multi-decade tuning range and gigahertz maximum speed. IEEE J. Solid-State Circuits **23**, 1386–1393 (1988)
4. Shrivastava, A., Saxena, A., Akashe, S.: High performance of low voltage-controlled ring oscillator with reverse body bias technology. Front. Optoelectron. **6**(3), 338–345 (2013)
5. Tao, R., Berroth, M.: 5 GHz voltage-controlled ring oscillator using source capacitively coupled current amplifier. In: Proceedings of IEEE Topic Meeting on Silicon Monolithic Integrated Circuits in RF System, pp. 45–48 (2003)

6. Jamal, M., Kazemeini, M.H., Naseh, S.: Performance characteristics of an ultra-low power VCO. In: ISCAS'03 Proceedings of the 2003 International Symposium on Circuits and Systems, vol. 1. IEEE (2003)
7. Niaboli-Guilani, M.: A low power low phase CMOS voltage controlled oscillator. In: 17th IEEE International Conference on Electronics, Circuits, and Systems (ICECS) (2010)
8. Prajapati, P.P.: Analysis of voltage controlled oscillator using 45nm CMOS technology. Int. J. Adv. Res. Electr. Electron. Instrum. Eng. **3**(3), (2014)
9. Razavi, B.: Design of analog CMOS integrated circuits, international edn. The McGraw-Hill Companies, Inc. (2001)
10. Bistritskii, S.A., Klyukin, V.I., Bormontov, E.N.: Ring voltage-controlled oscillator for high-speed PLL systems. Russ. Microlectron. **43**(7), 472–476 (2014)
11. Yadav, N., Gupta, S.: Design of low power voltage controlled ring oscillator using MTCMOS technique. Int. J. Sci. Res. **3**(7), 845–851 (2014)
12. Neihart, O.D., Neihart, M., Allstot, D.J.: differential VCO and passive frequency doubler in 0.18 μm CMOS for 24 GHz applications
13. Yan, N., Zhang, C., Hou, X.: Design of a LC-VCO in 65 nm CMOS technology for 24 GHz
14. Liu, Y., Li, Z., Gao, H.: A 24 GHz PLL with low phase noise for 60 GHz sliding-IF transceivers in a 65-nm CMOS. https://doi.org/10.1016/j.mejo.2021.105106
15. Behera, P., Siddique, A., Delwar, T.S., Biswal, M.R, Choi, Y. Ryu, J.-Y.: A novel 65 nm active-inductor-based VCO with improved Q-factor for 24 GHz automotive radar applications. https://doi.org/10.3390/s22134701
16. Rout, S.S., Acharya, S., Sethi, K.: A low phase noise gm-boosted DTMOS VCO design in 180 nm CMOS technology. https://doi.org/10.1016/j.kijoms.2018.03.001

Chapter 16
Detailed Performance Study of Data Balancing Techniques for Skew Dataset Classification

Vaibhavi Patel and Hetal Bhavsar

Abstract Many real-world classification problems involve changing events where one class has comparatively fewer samples called minority class which is more important to detect. Consequently, the dataset is often unbalanced and shows significantly skewed data. Since the majority class dominates the learning process and tends to sketch all predictions, the conventional classification model leads to biased results where it may easily display excellent performance in the dominant class and bad performance in the minority class. Additionally, the traditional accuracy score is inaccurate since it assigns equal weight to actual positives and actual negatives. This study is aimed to present an empirical analysis of the data imbalance effect on classification algorithms. Six popular and effective data balancing techniques are applied to eight benchmark skewed datasets from the KEEL and Kaggle repositories to analyze and compare the performance.

Keywords Keywords · Binary class · Data imbalance · Multi-class · Cost-sensitive · Class distribution · Under-sampling · Skew data · Big data · Hybrid method · Ensemble method · Over-sampling

16.1 Introduction

The data imbalance issue is a well-known problem in the field of data mining. In the binary class training set, it occurs when occurrences of one class (majority class) are significantly overwhelmed by samples of the other class (minority class) as shown in Fig. 16.1. In the multi-class dataset, this problem is more critical where it may possible that one class is minor in comparison with some classes and the same class might be the majority class in comparison with other classes. Imbalance substantially distorts the process of learning and harms the effectiveness of classification algorithms

V. Patel (✉) · H. Bhavsar
Faculty of Technology and Engineering, The Maharaja Sayajirao University of Baroda, Vadodara, Gujarat, India
e-mail: vkpatel93@gmail.com

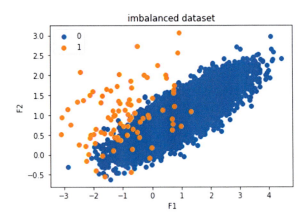

Fig. 16.1 Imbalanced binary class dataset

since the classifier usually does not have enough instances of the minority class for proper model training.

It is a key challenge in a wide variety of vital application domains, including computer vision [1], fraud detection, medical science [2], direct marketing, text classification, fire alarm systems, and any circumstance in which existing instances of one class greatly outweigh instances of other classes. The limited data in the minority class could be due to an extraordinarily costly or impossible-to-uniform data collection method, naturally uncommon, skewed, limited resources, errors, and so on. But for the data analyst, uncommon cases are usually the most valuable and intriguing.

The subject has attracted a lot of interest, leading to the suggestion of numerous works with a focus on unbalanced learning. These methods have enhanced classifier performance to some extent. External and internal strategies are the two main methods employed to address the problem of data imbalances. Internal techniques modify the current algorithms to lessen their sensitivity to class discrepancies, whereas external ones preprocess the training data to balance them. The most popular external approaches are sampling, bagging, and boosting, while internal strategies include algorithm-level adjustments and cost-sensitive solutions. Few methods have been proposed by researchers as a component of their internal methods.

This paper presents a detailed analysis of skew dataset effects on supervised learning approaches. To analyze and compare the performance of data balancing techniques, six popular such techniques are applied to eight benchmark imbalanced datasets from the KEEL and Kaggle repositories.

The remaining sections of this paper are organized as follows. The review of various methods to deal with imbalance issues is discussed in the following section. The experimental setup and the use of matrices to evaluate the efficiency of the classification method are explained in Sect. 16.3. Experiment results are recorded in Sect. 16.4. Section 16.5 discusses the experimental study remarks, and Sect. 16.6 is devoted to concluding remarks.

16.2 Literature Survey

Imbalance happens in real-world data because there are frequently fewer samples of a class that is more crucial to identify. The majority class typically outperforms the minority class when a classifier is developed utilizing skew datasets. Over the years, a lot of academics and researchers have effectively enhanced the unbalanced dataset. As illustrated in Fig. 16.2, the proposed strategies can be divided into two categories: external and internal.

External Approaches

The external approach mainly preprocesses the training data to balance the imbalance in the dataset. These techniques are known as external because they do not modify the internal working of classification algorithms. As a result, this technique has the benefit of being independent of the underlying classification algorithm [3]. Before applying the classification method, it changes the data distribution and balances it. The most popular external method is the data-level approach.

Data level Approaches: It is also known as a sampling method which is of three types: over-sampling, under-sampling, and hybrid.

The over-sampling method produces a superset of the initial dataset by repeating certain instances or making new examples from existing ones from the minority class. According to some authors, this method produces exact clones of current instances raising the possibility of overfitting. A number of over-sampling methods have been developed including random over-sampling, SMOTE [4], MSMOTE [5], adaptive synthetic sampling (ADASYN) [6], SNOCC [7], etc.

The under-sampling method creates a subset of the original dataset by removing the samples (typically samples from the majority class). It is a non-heuristic method for balancing the distribution of classes by arbitrarily removing samples from the dominant class. The major drawback is that it can remove information that could be useful for classification. Such under-sampling methods are evolutionary under-sampling [2], cluster-based sampling (SBC) [8], ACOSampling [9], one-sided

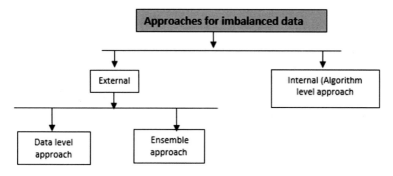

Fig. 16.2 Data balancing methods

selection [10], granular SVM-repetitive under-sampling (GSVM-RU) [11], and many more have been developed in past years.

The last category of data sampling method is hybrid which combines both, over- and under-sampling methods like selective preprocessing of imbalanced data (SPIDER) [12].

Ensemble Approaches: The ensemble approach's main objective is to attempt to improve the effectiveness of individual classifiers by creating various classifiers and combining them to create a new model that enhances each classifier individually. In order to improve generalization ability, the primary idea is to create many classifiers from the dataset and then integrate their predictions when new samples are presented [2]. The skew issue can also be resolved using ensemble approaches. Boosting and bagging are the two ensemble learning techniques that are most frequently utilized [3].

In bootstrap aggregating (bagging), classifiers are trained using bootstrapped duplicates of the initial training set of data. A majority of the weighted vote is then used to infer the class when a new instance is provided to each classifier. In contrast, boosting trains each weak classifier progressively using the complete dataset, and after each step, it assigns greater weight (importance) to poorly classified samples in order to correctly classify them in the subsequent iteration [13, 14].

The ensemble model provides two options for addressing the issue of data imbalance: cost-sensitive ensemble and ensemble with preprocessing (hybrid). Cost-sensitive ensembles are the same as the cost-sensitive technique, but in this case, the ensemble (boosting) approach can be used to direct the cost adjustment. Numerous suggestions have been made to alter the boosting algorithm's weight update. These concepts typically vary in how they modify the weight update rule. Within this family, AdaCost [15], CSB1, CSB2 [16], RareBoost [17], AdaC1, AdaC2, and AdaC3 [18] are the most widely used approaches.

One more is hybrid, which integrates an ensemble strategy with data preprocessing to address data imbalance. As demonstrated by many studies on the EUSBoost [2], RUSBoost [19], SMOTEBoost [14], MSMOTEBoost [4], and DataBoost-IM [20] algorithms, boosting can be done with either over- or under-sampling. OverBagging [14], REA [21], UnderBagging [13], asBaggingFSS [22], SMOTEBagging [23], IEFS [24], Regularized Ensemble Framework [25], and UnderOverBagging [26] are the most well-known of these algorithms.

Internal Approaches

This method involves changes in the learning process which makes it more sensitive to minority classes to naturally consider the class imbalance. To reduce the bias against the majority class, techniques are frequently adjusted to add a class penalty [27], class weight [28], or the decision threshold. Therefore, numerous algorithms have been put up by researchers, including post-boosting (PBI) [29], GM-based post-boosting (PBG) [30], IVTURSR [31], FURIACS [32], weighted regularized least squares (WRLSC) [33], deep neural network cost-sensitive (CoSen) [34], Can-CSC-GBE [35], One-class [36], weighted SVM [37], weighted Lagrangian twin SVM

(WLTSVM)[28], deep neural networks with imbalanced dataset [38], asymmetric kernel scaling (AKS) [39], etc.

A variety of techniques have been developed for the classification of imbalanced data. But most of the studies examined deal with issues of binary class inequality. A few studies are based on multi-class imbalanced learning, which converges to the issues of both multi-class imbalances with big data learning. No such efficient algorithm-level technique has been suggested for handling the multi-class massive data imbalance problem. In the future, there is a scope to extend this work. This paper is comparing different efficient existing techniques to analyze the performance. After the simulation, few conclusions are made.

16.3 Experimental Framework

The performance of various imbalance-handling techniques is analyzed on various datasets to understand the behavior. The framework for the experiments is outlined in this section. In the first subsection, real-world datasets are explained that were used in the experiments. The studied imbalance approaches, classification techniques, and other experimental settings are then briefly described in the following subsection. Finally, in the last subsection, all the evaluation parameters to compare the performance of algorithms are explained.

16.3.1 Datasets

The performance of various imbalance-handling techniques is analyzed using eight binary class datasets of varying sizes and degrees of imbalance from the KEEL and Kaggle repositories. Table 16.1 outlines the properties of these datasets, such as the number of attributes, occurrences, and imbalance ratio between majority and minority samples. The datasets range in size from 214 examples (glass1) to 43,400 examples (Stroke-Prediction). The datasets in Table 16.1 are sorted from most imbalanced to least imbalanced.

16.3.2 Experimental Setting

All the approaches are implemented using the Python framework. Fivefold stratified cross-validation is used in all the experiments due to the limited number of minority samples in all the datasets. To obtain generalized outcomes, the entire cross-validation [40] procedure is repeated 10 times, and the outcomes are the average of these 10 runs. A grid search process was used to identify the parameter values, and their performance is expected to be ideal. In decision trees and random forest algorithms,

Table 16.1 Summary of datasets

Dataset	Features (R/I/N)	Instances	Imbalance ratio
abalone19	8 (7/0/1)	4174	129.437500
KRI-vs-back	41 (26/0/15)	2225	100.14
Stroke-prediction	10 (2/4/4)	43,400	54.427842
Yeast1	8 (8/0/0)	1484	32.72
vowel0	13 (10/3/0)	988	9.97
new-thyroid1	5 (4/1/0)	215	6.10
vehicle1	18 (0/18/0)	846	2.89
glass1	9 (9/0/0)	214	1.10

In features R, I, and N represent real, integer, and nominal values, respectively

the Gini criterion [41] is used to reduce the computational power and to select the best feature. In SVM, kernel function RBF is used because when there is no prior data knowledge, it makes proper separation [42].

This study is performed over different classification algorithms including Decision tree (C4.5) [45] with criteria "Gini," K-nearest neighbor (KNN) where k = square root of N with uniform weight, Logistic regression with max-iteration 100, Naïve Bayes (GaussianNB), support vector machine with "RBF" kernel function and random forest with a number of estimators = 100 and "Gini" Criterion.

The first simulation runs over imbalanced datasets. After that several existing well-known and efficient data balancing methods are used to classify the imbalanced dataset. Those methods are synthetic minority over-sampling technique (SMOTE) [4], random majority under-sampling (RUS) [21], NearMiss (NM) [43], cost-sensitive learning (CS) [27], and BaggingEnsemble. In BaggingEnsemble method, SMOTE is used to balance the dataset before bootstrapping, and previously mentioned classification algorithms are used as a base estimator.

16.3.3 Performance Metrics

This paper analyzes the models developed for our experiment using four different measures of performance. Skew-insensitive measures are required to evaluate the performance of techniques when a dataset is imbalanced. Some of the effective metrics for an imbalanced dataset are: recall which evaluates the algorithm's ability to recognize positive samples, and precision which is the ratio of correctly predicted positive instances to total predicted positive instances [44]. With these two measures, it is difficult to compare the algorithms; as a result, F_1-score is the better choice of measure which is a harmonic mean of precision and recall [3]. When the dataset is skewed, accuracy is a good metric to use, though it is considered in this study because after applying the balancing technique overall accuracy is important. This experiment was performed 10 times, and the final value of accuracy, recall, precision, and F_1-score was calculated by taking an average of all individual experiments.

16.4 Experimental Results

This section shows the experimental results obtained by our experiment. Tables 16.2 and 16.3 contain all the classification results. Using accuracy as a metric to evaluate classifiers on highly imbalanced datasets is nearly useless as mentioned in the previous section. As a result, in Table 16.2 F_1-scores for a different dataset with different balancing techniques on different classification algorithms are recorded. Once the dataset is balanced by the balancing technique, we need to consider the overall accuracy of the model. As a result, in Table 16.3 accuracy is recorded. The experimental study has been discussed in the following section.

16.5 Result Analysis

Tables 16.2 and 16.3 show the performance comparison of LogisticRegression, Naive-Bayes, decision tree, SVM, KNeighbors, and RandomForest algorithms with and without taking into consideration the class skew problem. By analyzing this study, we can point out the following:

- It is observed from Fig. 16.3a, b that without balancing the dataset, F1-score is lowest and accuracy is highest except for the glass1 dataset results. It shows that the dataset is highly imbalanced and results are biased toward the majority class. F1-score for the abalone19, KRI-vs-back, and Stroke-Prediction is very poor and accuracy is very high, near 1. As a result, it is concluded that as the imbalance ratio increases performance degrades in terms of minority class sensitivity.
- As shown in Fig. 16.4 after applying the data balancing technique, overall accuracy is getting reduced because the model is now becoming sensitive to minority class and some of the majority instances are also classified as positive; hence, false positive rate (FPR) is getting increased. But at the same time, the true positive rate (TPR) is also getting increasing. The classification model is good if it gives good accuracy even after applying the data balancing technique.

External approaches

- From Fig. 16.4, it is concluded that random over-sampling gives comparatively higher accuracy and recall value because of the overfitting problem. But F_1-score is not good because of the low precision. Whereas in the case of random under-sampling F1-score as well as accuracy is also low due to the underfitting problem.
- Figure 16.4 also shows that the performance of synthetic minority over-sampling (SMOTE) gives a better result as compared to RUS and ROS but the distribution space of the original samples cannot be fully covered by SMOTE over-sampling. The NearMiss approach has not proven to be efficient in balancing the dataset because of the poor performance of all the datasets. So, among all the examined data-level approaches, SMOTE gave better results.

Table 16.2 Results on seven datasets in terms of f-measure and balancing approach

Datasets	Algorithm	Without balancing	ROS	RUS	SMOTE	NearMiss	Costsensitive	Ensemble
abalone19	LogisticRegression	0	0.05696	0.02706	0.05756	0.02538	0.0578	0.03082
	Naive-bayes	0.0374	0.02936	0.02552	0.02905	0.01249	NA	0.01553
	DecisionTree	0.062	0.76668	0.04203	0.77275	0.01805	0.07601	0.03005
	SVM	0	0.05886	0.02486	0.05932	0.01352	0.0559	0.02946
	KNeighbors	0	0.06698	0.02067	0.06898	0.03693	0.0691	0.03803
	RandomForest	0	0.8834	0.04505	0.88304	0.01843	0	0
yeast	LogisticRegression	0	0.45395	0.32572	0.54132	0.27373	0.64	0.58724
	Naive-bayes	0.6053	0.59281	0.61473	0.62403	0.30692	NA	0.61268
	DecisionTree	0.59001	0.8682	0.79226	0.90271	0.72768	0.97532	0.67575
	SVM	0	0.49644	0.32691	0.54919	0.22053	0.5567	0.6006
	KNeighbors	0.6038	0.68777	0.60864	0.71389	0.58309	0.7818	0.59904
	RandomForest	0.70606	0.90495	0.84185	0.91316	0.71608	0.97229	0.69739
new-thyroid1	LogisticRegression	0	0.86131	0.65428	0.9119	0.76161	0.9474	0.95266
	Naive-bayes	0.7524	0.84287	0.89781	0.88833	0.78429	NA	0.83723
	DecisionTree	0.59333	0.99333	0.93819	1	0.93623	1	0.99524
	SVM	0.21	0.78286	0.72357	0.79413	0.49907	0.85	0.82009
	KNeighbors	0.7933	0.94263	0.94369	0.94507	0.96358	1	0.8933
	RandomForest	0.69524	0.99333	0.94486	1	0.99333	0.99524	0.95716
vehicle1	LogisticRegression	0	0.49709	0.49807	0.48841	0.27998	0.4951	0.49388
	Naive-bayes	0	0.49709	0.49807	0.48841	0.27998	0.4951	0.49388
	DecisionTree	0.5761	0.58023	0.58048	0.58116	0.29558	NA	0.61232
	SVM	0	0.51685	0.51082	0.59015	0.25542	0.7904	0.50417
	KNeighbors	0.4298	0.58974	0.56165	0.59433	0.46303	0.6533	0.58957
	RandomForest	0.51123	0.87671	0.79662	0.89918	0.69033	0.91142	0.61032

(continued)

16 Detailed Performance Study of Data Balancing Techniques … 195

Table 16.2 (continued)

Datasets	Algorithm	Without balancing	ROS	RUS	SMOTE	NearMiss	Costsensitive	Ensemble
vowel0	LogisticRegression	0.2105	0.61219	0.57655	0.64604	0.34865	0.9875	0.7155
	Naive-bayes	0.6774	0.60013	0.57697	0.61997	0.19968	NA	0.66092
	DecisionTree	0.75284	0.95173	0.68382	0.97221	0.58912	0.9651	0.82452
	SVM	0.2312	0.69756	0.59462	0.7056	0.35115	0.7733	0.75667
	KNeighbors	0	0.60338	0.55864	0.60757	0.35048	0.7415	0.67678
	RandomForest	0.78301	0.99029	0.75599	0.99714	0.33686	0.94549	0.85922
glass1	LogisticRegression	1	1	1	1	1	1	1
	Naive-bayes	1	1	1	1	1	NA	1
	DecisionTree	1	1	1	1	0.94	1	1
	SVM	1	1	1	1	1	1	1
	KNeighbors	1	1	1	1	1	1	1
	RandomForest	1	0.996	0.90888	0.99765	1	1	1
KR1-vs-back	LogisticRegression	0.4837	0.99091	0.5834	0.99412	0.9	0.9991	0.9538
	Naive-bayes	0.5321	1	0.56651	1	0.9	NA	0.9546
	DecisionTree	0.2341	1	0.39656	1	0.34082	1	1
	SVM	0.1291	0.936	0.61609	0.94888	0.85758	0.9696	0.9345
	KNeighbors	0	0.71809	0.57964	0.76889	0.85556	0.8734	0.9635
	RandomForest	0.4474	1	0.52316	1	0.12267	0.9474	1
Stroke-Prediction	LogisticRegression	0	0.07955	0.04672	0.08415	0.04141	0.0699	0.07411
	Naive-bayes	0	0.03582	0.06099	0.03547	0.03792	NA	0.03655
	DecisionTree	0.09059	0.76704	0.09617	0.65985	0.04099	0.04913	0.01755
	SVM	0	0.09096	0.06025	0.09431	0.04536	0.087	0.0271
	KNeighbors	0	0.0838	0.06145	0.08665	0.04345	NA	0.03597
	RandomForest	0.0037	0.84909	0.10438	0.7994	0.03979	NA	0

Table 16.3 Results on seven datasets in terms of accuracy and balancing approach

Datasets	Algorithm	Without balancing	ROS	RUS	SMOTE	NearMiss	Costsensitive	Ensemble
abalone19	LogisticRegression	0.9942	0.78209	0.6589	0.80661	0.6271	0.8968	0.81741
	Naive-bayes	0.9253	0.54466	0.54463	0.55047	0.12961	NA	0.55855
	DecisionTree	0.9844	0.99617	0.66647	0.99804	0.13774	0.99585	0.98751
	SVM	0.9942	0.77338	0.53027	0.80739	0.43582	0.8816	0.84667
	KNeighbors	0.9942	0.80287	0.49953	0.89398	0.85394	0.9742	0.86401
	RandomForest	0.9942	0.99836	0.70527	0.99964	0.1452	0.9942	0.9894
yeast	LogisticRegression	0.6479	0.47592	0.4963	0.52962	0.43704	0.6197	0.58871
	Naive-bayes	0.5775	0.59073	0.5963	0.59815	0.47036	NA	0.58871
	DecisionTree	0.70282	0.91482	0.83706	0.92778	0.74259	0.94648	0.77042
	SVM	0.6479	0.49814	0.4963	0.52593	0.60925	0.6479	0.54084
	KNeighbors	0.7042	0.77593	0.72037	0.77594	0.66483	0.7761	0.67045
	RandomForest	0.79998	0.93888	0.87963	0.93517	0.72037	0.98167	0.77465
new-thyroid1	LogisticRegression	0.8592	0.95927	0.91111	0.96852	0.94999	0.9859	0.98731
	Naive-bayes	0.9859	0.93887	0.96666	0.96113	0.92961	NA	0.94509
	DecisionTree	1	0.99815	0.98149	1	0.97408	1	0.99859
	SVM	0.8873	0.94073	0.9389	0.96482	0.90185	0.9737	0.95772
	KNeighbors	1	0.9815	0.98149	0.97779	0.9889	0.98	0.96616
	RandomForest	0.99859	0.99815	0.98334	1	0.99815	0.99859	0.98731
vehicle1	LogisticRegression	0.7179	0.66273	0.65472	0.68481	0.33444	0.6986	0.62824
	Naive-bayes	0.7214	0.72547	0.7099	0.77038	0.42925	NA	0.73822
	DecisionTree	0.76751	0.93443	0.83868	0.94056	0.76651	0.9857	0.78465
	SVM	0.7179	0.65942	0.65613	0.68586	0.36084	0.6914	0.62787
	KNeighbors	0.7536	0.70707	0.67972	0.76226	0.60614	0.7771	0.67964
	RandomForest	0.77072	0.93867	0.86275	0.94716	0.78443	0.96892	0.78714

(continued)

16 Detailed Performance Study of Data Balancing Techniques ...

Table 16.3 (continued)

Datasets	Algorithm	Data Balancing approach							
		Without balancing	ROS	RUS	SMOTE	NearMiss	Costsensitive	Ensemble
vowel0	LogisticRegression	0.9083	0.89715	0.8757	0.91255	0.7575	0.9083	0.92048
	Naive-bayes	0.9388	0.89431	0.885	0.90526	0.4263	NA	0.91131
	DecisionTree	0.95657	0.99233	0.91375	0.99556	0.86155	0.99415	0.9682
	SVM	0.945	0.92348	0.88219	0.93038	0.72752	0.9991	0.93363
	KNeighbors	0.896	0.88461	0.86276	0.89189	0.75182	0.8874	0.90062
	RandomForest	0.96207	0.9984	0.93807	0.9996	0.60285	0.99688	0.97369
glass1	LogisticRegression	1	1	1	1	1	1	1
	Naive-bayes	1	1	1	1	1	NA	1
	DecisionTree	1	1	1	1	0.98949	1	1
	SVM	1	1	1	1	1	1	1
	KNeighbors	1	1	1	0.9865	1	1	1
	RandomForest	1	0.99973	1	0.99973	0.9903	1	1
KR1-vs-back	LogisticRegression	0.9993	0.9921	1	0.9982	0.99172	0.9973	1
	Naive-bayes	0.8901	1	1	1	0.9811	NA	1
	DecisionTree	0.9835	1	0.95863	1	0.66441	1	1
	SVM	0.8765	0.99856	0.99898	0.9982	0.99865	0.9959	1
	KNeighbors	0.9994	0.99316	0.99944	0.9937	0.9921	0.9975	1
	RandomForest	0.9986	1	0.9838	0.9986	0.83605	1	1
Stroke-Prediction	LogisticRegression	0.9814	0.83014	0.86792	0.83389	0.76036	0.6468	0.65303
	Naive-bayes	0.9814	0.05429	0.59512	0.05536	0.45483	NA	0.0596
	DecisionTree	0.96189	0.99153	0.68678	0.98547	0.1797	0.96649	0.97681
	SVM	0.9814	0.73989	0.88777	0.73085	0.47272	0.7627	0.0271
	KNeighbors	0.9814	0.65886	0.51733	0.66003	0.35375	0.9814	0.03597
	RandomForest	0.98133	0.99514	0.70456	0.9931	0.15554	0.98132	0

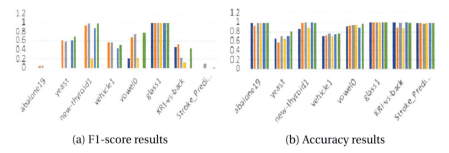

(a) F1-score results　　　　(b) Accuracy results

Fig. 16.3 Comparison of results without data balancing approaches

Fig. 16.4 Comparison of average F_1-score and accuracy over different dataset and classification techniques

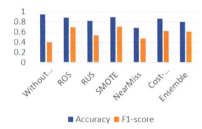

- The cost-sensitive approach where cost, the penalty associated with an incorrect prediction is different for minority and majority samples. This requires a modification that is specific to each algorithm and takes a long time to build. These algorithms are giving the best results compared to data-level approaches. We can conclude that algorithm-level changes for balancing the dataset are more efficient.
- Based on the results displayed in Fig. 16.5, it is stated that the cost-sensitive approach is proved better in all the datasets except the datasets having a large number of samples. The overall performance of these datasets is also low, and the time taken is high. So, we can say that there is a need for extension in data balancing approaches for big data.
- From the exceptional results of the glass1 dataset in Fig. 16.5, we can state that if the dataset is not biased, then the F1-score is not affected. Glass1 dataset is having lowest imbalance ratio.
- The SVM algorithm is more affected by the imbalance compared to other algorithms due to the more inclined hyperplane toward the minority class. Even after applying SMOTE and making it more cost-sensitive, performance is not getting increased which needs more efficient algorithms level changes.

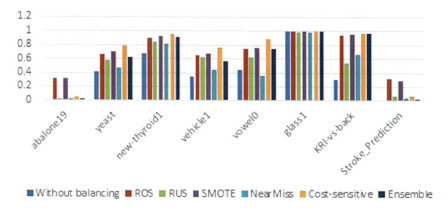

Fig. 16.5 Comparison of average F_1-score overall classification methods for all datasets with different balancing approaches

16.6 Conclusion

Many conventional machine learning techniques are less effective in predicting minority classes when the data distribution is skewed. This detailed study has provided an illustration of the problem of an unbalanced dataset using binary classification and multi-class classification. Additionally, this study provided several solutions for addressing the issue, including machine learning tools and evaluation matrices. We conducted an empirical analysis of the algorithms and explored all the significant research works with the performance comparison. It goes into greater detail about how class distribution affects learning.

Despite excellent work in this field, there are still numerous inadequacies in current approaches and problems that need to be handled; therefore, there is still an opportunity to improve on the prior work for improved outcomes, especially in algorithm-level approaches. Although approaches for binary skewed class classification are now quite developed, multi-class imbalance learning remains an unresolved problem. In the future, new algorithm-level approaches will be developed which can be further extended for the multi-class classification involving big data issues with imbalance issues.

References

1. Kaur, H., Pannu, H.S., Malhi A.K.: A systematic review on imbalanced data challenges in machine learning. ACM Comput. Surv. **52**(4) (2019)
2. Galar, M., Fern'andez, A., Barrenechea, E., Herrera, F.: EUSBoost: enhancing ensembles for highly imbalanced data-sets by evolutionary undersampling. Pattern Recogn. **46**(12), 3460–3471 (2013)

3. Patel, V., Bhavsar, H.: An empirical study of multi-class imbalance learning algorithms. In: ICT Systms and Sustainability, pp. 161–174. Springer, Singapore (2023)
4. Chawla, N.V., et al.: SMOTE: synthetic minority over-sampling technique. J. Artif. Intell. Res. **16**, 321–357 (2002)
5. S. Hu, Y. Liang, L. Ma, and Y. He, "MSMOTE: Improving classification performance when training data is imbalanced," in Proc. 2nd Int. Workshop Comput. Sci. Eng., 2009, vol. 2, pp. 13-17
6. He, H., Bai, Y., Garcia, E.A, Li, S.: ADASYN: adaptive synthetic sampling approach for imbalanced learning. In: Proceedings of the IEEE International Joint Conference on Neural Networks (IJCNN'08) (IEEE World Congress on Computational Intelligence), pp. 1322–1328. IEEE (2008)
7. Zheng, Z., Cai, Y., Li, Y.: Oversampling method for imbalanced classification. Comput. Info. **34**(5), 1017–1037 (2016)
8. Yen, S.-J., Lee, Y.-S.: Cluster-based under-sampling approaches for imbalanced data distributions. Expert Syst. Appl. **36**(3), 5718–5727 (2009)
9. Yu, H., Ni, J., Zhao, J.: ACOSampling: an ant colony optimization-based undersampling method for classifying imbalanced DNA microarray data. Neurocomputing **101**(2013), 309–318 (2013)
10. Miroslav Kubat, Stan Matwin, et al. Addressing the curse of imbalanced training sets: one-sided selection. In: Proceedings of the International Conference on Machine Learning (ICML'97), vol. 97. pp. 179–186 (1997)
11. Tang, Y., Zhang, Y.-Q., Chawla, N.V., Krasser, S.: SVMs modeling for highly imbalanced classification. IEEE Trans. Syst. Man Cybernet. Part B Cybernet. **39**(1), 281–288 (2009)
12. Stefanowski, J., Wilk, S.: Improving rule based classifiers induced by modlem by selective pre-processing of imbalanced data. In: Proceedings of the RSKD Workshop at ECML/PKDD, Warsaw, pp. 54–65. Citeseer (2007)
13. Sun, B., Chen, H., Wang, J., Xie, H.: Evolutionary under-sampling based bagging ensemble method for imbalanced data classification. Front. Comput. Sci. (2017). https://doi.org/10.1007/s11704-016-5306-z
14. Chawla, N.V., et al.: SMOTEBoost: Improving prediction of the minority class in boosting. In: European Conference on Principles of Data Mining and Knowledge Discovery. Springer, Berlin, Heidelberg (2003)
15. W. Fan, S. J. Stolfo, J. Zhang, and P. K. Chan, "Adacost: Misclassification cost-sensitive boosting," presented at the 6th Int. Conf. Mach. Learning, pp. 97-105, San Francisco, CA, 1999
16. K.M. Ting, "A comparative study of cost-sensitive boosting algorithms," in Proc. 17th Int. Conf. Mach. Learning, Stanford, CA, 2000, pp. 983-990
17. Joshi, M., Kumar, V., Agarwal, Evaluating boosting algorithms to classify rare classes: comparison and improvements. In: Proceedings of IEEE international conference on data mining, 2001, pp 257–264
18. Sun, Y., Kamel, M.S., Wong, A.K., Wang, Y.: Cost-sensitive boosting for classification of imbalanced data. Pattern Recog. **40**(12), 3358–3378 (2007)
19. Seiffert, C., et al. RUSBoost: a hybrid approach to alleviating class imbalance. IEEE Trans. Syst. Man Cybern.-Part A: Syst. Humans **40**(1), 185–197 (2009)
20. Guo, H., Viktor, H.L.: Learning from imbalanced data sets with boosting and data generation: the databoost-IM approach. SIGKDD Expl. Newsl. **6**, 30–39 (2004)
21. Qian, Y., Yanchun, L., Li, M., Feng, G., Shi, X.: A resampling ensemble algorithm for classification of imbalance problems. Neurocomputing **143**(2014), 57–67 (2014)
22. Yu, H., Ni, J.: An improved ensemble learning method for classifying high-dimensional and imbalanced biomedicine data. IEEE/ACM transactions on computational biology and bioinformatics **11**(4), 657–666 (2014)
23. Ahmed, S., et al.: Hybrid methods for class imbalance learning employing bagging with sampling techniques. In: 2017 2nd International Conference on Computational Systems and Information Technology for Sustainable Solution (CSITSS). IEEE (2017)

24. Yang, J., Zhou, J., Zhu, Z., Ma, X., Ji, Z.: Iterative ensemble feature selection for multiclass classification of imbalanced microarray data. J. Biol. Res. Thessaloniki **23**(1), 13 (2016)
25. Yuan, X., Xie, L., Abouelenien, M.: A Regularized ensemble framework of deep learning for cancer detection from multi-class, imbalanced training data. Pattern Recogn. (2017), S0031320317305034. https://doi.org/10.1016/j.patcog.2017.12.017
26. Wang, S., Yao, X.: Diversity analysis on imbalanced data sets by using ensemble models. In: IEEE Symposium on Computational Intelligence and Data Mining, pp. 324–331 (2009)
27. Hengyu, Z.: A new cost-sensitive SVM algorithm for imbalanced dataset. In: 2021 IEEE international conference on consumer electronics and computer engineering (ICCECE). IEEE (2021)
28. Shao, Y.-H., Chen, W.-J., Zhang, J.-J., Wang, Z., Deng, N.Y.: An efficient weighted Lagrangian twin support vector machine for imbalanced data classification. Pattern Recogn. **47**(9), 3158–3167 (2014)
29. Jie, D., Vong, C.-M., Pun, C.-M., Wong, P.-K., Ip, W.-F.: Post-boosting of classification boundary for imbalanced data using geometric mean. Neural Netw. **96**, 101–114 (2017)
30. Vong, C.-M., Du, J., Wong, C.-M., Cao, J.-W.: Postboosting using extended G-Mean for online sequential multiclass imbalance learning. IEEE Trans. Neural Netw. Learn, Syst (2018)
31. Sanz, J.A., Bernardo, D., Herrera, F., Bustince, H., Hagras, H.: A compact evolutionary interval-valued fuzzy rule-based classification system for the modeling and prediction of real-world financial applications with imbalanced data. IEEE Trans. Fuzzy Syst. **23**(4), 973–990 (2015)
32. Palacios, A., Trawiński, K., Cordón, O., Sánchez, L.: Cost-sensitive learning of fuzzy rules for imbalanced classification problems using FURIA. Int. J. Uncertain. Fuzz. Knowl.-Based Syst. **22**(05), 643–675 (2014)
33. Vo, N.K., Won, Y.: Classification of unbalanced medical data with weighted regularized least squares. In: Proceedings of the conference on frontiers in the convergence of bioscience and information technologies (FBIT'07), pp. 347–352. IEEE (2007)
34. Khan, S.H., Hayat, M., Bennamoun, M., Sohel, F.A., Togneri, R.: Cost-sensitive learning of deep feature representations from imbalanced data. IEEE Trans. Neural Netw. Learn, Syst (2017)
35. Ali, S., Majid, A., Javed, S.G., Sattar, M.: Can-CSC-GBE: Developing cost-sensitive classifier with gentleboost ensemble for breast cancer classification using protein amino acids and imbalanced data. Comput. Biol. Med. **73**(2016), 38–46 (2016)
36. Raskutti, B., Kowalczyk, A.: Extreme re-balancing for SVMs: a case study. ACM SIGKDD Explor. Newslett. **6**(1), 60–69 (2004)
37. Huang, Y.-M., Du, S.-X.: Weighted support vector machine for classification with uneven training class sizes. In: Proceedings of 2005 International Conference on Machine Learning and Cybernetics, vol. 7, pp. 4365–4369. IEEE (2005)
38. Wang, S., Liu, W., Wu, J., Cao, L., Meng, Q., Kennedy, P.J.: Training deep neural networks on imbalanced data sets. Int Joint Conf Neural Networks (IJCNN) **2016**, 4368–4374 (2016). https://doi.org/10.1109/IJCNN.2016.7727770
39. Chawla, N.V., Bowyer, K.W., Hall, L.O., Kegelmeyer, W.P.: Smote: synthetic minority over-sampling technique. Journal of artificial intelligence research **16**, 321–357 (2002)
40. Ramezan, A.C., Warner, T.A., Maxwell, A.E.: Evaluation of sampling and cross-validation tuning strategies for regional-scale machine learning classification. Remote Sens. **11**(2), 185 (2019)
41. Boulesteix, A.-L., et al.: Random forest Gini importance favours SNPs with large minor allele frequency: impact, sources and recommendations. Briefings Bioinform. **13**(3), 292–304 (2012)
42. Bhavsar, H., Panchal, M.H.: A review on support vector machine for data classification. Int. J. Adv. Res. Comput. Eng. Technol. (IJARCET) **1**(10), 185–189 (2012)
43. Mqadi, N.M., Naicker, N., Adeliyi, T.: Solving misclassification of the credit card imbalance problem using near miss. Math. Probl. Eng. **2021**, 1–16 (2021)
44. Ferri, C., Hernández-Orallo, J., Modroiu, R.: An experimental comparison of performance measures for classification. Pattern Recogn. Lett. **30**(1), 27–38 (2009)
45. Hssina, B., et al. A comparative study of decision tree ID3 and C4. 5. Int. J. Adv. Comput. Sci. Appl. **4**(2), 13–19 (2014)

Chapter 17
Compact Dual-Band Printed Folded Dipole for WLAN Applications

Abhishek Javali

Abstract Miniaturization and multi-band operability have been the key desirable characteristic features of smart antennas designed in wireless communication. Printed folded dipole antennas find their applications in radio frequency identification tags. In this paper, a printed planar folded dipole antenna operating at 2.4 GHz is investigated to meet the requirements of miniaturization and multi-band functionality at 2.6 and 5.6 GHz. Altair Feko is used for the simulation of the antenna. The parametric study is performed to study the change in impedance bandwidth for varying dipole trace width and substrate thickness. The proposed antenna is found to be potential candidate for WLAN applications with dual-band functionality and omnidirectional radiation pattern.

Keywords Printed folded dipole · Microstrip patch antenna · Gain · Reflection coefficient · VSWR · Return loss

17.1 Introduction

Dramatic changes taking place in the wireless technology have witnessed novel design methods for low profile antenna configurations. The possibility of integrating multiple communication standards and protocols in the mobile system for performance enhancement has been the need of the hour. It is obvious that miniature antenna design with enhanced performance in terms of impedance bandwidth and realized gain is pivotal. In addition to the portable antenna design, multi-band operability is desirable to meet the increasing demands [1, 2].

Microstrip-based antennas have revolutionized the antenna world with their many advantages such as cost-effectiveness, low profile design, less bulky geometry, robustness and ease of mounting to mention a few. There have been many tech-

---

A. Javali (✉)
CMR Institute of Technology, Bengaluru, India
e-mail: abhishek.j@cmrit.ac.in

niques proposed in the literature to enable multi-band operation of the microstrip antennas [3–5].

Antennas for wireless local area network (WLAN) have witnessed paradigm shifts in the last couple of decades. With the evolution of smart antennas and mobile networks with high data transfer, smart antennas for WLAN applications have received paramount attention. In order to meet the requirements of the IEEE 802.11$a/b/g/n$ standards, it is highly desirable to have multi-band operability at frequency bands in particular at 2.4 and 5 GHz. Hence, the combination of miniaturization and multi-band functionality has become hot research areas in the last couple of decades [6, 7].

A folded dipole antenna configuration having a closed loop antenna set up was simulated [8]. Two equally potential channels were established using 2 × 2 Multiple Input Multiple Output (MIMO) using similar folded dipole antennas operating at 5.6 GHz for WLAN applications [9]. A simple folded antenna with specific impedance was used in radio frequency identification (RFID) tag antenna [10]. A folded dipole antenna configuration operating at 2.45 GHz was simulated for RFID tag antenna prototype and verified with measurements [11]. A compact folded dipole antenna was proposed at ultra-high frequency for RFID tag antenna with C-shaped resonators for impedance matching [12]. A unique folded dipole antenna with dual-band operability at 2.4 GHz was designed for WLAN applications [13]. A low profile ultra-high-frequency RFID tag antenna with enhanced inductive input reactance was presented with good wake-up sensitivity at a frequency band 908–914 MHz [14].

In this paper, a unique miniature dual-band printed folded dipole antenna for WLAN applications is proposed and investigated. First, the printed folded dipole is studied at a center frequency of 2.4 GHz to get an impedance bandwidth of 329.335 MHz. Later, the same antenna is studied for its multi-band operability at 2.6 GHz and 5.6 GHz. The effect of varying the dipole trace width and substrate height on the impedance bandwidth as part of the parametric analysis is addressed. The antenna is simulated with cost-effective FR4 substrate.

The remainder of the paper is organized as follows. The antenna design aspects are enumerated in Sect. 17.2. The simulation results and discussions are presented in Sect. 17.3. The work is concluded in Sect. 17.4.

17.2 Antenna Design

The printed folded dipole antenna is designed at an operating frequency of 2.4 GHz. The antenna geometrical configuration is shown in Fig. 17.1. The constructional details of the antenna design are listed as follows.

Airbox length = 76.019 mm
Airbox thickness = 9.93 mm
Airbox width = 24.7 mm
Effective wavelength = $\lambda_{\text{eff}} = \frac{\lambda}{\sqrt{\epsilon_{\text{eff}}}} = 76.019$ mm

17 Compact Dual-Band Printed Folded Dipole for WLAN Applications

Fig. 17.1 Printed folded dipole antenna

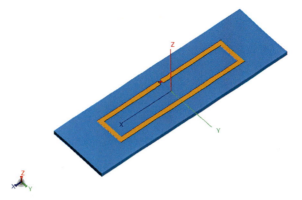

Effective permittivity = $\frac{1+\epsilon_r}{2} = 2.7$
Gap size at the port = $\frac{\lambda_{eff}}{90} = 0.844$ mm
λ = Wavelength at the center frequency = 124.91 mm
Length of the folded element = $\frac{\lambda}{1.8} = 42.23$ mm
Mesh = $\frac{width}{3} = 0.46$ mm
FR4 (substrate) permittivity = $\epsilon_r = 4.4$
Distance between the outer edges of the folded arms = $\frac{\lambda_{eff}}{8} = 9.5$ mm
Substrate length = 1.5 × length = 63.34 mm
Substrate thickness = $\frac{\lambda_{eff}}{100} = 1.249$ mm
Substrate width = 2 × folded dipole spacing = 19 mm
Substrate tan delta = 0.01 mm
Dipole trace width = $\frac{\lambda_{eff}}{55} = 1.382$ mm

17.3 Simulation Results and Discussions

Simulation is carried out at a center frequency of 2.4 GHz. The three-dimensional radiation pattern of the antenna is shown in Fig. 17.2. The 3D pattern shows a omnidirectional radiation pattern.

The return loss curve for the printed folded dipole antenna is shown in Fig. 17.3. The plot shows that an impedance bandwidth of 329.335 MHz was found. The voltage standing wave ratio (VSWR) was plotted in Fig. 17.4. The plot shows a VSWR of 1.11 which indicates good impedance matching between the feed and the input impedance of the antenna.

As multi-band functionality is desirable, the printed folded antenna is considered next for the same. A wide range of 2–7 GHz is selected to test the impedance bandwidth of the antenna by observing the 10 dB reflection bandwidth. It is found that the test antenna has the dual-band functionality at 2.6 and 5.6 GHz as shown in Fig. 17.5. The test printed folded dipole antenna has exhibited good return loss of

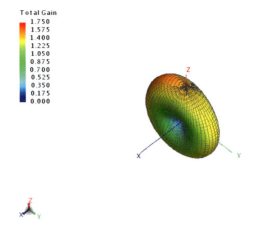

Fig. 17.2 3D radiation pattern of printed folded dipole antenna

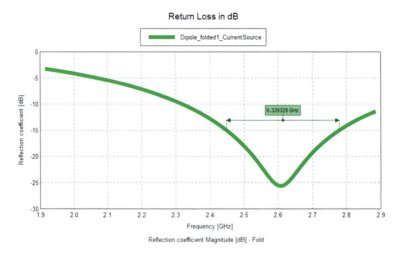

Fig. 17.3 Return loss in dB as a function of frequency showing the impedance bandwidth

−36.58 dB at 2.6 GHz for dipole trace width of 1.169 mm. For the other frequency band, it shows an optimum return loss of −10.84 dB at 5.6 GHz as shown in Fig. 17.5.

As part of the parametric analysis, the effect of varying dipole trace width and substrate thickness on the impedance bandwidth is studied using Altair Feko. Three different dipole trace widths are considered for the comparison. Dipole trace width of $\frac{\lambda}{55} = 1.382$ mm, $\frac{\lambda}{65} = 1.169$ mm and $\frac{\lambda}{75} = 1.013$ mm is taken into account, the corresponding impedance bandwidths are measured, and it is shown in Fig. 17.5. It can be seen that as dipole trace width is decreased, the impedance bandwidth of the proposed antenna increases.

17 Compact Dual-Band Printed Folded Dipole for WLAN Applications

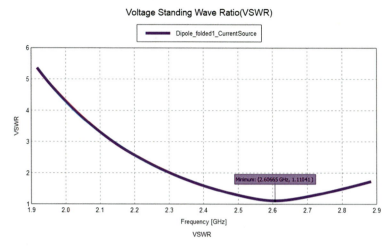

Fig. 17.4 Voltage standing wave ratio

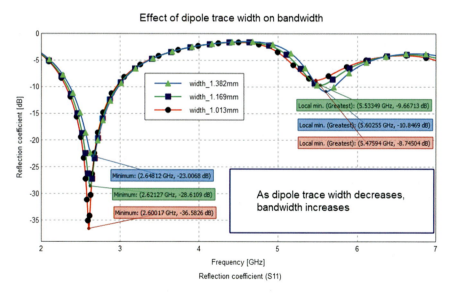

Fig. 17.5 Effect of dipole trace width on bandwidth

Similarly, three different substrate thicknesses are considered for the comparison. Substrate thickness of $\frac{\lambda}{100} = 1.249$ mm, $\frac{\lambda}{75} = 1.665$ mm and $\frac{\lambda}{50} = 2.498$ mm are taken into account and the corresponding impedance bandwidths is measured, and it is shown in Fig. 17.6. It can be seen that as substrate thickness is increased, the impedance bandwidth of the proposed antenna decreases. The demonstrated printed folded dipole antenna is suitable for dual-band operation for WLAN applications at 2.6 and 5.6 GHz.

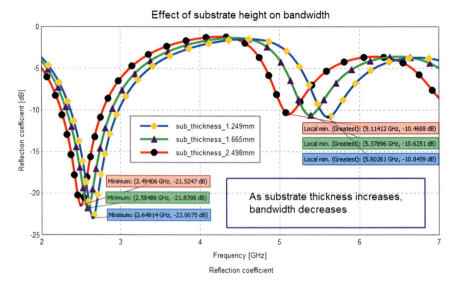

Fig. 17.6 Effect of substrate thickness on bandwidth

17.4 Conclusion

A miniature dual-band printed planar folded dipole antenna was designed and investigated. The proposed antenna is suitable for operation at 2.6 GHz and 5.6 GHz for WLAN applications. The proposed antenna is hence suitable for dual-band functionality. It was found from the simulation results that the as dipole trace width is decreased, impedance bandwidth increases, and as substrate thickness is increased, the impedance bandwidth of the proposed antenna decreases. This investigation helps in finding the suitability of the printed folded dipole antenna for its selectivity for WLAN applications.

References

1. Zuazola, I.G., Batchelor, J.C.: Compact multiband PIFA type antenna. Electron. Lett. **45**(15), 768–769 (2009)
2. Zhou, S., Guo, J., Huang, Y., Liu, Q.: Broadband dual frequency sleeve monopole antenna for DTV/GSM applications. Electron. Lett. **45**(15), 766–768 (2009)
3. Javali, A., et al.: A comparitive study on the performance of circular patch antenna using low cost substrate for S-band applications. In: 2018 Second International Conference on Advances in Electronics, Computers and Communications (ICAECC). IEEE (2018)
4. Wu, P., Kuai, Z., Zhu, X.: Multiband antennas comprising multiple frame-printed dipoles. IEEE Trans. Antennas Propag. **57**(10), 3313–3316 (2009)
5. Zhang, J., Zhang, X.M., Liu, J.S., Wu, Q.F., Ying, T., Jin, H.: Dual-band bidirectional high gain antenna for WLAN 2.4/5.8 GHz applications. Electron. Lett. **45**(1), 6–7 (2008)

6. Jolani, F., Dadgarpour, A., Hassani, H.R.: Compact M-slot folded patch antenna for WLAN. Progr. Electromag. Res. Letters **3**, 35–42 (2008)
7. Ren, W.: Compact dual-band slot antenna for 2.4/5GHz WLAN applications. Progr. Electromag. Res. **8**, 319–327 (2008)
8. Chen, S.L., Lin, K.H.: A folded dipole with a closed loop antenna for RFID applications. In: 2007 IEEE Antennas and Propagation Society International Symposium, pp. 2281–2284. IEEE (2007)
9. Maricar, M.I., Gradoni, G., Greedy, S., Ivrlac, M.T., Nossek, J.A., Phang, S., Creagh, S.C., Tanner, G.,Thomas, D.W.: Analysis of a near field MIMO wireless channel using 5.6 GHz dipole antennas. In: 2016 ESA Workshop on Aerospace EMC (Aerospace EMC), pp. 1–3. IEEE (2016)
10. Yang, B., Feng, Q.: A folded dipole antenna for RFID tag. In: 2008 International Conference on Microwave and Millimeter Wave Technology, vol. 3, pp. 1047–1049. IEEE (2008)
11. Qing, X., Yang, N.: A folded dipole antenna for RFID. In: IEEE Antennas and Propagation Society Symposium, vol. 1, pp. 97–100. IEEE (2004)
12. Erman, F., Hanafi, E., Lim, E.H., Mahyiddin, Wan Mohd, W.A., Harun, S.W., Umair, H., Soboh, R., Makmud, M.Z.H.: Miniature compact folded dipole for metal mountable UHF RFID tag antenna. Electronics **8**(6), 713 (2019)
13. Guo, Y.Y., Zhang, X.M., Ning, G.L., Zhao, D., Dai, X.W., Wu, Q.: Miniaturized modified dipoles antenna for WLAN applications. Progr. Electromag. Res. Lett. **24**, 139–147 (2011)
14. Choi, Y., Kim, U., Kim, J., Choi, J.: Design of modified folded dipole antenna for UHF RFID tag. Electron. Lett. **45**(8), 387–389 (2009)

Chapter 18
Performance Analysis of Various Feature Extraction Methods for Classification of Pox Virus Images

K. P. Haripriya and H. Hannah Inbarani

Abstract After the COVID-19 pandemic, people began to fear that the monkeypox virus will be the next epidemic. The World Health Organization (WHO) has reported that monkeypox outbreaks have taken place in different regions of Central and West Africa throughout the years, with the latest outbreak being reported in Nigeria in May 2021. Fever, swollen lymph nodes, dry cough, and red rashes all manifest as signs of the monkeypox virus. Most of the symptoms of Measles and chickenpox are comparable. These disorders are given only methodical treatment by the doctor. In Image processing, feature extraction techniques are used to transform raw pixel values of an image into a set of features that can be used for further analysis, such as object recognition, image classification, and image retrieval. Several feature extraction techniques are employed to determine the disease from the images in order to determine which technique works best for this collection of monkeypox skin images (MSID). Wavelets fused with gray-level co-occurrence matrix (GLCM), Haralick features, and local binary pattern are the various feature extraction techniques (LBP) applied in this work. For the classification of various pox virus diseases such as measles, chicken pox, and monkeypox, various classification algorithms such as Random Forest Classification (RF), Naive Bayes (NB), K-Nearest Neighbor algorithms (KNN), Support Vector Machine (SVM), Ada Boosting (AB), and Gradient Boosting (GB) are used in this work. In this paper, four evaluation metrics are used to determine the best feature extraction method for the monkeypox, chickenpox, and measles datasets. Wavelets fused with GLCM produce the highest accuracy (84.41% for gradient boosting and 83.87% for random forest) when extracting features from Monkeypox Skin Image Datasets (MSID).

Keywords Classification algorithm · RF · NB · KNN · SVM · AB · GB · Pox virus images · Feature extraction methods: wavelets with GLCM · LBP · Haralick features

---

K. P. Haripriya · H. Hannah Inbarani (✉)
Department of Computer Science, Periyar University, Salem, India
e-mail: hhinba@periyaruniversity.ac.in

© The Author(s), under exclusive license to Springer Nature Singapore Pte Ltd. 2024
P. K. Jha et al. (eds.), *Proceedings of Congress on Control, Robotics, and Mechatronics*, Smart Innovation, Systems and Technologies 364,
https://doi.org/10.1007/978-981-99-5180-2_18

18.1 Introduction

The monkeypox virus is a disease that is caused by a member of the orthopoxvirus [1]. Monkeypox started in West and Central Africa in 1970 [2]. Later, it started to spread around the UK, Spain, and other states. Recently, in India, monkeypox diseases were discovered in some people. Generally, monkeypox [2] takes 6–13 days to breed, but in the real world, it takes around 5–21 days to show its symptoms. The symptoms of monkeypox disease are fever, swollen lymph nodes, rectal pain, sore throat, oral lesions, and swollen tonsils. Hence, it is necessary to avoid close contact with a person who is already infected with the monkeypox virus. Later, the European Government named the monkeypox virus as the MPOX virus [3].

Chickenpox [4] is the most common disease; the infection occurs in people of different age groups. Chickenpox symptoms are very similar to monkeypox symptoms. For children, chickenpox can be easily spread from one to another. Measles [5] is a disease caused by a virus that is easily transmitted to children. The measles will look like red rashes around the body where it gets affected. The virus exposes its symptoms after 10–14 days. The common symptoms of measles are fever, sore throat, running nose, and dry cough. This disease can be easily spread from one person to another. Early diagnosis of the disease will help people recover fast, and for the doctor, it will be very useful to identify which disease comes from the human being.

The Monkeypox Skin Image Datasets (MSID) are taken from Kaggle [6], and it contain images for various diseases like chickenpox, measles, monkeypox, and normal skin images. The dataset contains 770 images, and from those images, rashes and infected parts are obtained from every part of the human body. To identify the rash and infected parts of the images first, the color images are converted into grayscale images and feature extraction is performed from grayscale images. For feature extraction, different methods like wavelets, Gray-Level Co-occurrence Matrix (GLCM), Haralick, and Local Binary Pattern (LBP) are used. Later, machine learning classification algorithms are applied to calculate the accuracy, precision, recall, and the F1-score to measure which feature extraction methods are suitable for these Monkeypox Skin Images Datasets (MSIDs). The highest accuracy and the F1-score are used to predict which method is suitable for this dataset. The sample images of the different diseases are described in Fig. 18.1a the chickenpox image, (b) the Measles image, (c) the monkeypox image, and (d) the normal skin images.

In Fig. 18.1, sample images of different diseases are presented. The first image is a chickenpox image, followed by measles, monkeypox, and the normal images. The images show that red rashes are common among all of the diseases. The remaining part of this paper is described as follows: In Sect. 18.2, the various feature extraction methods and classification algorithms used in this work are discussed. In this work, the most frequently used feature extraction methods are chosen. Section 18.3 is a detailed discussion of the datasets, feature extraction methods, and methodology for this paper. In Sect. 18.4, various methods are used to extract the features from the image, and various machine learning algorithms are used to determine which produce

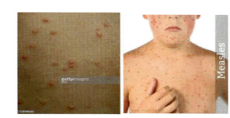 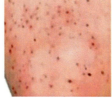 

Fig. 18.1 Sample images of different diseases

the highest accuracy. In Sect. 18.5, the conclusion and future enhancement of this paper are discussed.

18.2 Literature Survey

This section provides a survey of various methods of feature extraction for the images like Texture, Color, Morphological, Haralick, Wavelets, Ranklets, HOG, LBP, methods, etc. and different classification algorithms applied for many types of images as shown in Table 18.1.

The drawbacks and the issues faced by the above papers are described below: Aruraj et al. [7] say that the proposed methodology is only implemented on small datasets, and to improve the classification accuracy, different texture and color features have to be extracted. Hussian et al. [8] applied the texture and morphological features for machine learning-based classification. In future, deep learning techniques are applied. Perumal et al. [9] needed more data assimilation to strengthen their results. For the proposed segmentation, Azevedo Tosta et al. [10] discovered false positives and negatives for neoplastic nuclei. Goyal et al. [11] the proposed model can be improved further in terms of adaptive model construction and severity analysis for additional datasets with multiple classes (more than 5) and deep models are reported. Sahin et al. [13] used comparably very small datasets, and AI approaches based on Dl models are the issues.

18.3 Research Methodology

In Fig. 18.2, the different diseases like chickenpox, measles, and monkeypox and the normal skin images are taken as input. The next stage is image acquisition; the preprocessing of images is done under this section. To remove the noise from the image and to find the affected region and edges of images, the Gaussian filter is applied. The next step is feature extraction, and three various methods are used to extract the features of the images. The three different feature extraction methods are: wavelets fused with GLCM, Haralick features, and local binary pattern (LBP). Each

Table 18.1 Related works for this study

Authors	Year	Dataset used	Feature extraction methods	Classification algorithms	Best algorithm and accuracy
Aruraj et al. [7]	2019	Plant village dataset	Texture and color, LBP	SVM, KNN	SVM—89.4, 90.9%
Hussain, et al. [8]	2020	COVID-19 and non-COVID and normal chest CXR images	Texture and Morphological features	XGB-Linear, XGB-Tree, CART, KNN, NB	XGB-linear, XGB-tree, CART, NB—100%
Perumal et al. [9]	2021	COVID-19, ChestX-ray, Pneumonia, and others	Haralick feature extraction	VGG-16, ResNet50, InceptionV3, VGG-16 using transfer learning	VGG-16 using transfer learning—93%
Azevedo Tosta et al. [10]	2021	AxioCam MR5 CCD color camera-coupled with a Zeiss Axioscope microscope	Wavelets, Ranklets, Statistical descriptors, Haralick features	SVM, KNN, DT, RF, NN, LR	KNN—97.30% (Wavelet) LR—99.10% (Ranklet) NN and KNN—97.51% (Wavelet)
Goyal, et al. [11]	2021	COVID-19 radiography Database (C19RD) collected from Kaggle and others	HOG, texture, intensity features, and geometric invariant features	ANN, SVM, KNN, ensemble classifier, RNN with LSTM	RNN with LSTM—9.31%
Tavakoli, et al. [12]	2021	Raabin-WBC15, LISC7, and BCCD16	Shape, and color feature extraction	ResNet50, ResNext50, MobileNetV2, MnasNet1, ShuffleNet-V2, proposed (SVM)	Proposed (SVM)—96.04, 94.65, 50.97%
Sahin, et al. [13]	2022	Monkeypox Skin lesion dataset (MSLD)	—	ResNet18, GoogleNet, EfficientNetb0, NasnetMobile, ShuffleNet, MobileNetv2	MobileNetv2, EfficientNetb0—91.11%

(continued)

18 Performance Analysis of Various Feature Extraction Methods ... 215

Table 18.1 (continued)

Authors	Year	Dataset used	Feature extraction methods	Classification algorithms	Best algorithm and accuracy
Sitaula, et al. [1]	2022	Monkeypox image dataset	–	VGG-16, VGG-19, ResNet50, ResNet-101, IncepResNetv2, MobileNetV2, InceptionV3, Xception, EfficientNet-B0, EfficientNet-B1, EfficientNet-B2, DenseNet-121, DenseNet-169, ensemble approach	Ensemble approach—87.13%
Haque, et al. [14]	2022	Monkeypox skin lesion dataset (MSLD)	–	VGG-19-CBAM-Dense, Xception CBAM-Dense, DenseNet-121-CBAM-Dense architecture, MobileNetV2-CBAM-Dense architecture, and lastly the EfficientNetB3-CBAM-Dense	XceptionCBAM-Dense, 83.89%

Abbreviation used in Table 18.1: *LBP* Local Binary Pattern, *2DGF* 2D Gabor Filtering, *CART* Classification and Regression Tree, *KNN* K-Nearest Neighbor, *NB* Naive Bayes, *ANN* Artificial Neural Network, *SVM* Support Vector Machine, *RNN* Recurrent Neural Network, *LSTM* Long Short-Term Memory, *CBAM* Convolutional Block Attention Module, *VGG* Visual Geometry Group

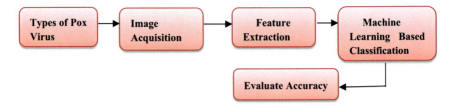

Fig. 18.2 Proposed model

Table 18.2 Detailed information about the datasets

Category	Chickenpox	Measles	Monkeypox	Normal	Total
#	107	91	279	293	770

and every method produces different features. After extracting the features, various machine learning classification algorithms are applied and accuracy, precision, recall, and the F1-score are used to measure the accuracy of classification. In the end, the result is compared to show which feature extraction method is suitable for this monkeypox skin image dataset.

18.4 Materials and Methods:

18.4.1 The Source of the Datasets

The benchmark datasets for the Monkeypox Skin Image Datasets (MISD) are gathered from Kaggle [6]. The MISD datasets contain four folders, and each folder contains different disease images like chickenpox, measles, and monkeypox, as well as normal images. Each image has a dimension of about 224 by 224. In Table 18.2, the entire information about the datasets is described.

18.4.2 Image Acquisition

Image acquisition is the first stage in digital image processing techniques. In this stage, the preprocessing of the image is done. In this work, the RGB (Red, Green, and Blue) images are converted into grayscale images. Later, Gaussian filters are applied to the images to reduce the noise. To predict the region of interest for the image, Gaussian filters are used.

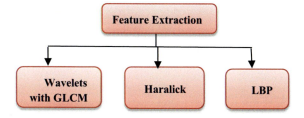

Fig. 18.3 Feature extraction methods

18.4.3 Feature Extraction Methods

In Fig. 18.3, the various methods of feature extraction are explained. The partitioning of images into regions of interest and the classification of those regions are called texture features. It measures the intensities of an image or an arrangement of colors. The texture features used in this work are Gray-Level Co-occurrence Matrix (GLCM) and Local Binary Pattern (LBP). The multi-scaled and other features are Haralick features and wavelets.

Wavelets with GLCM

A wavelets' transformation is a multi-scale and multi-resolution transformation to extract the features of high-resolution to low-resolution components.

In Fig. 18.4, the wavelets are used in four different directions. The LL talks about the approximation and tells about the low-frequency components of an image. The LH talks about the horizontal details, and it depicts the horizontal edge components. The HL talks about the "vertical details," which means that it detects all vertical edge information. The HH talks about "diagonal details," which means it detects all diagonal edge information.

The Gray-Level Co-occurrence Matrix (GLCM) is a texture feature [8] that covers the image spatial correlations. Generally, GLCM has N number of properties, but most of the research works used only these four properties as the main ones. The wavelet uses the Daubechies wavelet family. The wavelets use different angles, like horizontal, vertical, and diagonal, for the images. From these four angles, the GLCM

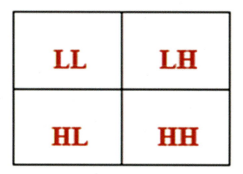

Fig. 18.4 Wavelets' angle samples

is applied for color images. So, the color images produce the three channels Red, Green, and Blue, and then with the help of gray-level properties, the four important features are extracted from the image. The four features are contrast, energy, dissimilarity, and homogeneity. So, 4 * 3 * 4 = 48 features are extracted from the image.

Haralick Features

The Haralick [9] features are part of GLCM. The Haralick features talk about the relationship between the pixel intensities that are adjacent to each other. The Haralick features produce 14 features that are extracted from GLCM. For each image, 14 features are extracted to form a new interrelationship. The GLCM is usually calculated at different angles like 0°, 90°, 180° and 270° of the images.. The Haralick features produce different features, namely, Homogeneity (or) Angular Second Moment (ASM), Contrast, correlation, Variance (or) Sum of Squares, Inverse Difference Moment (IDM), Sum Average, Sum Variance, Sum Entropy, Entropy, Difference Variance, Difference Entropy, First Information Measure, and Second Information.

Local Binary Pattern (LBP)

The texture features include the Local Binary Pattern (LBP) [7]. This technique provides a uniform and efficient texture pattern for the image, which describes the local texture patterns. The LBP [15] measures both the shape and texture features of the image. LBP offers information about the neighborhood pixels of the image, which provides the combination of binary code for the image pixels. From Fig. 18.5, [7] the image of pixels is given, and from that threshold value, the rest is calculated. For my datasets, 24 is my degree and 8 is the radius. The mathematical expression of the LBP code is given below [15].

$$\text{LBP}_{P,R}(x_c, y_c) = \sum_{p=0}^{p-1} S(g_p - g_c)^2.$$

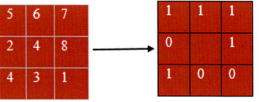

Fig. 18.5 Example of LBP operation [7]

The value is 11110010

The center value is taken as a threshold, and then the threshold value is compared to the neighborhood pixels. If the value is greater than or equal to the threshold, the value is assigned as 1. Otherwise, the value is assigned as 0. For example, if a 3 * 3 matrix is given at the center point, it acts as a threshold, and the remaining part acts as neighborhood values.

18.5 Experimental Analysis

Naïve Bayes [8] algorithms work based on the Bayes theorem, and they are suitable for problems with higher dimensionality and for some independent variables that have categorical or continuous values. This algorithm provides minimal computational time to construct the model. The probability of computational error is large in Naïve Bayes techniques. To overcome this problem, the probability of valuation error is reduced in Bayes' methods. These techniques need a large number of parameters for training the models. The KNN [8] algorithm is widely used in machine learning for classification purposes. KNN is also called a lazy learner algorithm because it does not have any defined function to perform. Instead, it trains the model by memorizing training data. The K components are used to calculate the nearest neighbor in the datasets. Random Forest [16] is a collection of many independent decision trees that train independently on random subsets of data. The final prediction will be made with the help of "voting". This means that each decision tree class will vote for the output class (the class with the most votes is chosen as the final prediction). SVM [17] considers itself to be one of the most well-known and practical techniques for dealing with data classification, learning, and prediction problems. The data point closest to the decision surface is the support vector. The simplest binary classification problem works with linearly separable training data. AdaBoost [18] is a method in which several weaker models are trained independently and their predictions are combined in some way to produce the overall prediction. Gradient boosting [19] begins with a single leaf with weights for all attributes. When predicting a continuous value, the first guess value is used as the average value. For all the algorithms, sensitivity analysis is done to choose the best hyperparameters. To get good accuracy values, the following work is done:

In Table 18.3, the accuracy values of wavelets and GLCM features with various machine learning algorithms are depicted. In that, various machine learning algorithms are used, like KNN, Naïve Bayes, Random Forest, SVM, Gradient Boosting, and AdaBoosting. From the table, it can be seen that the gradient boosting got the highest accuracy and F1-score measures of 84.41% and 84.02% when compared to others. The Haralick feature extraction method is used in that gradient boosting, which produces the highest accuracy and F1-score values of 71.3% and 60.27%, respectively. When compared to other machine learning methods, it produces good values. In the table, A—Accuracy, P—Precision, R—Recall, and F1—F1-score.

In Table 18.3, the local binary pattern (LBP) feature extraction methods are used in that constant values are observed for RF, SVM, AB, and GB, but still, the evaluation of

Table 18.3 Wavelets with GLCM, Haralick, and LBP features with various ML algorithms

Model	Wavelets with GLCM				Haralick				LBP			
	A	P	R	F1	A	P	R	F1	A	P	R	F1
KNN	62.37	62.52	63.51	63.01	47.40	34.88	32.48	33.64	8.24	2.11	25	3.89
NB	50.53	55.63	52.80	54.18	57.79	52.92	48.14	50.42	16.88	4.22	25	7.22
RF	83.87	83.55	82.64	83.09	70.12	54.67	55.02	54.84	38.96	9.74	25	14.02
SVM	63.44	60.22	60.10	60.16	61.69	41.47	43.45	42.44	38.96	9.74	25	14.02
AB	64.52	63.38	56.15	59.55	70.78	53.65	49.29	51.38	38.96	9.74	25	14.02
GB	84.41	84.53	83.52	84.02	71.43	60.99	59.56	60.27	38.96	9.74	25	14.02

accuracy and F1-score are not better than Haralicks and wavelet fusion with GLCM. So, the conclusion can be made that LBP is not suitable for monkeypox skin image datasets.

In Fig. 18.6, the various machine learning algorithms are used with the feature extraction method, i.e., wavelets fused with GLCM. From the figure, the highest accuracy and F1-score value are attained by the Gradient Boosting method. In Fig. 18.7, the Haralick Features extraction method is used with various machine learning algorithms, and gradient boosting produces good accuracy and F1-score values when compared to others.

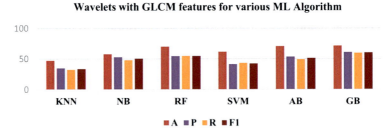

Fig. 18.6 Wavelets with GLCM features for various ML algorithms

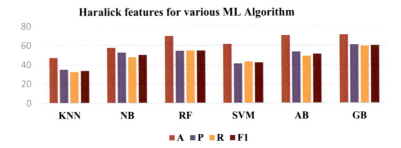

Fig. 18.7 Haralick features for various machine learning algorithms

18 Performance Analysis of Various Feature Extraction Methods … 221

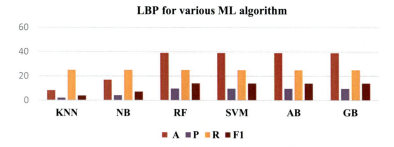

Fig. 18.8 LBP for various machine learning algorithm

In Fig. 18.8, the LBP feature extraction method produces the same values for all the four metrics. So, the LBP feature extraction method is not suitable for this dataset because it produces very low accuracy and F1-score values when compared to the other two methods.

In Table 18.4, different parameters are employed with different machine learning algorithms and various feature extraction approaches. The various feature extraction techniques and their associated characteristics are shown in Table 18.4. Table 18.4 depicts the accuracy, precision, recall, and F1-score values obtained using different parameter settings for various classification approaches. In this work, hyperparameters like n-estimators, neighbors, learning rate, and kernels are applied. This hyperparameter helps us find some constant value at some stage. From the Table, it can be observed that the wavelets fused with GLCM features produce the highest accuracy and F1-score when compared to other feature extraction methods. Based on experimental analysis, it has been observed that for these Monkeypox Skin Image Datasets (MSIDs), wavelets fused with GLCM are the best feature extraction method.

Table 18.4 Various parameters used in features extraction methods with various machine learning algorithms

Methods	Different methods	A	P	R	F1
KNN (neighbors = 7)	Wavelets and GLCM	61.29	60.01	60.02	60.02
RF (entropy, 12)	Wavelets and GLCM	82.79	82.36	81.17	81.76
AB (0.2)	Wavelets and GLCM	63.44	54.59	55.10	54.84
GB(0.1)	Wavelets and GLCM	83.87	82.97	82.52	82.74
KNN (neighbors = 10)	Haralick	47.40	34.88	32.48	33.64
RF (entropy, 15)	Haralick	69.48	55.69	54.56	55.12
AB (0.2)	Haralick	70.13	53.70	48.10	50.75
GB(0.1)	Haralick	68.18	52.72	50.67	51.67

18.6 Conclusion

In this work, various feature extraction methods are used to extract the features from the images. The different feature extraction methods, like wavelets with GLCM, Haralick features, and LBP, are used for feature extraction and various machine learning algorithms are applied on these extracted features for identification of type of pox virus. The wavelets with GLCM features used with various machine learning algorithms got the highest accuracy and F1-score values for gradient boosting at 84.1% and 84.02, respectively. The Haralick features are used with various machine learning algorithms and got the highest accuracy and F1-score values for gradient boosting at 71.43% and 60.27%, respectively. The LBP features used with various machine learning algorithms produce very low accuracy and F1-score values. From this, the observation in the above table says that the wavelets combined with GLCM got the highest accuracy and the F1-score value. So, the conclusion can be drawn that wavelets combined with GLCM is the best feature extraction method for these Monkeypox Skin Image Datasets (MISD). In this work, feature extraction is done directly for the images. After the segmentation of the affected part of the image, feature extraction has to be done to improve accuracy. Data augmentation has to be done. Still, the dataset contains only 770 images, and if the count has increased, the accuracy may be improved further.

References

1. Sitaula, C., Shahi, T.B.: Monkeypox virus detection using pre-trained deep learning-based approaches. J. Med. Syst. **46**(11), 1–9 (2022)
2. Riaz, A.: Monkeypox and Its outbreak. Pak BioMed J **5**(8), 02–02 (2022). https://doi.org/10.54393/pbmj.v5i8.792
3. Guarner, J., del Rio, C., Malani, P.N.: Mpox in 2022—What clinicians need to know. https://pubmed.ncbi.nlm.nih.gov/35696257/. JAMA. Published online 13 June 2022. Accessed 17 June 2022
4. Chen, B., Sumi, A., Wang, L., et al.: Role of meteorological conditions in reported chickenpox cases in Wuhan and Hong Kong, China. BMC Infect Dis **17**, 538 (2017). https://doi.org/10.1186/s12879-017-2640-1
5. Goldman, L., et al. (eds): Measles. In: Goldman-Cecil Medicine, 26th edn. Elsevier (2020). https://www.clinicalkey.com. Accessed 7 Feb 2022
6. Bala, D.: Monkeypox Skin Images Dataset (MSID). Kaggle (2022). https://doi.org/10.34740/KAGGLE/DSV/3971903
7. Aruraj, A., et al.:Detection and classification of diseases of banana plant using local binary pattern and support vector machine. In: 2019 2nd International Conference on Signal Processing and Communication (ICSPC). IEEE (2019)
8. Hussain, L., et al.: Machine-learning classification of texture features of portable chest X-ray accurately classifies COVID-19 lung infection. Biomed Eng Online **19**(1), 1–18 (2020)
9. Perumal, V., Narayanan, V., Rajasekar, S.J.S.: Detection of COVID-19 using CXR and CT images using transfer learning and Haralick features. Appl Intell **51**(1), 341–358 (2021)
10. Azevedo Tosta, T.A., et al.: Evaluation of statistical and Haralick texture features for lymphoma histological images classification. Comput. Methods. Biomech. Biomed. Eng. Imaging Vis. **9**(6), 613–624 (2021)

11. Goyal, S., Singh, R.: Detection and classification of lung diseases for pneumonia and Covid-19 using machine and deep learning techniques. J. Ambient Intell. Hum. Comput. 1–21 (2021)
12. Tavakoli, S., et al.: New segmentation and feature extraction algorithm for classification of white blood cells in peripheral smear images. Sci. Rep. **11**(1), 1–13 (2021).
13. Sahin, V.H., Oztel, I., Oztel, G.Y.: Human monkeypox classification from skin lesion images with deep pre-trained network using mobile application. J. Med. Syst. **46**(11), 1–10 (2022)
14. Haque, Md., et al.: Classification of human monkeypox disease using deep learning models and attention mechanisms. arXiv:2211.15459 (2022)
15. Dahea, W., Fadewar, H.S.: Finger vein recognition system based on multi-algorithm of fusion of Gabor filter and local binary pattern. In: 2020 Fourth International Conference on I-SMAC (IoT in Social, Mobile, Analytics and Cloud) (I-SMAC). IEEE (2020)
16. Sailasya, G., ArunaKumari, G.L.: Analyzing the performance of stroke prediction using ML classification algorithms. Int. J. Adv. Comput. Sci. Appl. **12**(6) (2021)
17. Soofi, A.A., Awan, A.: Classification techniques in machine learning: applications and issues. J. Basic Appl. Sci. **13**, 459–465 (2017)
18. Pandey, P., Prabhakar, R.: An analysis of machine learning techniques (J48 & AdaBoost)-for classification. In: 2016 1st India International Conference on Information Processing (IICIP). IEEE (2016)
19. Cahyana, N., Khomsah, S., Aribowo, A.S.: Improving imbalanced dataset classification using oversampling and gradient boosting. In: 2019 5th International Conference on Science in Information Technology (ICSITech). IEEE (2019)

Chapter 19
An Elucidative Review on the Current Status and Prospects of Eye Tracking in Spectroscopy

V. Muneeswaran, P. Nagaraj, L. Anuradha, V. Lekhana, G. Vandana, and K. Sushmitha

Abstract This study examined how the measurement of eye activity is done using various technologies. Eye tracking (ET) is dominant and prominent. It analyzes the gazing direction of people and measures the entire functioning of the eye. In this paper, we will come to know about various technologies for eye tracking and can able to choose the best approach. Irrespective of technology, as the eye is the most affected part, we can employ this technique in all fields. The main motto of this is to demonstrate a full-fledged review of diverse topics and techniques used in eye tracking. We will be seeing some interesting techniques like skin electrodes, contact lenses, head-mounted and remote systems, and the approach behind them. Most importantly we will learn what is pupil center corneal reflection technique (PCCR). ET gives numerous application which relates to the interaction between humans and computers. This paper also incorporates various elements which took part in the selection of a particular eye-tracking method.

Keywords Eye tracking · Pupil center corneal reflection technique · Skin electrode · Contact lens · Head-mounted · Remote system

---

V. Muneeswaran (✉) · P. Nagaraj · L. Anuradha · V. Lekhana · G. Vandana · K. Sushmitha
Kalasalingam Academy of Research and Education, Anand Nagar, Krishnankoil 626126, India
e-mail: munees.klu@gmail.com

P. Nagaraj
e-mail: nagaraj.p@klu.ac.in

L. Anuradha
e-mail: 9919005123@klu.ac.in

V. Lekhana
e-mail: 9919005290@klu.ac.in

G. Vandana
e-mail: 9919005129@klu.ac.in

K. Sushmitha
e-mail: 9919005322@klu.ac.in

19.1 Introduction

Eye tracking clearly says it tracks the positions and movement of the eyes. This method is mainly to understand the conscious and unconscious information using pupil center corneal reflection reported in 1901, and some of the techniques invasive such as using contact lenses and non-invasive techniques like remote trackers and the important application which is video-based combined pupil/corneal reflection can be used by the techniques such as fast image processing. In this paper, we can see what are all the things measured with an eye tracker, interpret eye gaze data, use eye tracking in various fields, and clear information about all the techniques involved in eye tracking and how this review will help future aims and modifications. One must know what is this eye tracking and why it has a specific significance. To understand this, simply let me give a real-time example. Assume we are in one room where we are giving some explanation on some topic and someone was delivering the content by keeping some kind of glasses for his eyes! Now my question is have you all been satisfied with such delivery content in which you aren't able to observe the receiver's eye movements and expressions? Need a genuine answer! Absolutely no, because among the total body, one must always be attracted by eyes as they will provide clear-cut information than those of listening.

So from this, we can say we are always interested to see the eye movements of others as they will say exactly what a person is gazing at whether it may be while watching television ads, browsing websites, and any other kind of activities. We can run eye tracking studies by using an ordinary webcam also whether it may be built into a laptop or an external one or by using a special device called an infrared eye tracker but it all depends on our purpose or needs and conditions. We can do a small activity on eye tracking using a cool tool that is available online. Eye trackers can measure observers gazing points with high resolution and accuracy using some optical sensors and projection patterns. I want to convey one interesting thing here we humans have peripheral vision and central vision. Both may vary based on the concentration of cones. To say that nearly our peripheral vision got 70 times less resolution than you do in your central vision which is a beautiful thing! Though one can see an object it does not mean that an observer can able to recognize entire objects. To understand this better, we will take a well-known example that everyone once in our childhood has done that activity, that is nothing but in newspapers or Funday books, we will see two similar images given and we need to recognize differences or simply we can say that we need to tell what all the things are present in that. After we have done this, we will come to know that for the first time, we will be able to specify some things and other things also we saw, but compared to the remembered things, other things are less concentrated by us. So based on eye gazing points things may vary. With this intro, we will proceed with further things which we are all interested to know like does eye tracking gets affected by our head movements? Does blinking have any effect? As said earlier we will come to know about fixations and saccades in detail and how the different approaches for eye tracking differ like normal processes, image processing, and using infrared radiation. What are the future scopes, their advantages, metrics, and some important applications and electrodes used for eye tracking? Before that one doubt, you may all have is does infrared light affect or harm our eyes. The answer is no it doesn't have any harmful effect if we see fire does it harm our eyes!! To know all this we need to know about Tobi, which is the Swedish top company for eye tracking, and the three persons of Swedish who found the potential of eye tracking with a lot more inventions. [1–5]

Significance: Our eyes and their viewing ability are the brain's primary things to get to know about the things or information happening around us. With eye tracking, one can able to find where the person is looking at. Eye tracking does not define brain choice but it can able to analyze and estimate the ultimate decision made by the brain.

Nowadays compared to human–human interaction, human–gadgets interaction became prominent and most electronic gadgets have touch screens using fingers as an input source. Now if we see to observe the changes or operations happening in software, we just need an eye. So, for every step eye movement is the key point.

As we know our pupils will always be in working condition. We call that stage a saccade. But to see or visualize a particular thing our pupil must concentrate on one point. This state is called fixation. Therefore, we can say that changing our pupil movement is changing our concentration. In this review, we can see how eye-tracking systems use the most available method called the corneal-reflection/pupil-center method.

19.2 Why Eye Tracking?

Though we have given clear-cut ideas on eye tracking, to make it much easier to understand its significance, we have given one image below! Have you recognized which image was shown in Fig. 19.1? Hope it is well known for all its boost! We can take anything in nature as an example. Now when we can buy any product in our care boost as soon as we buy a product our first focus is on MRP/Manufacturing date in most cases. Though we open our both eyes, we are seeing a whole image, but we can concentrate on one specific thing at a time, it does not mean other things are not seen, if we see carefully, we can indicate with different colors arrowed marks, based on that colors our focus may vary. So, to conclude though a person is looking at the whole image, to grasp his tracking where he is looking at it plays a very important role in psychology, biomedical in all other fields [6–10].

19.3 How Eye Tracking Works

One of the most used eye tracking in which infrared cameras are fixed with the sensor and LED when radiation is emitted that passes through the IR filter, which saves eyes from other effects. The process of eye tracking comes into play when IR radiation

Fig. 19.1 Illustration of Eye tracking

hits the center part of the eye called the pupil which produces corneal reflections in the cornea. Once the image processing unit identified the vector between the pupil and the location of corneal reflections, one can find the gazing points which are popularly known as the pupil center corneal reflection (PCCR) technique.

Figure 19.2 Illustrated how the eye tracking system works, based on the location of infrared light concerning the pupil one can find the gazing point (a) represents looking down and right of the camera, (b) looking straight at the camera, and (c) looking directly above the camera and left of the camera. But whatever the technique we are using for tracking eyes, we need one light source and one camera for recording eye movements. As we know when the pupil continuously moves, we cannot visualize things clearly which we call that state saccade, to see we need to fix our pupil position for getting focused on the things which we call fixation. When infrared light falls into our eyes, there were many layers present inside the eyes which can reflect that in consequence that we are getting corneal reflections. To understand we can observe Fig. 19.3

Mainly when we need to give rest to our hands or the people who cannot use their hands due to their physically challenging issues this is an excellent technique.

19 An Elucidative Review on the Current Status and Prospects of Eye ... 229

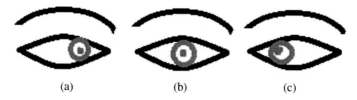

Fig. 19.2 Illustration of how eye tracking works

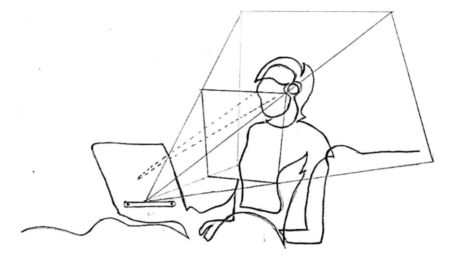

Fig. 19.3 Real-time example of Eye tracking

As seen in the image as soon as cameras have cameras with IR filter project their radiation into the user's eyes it starts finding the target and by using some algorithms, we can calculate the gazing points.

19.4 Eye Tracking and Metrics

As we understood the basic process involved in eye tracking, now we will try to know further information regarding eye tracking. As we said when the pupil is stable and focuses on a particular object, then only we will see the object called fixation and when it is moving continuously, we cannot see called saccade and the result of these two states is called scan path. To say one most knowable example which might be sarcastic that even in LinkedIn profiles based on this particular eye tracking, most of them will get the benefit for sure as many researchers say clearly. Eye tracking had undergone many rapid developments with good efficiency, stability, and many more pros; though those advancements gave good creation and better assumption techniques, some of the issues were traced out by the reviewers after many submissions on eye tracking. Among all the submissions hardly say only 7 were accepted by the reviewers after reviewing each successful paper. One of the papers was done by X. Zou et al. He has done his paper on task-oriented attention; similarly, one more paper has given an idea on reducing human interference in automating the dynamic process by McClung and Kang [11–13].

19.5 Types of Eye-Tracking Devices

As we have seen the process of eye tracking and the basic methodology is the same for all, we have different types of eye-tracking devices such as screen-based, wearable, and webcam, etc. [14].

To be clear when a receiver gets multimedia stimuli to record the output responses, screen-based device is used to increase information ability from deeper sides of the object. Similarly, when we are particular about some limitations like better contrast ratio, and a good resolution, we are going to track using a webcam device and mostly it can be done even at home without any requirement of the lab. Mainly two measurements were used in eye tracking as we mentioned earlier one is fixation and the other one is scanning path [15]. Scan path is used to view the object with less fixation where long-lasting fixation indicates less efficient scanning and other important eye-tracking skin electrodes; as we said earlier in this method, the electrodes are placed around the eye socket, and when the infrared light got struck, it will measure the retina and corneal reflections, and using this we can track both eyes of the viewer at a time but it is the particular limit but is simple to use. Next one is contact lens. It is fitted around the cornea (non-slipping lens), then tracking can be done by removing the magnetic coil around the cornea lens; compared to skin electrode, contact electrodes provide precise and accurate information, it is also limited to a specific range and it is not much comfort for the user and we have head-mounted electrode by the name a light source or camera is mounted around the head using helmet or headband, and mainly it does not cause restrictions to user head movement [16].

19.6 Applications Involving Eye-Tracking Methodology

19.6.1 Psychology and Neuroscience

As we say earlier, eye tracking has its specifications and applications in various fields, among them prominent psychology. In another way, we can say it shows the direct relationship between our eye movements, that is where we are looking and in correspondence to that how we are reacting. As eye tracking measures one's concentration, similarly, based on eye movements, one can study the insight process happening in the mind, that is our mental process [17].

19.6.2 Cognitive Process

To understand this better, let us speak with an example and the case is shown in Fig. 19.4: let us assume a person is driving a vehicle without any experience, then sudden if any critical situation occurs then? Hard to imagine! So here comes the eye-tracking usage. That is if a person is driving with no experience, if he wears eye-tracking glasses, then if any harm occurs, our subconscious mind can track the issue and give hints before itself [18].

19.7 Medical Research

To diagnose diseases like attention-deficit hyperactivity disorder (ADHD), autism spectrum disorder (ASD), obsessive compulsive disorder (OCD), etc., diseases were able to get a solution through eye tracking which will combine with some biosensors. To be simple, it can able to find the drowsiness of a person with this application. Like this, there were many applications including academic and scientific research, marketing, gaming, etc. [19]

Fig. 19.4 Eye-tracking glasses and their impact

19.8 Existing Technologies for Eye Tracking

19.8.1 Video-Based Eye Tracking

In this particular application as we said, there will be a camera embedded with LED and a sensor and when it focuses on one's eyes (either one or both), then the reflections can be seen in a cornea called corneal reflections and from that one can grab one's movements [20].

19.8.2 Interface Usability Eye Tracking

It will represent whether the person is looking at a particular thing or he is reading something or he is just having an overall scan on that, and if he is observing something, then at what concentration he is having on that content, and it also helps to find out whether the user needs a specific thing [21].

19.8.3 Interactive Applications for Eye Tracking

There were many interactive applications like accessibility, non-command-based systems, and virtual displays, and many among them accessibilities have their importance or specific dance. The name accessibility itself tells us it is giving access which in the sense of using eye tracking, one can communicate with others using their eyes! It seems normal but to be clear the persons who are facing brain injuries, spinal cord-related issues, or any strokes. Among such people, eye tracking is a boon we can simply say! [22–26]

19.8.4 PCCR Technique in Eye Tracking

The process of image acquisition is shown in Fig. 19.5. Though we discussed the PCCR technique known as pupil center corneal reflection, its idea works like an infrared camera possessing a filter and an LED source. When light falls on the camera, it passes through the filter to reduce harm to the eye; then once it filters, it hits the pupil in the cornea that gives rise to reflections in the cornea known as corneal reflections, thereby the monitor screen enables and eye tracking comes into play [27].

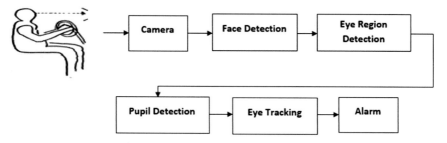

Fig. 19.5 Image Acquisition Scheme

19.9 Advantages of Using Eye Tracking

1. Compared to all other input media, eye tracking is faster which makes it more efficient and accurate.
2. For eye tracking like any other technology, it does not require any special training or coordination like normal users.
3. Eye tracking makes it easy to understand one's area of interest, which everyone likes to know!
4. Eye tracking plays a key role in user readability applications in finding out one's interaction with their environment.
5. As we said eye tracking is very much helpful for disabled or people with some injuries related to the brain, spinal, or any other.

19.10 Disadvantages of Using Eye Tracking

If we say advantages for any one particular technology for sure, there will be some disadvantages or consequences for the same technology. Similarly, eye tracking too has some disadvantages as we are saying their many advantages then definitely the cost and resources are not so reliable or less, so resources are much more expensive we are saying eye tracking can gaze at users' eye movements what about thoughts? As we said to calculate gaze points, we are using many algorithms, so interpretation is not so easy [28–30].

19.11 Future of Eye Tracking

Definitely, in the future, much more possibilities with better, precise, and accurate technologies will emerge for sure, and no doubt that eye tracking is an emerging technology in marketing industries as they are willing to increase the stock market by using eye-tracking technology to grab audience attention as if we take movies as the best example, the director can narrate the entire cinema in such a way that he will capture each scene by keeping audience perception in mind. Hope in the coming days this technology may replace many activity applications. As we said it is still an emerging technology and not quite there yet, but we can strongly assure you that eye tracking is the most exciting technology ever.

19.12 Conclusion

To sum up, all these points so far discussed, we can say that eye tracking has more importance in the upcoming days. Our eye movements are faster than computers. Through eye-tracking technology, we can make human–computer interaction. It is used to know that human visualization of a subject is being processed. It is useful in many fields, including developing video games, setting the gazing point at conferences, in graphics, and minimizing dangers in air traffic. It is very flexible and easy; through this technology, we get real-time feedback on eye movements. But not the thoughts of people. It can help people with disabilities operate through their eyes. Eye tracking can say whether a person is looking at the screen or just acting as if he is looking. It says a person is reading or searching for a specific word or just scanning. It shows where the person is looking at different parts on the screen or in front.

References

1. Hutton, S.: Cognitive control of saccadic eye movements. Brain Cogn. **68**(3), 327–340 (2008). https://doi.org/10.1016/j.bandc.2008.08.021
2. Jacob, R.J.K., Karn, K.S.: Eye tracking in human-computer interaction and usability research: ready to deliver the promises. In: The mind's eye: cognitive and applied aspects of eye movement research, pp. 573–605. Elsevier, Amsterdam (2003)
3. Just, M.A., Carpenter, P.A.: A theory of reading: From eye fixations to comprehension. Psychol. Rev. **87**(4), 329–355 (1980). https://doi.org/10.1037/0033-295X.87.4.329
4. Knight, B.A., Horsley, M., Eliot, M.: Eye-tracking and the Learning System: An Overview. In Current Trends in Eye-tracking Research, pp. 281–285. Springer (2014)
5. Liversedge, S., Findlay, J.: Saccadic eye movements and cognition. Trends Cogn. Sci. **4**(1), 6–14 (2000). https://doi.org/10.1016/S1364-6613(99)01418-7
6. Ghosh, S., Nandy, T., Manna, N.: Real-time eye detection and tracking method for driver assistance system. In: Gupta, S., Bag, S., Ganguly, K., Sarkar, I., Biswas, P. (eds.) Advancements of Medical Electronics. Lecture Notes in Bioengineering. Springer, New Delhi (2015). https://doi.org/10.1007/978-81-322-2256-9_2
7. Wedel, M., Pieters, R.: A review of eye-tracking research in marketing. Rev. Mark. Res. 123–147 (2017)
8. Holmqvist, K., Nyström, M., Andersson, R., Dewhurst, R., Jarodzka, H., Van de Weijer, J.: (2011) Eye Tracking: A Comprehensive Guide to Methods and Measures. OUP, Oxford (2011)
9. Djamasbi, S.: Eye tracking and web experience. AIS Trans Human-Comput Interact **6**(2), 37–54 (2014)
10. Carter, B.T., Luke, S.G.: Best practices in eye tracking research. Int. J. Psychophysiol. **155**, 49–62 (2020)

11. Lai, M.-L., Tsai, M.-J., Yang, F.-Y., Hsu, C.-Y., Liu, , T.-C., Lee, , S.W.-Y., Lee, M.-H., Chiou, G.-L., Liang, J.-C., Tsai, C.-C.: A review of using eye-tracking technology in exploring learning from 2000 to 2012. Edu. Res. Rev. **10**, 90–115 (2013)
12. Majaranta, P., Bulling, A.: Eye tracking and eye-based human–computer interaction. In: Advances in Physiological Computing, pp. 39–65. Springer, London (2014)
13. Horsley, M., Eliot, M., Knight, B.A., Reilly, R. (eds.): Current Trends in Eye Tracking Research. Springer (2013)
14. Arthi, V., Murugeswari, R., Nagaraj, P.: Object detection of autonomous vehicles under adverse weather conditions. In: 2022 International Conference on Data Science, Agents & Artificial Intelligence (ICDSAAI), vol. 1, pp. 1–8. IEEE (2022)
15. Sudar, K.M., Nagaraj, P., Ganesh, M., Yadav, D.A., Kumar, K.M., Muneeswaran, V.: Analysis of seminary learner campus network behaviour using machine learning techniques. In: 2022 7th International Conference on Communication and Electronics Systems (ICCES), pp. 1117–1122. IEEE (2022)
16. Muneeswaran, V., Nagaraj, P., Rajasekaran, M.P., Reddy, S.U., Chaithanya, N.S., Babajan, S.: IoT based multiple vital health parameter detection and analyzer system. In: 2022 7th International Conference on Communication and Electronics Systems (ICCES), pp. 473–478. IEEE (2022)
17. Vignesh, K., Nagaraj, P.: Analysing the nutritional facts in Mc. Donald's menu items using exploratory data analysis in R. In: Emerging Technologies in Computer Engineering: Cognitive Computing and Intelligent IoT: 5th International Conference, ICETCE 2022, Jaipur, India, 4–5 Feb 2022, Revised Selected Papers, pp. 573–583. Springer, Cham (2022)
18. Deny, J., Perumal, B., Nagaraj, P., Alekhya, K., Maneesha, V., Reddy, S.A.: Detection of osteoarthritis by using multiple edge detections. In: 2022 6th International Conference on Intelligent Computing and Control Systems (ICICCS), pp. 581–588. IEEE (2022)
19. Nagaraj, P., Saiteja, K., Ram, K.K., Kanta, K.M., Aditya, S.K., Muneeswaran, V.: University recommender system based on student profile using feature weighted algorithm and KNN. In: 2022 International Conference on Sustainable Computing and Data Communication Systems (ICSCDS), pp. 479–484. IEEE (2022)
20. Perumal, B., Nagarai, P., Venkatesh, R., Muneeswaran, V., GopiShankar, Y., SaiKumar, A., Koushik, A., Anil, B.: Real time transformer health monitoring system using IoT in R. In: 2022 International Conference on Computer Communication and Informatics (ICCCI), pp. 1–5. IEEE (2022)
21. Sudar, K.M., Nagaraj, P., Muneeswaran, V., Swetha, S.K.J., Nikhila, K.M., Venkatesh, R.: An empirical data analytics and visualization for UBER services: a data analysis based web search engine. In: 2022 International Conference on Computer Communication and Informatics (ICCCI), pp. 1–6. IEEE (2022)
22. Nagaraj, P., Muneeswaran, V., Muthamil Sudar, K., Hammed, S., Lokesh, D.L., Samara Simha Reddy, V.: An exemplary template matching techniques for counterfeit currency detection. In: Second International Conference on Image Processing and Capsule Networks: ICIPCN 2021 2, pp. 370–378. Springer (2022)
23. Muneeswaran, V., Nagaraj, M.P., Rajasekaran, M.P., Chaithanya, N.S., Babajan, S., Reddy, S.U.: Indigenous health tracking analyzer using IoT. In: 2021 6th International Conference on Communication and Electronics Systems (ICCES), pp. 530–533. IEEE (2021)
24. Perumal, B., Deny, J., Devi, S., Muneeswaran, V.: Region based skull eviction techniques: an experimental review. In: 2021 5th International Conference on Intelligent Computing and Control Systems (ICICCS), pp. 629–634. IEEE (2021)
25. Muneeswaran, V., Nagaraj, P., Godwin, S., Vasundhara, M., Kalyan, G.: Codification of dental codes for the cogent recognition of an individual. In: 2021 5th International Conference on Intelligent Computing and Control Systems (ICICCS), pp. 1387–1390. IEEE (2021)
26. Vb, S.K.: Perceptual image super resolution using deep learning and super resolution convolution neural networks (SRCNN). Intell. Syst. Comput. Technol. **37**(3) (2020)
27. Muneeswaran, V., BenSujitha, B., Sujin, B., Nagaraj, P.: A compendious study on security challenges in big data and approaches of feature selection. Int. J. Control Autom **13**(3), 23–31 (2020)

28. Nagaraj, P., Rao, J.S., Muneeswaran, V., Kumar, A.S.: Competent ultra data compression by enhanced features excerption using deep learning techniques. In: 2020 4th International Conference on Intelligent Computing and Control Systems (ICICCS), pp. 1061–1066. IEEE (2020)
29. Vamsi, A.M., Deepalakshmi, P., Nagaraj, P., Awasthi, A., Raj, A.: IOT based autonomous inventory management for warehouses. In: EAI International Conference on Big Data Innovation for Sustainable Cognitive Computing: BDCC 2018, pp. 371–376. Springer (2020)
30. Sudar, K.M., Deepalakshmi, P., Ponmozhi, K., Nagaraj, P.: Analysis of security threats and countermeasures for various biometric techniques. In: 2019 IEEE International Conference on Clean Energy and Energy Efficient Electronics Circuit for Sustainable Development (INCCES), pp. 1–6. IEEE (2019)

Chapter 20
Speech Recognition and Its Application to Robotic Arms

V. P. Prarthana, G. Sahana, A. Sheetal Prasad, and M. S. Thrupthi

Abstract This paper focuses on the development of a lightweight and easy-to-use robotic arm that can be controlled using speech recognition technology. The arm can be controlled using live voice instructions, and the system has been designed to work with both English and Kannada voice commands. The image model used for speech recognition predicts audio labels based on spectrograms, and the system transfers the prediction to the microcontroller using a serial port connection. The microcontroller then moves the arm based on the instructions received. The system is designed to be accessible to people of all backgrounds, regardless of their educational level or other factors. The project has the potential to make a significant impact by enabling people to use robotics to perform tasks that would otherwise be difficult or impossible. The work describes the use of four predefined English and Kannada commands to assess the success rate of speech recognition using machine learning models, specifically the ANN and the CNN. The accuracy of the models was evaluated and compared for both the English and Kannada datasets.

Keywords Speech recognition · Spectrogram · Serial communication · Native language

20.1 Introduction

In this developing world where everything is automated, robotics plays an important role. Despite the goal of making lives easier for everyone, some robots make everyone's job easy. Autonomous robots in industrial companies, on the other hand, are more likely to perform complex jobs. Not all robotics takes the form of a human person. A good example is a robotic arm. They are tough, adaptable, and accurate

V. P. Prarthana (✉) · A. Sheetal Prasad
Department of ECE, PES University, Bangalore, Karnataka, India
e-mail: prarthanavp31@gmail.com

G. Sahana · M. S. Thrupthi
Department of CSE, PES University, Bangalore, Karnataka, India

enough to handle a wide range of jobs. The first term to consider in voice control is speech reconnaissance, i.e., making the system understand human vocalization. Speech recognition is a technology where the system understands the words given by oration. Speech is a great way to control and communicate with robots. The advantage is the programmability of this autonomous speech recognition system. You may program and train the speech recognition circuit to recognize the specific words you want. The speech recognition model is simple to connect to the microcontroller. It will be easier to control by speaking to it. At its most basic level, speech recognition allows the user to complete many tasks simultaneously while working on the computer or appliance (i.e., hands and eyes are occupied elsewhere). Robotic arms are manipulators that have the appearance of a human arm and are capable of carrying out challenging things like welding, trimming, picking, and painting. Furthermore, their ability to operate in hazardous conditions that are inaccessible to human operators is their greatest advantage. This initiative's main goal is to help companies cut labor costs and employee errors while raising productivity. However, it can help the differently abled with daily duties with small structural alterations. The ability of robotic arms to operate in hostile situations and places that are inaccessible to people is their biggest benefit. There are many versions, including keypad controlled, voice controlled, gesture controlled, and more. Humans are replaced by robots and automation while doing mundane, dangerous, challenging, or hazardous work. In the high-tech era of today, automation significantly increases production capacities, enhances product quality, and lowers production costs. To program, monitor, and perform normal maintenance on the computer, just a small number of people are required.

20.2 Related Work

In this section, some of the research papers studied to build our arm have been cited. In [1], with the help of the microphone, speech commands are given to the voice recognition module. The voice recognition module receives the analog output of a microphone. It is capable of processing that signals and producing digital output. The signal is then passed to the ARM 7 controller, which passes it to the wireless transmission module for transmission. The ARM 7 controller sends a signal to the motor drive. The motor driving signal is sent to several motors that govern the robotic arm's mobility. Hence, robotic arm gives the movement according to the signal.

In [2], five predefined words which are "Turn Right"," Turn Left"," Stop", "Clamp", and "Release" were selected to determine the working of this robotic arm. The arm was built using servomotors and an Arduino microcontroller. The voice instruction was passed to the arm through the voice module available in Arduino IDE. The average accuracy was around 90%. In [3], four potentiometers were utilized to control the servomotors in this work. They constructed the robotic arm out of cardboard. They are recommended for applications requiring low speed, medium torque, and precise positioning. This model appears to be a robotic crane. The use of a servomotor is to create robotic arm joints and regulate them with a potentiometer.

The servomotors are controlled by an Arduino UNO board, and the Arduino's analog input is delivered to a potentiometer. The user can manage the digital values of Arduino, which are used to alter the servomotor position. By rotating these pots, we can move the joints of the robotic arm and pick, grab, or place any object.

In [4], the authors created a robot arm that uses Malayalam speech commands to control it. The word units are captured at 8 kHz, quantized at 16 bits, and then processed at 10 ms frames per second with a 25-millisecond overlapping Hamming window. The speech units are then parameterized using 12 Mel frequency cepstral coefficients. As features matching approaches in the next phase, DTW was applied. This algorithm compares two time series that differ in time or speed, as well as finds the best alignment between them if one of the time series is "warped" nonlinearly by expanding or contracting it along its time axis. The code for the spoken instruction is sent to the microcontroller for the appropriate arm motion, which is recognized by the SR unit. The motor driver IC in the microcontroller runs the motors in SCMRA in response to the received instructions.

In [5], using the HMM method, the suggested system detects continuous Kannada speech in speaker-dependent mode. The proposed method preprocesses the original Kannada speech signal before framing it every 20 milliseconds with a 6.5-ms overlapping interval. Second, the voiced section is identified by establishing a dynamic threshold based on signal size and short-time energy. Finally, in the voiced part of the signal, linear predictive coding coefficients are extracted and converted into real cepstrum coefficients. Fourth, the real cepstrum coefficients are processed through a k-means clustering algorithm with $k = 3$ and then through the Baum–Welch algorithm, which creates a three-state HMM model for each syllable, subword, or sentence.

In [6], in this paper, the author examines numerous strategies and techniques for designing and developing a robotic arm for agricultural use. A robotic arm with three degree of freedom with joints that resemble those of a human hand. To give precise movement and control over the arm, servo systems were used at each joint. A portable PC with Kinect SDK and an Arduino UNO is used to control the system. An inverse kinematics technique is used to calculate the angles at each joint required to completely characterize the location of the robot's joints. The robotic arm end effector moves to the desired location using a correct microcontroller board, which can be used to pluck a fruit or prune a branch.

In [7], in this paper, the authors used two motors. The ups and downs are controlled by motor one. The movement of the opening and closing is controlled by motor two. It can be halted here at any time the user wishes. Two of the robotic arm's joints can be moved. These joints are made by a four-pole EM 546 stepper motor (found in an old printer). The gripper joint can move from 0 to 100°, while motor 1 can move from 0 to 120°. The audio is transmitted to the computer interface via wireless technology. The audio is captured by the computer system's receiver driver. The program's sound to text converter subsystem subsequently converted the captured audio to a text string. For the exaction command, this text string was compared to the database. If a command is located in the text, the robot is given the implementation command to complete the task. This project's software is written in the C programming language. For speech rearrangement, Microsoft Speech SDK 5.1 is used.

Research Gap: Sometimes it cannot recognize command because of different pronunciation and different tone of different people.

In [8], this paper showcases an artificial neural network-based adaptive control technique for training a four-DOF robotic arm (ANN). At regular intervals, the technique comprises uniform sampling of the coordinates of the robot's joints and end effector in joint space. These coordinates are then preprocessed to create a Jacobian matrix, which is subsequently sent into a neural network to produce the upper and lower joint limits that are utilized to maneuver the arm. The architecture of this robotic arm's controller is unique in that it properly compensates for loading effects with an absolute inaccuracy of less than 1.

In [9], loudness, pitch, and quality are all characteristics of sound. Different sorts of parametric representations based on these features can be used to represent the voice signal. Mel frequency cepstral coefficients (MFCC) and Bark frequency cepstral coefficients are two available techniques for this (BFCC). The Mel scale is a perceptual scale based on signal "pitch" measurement. Mel scale cutoff frequencies represent nonlinear frequency perception in the human auditory system. Mel scale suggests a maximum sampling frequency of 14,000 Hz. The crucial bandwidth of the human hearing spectrum is divided into 24 non-overlapping critical bands. In the Bark scale, each crucial bandwidth corresponds to one bark. The Bark scale's 24 center frequencies are fixed. The Bark scale is based on the "loudness" of a voice signal. The Bark scale recommends a sampling frequency of 15,500 Hz.

In [10], the authors of this work proposed the DFCC feature extraction approach, which is based on diatonic scale frequencies. The overall classification accuracy of this approach is 95.20%, compared to 93.50% for the previous MFCC technique. Because diatonic scale cutoff frequencies can be generated for any sampling frequency value, DFCC pattern synthesis can be employed for a variety of applications such as voice recognition and music classification. We created a hybrid NF classifier in the second stage of the proposed study to recognize patterns generated by the DFCC algorithm. The accuracy of the DFCC and NF classifier combination is 99.23%, which is much greater than the accuracy of existing SVM and NN classification approaches. In the suggested work, the NF classifier is implemented in a fresh and unique way.

20.3 Proposed Work

Our main objective was to make it useful for the wider population as many of them do not know the English language. We found an English dataset that contained voice inputs of different speakers in 16-bit PCM WAV form. The voice inputs are only one-word commands in the English language. In the dataset where the voice data was used to recognize the words, we have considered the same dataset for controlling the robotic arm by just considering the words or commands the arm can perform. All the voice inputs are grouped under the label it belongs which is required to test the predictions. All voice data are unique; no two voice inputs of the same user

20 Speech Recognition and Its Application to Robotic Arms

Table 20.1 Data

English	Kannada	Action
Up	Mele	Arm will move upward
Down	Kelage	Arm will move downward
Left	Yadake	Arm will move toward left
Right	Balake	Arm will move toward right

are present in the data. Due to the non-availability of the Kannada dataset, we have created the dataset by recording the voice instructions manually. The recorded voice was not in format to preprocess, so we had to convert it to 16-bit PCM WAV form. All the voice data are unique. The data are speaker independent in which the system is trained to react to a word independently of who is speaking.

Table 20.1 shows the expected action of the arm when the specified input is passed in the form of live or recorded voice notes.

20.3.1 Robotic Arm Construction

The arm construction was focused on making it low cost, lightweight, and inexpensive. The arm has two degrees of freedom (DOF) which enables movement in four directions. The electronic components of the arm include servomotors, Arduino UNO board, UART, jumper wires, and breadboard. The main body of the arm is made up of cardboard. UART enables serial communication between the machine learning model and the Arduino board. Arduino is responsible for passing the command to the servomotors, where each direction acts as a separate function. One of the servomotors is involved in left to right and vice-versa movement of the arm. The other two are responsible for the up and down movement of the arm. The model mimics the human arm with the base acting as the shoulder, an elbow to move up and down followed by the gripper which is the wrist.

The Arduino board is dumped with the code which has different functions for different inputs (l, r, d, u). The output of machine learning model is decoded and sent to the microcontroller, which has functions defined for it. The specific function is called in which the amount of degrees the arm should move is defined.

Figure 20.1 shows the robotic arm built and the connections that have been made to the Arduino UNO board. The servomotors are connected to two points on the breadboard which are high and low. The high point goes to the 5 V port; the low point goes to the 0 V on the Arduino. The ground wires are directly connected to the GND port in the Arduino board.

Fig. 20.1 Robotic arm

20.3.2 Machine Learning Model

Preprocessing: In this stage, the audio waveforms are converted as per requirements. We generate the label and waveform for each audio file and then transform the audio dataset into a waveform dataset.

Feature extraction: During feature extraction, waveforms are transformed into spectrograms. The dataset's waveforms are displayed in the time domain, by computing the short-time Fourier transform (STFT). We divide the signal into time windows, perform a Fourier transform on each window, keep some time information, and return a 2D tensor. We convert the waveforms from the time-domain signals into the time-frequency-domain signals. The waveforms must be the same length in order for the resulting spectrograms to have equal dimensions, and simply zero-padding the audio segments that are under a second in length will accomplish this. An array of complex integers that represent magnitude and phase is generated by the STFT, but we only use the magnitude.

Feature recognition: Since we converted the audio data into spectrogram images, we built a simple convolution neural network (CNN) model for feature recognition. The model will use the following Keras preprocessing layers. Resizing and normalization of convolution neural networks perform better due to three extremely unique processes: convolution, pooling, and flattening. Two convolutional layers with ReLU activation function and 64 and 32 kernels and a kernel size of 3. Convolution is just searching an image by moving a kernel-based filter across it to find various aspects of the picture. Kernels are merely 2D matrices with various weights. Basically, as this kernel passes across the image, the pixel values are replaced with the average of the weighted sum of their weight for that particular section of the image. These kernels are an incredible tool for locating the image's key details. We must pool the features after adding the convolution layer to our model. Pooling merely shrinks the image without erasing the information we discovered during convolution. We employ a single MaxPooling layer, whose method returns the larger value within a given range and takes the form of a matrix. By doing this, we may compress the image without sacrificing it. The final stage in processing the image is to apply a flattening layer, which is nothing more than turning a 3D or 2D matrix into a 1D input for the model. A fully connected dense layer is used for classification. Additionally, we employ the Adam optimizer for compilation. We compute the loss, accuracy, and data fitting for the metrics and, also, train the model across ten iterations.

20.4 Results and Discussion

In this work, four predefined English commands (up, down, right, and left) and four predefined Kannada commands (mele, kelage, yedake, and balake) were utilized to assess the success rate of speech recognition. Speech recognition functions by examining the variations in speech patterns among speakers. Every person has a distinct speech pattern that is influenced by their anatomy (the size and shape of their mouth and throat) and behavioral habits (their speaking style, accent, and so on). For voice recognition, we used two machine learning models: the ANN and the CNN.

Table 20.2 represents the accuracy of the robotic arm movement when the commands were passed in English and Kannada. The table also sheds light on the accuracy obtained by employing different machine learning models like CNN and ANN.

The accuracy attained for each model is displayed in Table 20.2. For the English dataset, we obtained an accuracy of 94% using CNN and 75% using ANN, while for the Kannada dataset, we obtained an accuracy of 78% using CNN and 56% using ANN. Comparing both the models, we obtained maximum accuracy using CNN model. CNN models have better accuracy than ANN models. However, the accuracy of machine learning models for speech recognition is highly dependent on the quality and quantity of the data used to train them. A larger dataset with diverse speakers and speaking styles can improve the accuracy of the models, as demonstrated by the higher accuracy obtained for the larger English dataset (4000 voice commands) compared to the smaller Kannada dataset (400 voice commands). Additionally, the presence of noise or other factors that may affect speech patterns can reduce the accuracy of the models, as observed in the lower accuracy obtained for the Kannada dataset. By incorporating Kannada commands into the speech recognition model, the researchers have demonstrated the possibility of developing more inclusive and accessible technologies that can cater to a wider range of users.

Table 20.2 Accuracy table

	CNN	ANN
English	94	75
Kannada	78	56

20.5 Conclusion

In conclusion, the paper has explored various aspects of speech recognition and robotic arm technology, with a focus on developing a model that can process voice commands and perform physical tasks using a cardboard-based robotic arm. The study highlights the potential of machine learning models for speech recognition and the importance of dataset size and quality in achieving accurate results. It also highlights the need to consider inclusivity and accessibility in developing such technologies by incorporating languages and dialects spoken by a diverse population. The proposed model has some constraints and assumptions, including the limited degrees of freedom of the robotic arm and the compatibility of the components used. However, the model has several advantages, including the ability to process instructions in both Kannada and English, the use of preprocessed and live voice notes, and the ability to train the arm to perform human actions. The model can also be expanded to support other regional languages and to increase the complexity of the arm for enhanced usability. Overall, the study demonstrates the potential of integrating speech recognition and robotic arm technologies to develop new and innovative solutions for physical tasks. The findings of this study can be useful for developing new models and improving existing ones, with a focus on enhancing accessibility, usability, and versatility.

References

1. Miss. Gayatri, M., Miss. Poonam, M., Miss. Deepti, L., Mr. Bipin, B.: Voice and gesture controlled robotic ARM (2016)
2. Naik, A., Abraham, A.: Arduino based voice controlled robotic arm (2020)
3. Mishra, P., Patel, R., Upadhyaya, T., Desai, A.: Development of robotic arm using Arduino UNO (2017)
4. Lajish V.L., Vivek P., Sunil Kumar, R.K.: Malayalam speech controlled multipurpose Robotic arm (2014)
5. Dr. Punitha, P., Hemakumar, G.: Speaker dependent continuous Kannada speech recognition using HMM (2014)
6. Megalingam, R.K., Vivek, G.V., Bandyopadhyay, S., Rahi, M.J.: Robotic arm design, development and control for agriculture applications (2017)
7. Al Ahasan, M.A., Awal, M.A., Mostaf, S.S.: Implementation of speech recognition based robotic system (2011)
8. Ligutan, D.D., Abad, A.C., Dadios, E.P.: Adaptive robotic arm control using artificial network. Gokongwei College of Engineering, De La Salle University, Manila (2018)
9. Gupta, D., Bansal, P., Choudhary, K.: The state of the art of feature extraction techniques in speech recognition. In: Agrawal, S., Devi, A., Wason, R., Bansal, P. (eds.) Speech and Language Processing for Human-Machine Communications (2018)
10. Kondhalkar, H., Mukherji, P.: Speech recognition using novel diatonic frequency cepstral coefficients and hybrid neuro fuzzy classifier

Chapter 21
Machine Learning Robustness in Predictive Maintenance Under Adversarial Attacks

Nikolaos Dionisopoulos, Eleni Vrochidou ⓘ, and George A. Papakostas ⓘ

Abstract Predictive maintenance (PdM) techniques can increase industrial productivity and reduce maintenance costs by predicting the remaining useful life (RUL) of complicated machines. However, PdM systems involve industrial internet of things (IIoT) devices and machine learning (ML) algorithms, which are prone to adversarial attacks. In this work, first PdM is developed based on four ML classification models: Random Forest (RF), Light Gradient-Boosting Machine (LGBM), Gaussian Naive Bayes (GNB), and Adaptive Boosting (AdaBoost) classifier. Second, the robustness of the ML models under three adversarial attacks is evaluated, using the NASA turbofan engine dataset: Zeroth Order Optimization (ZOO), Universal Adversarial attack, and HopSkipJump attack. Results indicate RF as the most efficient classifier, reaching 96.35% of classification accuracy. Moreover, RF is proven to be the most robust of the examined classifiers under the considered attacks, displaying comparative resilience of up to 83.58% higher, compared to other models.

Keywords Predictive maintenance · Remaining useful life · Adversarial attacks · Robustness · Machine learning · Robustness

21.1 Introduction

The Industrial Internet of Things (IIoT) is a significant part of the Internet of Things that is currently on the focus of manufacturing industries [1]. The IIoT uses computer networks to accumulate data from connected industrial machines and translates them into useful information for their remaining useful life (RUL) [2]. This data can be exploited by machine learning (ML) models to perform data-driven predictive maintenance (PdM) [3, 4]. PdM is currently becoming a common practice in industry to timely prevent machine failures [5]. ML algorithms such as convolutional neural

---

N. Dionisopoulos · E. Vrochidou · G. A. Papakostas (✉)
MLV Research Group, Department of Computer Science, International Hellenic University, 65404 Kavala, Greece
e-mail: gpapak@cs.ihu.gr

networks (CNNs) [6, 7], recurrent neural networks (RNN) [8], and long short-term memory (LSTM) [9, 10], have reported high prediction accuracies in RUL prognostics tasks.

However, several IIoT devices and ML algorithms are vulnerable to potential cyber-attacks. Attacks can affect the performance of ML models and can lead to serious PdM performance loss with threatening consequences such as undetected failures or incorrect maintenance assessments. Robust machine learning algorithms under adversarial attacks for data-driven RUL prediction are therefore needed [11, 12]. The design of adversarial robust PdM systems to defense ML models is an active topic of the recent literature [13, 14]. The adversarial robustness of neural networks (NNs) was studied in [15] by proposing an optimization-based defense method. A novel defend method for recurrent neural networks (RNNs) was presented in [16]. The effect of adversarial attacks in ML regression tasks was investigated in [17].

To this end, in this work, the impact of three adversarial attacks is evaluated in four well-known ML classification models: Random Forest (RF), Light Gradient-Boosting Machine (LGBM), Gaussian Naive Bayes (GNB), and Adaptive Boosting (AdaBoost) classifier. First, the performance of the models in PdM is evaluated on data from the NASA turbofan engine dataset. Second, the impact of three adversarial attacks on the ML models is investigated, to highlight comparatively the most resilient model. Figure 21.1 illustrates the conceptual flow of this work.

The rest of the paper is organized as follows: Sect. 21.2 includes details regarding used materials and methods. Section 21.3 presents and discusses the main results of the study, including PdM classification and robustness under adversarial attacks. Section 21.4 concludes this chapter.

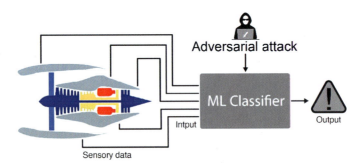

Fig. 21.1 PdM concept under ML adversarial attacks; sensory data from the turbofan engine are used as input to ML models for PdM, while cyber-attacks are affecting the models' output

21.2 Materials and Methods

In this section, the selected ML algorithms for PdM and the adversarial attacks are briefly presented. All evaluation experiments in this study are conducted with NASA's Turbofan engine dataset (Commercial Modular Aero-Propulsion System Simulation Dataset), also known as C-MAPSS [18], as this is the most significant part of an aircraft. The C-MAPSS dataset comprises four sub-datasets, each one containing training and testing data. More specifically, each dataset includes: three environmental variables, 21 sensory signal variables, the turbofan engine number, and the time cycle. Gaussian noise is added to the data to simulate real environmental conditions. In this work, the first sub-dataset (FD001) is only considered, including 100 training (train_FD001.txt) and 100 testing instances (test_FD001.txt), along with the actual RUL of the engine (RUL_FD001.txt).

21.2.1 ML Classification Models for Predictive Maintenance

Binary classification is extensively used for PdM, as a viable way to estimate the probability of a machine's failure over time. Two types of outputs are identified in binary classification for PdM: *positive* and *negative*. Positive types indicate malfunctions, while negative ones denote normal operations. The main scope is to identify the most accurate ML model to classify the probability of each new data to operate normally or fail, within the next time unit. Four well-known classifiers were selected for this task due to their outstanding performance on PdM applications and the lack of their comparative performance study in the bibliography for the problem under study. Moreover, all selected ensemble methods comprise similar steps and design decisions, therefore, could be fairly compared on the same data. In the rest of the section, a brief overview of the different selected ML models takes place.

Random Forest (RF). RFs are a collection of decision trees that rank the classification result based on the votes of each tree with majority voting. RFs outstand due to their high reported classification accuracies, their ability to handle big data with multiple variables and imbalanced datasets. RFs have been already tested for their efficiency in PdM related applications [19–21].

Light Gradient-Boosting Machine (LGBM). LGBMc is a gradient-boosting framework for ML, based on decision trees. light GBM is fast and reported high performances in the literature in PdM [22, 23].

Gaussian Naive Bayes (GNB). GNB assumes that each class follows a Gaussian distribution. It can deal with both continuous and discrete data and it is highly scalable with the number of predictors and data points. It is fast and it can be successfully applied to real-time predictions. GNB tested their efficiency in PdM problems [24].

Adaptive Boosting (AdaBoost). Boosting is an ensemble learning algorithm that combines the prediction of several week classifiers toward a strong one with enhanced robustness. AdaBoost classifier starts by fitting a classifier on the dataset and subsequently tries to fit copies of the classifier on the same dataset at the points where misclassified patterns were located. AdaBoost has also been investigated in PdM tasks [25, 26].

21.2.2 Adversarial Machine Learning Attacks

ML models are trained on large datasets. Malicious adversarial ML attacks can manipulate the input data; thus, the ML models cannot interpret them correctly leading to classification errors. Adversarial machine learning attacks can be either (1) misclassification inputs, when the malicious content is in the filters of the ML algorithm, or (2) data poisoning, when inaccurate data are inserted in the dataset to affect the models' performance [27]. In this work, three common adversarial attacks were employed to impact the performance of ML-based PdM methods, toward identifying the most resilient individual ML method.

Zeroth Order Optimization (ZOO) Attack. ZOO attack is a black-box attack that needs access only to the input and output of a ML model and not to its internal configurations [28].

Universal Adversarial Attack. Adversarial training is employed to enhance the robustness of deep NNs. In the universal adversarial attack, the input vector is reconstructed based on the adversarial sample so that ML models to conclude in misjudgment [29].

HopSkipJump Attack. HopSkipJump attack is a family of algorithms that are based on an estimate of gradient direction at the decision boundary based only on access to model's decision [30].

21.3 Experimental Results and Discussion

The parameters of the models are set by cross-validation hyperparameter selection procedure. Information regarding the used parameters is included in Table 21.1. All experimental results correspond to the average of tenfold cross-validation. Experiments are conducted on a PC Intel(R) Xeon(R) CPU, 12 GB RAM, GPU NVIDIA Quadro 2000 and in Python programming environment.

Accuracy is the most common metric to evaluate classification performance, especially for balanced data. Alternative metrics, however, are also considered so as to better understand the classifiers' performance beyond accuracy. Therefore, in this

Table 21.1 Models' hyperparameters

Model	Model parameters
LGBM	learning_rate = 0.01, n_estimators = 5000, num_leaves = 100, objective = 'binary', metrics = 'auc', random_state = 50, n_jobs = − 1
RF	n_estimators = 100, min_samples_leaf = 1, max_depth = 5
GNB	Priors = None, var_smoothing = 1e−9
AdaBoost	n_estimators = 100, random_state = 0

work, the performance of the classifiers is evaluated in terms of five selected metrics: *accuracy, precision, recall, F1-score,* and *area under curve (AUC)*.

21.3.1 ML Performance for Predictive Maintenance

This first step of the research is focused on finding the optimal ML classifier to better classify the status of a part of the aircraft engine in two labels. The classification performance of the four examined ML models is summarized in Table 21.2. The best performance for each metric is marked in bold in Table 21.2. It can be observed that RF gives the highest overall performance compared to the other selected models. LGBM also displays a similar performance. It should be noted that both RF and LGBM are decision tree-based classifiers.

Therefore, it could be concluded that RF and LGBM can classify data more precisely, and they display similar performance in the case of small datasets, as in our case, due to the fact that both models work on the same principles. However, from Table 21.2 it can be noticed that recall and AUC metrics are higher for the GNB model. While accuracy tells how many times the ML model was overall correct, recall tells how many times the model was able to detect a specific category, i.e., calculates the proportion of actual positives that were identified correctly. Therefore, in the case of the GNB model, the true positive rate was greater than RF, regardless of the fact that the model displayed lower accuracy. The same is for the AUC metric since it is directly related to the true positive rate.

Table 21.2 Comparative performance of ML models (higher metrics are marked in bold)

	Accuracy (%)	Precision	Recall	F1-score	AUC
LGBM	96.25	0.8909	0.8430	0.8669	0.9128
RF	**96.35**	**0.8973**	0.8430	**0.8693**	0.9134
GNB	93.15	0.6887	**0.9573**	0.8011	**0.9422**
AdaBoost	69.25	0.8819	0.8542	0.8678	0.9175

21.3.2 ML Performance Under Adversarial Attacks

In the second step of this research, the robustness of the PdM models under adversarial attacks is investigated. The parameters used in this work for all attacks are included in Table 21.3.

Performance results of the models after ZOO, Universal Adversarial, and HopSkipJump attacks are included in Tables 21.4, 21.5, and 21.6, respectively. The higher performance for each metric is marked in bold in all Tables.

It can be first observed that ML models' performance is impacted poorly by the Universal Adversarial attack, while the rest two attacks have a greater impact. Note that the resiliency of ML methods is strongly related to the used dataset. Different data would conclude to different robustness for the same models under the same attacks.

Table 21.3 Parameters of the attacks

	Parameters
ZOO	Confidence = 1.5, targeted = False, learning_rate = 1.5, max_iter = 30, binary_search_steps = 1, initial_const = 1.5, abort_early = True, use_resize = True, use_importance = True, nb_parallel = 10, batch_size = 1, variable_h = 1.5
Universal Adversarial	max_iter = 10, finite_diff = 3., eps = 3., batch_size = 32
HopSkipJump	batch_size = 16, targeted = False, norm = 2, max_iter = 6, max_eval = 6, init_eval = 3, init_size = 3

Table 21.4 Comparative performance of ML models after ZOO attack (higher metrics are marked in bold)

	Accuracy (%)	Precision	Recall	F1-score	AUC
LGBM	37.80	0.0930	0.3789	0.1493	0.3784
RF	**54.43**	**0.1675**	0.4921	**0.2499**	**0.5401**
GNB	37.31	0.1458	**0.6894**	0.2407	0.5046
AdaBoost	6.42	0.0542	0.3340	0.0933	0.1764

Table 21.5 Comparative performance of ML models after Universal Adversarial attack (higher metrics are marked in bold)

	Accuracy (%)	Precision	Recall	F1-score	AUC
LGBM	78.64	0.3882	0.8374	0.5305	0.8076
RF	**89.58**	**0.6006**	0.8262	**0.6956**	**0.8668**
GNB	39.96	0.1849	**0.9293**	0.3085	0.6199
AdaBoost	49.58	0.2022	0.8486	0.3266	0.6425

21 Machine Learning Robustness in Predictive Maintenance Under …

Table 21.6 Comparative performance of ML models after HopSkipJump attack (higher metrics are marked in bold)

	Accuracy (%)	Precision	Recall	F1-score	AUC
LGBM	71.51	0.3257	0.7461	0.4535	0.7277
RF	**87.35**	**0.5808**	0.7247	**0.6448**	**0.8131**
GNB	60.02	0.2489	**0.7555**	0.3745	0.6632
AdaBoost	12.74	0.0592	0.3027	0.0990	0.1985

Table 21.7 Degradation of accuracy of ML models for all attacks (lower degradation is marked in bold)

	ZOO (%)	Universal Adversarial (%)	HopSkipJump (%)
LGBM	60.72	18.29	25.70
RF	**43.50**	**7.02**	**9.34**
GNB	59.94	57.10	35.56
AdaBoost	90.72	28.40	81.60

The degradation of accuracy for all models under the attacks is included in Table 21.7. The lower degradation of accuracy is marked in bold in Table 21.7. Results show that RF performs well in the presence of all cyber-attacks, being the most resilient among the selected ML models. The latter can be attributed to the architecture of the model that is not highly sensitive to attacks. This can be also verified by the fact that LGBM that is based on the same architecture, displays quite similar behavior.

From Table 21.7 it can be calculated comparatively the robustness of each model. To this end, it can be concluded that RF is 30.48, 29.09, and 83.58% more resilient in ZOO attack, compared to LGBM, GNB, and AdaBoost, respectively. Regarding the Universal Adversarial attack, RF is more resilient by 12.12, 53.85, and 22.93% compared to LGBM, GNB, and AdaBoost, respectively. Finally, RF has 18.04, 28.92, and 79.70% higher resilience in HopSkipJump attack compared to LGBM, GNB, and AdaBoost, respectively.

21.4 Conclusions

In this paper, first, PdM based on four different machine learning models on the NASA turbofan dataset is presented. PdM is a strategy viable adopted when dealing with machine maintenance, considering the growing demand for minimizing downtime and related costs. Results indicated a proper behavior of the RF model in predicting two different machine states with the highest reported accuracy of 96.35% on the testing data attributed to the RF model. In the future, additional ML models can be investigated in this direction using more extended datasets.

Second, the impact of three adversarial attacks on the PdM models was investigated. RF was proven to be the most robust ML model under all three attacks, displaying higher comparative resilience of up to 83.58%. Therefore, the RF ensemble model can perform well under adversarial attacks, leading to more accurate replacement and maintenance decisions even under cyber-attacks. Future work will focus on having more robust datasets and more ML models, as well as investigating alternative attacking scenarios and defense strategies. Future potential research could also include training with historical data and the investigation of the robustness of algorithms on live testing data.

Acknowledgements This work was supported by the MPhil program "Advanced Technologies in Informatics and Computers", hosted by the Department of Computer Science, International Hellenic University, Kavala, Greece.

References

1. Boyes, H., Hallaq, B., Cunningham, J., Watson, T.: The industrial internet of things (IIoT): an analysis framework. Comput. Ind. **101**, 1–12 (2018). https://doi.org/10.1016/j.compind.2018.04.015
2. Gungor, O., Rosing, T., Aksanli, B.: STEWART: stacking ensemble for white-box adversarial attacks towards more resilient data-driven predictive maintenance. Comput. Ind. **140**, 103660 (2022). https://doi.org/10.1016/j.compind.2022.103660
3. Maher, Y., Danouj, B.: Survey on deep learning applied to predictive maintenance. Int. J. Electr. Comput. Eng. **10**, 5592 (2020). https://doi.org/10.11591/ijece.v10i6.pp5592-5598
4. Samatas, G.G., Moumgiakmas, S.S., Papakostas, G.A.: Predictive maintenance—bridging artificial intelligence and IoT. In: 2021 IEEE World AI IoT Congress (AIIoT), pp. 0413–0419. IEEE (2021). https://doi.org/10.1109/AIIoT52608.2021.9454173
5. Çınar, Z.M., Abdussalam Nuhu, A., Zeeshan, Q., Korhan, O., Asmael, M., Safaei, B.: Machine learning in predictive maintenance towards sustainable smart manufacturing in industry 4.0. Sustainability **12**, 8211 (2020). https://doi.org/10.3390/su12198211
6. Yang, B., Liu, R., Zio, E.: Remaining useful life prediction based on a double-convolutional neural network architecture. IEEE Trans. Ind. Electron. **66**, 9521–9530 (2019). https://doi.org/10.1109/TIE.2019.2924605
7. Liu, L., Song, X., Zhou, Z.: Aircraft engine remaining useful life estimation via a double attention-based data-driven architecture. Reliab. Eng. Syst. Saf. **221**, 108330 (2022). https://doi.org/10.1016/j.ress.2022.108330
8. Zhang, X., Dong, Y., Wen, L., Lu, F., Li, W.: Remaining useful life estimation based on a new convolutional and recurrent neural network. In: 2019 IEEE 15th International Conference on Automation Science and Engineering (CASE), pp. 317–322. IEEE (2019). https://doi.org/10.1109/COASE.2019.8843078
9. Park, K., Choi, Y., Choi, W.J., Ryu, H.-Y., Kim, H.: LSTM-based battery remaining useful life prediction with multi-channel charging profiles. IEEE Access. **8**, 20786–20798 (2020). https://doi.org/10.1109/ACCESS.2020.2968939
10. Li, J., Li, X., He, D.: A directed acyclic graph network combined with CNN and LSTM for remaining useful life prediction. IEEE Access. **7**, 75464–75475 (2019). https://doi.org/10.1109/ACCESS.2019.2919566
11. Martins, N., Cruz, J.M., Cruz, T., Henriques Abreu, P.: Adversarial machine learning applied to intrusion and malware scenarios: a systematic review. IEEE Access **8**, 35403–35419 (2020). https://doi.org/10.1109/ACCESS.2020.2974752

12. Apostolidis, K.D., Papakostas, G.A.: A survey on adversarial deep learning robustness in medical image analysis. Electronics **10**, 2132 (2021). https://doi.org/10.3390/electronics1017 2132
13. Siddique, A., Kundu, R.K., Mode, G.R., Hoque, K.A.: RobustPdM: designing robust predictive maintenance against adversarial attacks. arXiv Prepr. arXiv:2301 (2023). https://doi.org/10.48550/arXiv.2301.10822 Focus to learn more
14. Chakraborty, A., Alam, M., Dey, V., Chattopadhyay, A., Mukhopadhyay, D.: A survey on adversarial attacks and defences. CAAI Trans. Intell. Technol. **6**, 25–45 (2021). https://doi.org/10.1049/cit2.12028
15. Madry, A., Makelov, A., Schmidt, L., Tsipras, D., Vladu, A.: Towards deep learning models resistant to adversarial attacks. 6th Int. Conf. Learn. Represent. ICLR 2018—Conf. Track Proc. (2017)
16. Rosenberg, I., Shabtai, A., Elovici, Y., Rokach, L.: Sequence Squeezing: A defense method against adversarial examples for API call-based RNN VARIANTS. In: 2021 International Joint Conference on Neural Networks (IJCNN), pp. 1–10. IEEE (2021). https://doi.org/10.1109/IJCNN52387.2021.9534432
17. Mode, G.R., Anuarul Hoque, K.: Crafting adversarial examples for deep learning based prognostics. In: 2020 19th IEEE International Conference on Machine Learning and Applications (ICMLA), pp. 467–472. IEEE (2020). https://doi.org/10.1109/ICMLA51294.2020.00079
18. Saxena, A., Goebel, K., Simon, D., Eklund, N.: Damage propagation modeling for aircraft engine run-to-failure simulation. In: 2008 International Conference on Prognostics and Health Management, pp. 1–9. IEEE (2008). https://doi.org/10.1109/PHM.2008.4711414
19. Allah Bukhsh, Z., Saeed, A., Stipanovic, I., Doree, A.G.: Predictive maintenance using tree-based classification techniques: a case of railway switches. Transp. Res. Part C Emerg. Technol. **101**, 35–54 (2019). https://doi.org/10.1016/j.trc.2019.02.001
20. Bakdi, A., Kristensen, N.B., Stakkeland, M.: Multiple instance learning with random forest for event logs analysis and predictive maintenance in ship electric propulsion system. IEEE Trans. Ind. Inform. **18**, 7718–7728 (2022). https://doi.org/10.1109/TII.2022.3144177
21. Kizito, R., Scruggs, P., Li, X., Kress, R., Devinney, M., Berg, T.: The application of random forest to predictive maintenance. In: IISE Annual Conference and Expo 2018, pp. 354–359 (2018)
22. Aziz, N., Akhir, E.A.P., Aziz, I.A., Jaafar, J., Hasan, M.H., Abas, A.N.C.: A study on gradient boosting algorithms for development of AI monitoring and prediction systems. In: 2020 International Conference on Computational Intelligence (ICCI), pp. 11–16. IEEE (2020). https://doi.org/10.1109/ICCI51257.2020.9247843
23. Singh, D., Kumar, M., Arya, K.V., Kumar, S.: Aircraft engine reliability analysis using machine learning algorithms. In: 2020 IEEE 15th International Conference on Industrial and Information Systems (ICIIS), pp. 443–448. IEEE (2020). https://doi.org/10.1109/ICIIS51140.2020.9342675
24. Carvalho, T.P., Soares, F.A.A.M.N., Vita, R., Francisco, R. da P., Basto, J.P., Alcalá, S.G.S.: A systematic literature review of machine learning methods applied to predictive maintenance. Comput. Ind. Eng. **137**, 106024 (2019). https://doi.org/10.1016/j.cie.2019.106024
25. Vasilic, P., Vujnovic, S., Popovic, N., Marjanovic, A., Durovic, Z.: Adaboost algorithm in the frame of predictive maintenance tasks. In: 2018 23rd International Scientific-Professional Conference on Information Technology (IT), pp. 1–4. IEEE (2018). https://doi.org/10.1109/SPIT.2018.8350846
26. Bahad, P., Saxena, P.: Study of Adaboost and gradient boosting algorithms for predictive analytics (2020). https://doi.org/10.1007/978-981-15-0633-8_22
27. Goldblum, M., Tsipras, D., Xie, C., Chen, X., Schwarzschild, A., Song, D., Madry, A., Li, B., Goldstein, T.: Dataset security for machine learning: data poisoning, backdoor attacks, and defenses. IEEE Trans. Pattern Anal. Mach. Intell. **45**, 1563–1580 (2023). https://doi.org/10.1109/TPAMI.2022.3162397

28. Chen, P.-Y., Zhang, H., Sharma, Y., Yi, J., Hsieh, C.-J.: ZOO: zeroth order optimization based black-box attacks to deep neural networks without training substitute models. AISec 2017—Proc. 10th ACM Work. Artif. Intell. Secur. co-located with CCS 2017 (2017). https://doi.org/10.1145/3128572.3140448
29. Xie, C., Zhang, L., Zhong, Z.: Virtual adversarial training-based semisupervised specific emitter identification. Wirel. Commun. Mob. Comput. **2022**, 1–14 (2022). https://doi.org/10.1155/2022/6309958
30. Chen, J., Jordan, M.I., Wainwright, M.J.: HopSkipJumpAttack: a query-efficient decision-based attack. In: 2020 IEEE Symposium on Security and Privacy (SP), pp. 1277–1294. IEEE (2020). https://doi.org/10.1109/SP40000.2020.00045

Chapter 22
Modelling and Grasping Analysis of an Underactuated Four-Fingered Robotic Hand

Deepak Ranjan Biswal, Alok Ranjan Biswal, Rasmi Ranjan Senapati, Abinash Bibek Dash, Shibabrata Mohapatra, and Poonam Prusty

Abstract In the field of industrial environment, prosthetics, rehabilitation, space applications, medical applications, etc., the implementation and use of the robotic hand is of significant prominence to achieve reasonable accuracy and improve production. For perfect grasping, dexterity manipulation and shape adaption the concept of underactuation is an appropriate approach. This paper implements the modelling of a four-fingered robotic hand and the analysis of the material used in the hand during grasping. The proposed robotic hand consists of a four underactuated fingers that are alike to each other. Each finger consists of three phalynx and three joints. Under actuation is carried out by tendon and pulleys. The proposed robotic hand has four fingers with a total of twelve numbers of joints with a total of sixteen degrees of freedom. Solid work platform is used to model the hand and a finite element-based analysis is performed to analyse the various mechanical parameters based on the materials used in the model. The robotic hand holds a cuboid with its fingers. The analysis is accomplished concerning three types namely deformation, stress and strain test. The analysis provides significant data about the parts of the robotic hand that can be destroyed if subjected to greater force. Therefore, this could be useful for a major design upgrading and confirming the suitability of the robotic hand in actual application.

Keywords Finite element analysis · Four fingered · Grasping · Robotic hand · Tendon driven · Underactuation

D. R. Biswal (✉) · A. R. Biswal · R. R. Senapati · A. B. Dash · S. Mohapatra · P. Prusty
Department of Mechanical Engineering, DRIEMS (Autonomous), Cuttack, India
e-mail: deepak.biswal@driems.ac.in

22.1 Introduction

The human hand is considered as the utmost intellectual outer extremity, capable of carrying out a wide-ranging of tasks but with certain limitations. Robotic hands have the advantage of exceeding human limitations in difficult, dangerous, repetitive and gloomy jobs. With the decreasing number of trained labourers in the assembly industry, an ideal robot with resolution, flexibility and acumen is anticipated to improve quality of production. Automated machines can contribute to the result of building machines to do human-like tasks. However, if there are significant changes in the manufacturing process, the peripheral equipment must be rebuilt [1]. Meanwhile, industrial manipulator and robots typically have three to four movability. The end effector of these manipulators is crucial in directly manipulating the products. Suction cups and grippers are the most commonly utilised end effectors. However, these are restricted to the dimension and weight of the body being manipulated. In the same context a robot hand is projected to aid the business because it can mimic human hands to perform more complex tasks. Some significant contributions are presented. Researchers at the German Aerospace Centre developed DLR Hand II [2] an upgraded form of the original DLR Hand that has the appearance and dimensions of the hand of a human. It is made up of a moveable palm, five articulated fingers and have powerful tendons that can withstand 30 N load at the fingertip. In order to achieve more flexible grasps, General Motors and NASA developed Robonaut Hand-1 [3] and Robonaut Hands-2 [4]. These hands possesses driving elements having lower frictional material, more durability, and have capability of sensing. A robotic hand with four fingers and the ability to perform grasping tasks along with various object manipulations is mentioned [5]. RBO Hand-2, is proposed which is an underactuated one and is similar to human hand and is highly compliant and robust [6]. Lu et al. [7] presented a Robotic gripper which comprises a touch sensor and having reversible mechanism present in the fingers and palm. A robotic hand with four fingers is designed in which vacuum suction nozzle is implemented for sucking of tomatoes [8]. Higashimori et al. [9] planned a robot hand comprising four finger with dual turning mechanism in which fingers can autonomously rotate in circles with a common centre. Laliberte et al. [10] presented the technique of under actuation in robot grasping hand. A six degrees of freedom and five actuator operated prosthetic hand named I-LIMB hand was developed. It has one actuator for every finger and a total of five actuators in the hand are present which directly activates the flexion of each metatarsophalangeal joint [11]. Be Bionic Hand a 6-Degrees of Freedom (DOF), 5-Degrees of Actuation (DOA) prosthetic hand was produced. There is one actuator for each finger (a total of five actuators), but unlike the I-LIMB hand, the motors are located in the middle of the hand (metacarpus) [12]. A 12-DOF, 5 actuator-based Meka H2 hand was proposed which is a compliant, four-fingered one. Here each finger has a separate actuator and the thumb's abduction and adduction have their own individual actuators, i.e. a total 5 actuators [13]. A hand with only one actuator and five fingers called ADAM'S Hand was created for prosthetic use. A fixed frame is there in this hand alike a human palm where all the fingers are interconnected through revolute joints [14]. An underactuated Valkyrie hand driven by pulleys and tendons is proposed for design and analysis in which there are a total of four fingers, thirteen joints and six DOA in this hand [15]. For heavy-duty tasks required in the field or rough terrain, Ko et al. [16] suggested an under actuated four-fingered hand with five electro hydrostatic actuators clustered together. An anthropomorphic hand with hydrostatic motion is created. Ohol et al. [17] described on FEA-based study on robot hand design in which he used a shape optimization algorithm for a multi-fingered robotic gripper. A robotic hand with three fingers and its analysis is described by Azri and Shauri [18]. Based on the finite element method, Zhang et al. [19] proposed stress analysis of significant finger rehabilitation exoskeleton robot components where the author examines the flexion and extension of three joints on a single mechanical exoskeleton finger. Grasping aptitude and motion interactions in affordable tendon-driven prosthetic hands controlled by Able-Bodied Subjects is presented in [20]. An underactuated hand of anthropomorphic type with fifteen DOF and one DOA is proposed by Lalibert [21]. Key components' stress was examined by Zhang [22] using the finite element method, which included calculating and examining its stress and strain. In the present work a four-fingered robotic hand is proposed. All the four fingers are placed symmetrically at the periphery of the palm and all the fingers are underactuated. The process of underactuation is accomplished by tendon and pulley arrangement. Each finger consists of 3 phalanges names as: distal phalynx (P_d), middle phalynx (P_m), proximal phalynx (P_p). Three joints are there in each finger. The joint between distal and middle phalynx is distal inter phalangeal joint (J_{dip}), the joint between middle and proximal phalynx is proximal inter phalangeal joint (J_{pip}) and the joint between proximal phalynx and thumb or base is Metacarpophalangeal joint (J_{mcp}). All the three joints are underactuated that improves the self-adaption. Eight separate motors are used near the four J_{mcp}. Out of which four motors are used for flexion/ extension and rest fours are used for abduction/adduction operation.

22.2 Materials and Methods

Stainless steel and Aluminium alloy are the materials used individually for the proposed hand model and for grasping the cuboidal object. In the present model four fingers are used to grasp the cuboidal shaped object. The material properties are shown in Table 22.1.

Table 22.1 Properties of material

Properties	Stainless steel	Aluminium alloy
Density	7750 kg/m^3	2270 kg/m^3
Yield strength (tensile)	2.07E + 08	2.8E + 08
Yield strength (compressive)	2.07E + 08	2.8E + 08
Ultimate strength (tensile)	5.86 E + 08	3.1E + 08

22.2.1 Modelling of the Hand

The proposed hand model comprises of four fingers which are placed at an angle of 90° to each other and are placed symmetrically. Each finger has three phalanges, i.e. P_p, P_m and P_d. The three phalanges are connected with each other by intermediate joints. The joint J_{mcp} joins palm of the hand with P_p. In between the P_p and P_m the J_{pip} is present. The joint J_{dip} is present between the phalynx P_d and P_m. All the joints are rotational joints. The inward movement of the phalanges towards the palm are accomplished by the tendon and pulleys. Springs are present at the backside of the joints for retraction and to position the phalanges to its original position. At each joint pulleys are present and are free to rotate about the joint axes [23]. Near the J_{mcp} of each finger upon the palm individual motors are present. Tendons are wounded around the pulleys and are attached to the motor shaft. By the rotation of the motor shaft in one direction the three digits of the fingers move inward making all the springs extended. A total of eight motors are used in the proposed hand model and the total degrees of freedom of the hand is sixteen.

When all the fingers closes to each other they simultaneously holds the cuboid. When the motor rotate in opposite direction the springs got contracted and the motor shaft releases the tendon up to the three phalanges come to the original position.

The layout of the hand model is presented in Fig. 22.1. The fingers are operated by tendon and pulley. In a single finger two actuation is there to operate the movement of the fingers. The contraction of the finger is done by the tendon and the expansion of all the three digits of the finger is accomplished by the springs. In this present work the proposed hand comprises of four fingers. Each finger consists of three phalanges, i.e. P_p, P_m and P_d. The dimensions of the hand, palm and the cuboid to be hold are presented in Table 22.2. The model of the proposed hand is mentioned in the Fig. 22.2 where the hand model is drawn in solid work platform.

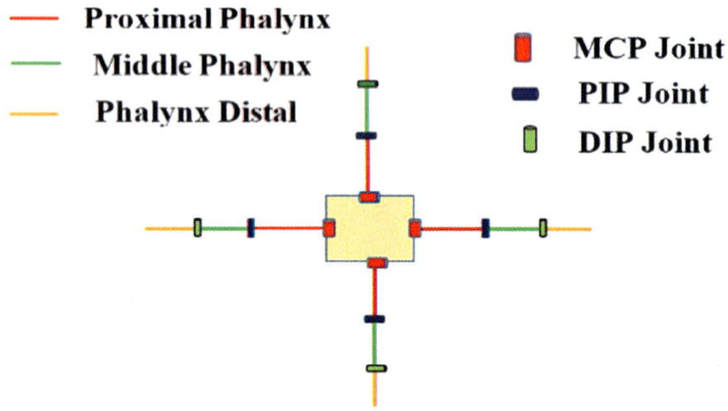

Fig. 22.1 Layout of the hand

Table 22.2 Dimension detail of the proposed hand

Name	Dimension (L × W × T) in mm
Proximal phalynx	52 × 10 × 5
Middle phalynx	30 × 10 × 5
Distal phalynx	20 × 10 × 5
Palm	80 mm side
Cuboid	80 mm side

Fig. 22.2 Model of the hand

22.2.2 Grasping

During grasping the desired object is held firmly by the fingers in the hand. Fingers present in the hand move in order to grab an object. There are different ways to grasp anything, and in this paper, the fingers are actuated by a tendon pulley mechanism.

The four-fingered hand holds the cuboid of size of 80 mm each side presented in Fig. 22.3. The movement of the hand is done by the tendon and pulley arrangement as the hand is an underactuated hand. Figure 22.4 shows a representation of tendon pulley planning with contact forces.

22.2.3 Contact Force Calculation

Schematic layout of a tendon pulley setup with contact forces acting on the phalynx is presented in Fig. 22.4. The contact force vector expressed as

$$f = [f_1 \, f_2 \, f_3]^T \tag{22.1}$$

Fig. 22.3 Model of the hand grasping a cuboid

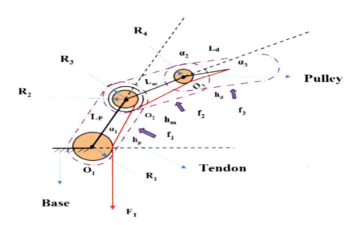

Fig. 22.4 Representation of forces on the phalynx

where f_1 is the contact force at the proximal phalynx, f_2 and f_3 are the contact forces at the proximal and distal phalanges. Mathematical formulation of f_1, f_2 and f_3 are mentioned in the Eqs. 22.2, 22.3 and 22.4

$$f_1 = \frac{\Gamma T_a}{h_p h_m h_d R_1 R_3} \qquad (22.2)$$

where $\Gamma = E + H + M$; where $E = L_P R_2 \cos\alpha_2 (R_4 L_m \cos\alpha_3 + (R_4 - R_3)h_d)$; $H = -L_p R_2 R_4 h_m \cos(\alpha_2 + \alpha_3)$; $M = (R_1 - R_2) h_m h_d R_3$

Middle phalynx force is as follows:

$$f_2 = \frac{-R_2(-h_d R_3 + R_4 L_m \cos\alpha_3 + R_4 h_d) T_a}{h_m h_d R_1 R_3} \tag{22.3}$$

Distal phalynx force is mentioned as follows:

$$f_3 = \frac{R_2 R_4 T_a}{h_d R_1 R_3} \tag{22.4}$$

22.3 Results and Discussion

Contact forces at the point of contact are obtained by taking the dimensions of the phalanges, radius of the pulleys, distance of the contact points and phalynx angles. The plot for the contact force at the distal, middle and proximal phalanges are presented. The proposed four-fingered hand holds a cuboid with all its fingers. The finite element-based analysis is carried out by taking two different materials. Force is applied vertically in upward and downward direction separately. The mechanical parameters such as deformation, stress and strain are analysed by increasing the magnitude of the forces and the plot is obtained from Ansys work bench.

22.3.1 Contact Force Graphs

The graphs of the contact forces for the fingers are presented. The variation of contact forces f_1 with respect to θ_2 and θ_3 is shown in Fig. 22.5. The figure indicates that as the magnitude of the joint angles increases the contact force between the proximal phalynx and the grasped body also increases up to a certain value and then the magnitude of the force decreases. The contact force graph of f_2 with angle θ_2 and θ_3 is shown in Fig. 22.6. The figure represents that as the values of joint angles increases the magnitude of the contact forces gradually increases. Referring to the plot it is viewed that as the value of thita-3 increases with constant value of thita-2 the magnitude of the contact force increases. The value of the force remains constant as the magnitude of the thit-2 increases keeping thita-3 constant.

The variation of contact force f_3 with angle θ_2 and θ_3 is shown in Fig. 22.7 which presents that with the increase in the values of angles in *x*-axis and *y*-axis the value of the contact force remains constant. The plotted figure is a plane which indicates that as the magnitude of the angles increases the value of the force remains constant and is fixed. It represents that as the joint angles increases the magnitude of the contact force between the object and the distal phalynx remains constant at a particular value.

Fig. 22.5 Contact force of proximal phalynx

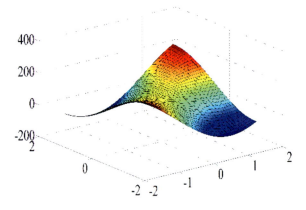

Fig. 22.6 Contact force of middle phalynx

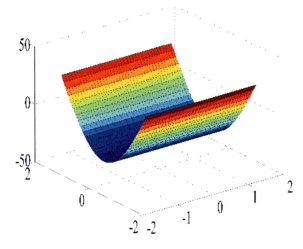

Fig. 22.7 Contact force of distal phalynx

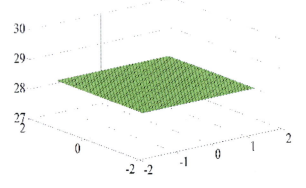

22.3.2 Analysis for Mechanical Properties Through Ansys

There are various kinds of structural analysis techniques. The classical approach is one of them, along with material strength analysis. It is typically employed for geometrically straightforward regions and loadings. FEA essentially makes use of nodes, which are chosen finite points connected to one another to create a mesh-like grid. The accuracy of the analysis is based on the number of nodes in each element. The structural modelling programme ANSYS utilises an automatic mesh generation algorithm. When subjected to some degree of external force, a body will indeed experiences some deformation. The cohesion of the molecules however helps the body to resist deformation. In general, stress or strain can be used to classify this resistance, which is known as material strength [24]. Strain is the change from the initial dimension. Various strains includes shear, tensile, compressive and volumetric strains [25]. Stress, meanwhile, is described as a body's ability to resist deformation on a unit area basis. Four cases are studied in the present analysis. In case-1 the material for hand and cuboid is stainless steel and force direction is vertically downward on the cuboid.

In case-2 the material for hand and cuboid is stainless steel for both hand and cuboid and force act vertically upward on the cuboid. In case-3 the material for hand and cuboid is taken as Aluminium alloy and force acting on the cuboid is vertically downward on the cuboid. In case-4 the material for hand and cuboid is taken as Aluminium alloy and force acting on the cuboid is vertically upward on the cuboid forces are applied separately in downward and upward direction. The magnitude of the force is increasing from 10 to 100 N in both cases. The meshing of the hand holding a cuboid is shown in the Fig. 22.8. Number of nodes is 17249 and number of elements is 5279 in the Ansys workbench. The meshing for all the four fingers and the palm is triangular meshing and the meshing for the cuboid is square meshing.

In the Fig. 22.9 the force direction is acting downward and the direction of force is upward as shown in Fig. 22.10. All the four fingers are in touch with the cuboidal shaped object for firm grasping of the object. The analysis is carried out in Ansys work bench by applying load as referred to the figures and the magnitude gradually increases. In all the cases the magnitude of the force acting on the surface of the cuboid varies from minimum value of 10 N to maximum value of 100 N. The mechanical parameters analysed are Total deformation, Maximum principal stress σ_{mp} and Maximum principal elastic strain ζ_{mp}. The analysis for case-1 is done in Ansys work bench and the result is presented in the Table 22.3. Here the magnitude of the force increases from 10 to 100 N. The values of d_t, σ_{mp} and ζ_{mp} is analysed. It has been observed that as the magnitude of the force increase from 10 to 100 N the mechanical parameters also increases.

The analysis for the total deformation for case-1 is presented in Fig. 22.11. The minimum value is presented in blue colour and the maximum value is presented in red colour. From Fig it has been observed that at 50 N of force when acting downward on the cuboid for the material of stainless steel the values of total deformation

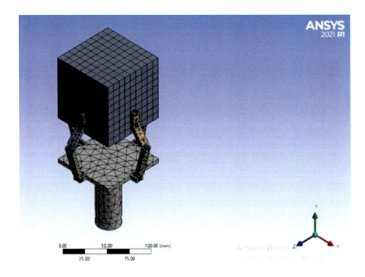

Fig. 22.8 Meshing of the proposed hand

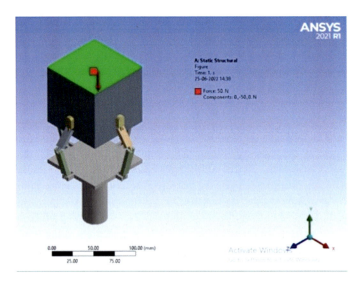

Fig. 22.9 Direction of force acting downward

is 3.08E−02, the maximum principal stress 60.514 MPa and value of maximum principal elastic strain is 4.05E−04 mm/mm.

The analysis of the maximum principal stress is done and is presented in Fig. 22.12. It has been found that the maximum value at the force of 50 N is 60.514 Mpa and the minimum value is − 26.377 Mpa. It is observed that the maximum stress occurs at the finger joints.

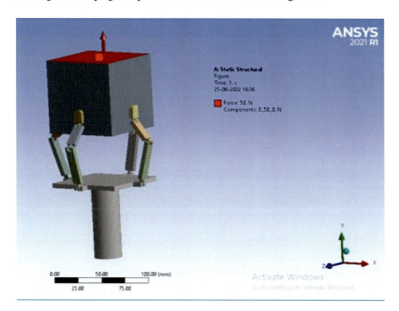

Fig. 22.10 Direction of force acting upward

Table 22.3 Analysis for case-1

F(N)	d_t (mm)	σ_{mp} (Mpa)	ζ_{mp} (mm/mm)	f (N)	d_t (mm)	σ_{mp} (Mpa)	ζ_{mp} (mm/mm)
10	6.15E−03	12.103	8.11E−05	60	3.69E−02	72.617	4.86E−04
20	1.23E−02	24.206	1.62E−04	70	4.31E−02	84.719	5.68E−04
30	1.85E−02	36.308	2.43E−04	80	4.92E−02	96.822	6.49E−04
40	2.46E−02	48.411	3.24E−04	90	5.54E−02	108.92	7.30E−04
50	3.08E−02	60.514	4.05E−04	100	6.15E−02	121.03	8.11E−04

Analysis result of maximum principal elastic strain is presented in Fig. 22.13. It has been found that the maximum value of the elastic strain at 50 N is 4.05E−04 and the minimum value is − 3.6621E−8. In case-2 the magnitude of the force increases from 10 to 100 N with an interval of 10 N and the material taken for the hand and the cuboid is stainless steel. Analysis result for case-2 is presented in Table 22.4. At 10 N force all the three parameters d_t, σ_{mp} and ζ_{mp} are 6.15E−03 mm, 15.426 Mpa and 8.24E−05 mm/mm and the values of the above parameters for 100 N force is 6.15E−02 mm, 154.26 Mpa and 8.24E−04 mm/mm.

Analysis for case-3 is presented in Table 22.5. It has been observed that the value of d_t, σ_{mp} and ζ_{mp} for 10 N force is 1.69E−02 mm, 12.121 Mpa and 2.23E−04 mm/mm and the values of the above parameters for 100 N force is 1.69E−01 mm, 121.21 Mpa and 2.23E−03 mm/mm.

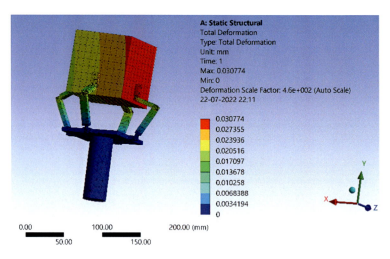

Fig. 22.11 Total deformation at 50 N force for case-1

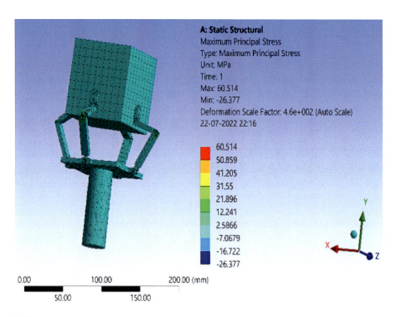

Fig. 22.12 Maximum principal stress at 50 N force for case-1

Analysis for case-4 is presented in Table 22.6. It has been observed that the value of d_t, σ_{mp} and ζ_{mp} for 10 N force is 1.69E−02 mm, 16.211 Mpa and 2.27E−04 mm/mm and the values of the above parameters for 100 N force is 1.69E−01 mm, 162.11 Mpa and 2.27E−03 mm/mm.

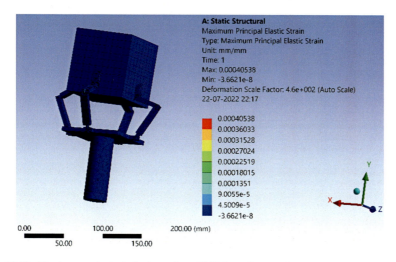

Fig. 22.13 Maximum principal elastic strain at 50 N force for case-1

Table 22.4 Analysis report for case-2

f (N)	d_t (mm)	σ_{mp} (Mpa)	ζ_{mp} (mm/mm)	f (N)	d_t (mm)	σ_{mp} (Mpa)	ζ_{mp} (mm/mm)
10	6.15E−03	15.426	8.24E−05	60	3.69E−02	92.553	4.95E−04
20	1.23E−02	30.851	1.65E−04	70	4.31E−02	107.98	5.77E−04
30	1.85E−02	46.277	2.47E−04	80	4.92E−02	123.4	6.59E−04
40	2.46E−02	61.702	3.30E−04	90	5.54E−02	138.83	7.42E−04
50	3.08E−02	77.128	4.12E−04	100	6.15E−02	154.26	8.24E−04

Table 22.5 Analysis report for case-3

f (N)	d_t (mm)	σ_{mp} (Mpa)	ζ_{mp} (mm/mm)	f (N)	d_t (mm)	σ_{mp} (Mpa)	ζ_{mp} (mm/mm)
10	1.69E−02	12.121	2.23E−04	60	1.01E−01	72.727	1.34E−03
20	3.38E−02	24.242	4.45E−04	70	1.18E−01	84.848	1.56E−03
30	5.07E−02	36.364	6.68E−04	80	1.35E−01	96.97	1.78E−03
40	6.76E−02	48.485	8.90E−04	90	1.52E−01	109.09	2.00E−03
50	8.45E−02	60.606	1.11E−03	100	1.69E−01	121.21	2.23E−03

Comparing the deformation of case-1 and case-2 it is found that the deformation for upward and downward movement is approximately same. The maximum principal stress for case-1 and case-3 are nearly same. The total deformation for case-3 and case-4 are nearly same. The comparison of maximum principal stress for case-1 and case-2 is as follows.

Table 22.6 Analysis report for case-4

f (N)	d_t (mm)	σ_{mp} (Mpa)	ζ_{mp} (mm/mm)	f (N)	d_t (mm)	σ_{mp} (Mpa)	ζ_{mp} (mm/mm)
10	1.69E−02	16.211	2.27E−04	60	1.01E−01	97.264	1.36E−03
20	3.38E−02	32.421	4.55E−04	70	1.18E−01	113.48	1.59E−03
30	5.07E−02	48.632	6.82E−04	80	1.35E−01	129.69	1.82E−03
40	6.76E−02	64.843	9.10E−04	90	1.52E−01	145.9	2.05E−03
50	8.45E−02	81.054	1.14E−03	100	1.69E−01	162.11	2.27E−03

As shown in the Fig. 22.14 as the force increases the value of the maximum principal stress increases. For the same load the value of the stress for case-2 is more as compared to case-1. As the load increases it is found that the stress value for case-2 is increases more as compared to case-1. As shown in Fig. 22.15 stress diagram for case-3 and case-4 is presented. From the figure it is viewed that as the force increases the value of the stress also increase. For the same load the stress value of case-4 is more as compared to case-3 referring to the obtained figure.

The maximum principal strain diagram for case-1 and case-3 is shown in Fig. 22.16. As the force increases the value of maximum principal strain also increases. For the same load the strain value of case-1 is more as compared to case-3. The maximum principal strain diagram for case-2 and case-4 is shown in Fig. 22.17. As the force increases the value of maximum principal strain also increases. For the same load the strain value of case-4 is more as compared to case-2.

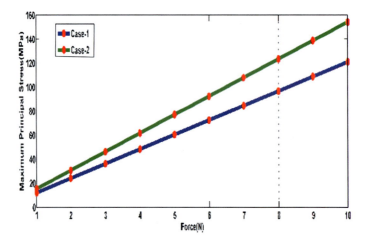

Fig. 22.14 Comparison of stress diagram between case-1 and case-2

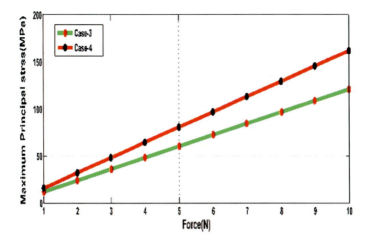

Fig. 22.15 Comparison of stress diagram between case-3 and case-4

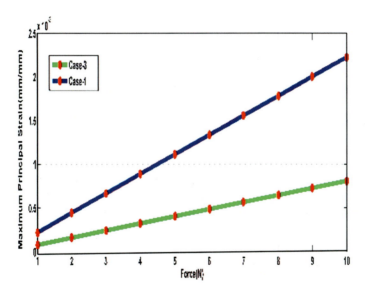

Fig. 22.16 Comparison of strain diagram between case-1 and case-3

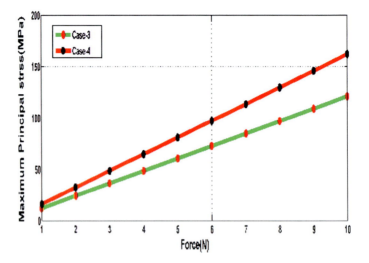

Fig. 22.17 Comparison of strain diagram between case-2 and case-4

22.4 Conclusion

The present work deals with the modelling of a robotic hand which comprises of four fingerers and a base or palm. Grasping analysis of the proposed robotic hand in which the hand grasp the cuboid with all its four fingers is analysed. The proposed hand is underactuated one which is operated through pulleys and tendons. The modelling has been carried out in modelling software. For analysis of the proposed robotic hand which grasp the cuboid, two types of material, i.e. stainless steel and Aluminium alloy are taken separately. The analysis has been carried out for the proposed hand model during the grasping. Four separate cases are analysed in Ansys work bench. Mechanical parameters like the maximum deformation, maximum principal stress and maximum principal elastic strain have analysed in the Ansys environment with variation of application of load at the surface of the cuboidal shaped object in downward and upward directions separately. The analysis is carried out keeping in view of the object handling in industrial environment. It has been found that the proposed hand model can be suitable for industrial applications.

References

1. Ishiguro, H.: Scientific issues concerning androids. Int. J. Rob. Res. **26**(1), 105–117 (2007)
2. Butterfaß, J., Grebenstein, M., Liu, H., Hirzinger, G.: DLR-Hand II: next generation of a dextrous robot hand. Proc. IEEE Int. Conf. Robot. Autom. **1**, 109–114 (2001)
3. Lovchik, C.S., Diftler, M.A.: The robonaut hand: a dexterous robot hand for space. In: Proceedings, IEEE International Conference on Robotics and Automation (Cat. No. 99CH36288C), vol. 2, pp. 907–912

4. Diftler, M.A.: Robonaut 2-the first humanoid robot in space. IEEE International Conference on Robotics and Automation, pp 2178–2183 (2011)
5. Neha E., Suhaib, M., Mukherjee, S., Shrivastava, Y.: Kinematic analysis of four-fingered tendon actuated robotic hand (2021). https://doi.org/10.1080/14484846.2021.1876602
6. Deimel, R., Brock, O.: A novel type of compliant and underactuated robotic hand for dexterous grasping. Int. J. Rob. Res. **35**(1–3), 161–185 (2016)
7. Lu, Z., Guo, H., Zhang, W., Yu, H.: GTac-Gripper: a reconfigurable under-actuated four-fingered robotic gripper with tactile sensing. IEEE Robot. Autom. Lett. (2022)
8. Vu, Q., Ronzhin, A.: A model of four-finger gripper with a built-in vacuum suction nozzle for harvesting tomatoes. In: Proceedings of 14th International Conference on Electromechanics and Robotics Zavalishin's Readings, pp. 149–160 (2020)
9. Higashimori, M., Jeong, Ishii, H.I., Kaneko, M., Namiki A., Ishikawa, M.: A new four-fingered robot hand with dual turning mechanism. In: Proceedings of the 2005 IEEE International Conference on Robotics and Automation, pp. 2679–2684 (2005)
10. Laliberte, T., Birglen, L., Gosselin, C.: Underactuation in robotic grasping hands. Mach. Intell. Robot. Control. **4**(3), 1–11
11. Van Der Niet Otr, O., Reinders-Messelink, H.A., Bongers, R.M., Bouwsema, H., Van Der Sluis, C.K.: The i-LIMB hand and the DMC plus hand compared: a case report. Prosthet. Orthot. Int. **34**(2), 216–220 (2010)
12. Tavakoli, M., Enes, B., Marques, L., Feix, T.: Actuation configurations of bionic hands for a better anthropomorphism index. J. Mech. Robot. **8**(4) (2016)
13. Tavakoli, M., de Almeida, A.T.: Adaptive under-actuated anthropomorphic hand: ISR-SoftHand. In: 2014 IEEE/RSJ International Conference on Intelligent Robots and Systems, pp. 1629–1634 (2014)
14. Zappatore, G.A., Reina, G., Messina, A.: Analysis of a highly underactuated robotic hand. Int. J. Mech. Control. **18**(2), 17–24 (2017)
15. Guo, R.: (2016) DETC2014-35069, pp 1–10
16. Ko, T., Kaminaga, H., Nakamura, Y.: Underactuated four-fingered hand with five electro hydrostatic actuators in cluster. Proc. IEEE Int. Conf. Robot. Autom. 620–625 (2017)
17. Ohol, S.S., Karade, S., Kajale, S.R.: Optimization of four finger robotic hand using FEA. Int. J. Technol. Knowl. Soc. **6**(6) (2010)
18. Bin Mohamed Azri, M. H., Shauri, R. L. A,(2014), Finite element analysis of a three-fingered robot hand design. In: 2014 IEEE 4th International Conference on System Engineering and Technology (ICSET), vol. 4, pp. 1–6 (2014)
19. Wen, L., Li, Y., Cong, M., Lang, H., Du, Y.: Design and optimization of a tendon-driven robotic hand. In: 2017 IEEE International Conference on Industrial Technology (ICIT), pp. 767–772 (2017)
20. Llop-Harillo, I., Pérez-González, A., Andrés-Esperanza, J.: Grasping ability and motion synergies in affordable tendon-driven prosthetic hands controlled by able-bodied subjects. Front. Neurorobot. **14**, 57 (2020)
21. Lalibert, T.: An anthropomorphic underactuated robotic hand with 15 dofs and a single actuator, pp. 749–754 (2008)
22. Zhang, Q., Lu, G., Chen, Y.G., Han, T.Y., Qie, T.: Stress analysis of key components of finger rehabilitation exoskeleton robot based on finite element method. J. Phys. Conf. Ser. **2216**(1), 12001 (2022)
23. Birglen, L., Laliberté, T., Gosselin, C.: Underactuated robotic hands (2007)
24. Collins, J.A., Busby, H.R., Staab, G.H.: Mechanical design of machine elements and machines: a failure prevention perspective. Wiley (2009)
25. Dupen, B.: Applied strength of materials for engineering technology. Purdue University Fort Wayne, School of Polytechnic (2019)

Chapter 23
Design and Development of Six-Axis Robotic Arm for Industrial Applications

D. Teja Priyanka, G. Narasimha Swamy, V. Naga Prudhvi Raj, E. Naga Lakshmi, M. Maha Tej, and M. Purna Jayanthi

Abstract Automation is currently necessary in many industries including the core. After studying issues in various industries, we discovered that the majority of errors take place in tasks that require human intervention; in fact, the product's quality and turnaround time must be maintained. By automating the industry completely or partially, these can be maintained. Our main aim is to create a six-axis robotic arm that can fully or partially automate industry and provide a production line with a variety of benefits. They have better wrist movements and are more elastic, which helps to improve movement and the production line. We utilized an Arduino Nano for automatic control and an HC05 Bluetooth module for manual operation to complete this project.

Keywords Android studio · Embedded C · Arduino nano board · PCB board · PCA9685 controller · MG996R servo motors · MG958 servo motors

23.1 Introduction

23.1.1 What Is Robotic Arm?

A robotic arm with rotational couplings is referred to as articulated robotic arm. A wide range of automatic operations will be carried out with articulated robots. The forearm, wrist, upper arm, shoulder, and case together make up the articulated robot arm. The architecture and spontaneous movements of articulated robots are very

---

D. Teja Priyanka (✉) · E. Naga Lakshmi · M. Maha Tej · M. Purna Jayanthi
Velagapudi Ramakrishna Siddhartha Engineering College, Kanuru 520007, India
e-mail: priyankaammulu666@gmail.com

G. Narasimha Swamy
Andhra University, Visakhapatnam 530003, India

V. Naga Prudhvi Raj
Madras University, Chennai 600005, India

similar to those of a human arm. The arm itself can be made up of two to ten rotary couplings that serve as axes, with each joint or axis allowing for a greater range of motion.

23.1.2 Laws of Robot

The term robot first came out in 1921 but was not a specialized term. In, 1942, Isaac Asimov authored a short sci-fi story in which the word robotics was first applied and he presented three laws of robotics [1]. They are:

Law 1: Human beings should not be injured by robots or become injured through inactivity by robots.

Law 2: Unless similar orders conflict with the first law, robots are required to cleave the arrangements made by humans.

Law 3: In accordance with the first law, robots must defend their own actuality.

Laws-Asimov's robotics laws are not scientific laws; rather, they are instructions incorporated into every robot in his stories to protect them from malfunctioning in a dangerous fashion. The first rule is that a robot must not harm a human or enable a human to be harmed by inaction. The second law states that a robot must obey any human-given order, and the third law states that a robot must avoid actions or situations that could lead it to hurt itself. Where these principles clash, the first law takes precedence, followed by the second, with the robot's self-preservation coming last. For instance, if a person told a robot to attack another human, the robot would not comply (the first law taking precedence over the second), but it would comply if the human told the robot to destroy itself (the second law taking precedence over the third).

Robotic arm mainly consists of manipulator, end-effector actuators, sensors, controller, processor, and eventually software. The degrees of freedom can be calculated as

$$n_{dof} = \beth(n-1) - \sum_{i=1}^{k}(\beth - f_i), \qquad (23.1)$$

where n is the number of links, k is the number of joints, f_i is the number of degrees of freedom of ith joint, and $\beth$ is 3 for planar mechanisms and 6 for spatial mechanisms.

In robotics, both direct and inverse kinematics can be used for calculation. Here, we used inverse kinematics. Steps to calculate:

1. Determine the matrix's determinant.
2. Matrix transposition.
3. The transposed matrix should be replaced with the minor of each element.

23.2 Related Work

Agbaraji et al. proposed the modeling of a 3-DOF articulated manipulator rested on base-dependent collaborative artifice and the decision of thick damping coefficient for the common torque-controlled model. This path was thus recommended for robust controlled of automated manipulators under misgivings [2]. Serrezuela, Ruthber rodríguez had proposed the evolution of the design and application of a kinematic model for a four-degree-of-freedom manipulator robot armed typed. The main demerits were packed to achieve message between MATLAB and Arduino were dlls [3]. Mouli et al. proposed the design and implementation of a robotic arm using LabVIEW and embedded system tools. An optimization procedure using neural network models and LabVIEW for simulation was conducted to improve overall inverse kinematics issue of a robotic arm [4]. Proposed a robotic arm using LabVIEW software. LabVIEW was used to create the lab clients and trial machines. A webcam was used to deliver optical feedback from the lab to the client [5]. Limb and Brian proposed the robot controlling using LabVIEW. Utilizing LabVIEW, they have programmed a robotic arm of five degrees of freedom to displace a rustic range from one rustic cut to another, bluffing a normal wafer-addressing job [6]. Tomas et al. proposed implementation of robotic arm using plc. The authors demonstrate the benefits of employing fuzzy logic in situations where the mathematical model of the controlled system was unavailable. Fuzzy control++ had been used successfully to create fuzzy control in the plc sematic s7-300 [7]. Kruthika et al. proposed the advancement of a robotic arm with five degrees of freedom (DOF), which was used to provide elderly or particularly challenged people. The robotic arm was controlled by using the principles of robotic kinematics and MATLAB. The Arduino MEGA2560 I/O board serves as the primary processing unit for this project and links to the graphical user interface, motors, and sensors. Compim, proteus, the Arduino IDE, and processing communicate serially was used [8]. Deshpande and George used kinematic modeling and implemented 5-DOF robotic arm. In this article, the forward and inverse kinematics of a 5-DOF automated armed for straightforward pick and place operation was modeled [9].

23.3 Need for Six-Axis Robotic Arm

This –six-axis robotic arm is very useful for complex applications which cannot be done by –five-axis, four-axis, and so on. The functionality of six-axis robotic arm is suitable for complex movements that simulate a human arm [10–15]. Better wrist action and elasticity, robot software and programming qualifications, and numerous mounting add-ons are just some of the advantages these six-axis robots possess to give. There is also a bulky range of robot sizes, loads, and preferences to opt from. The work envelope is more for 6-DOF robotic arm when compared to other arms. By

Table 23.1 Comparison table for degrees of freedom

S. No.	Features	2-DOF	3-DOF	4-DOF	5-DOF
1	Links	2	3	4	5
2	Joints	1	2	3	4
3	DOF	2	3	4	5
4	No. of motors	2	3	4	5
5	Range (angle)	90°	±90°	180°	±180°

using six-axis robotic arm, the production line will increase and avoid unnecessary costs. For example, if an object is picked, need to change its orientation and placed to be in new destination, which cannot be done by three-axis or five-axis robotic arm because its end-effector does not have 360° rotation, so six-axis robotic arm is suitable for such kind of complex applications. The comparison table for different features of robotic arm for different degrees of freedom is shown in Table 23.1.

23.4 Implementation of 6-DOF Robotic Arm

23.4.1 Construction Principle

Axis 6 is the wrist of an artificial robot. This axis is accountable for the entire 360-degree gyrations of the wrist. The 6th axis gives artificial robots the capacity to remake an allotment's aspect in the x, y and z airplanes with roll, pitch, and yaw motions.

23.4.2 Mathematical Expressions

The step-by-step frame assignment for each joint link of the manipulator according to algorithm is shown in Fig. 23.1 and is explained here. Beginning with joint 1, which connects links 0 and 1, the six joints are numbered from 1 to 6. Each joint axis's orientation and each joint's variable are noted and labeled in accordance with the home position. The joint-link parameters for standard manipulator are shown in Table 23.2.

23 Design and Development of Six-Axis Robotic Arm for Industrial ...

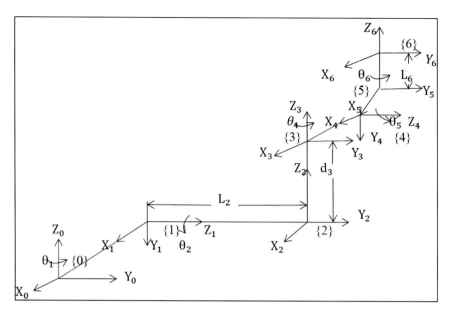

Fig. 23.1 Frame assignment for 6-DOF manipulator

Table 23.2 Joint-link parameters for standard manipulator

Link i	a_i	α_i	d_i	θ_i	q_i	$C\theta_i$	$S\theta_i$	$C\alpha_i$	$S\alpha_i$
1	0	$-90°$	0	θ_1	θ_1	C_1	S_1	0	-1
2	0	$90°$	L_2	θ_2	θ_2	C_2	S_2	0	1
3	0	0	d_3	0	d_3	1	0	1	0
4	0	$-90°$	0	θ_4	θ_4	C_4	S_4	0	-1
5	0	$90°$	0	θ_5	θ_5	C_5	S_5	0	1
6	0	0	L_6	θ_6	θ_6	C_6	S_6	1	0

The six link transformation matrices are, thus

$${}^{1}_{2}T(\theta_2) = \begin{bmatrix} C_2 & 0 & S_2 & 0 \\ S_2 & 0 & -C_2 & 0 \\ 0 & 1 & 0 & L_2 \\ 0 & 0 & 0 & 1 \end{bmatrix},$$

$${}^{2}_{3}T(d_3) = \begin{bmatrix} 1 & 0 & 0 & 0 \\ 0 & 1 & 0 & 0 \\ 0 & 0 & 1 & d_3 \\ 0 & 0 & 0 & 1 \end{bmatrix},$$

$$^3_4T(\theta_4) = \begin{bmatrix} C_4 & 0 & -S_4 & 0 \\ S_4 & 0 & C_4 & 0 \\ 0 & -1 & 0 & 0 \\ 0 & 0 & 0 & 1 \end{bmatrix},$$

$$^4_5T(\theta_5) = \begin{bmatrix} C_5 & 0 & S_5 & 0 \\ S_5 & 0 & -C_5 & 0 \\ 0 & 1 & 0 & 0 \\ 0 & 0 & 0 & 1 \end{bmatrix},$$

$$^5_6T(\theta_6) = \begin{bmatrix} C_6 & -S_6 & 0 & 0 \\ S_6 & C_6 & 0 & 0 \\ 0 & 0 & 1 & L_6 \\ 0 & 0 & 0 & 1 \end{bmatrix}.$$

Lastly, by swapping the individual transform matrix, you can obtain the transformation of the tool frame with respect to the base frame:

$$^0_6T = {^0_1}T\,{^1_2}T\,{^2_3}T\,{^3_4}T\,{^4_5}T\,{^5_6}T,$$

$$T = \begin{bmatrix} n_x & o_x & a_x & d_x \\ n_y & o_y & a_y & d_y \\ n_z & o_z & a_z & d_z \\ 0 & 0 & 0 & 1 \end{bmatrix}.$$

The displacement of prismatic joint d_3 is always positive.

$$d_3 = \sqrt{(d_x - L_6 a_x)^2 + (d_y - L_6 a_y)^2 + (d_z - L_6 a_z)^2 - L_2^2}. \qquad (23.2)$$

By solving, we get joint displacement θ_2 as

$$\theta_2 = A\tan 2(\pm\sqrt{(d_x - L_6 a_x)^2 + (d_y - L_6 a_y)^2 + (d_z - L_6 a_z)^2}). \qquad (23.3)$$

Similarly, by solving we get joint displacement θ_1 as

$$\theta_1 = A\tan\left(\frac{S_2 d_3}{\sqrt{S_2^2 d_3^2 + L_2^2}}, \frac{L_2}{\sqrt{S_2^2 d_3^2 + L_2^2}}\right)$$

$$- A\tan 2\left(\frac{d_x - L_6 d_x}{\sqrt{S_2^2 d_3^2 + L_2^2}}, \frac{d_y - L_6 d_y}{\sqrt{S_2^2 d_3^2 + L_2^2}}\right). \qquad (23.4)$$

Path 1 frame {3} → frame {4} → frame {5} → frame {6}

Along this path, the transformation ^3_6T can be obtained as

$$^3_6T = {^3_4T}{^4_5T}{^5_6T},$$

$$^3_6T = \begin{bmatrix} r_{11} & r_{12} & r_{13} & r_{14} \\ r_{21} & r_{22} & r_{23} & r_{24} \\ r_{31} & r_{32} & r_{33} & r_{34} \\ 0 & 0 & 0 & 1 \end{bmatrix}.$$

The arm point $\theta_4, \theta_5, \theta_6$ can be found by equating the corresponding elements of matrices.

$$S_5 C_6 = r_{31},$$
$$-S_5 S_6 = r_{32}$$
$$C_5 = r_{33}.$$

Squaring Eqs. $S_5 C_6$ and $-S_5 S_6$, adding and dividing the result by C_5 give solution for θ_5:

$$\theta_5 = A\tan 2(\pm\sqrt{(r_{31}^2 + r_{32}^2)}, r_{33}). \tag{23.5}$$

Similarly, θ_4 and θ_6 can be expressed as

$$\theta_4 = A\tan 2\left(\frac{r_{24}}{S_5}, \frac{r_{14}}{S_5}\right), \tag{23.6}$$

$$\theta_6 = A\tan 2\left(-\frac{r_{32}}{S_5}, \frac{r_{31}}{S_5}\right). \tag{23.7}$$

23.4.3 Components Used

- Robotic arm metal chassis.
- MG996R servo motors.
- MG958 servo motors.
- Arduino Nano.
- PCA9685 Controller.
- Arduino IDE wire.
- HC05 Bluetooth module.

To finish this project, we have chosen a metal robotic arm chassis, MG996R and MG958 servo motors, an Arduino Nano, a PCA9685 controller, Arduino IDE wire, and an HC05 Bluetooth module. In order to upload programs and interact with the Arduino hardware, we used the Arduino studio, which connects to the hardware. The Arduino Nano code was written in embedded C.

23.4.4 Flowchart and Operation

The operation of flowchart shown in Fig. 23.2 is explained below.

Step 1: Start.

Step 2: Initialize the servo motor and Bluetooth.

Step 3: If the condition if (SW) is YES, the manual mode operation will take place, otherwise automatic mode operation will take place.

Step 4: In manual mode, if the condition If (CMD from BT) is YES, then the operation will continue otherwise stops.

Step 5: Servo motors will rotate when their respective conditions are YES, otherwise need to check and replace the servo motor if it is damaged.

Step 6: In automatic mode when the condition if (SW) is NO, the servo motors 1 to 6 will rotate according to their conditions until the power supply is disconnected.

Note: Robotic arm will halt only when the supply is disconnected.

23.5 Results and Discussion

A robotic arm that can function in both manual and automatic modes is developed because automation is required in many industries to decrease errors, enhance the production line, and enhance the quality of the product. A six-axis robotic arm allows us to reduce the number of robotic arms we need to utilize, which implies that in some applications, we can use just one six-axis robotic arm rather than two three-axis robotic arms. Six-axis robotic arm is capable of reaching a longer range when compared to the five-axis, four-axis, and so on. The prototype of the developed robotic arm is shown in Fig. 23.3. Fully automated robotic arm for palletizing and depalletizing application; partially automated to control the robotic arm when obstacles occur during the automatic operation—both have been implemented. The jerks that occurred during operation have been reduced with the help of reducing the speed in the code. The chassis is controlled by high-torque motors such as the MG996R and MG958 servo motors. The comparison of the proposed solution with the existing solutions with respect to different parameters is shown in Table 23.3.

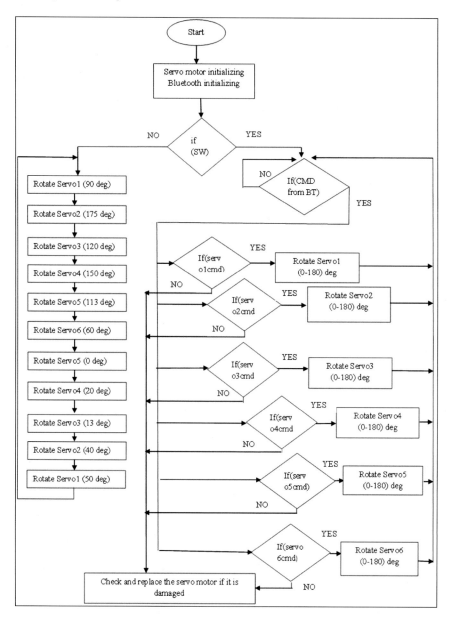

Fig. 23.2 Flowchart for operation

Fig. 23.3 Implemented 6-DOF robotic arm prototype

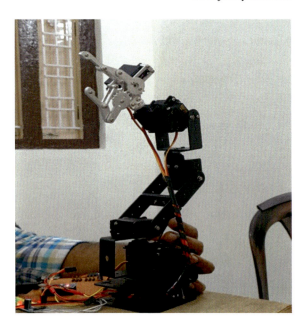

Table 23.3 Comparison table for proposed and existing solutions

S. No	Parameters	Existing solutions	Proposed solution
1	Range	$< 360°$	$\pm 360°$
2	Man power	Required more	Required less
3	Time	More	Less
4	Speed of operation	Speed of operation is low	Speed of operation is high

23.6 Conclusion

The proposed model is developed to carry out a single continuous application in any industry. Same model can be used for different applications by changing code for servo motors and end-effectors. Instead of man doing the same work continuously, we can replace the person with a robotic arm for such tasks which are boring and hazardous. By doing this, the speed of operation as well as quality increases and there will be a good productivity line. But disadvantage in our proposed model is, the dimensions of the object need to be entered frequently if the object is replaced in the application and also need to add proximity sensor for detecting an object itself.

References

1. Parhi, D.R., et al.: Forward and inverse kinematic models for an articulated robotic manipulator. Int. J. Artif. Intell. Comput. Res. **4.**2, 103–109 (2012)
2. Agbaraji, E.C., Inyiama, H.C., Okezie, C.C.: Dynamic modeling of a 3-DOF articulated robotic manipulator based on independent joint scheme. Phys. Sci. Int. J. **15**(1), 1–10 (2017)
3. Serrezuela, R.R. et al.: Kinematic modelling of a robotic arm manipulator using Matlab. ARPN J. Eng. Appl. Sci. **12.**7, 2037–2045 (2017)
4. Mouli, C.C., Jyothi, P., Nagabhushan Raju, K., Nagaraja, C.: Design and implementation of robot arm control using labview and arm controller. IOSR J. Electri. Electron. Eng. **6**(5), 80–84 (2013)
5. Akinwale, O., Kehinde, L., Ayodele, K.P., Jubril, A.M., Jonah, O. P., Ilori, S., Chen, X.: A labview based on line robotic arm for students' laboratory. In: 2009 Annual Conference and Exposition, June, pp. 14–39. (2009)
6. Lim, B.: LabVIEW programming for robot control (2004)
7. Tomas, S., Michal, K., Alena, K.: Fuzzy control of robotic arm implemented in PLC. In: 2013 IEEE 9th International Conference on Computational Cybernetics (ICCC). IEEE (2013)
8. Kruthika, K., Kiran Kumar, B.M., Lakshminarayanan, S.: Design and development of a robotic arm. In: 2016 International Conference on Circuits, Controls, Communications and Computing (I4C). IEEE, 2016.R. Nicole, "Title of paper with only first word capitalized," J. Name Stand. Abbrev., in press
9. Deshpande, V., George, P.M.: Kinematic modelling and analysis of 5 DOF robotic arm. Int. J. Robot. Res. Developm. (IJRRD) **4.**2, 17–24 (2014)
10. Ahmed, A., Yu, M., Chen, F.: Inverse kinematic solution of 6-DOF robot-arm based on dual quaternions and axis invariant methods. Arab. J. Sci. Eng. 1–16 (2022)
11. Brathikan, V.M., Prashanth, P., Vikash, S.: Design and development of 5-axis robot. Mater. Today: Proc. (2022)
12. Manolescu, V.D., Secco, E.L.: Design of an assistive low-cost 6 dof robotic arm with gripper. In: Proceedings of Seventh International Congress on Information and Communication Technology: ICICT 2022, London, vol. 1, Singapore, Springer Nature Singapore (2022)
13. Chen, F., Hehua, J., Liu, X.: Inverse kinematic formula for a new class of 6R robotic arms with simple constraints. Mech. Mach. Theory **179**, 105118 (2023)
14. Suwarno, I., et al.: Current trend in control of artificial intelligence for health robotic manipulator. J. Soft Comput. Explor. **4.**1 (2023)
15. Cornejo, J., et al.: Mechatronics design and kinematic simulation of UTP-ISR01 robot with 6-DOF anthropomorphic configuration for flexible wall painting. In: 2022 First International Conference on Electrical, Electronics, Information and Communication Technologies (ICEEICT). IEEE (2022)

Chapter 24
A Solution to Collinear Problem in Lyapunov-Based Control Scheme

Kaylash Chaudhary, Avinesh Prasad, Vishal Chand, Ahmed Shariff, and Avinesh Lal

Abstract Robots are widely used to carry out various tasks in different industries worldwide. The movement of a robot is necessary for any task accomplishment. While moving, a robot must prevent collisions with obstacles to reach its destination successfully. The motion control algorithm governs a robot's movement. One such method is Lyapunov-based control scheme (LbCS). LbCS is a popular method for controlling a robot's motion, but the technique suffers from a problem known as collinear. This problem occurs when a robot, an obstacle, and a target are in a linear position, which gets the method trapped into local minima. This paper tackles this problem using a heuristic-based method, ant colony optimization (ACO). The ACO will be activated when the LbCS gets trapped in local minima. This paper presents an algorithm, *ACO-LbCS*, that solves the collinear problem of LbCS. This hybrid algorithm has been strategically formulated using ACO and LbCS. The algorithm has been applied to multiple obstacle's environment. The results show that the problem of local minima has been solved by the proposed algorithm.

Keywords Robot · Optimization · Collinear · Motion · Control

24.1 Introduction

Today, robots are used in many industries due to their improved performance and ability to perform tasks. These industries include agriculture [11], mining [19], transportation [25], military [12], manufacturing, and civil engineering [11], to name a few. For a robot to complete a task, it must move safely from an initial location to a target location by preventing collisions with obstacles in its path. In any case, a robot should not compromise safety. This problem is known as a find path or robot navigation problem. In particular, this paper focuses on motion control.

K. Chaudhary (✉) · A. Prasad · V. Chand · A. Shariff · A. Lal
The University of the South Pacific, Suva, Fiji
e-mail: kaylash.chaudhary@usp.ac.fj

Different ways to tackle the motion control problem include heuristic, machine learning, deep learning, and classical. This paper will use a heuristic and classical approach to solve the motion control problem. Examples of heuristic approaches include ant colony optimization, firefly algorithm, and artificial bee colony, to name a few [20, 30]. Classical techniques comprise the artificial potential field [18], cell decomposition [13], virtual force field and road map [5]. This paper will focus on the artificial potential field approach, in particular, Lyapunov-based control scheme (LbCS). LbCS is a movement control method that has been widely used in literature [8, 14, 24-25]. Though LbCS has been used in many research studies, it has a problem of getting trapped in local minima. That is, a robot driven by LbCS will come to a complete stop when trapped in local minima and cannot complete its allocated task. According to the author's knowledge, this problem has not been addressed yet.

The focus of this paper will be to resolve the local minimum problem of LbCS using ACO and kinematic equations. The choice of ACO in the continuous domain has been made because of its popularity [26]. The reason behind choosing the ACO variant by Socha and Dorigo is its performance advantage over other variants [28]. In particular, this paper will present an algorithm named ACO-LbCS, a combination of classical and heuristic methods. The hybrid algorithm will start with LbCS and will only switch to ACO when there is a likelihood of a robot being caught at a location because the algorithm is in the local minima. The ACO will generate points which the kinematic equations will then use to move the robot to that point. The algorithm will then switch back to LbCS, which will continue to govern the motion of the robot.

This paper contributes the following to the literature:

- ACO-LbCS algorithm: This algorithm is a new hybrid specifically designed to rescue a trapped robot governed by LbCS. According to the author's knowledge, no such type of algorithm exists in the literature.
- Application: The proposed algorithm has been applied to two scenarios. The results show that the algorithm successfully solved the collinear problem of LbCS.

The next section will present an overview of the literature regarding motion control, optimization algorithm, and collinear problems. Section 24.3 presents the research objectives of this paper. Lyapunov-based control scheme is discussed in Sect. 24.4. Ant colony optimization algorithms with the multiobjective problem are discussed in Sect. 24.5. Section 24.6 presents and discusses the new hybrid algorithm. Results are discussed in Sect. 24.7. Section 24.8 discusses the conclusion and future work.

24.2 Related Work

Many motion control algorithms in the literature are either heuristics, classical, or hybrid. This section will summarize the motion control problems these algorithms solved.

Firstly, there are many applications of heuristic algorithms to path planning problems. For example, the research article by Akka and Khaber [3] used an improved version of ant colony optimization for mobile robot path planning. Likewise, an aging-based ant colony optimization has been applied to grid-based mobile robot path planning in both, dynamic and static environments [2]. The authors also compared the aging-based ant colony optimization with an artificial bee colony, genetic algorithm, and particle swarm optimization algorithms, proving to be superior to the three. An improved brainstorm optimization algorithm has also been applied to grid-based robot path panning [29]. Dewang et al. [10] conducted research on the path planning of robots using modified particle swarm optimization. The reader is referred to some other examples of heuristic algorithms used in robot path planning [4, 6, 20, 31].

Secondly, research on robot path finding and motion control has been carried out using the artificial potential field. The development of Lyapunov-based control scheme by [27] has been to solve many motion control problems by different researchers. For example, LbCS has been used in 3D environments for the mobile manipulator [22]. It has also been applied for controlling quadrotors, tractor-trailer robots, and navigating car-like robots [15-17, 21, 23-24].

Thirdly, some research is available on the hybrid algorithms used for robot path planning. For example, a firefly algorithm has been combined with the artificial potential field for optimized robot path planning [1]. Similarly, ant colony optimization has been combined with kinematic equations for robot motion control [9].

All the algorithms applied to different problems above do not address the collinear problem of LbCS. Therefore, the proposed algorithm in this research will be the first to solve the collinear problem of LbCS.

24.3 Research Objectives

This paper addresses the following objectives:

- Design and implement a hybrid algorithm consisting of Lyapunov-based control scheme (LbCS) and ant colony optimization algorithm to solve the local minimum problem of LbCS.
- Apply the proposed algorithm to different scenarios.

24.4 Lyapunov-Based Control Scheme

Lyapunov-based control scheme (LbCS) is a powerful technique developed by Sharma et al. in [27] for deriving control laws for autonomous systems. The LbCS is based on Lyapunov method and works within the framework of the artificial potential field method. As an example, consider a moving point-mass robot in a two-dimensional plane, whose kinematic equations are:

$$\dot{x} = u_1, \quad \dot{y} = u_2, \tag{24.1}$$

where (x, y) are the positions of the robot and (u_1, u_2) are the velocity components in the x and y directions, respectively. Suppose there are $m > 0$ fixed circular obstacles and $n > 0$ fixed line segment obstacles that the robot needs to avoid on its way. Let the center of the ith circular obstacle be (o_{i1}, o_{i2}) and radius ro_i. Similarly, assume that the kth line segment is a straight line from (a_{k1}, a_{k2}) to (b_{k1}, b_{k2}). If we require the robot to relocate from a starting location to a destination at (p_1, p_2) and avoid fixed circular and line obstacles, then a tentative Lyapunov function is

$$L(x, y) = V(x, y)\left(1 + \sum_{i=1}^{m} \frac{\alpha_i}{W_i(x, y)} + \sum_{k=1}^{n} \frac{\beta_k}{LO_k(x, y)}\right), \tag{24.2}$$

where $V(x, y) = \frac{1}{2}\left[(x - p_1)^2 + (y - p_2)^2\right]$ is the target attraction function;
$W_i(x, y) = \frac{1}{2}\left[(x - o_{i1})^2 + (y - o_{i2})^2 - ro_i^2\right]$ is the avoidance function for the ith circular obstacle;
$LO_k(x, y) = \frac{1}{2}\left[(x - a_{k1} - \gamma_k(a_{k2} - a_{k1}))^2 + (y - b_{k1} - \gamma_k(b_{k2} - b_{k1}))^2\right]$ is the avoidance function for the kth line obstacle; $\alpha_i > 0$ and $\beta_k > 0$ are control parameters and $\gamma_k \in [0, 1]$.

Differentiating the Lyapunov function and setting $\dot{L}(x, y) \leq 0$, we obtain

$$u_1 = -\frac{1}{\delta_1}\frac{\partial L}{\partial x} \quad \text{and} \quad u_2 = -\frac{1}{\delta_2}\frac{\partial L}{\partial y}. \tag{24.3}$$

where $\delta_1 > 0$ and $\delta_2 > 0$ are constants called convergence parameters.

The LbCS has been used in different domains by many researchers [7, 8, 14, 24, 27]. In a cluttered environment with varying types of obstacles, the LbCS is a powerful method for developing movement controllers for a robot so that it can move from an initial position to the target. The convergence in LbCS occurs because of an attractive force which pulls a robot to its target. The safety of a robot is ensured by a repulsive force, which enables it to avoid obstacles on its way.

When planning robot trajectories, artificial potential field methods, including LbCS, suffer from the problem of local minima. Local minima in robot motion control mean that the position of a robot, its destination, and an obstacle are collinear. An example of a collinear problem is shown in Fig. 24.1.

Figure 24.1 shows a collinear problem with a start (5, 5), an obstacle (15, 15), and a target (25, 25) locations. It also shows that the point-mass robot driven by LbCS could not move forward because it is trapped at local minima and could not escape by itself. This collinear problem will be addressed in this paper.

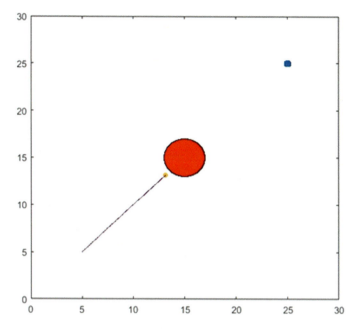

Fig. 24.1 Collinear problem example

24.5 Ant Colony Optimization

Ant colony optimization, also known as ACO, was inspired with natural ants that leave their nests to find food by moving randomly through the environment. Upon discovering and evaluating food, an ant transports it on its back. While returning to the nest, the ants leave trails of pheromones on the terrain, depending on the volume and quality of food. These trails left by an ant guide others to the food source from their nest. The ACO variant used in this paper is for the continuous domain known as ACOR, which has been proposed by Socha and Dorigo [28].

The two population types used by the ACOR algorithm are archive and new. Both populations represent solutions, and the archive population consists of the pheromone information. Each solution in the archive is evaluated by the fitness function and is stored in ascending order. A solution quality weight percentage is associated with each solution in the archive population, in which ACOR uses to make probabilistic choices. Each solution has an associated solution quality weight percentage. When LbCS gets trapped at local minima, the robot's next step is determined by the best ant in the population. The pheromone information determines the selection of the best ant. The robot then moves to that step using the equation of motion.

The steps are planned using the objective function, which considers a safe and short path. Even though the purpose of the ant colony optimization algorithm is to relieve the LbCS from local minima, the objective function still needs to consider a safe and short path because a point that ants will generate needs to be safe (away from obstacles) and closest to the target.

A robot's position from the target should always be at a minimum distance. To achieve this, the distance between an ant and the target can be calculated, as shown in Eq. (24.4).

$$d_{it} = \sqrt{(p_1 - xa_i)^2 + (p_2 - ya_i)^2}, \qquad (24.4)$$

where d_{it} is the distance between an ant and the target, (p_1, p_2) are the coordinates of the target, and (xa_i, ya_i) are the coordinates of the ith ant.

Since the workspace is cluttered with obstacles, a robot must avoid them to reach the target. The paper uses circular and line obstacles. The distance between an ant and an obstacle is calculated for circular obstacles, as shown in Eq. (24.5).

$$d_{il} = \sqrt{(o_{11} - xa_i)^2 + (o_{12} - ya_i)^2}, \qquad (24.5)$$

where d_{il} is the distance between an ant and an obstacle and $(o_{11} o_{12})$ are the coordinates of the lth obstacle.

Multiple points are generated on the line for line obstacles, and the distance between the ants and the points is calculated. The minimum distance technique considers the point with the smallest distance [6]. Again, the distance between an ant and a point on the line is calculated as shown in Eq. (24.6).

$$d_{ik} = \sqrt{(a_{k1} + \lambda_k(a_{k2} - a_{k1}) - xa_i)^2 + \lambda_k(b_{k1} + (b_{k2} - b_{k1}) - ya_i)^2}, \qquad (24.6)$$

where d_{ik} is the distance between an ant and a point on the kth line segment, $\lambda_k \in [0, 1]$, and (a_{k1}, a_{k2}) and (b_{k1}, b_{k2}) are the coordinates of the points on the line segment.

Since there are many line segments ($k = 1, 2, .. n$) and circular obstacles ($l = 1, 2, .. n$), the distances between an ant and obstacles are summed and calculated as follows:

$$f_l = \sum_{l=1}^{n} d_{il}, \quad f_k = \sum_{k=1}^{n} d_{ik}. \qquad (24.7)$$

The problem of releasing the LbCS from local minima is a minimization optimization problem. The fitness equation for each ant (f_i) is defined as follows:

$$f_i = a.\frac{1}{f_l} + b.\frac{1}{f_k} + c.d_{it}. \qquad (24.8)$$

The coordinates of the ant with minimum fitness value will be the robot's new position that has been stopped due to a collinear problem. The position of the fittest ant, calculated by Eq. (24.8), will be chosen as the robot's next step. This procedure will persist until the robot moves from its current location to another; hence, the robot is said to be out of the local minima. Then LbCS will then take control of the robot. Control parameters such as a, b, and c determine path safety and are set using the brute force technique.

24.6 Proposed Algorithm

The proposed algorithm is implemented by strategically combining the ant colony optimization, kinematic equations, and LbCS together and is known as *ACO-LbCS*. ACO with kinematic equations has been recently applied to robot motion control problems, hence the choice of combining with LbCS. There are various hybrid algorithms in the literature, but according to the author's knowledge, none of these algorithms solves the collinear problem of LbCS. The pseudocode of the ACO-LbCS is presented in Algorithm 1.

Algorithm 1: ACO-LbCS.

While (robot location < target)

　Phase1: LbCS
　Move the robot using LbCS
　Phase 2: ACO
　While (($\dot{x} < 0.01$ && ($\dot{x} > -0.01$) && ($\dot{y} < 0.01$ && ($\dot{y} > -0.01$))

　　Initialize archive population
　　Initialize the weights and selection probabilities
　　For all ants
　　　Compute the means
　　End For
　　For all ants
　　　Compute the standard deviation
　　End For
　　Generate a sample size population
　　For all ants
　　　Build solutions
　　End For
　　Assess sample size solutions
　　Combine the archive population with the sample population
　　Sort ants in ascending order in the combined population
　　Use the Kinematic equation to move the robot to the first ant's (best) position

> Revise the location of the robot with the best ants position
> **End While**
> **End While**

The algorithm starts with LbCS, which moves the robot toward the target while avoiding obstacles. If the robot driven by LbCS is trapped in local minima, the second phase of the algorithm starts with the generation of ants, and after processing, the robot moves to the best position determined by ants. This moves the robot to a new position, solving the problem of LbCS. The authors in this paper have used a range to determine the collinear problem, that is:

$$-0.01 \langle \dot{x} \rangle 0.01 \text{ and } -0.01 \langle \dot{y} \rangle 0.01,$$

where $\dot{x}$ and $\dot{y}$ are kinematic equations defined in Eq. (24.1).

24.7 Results

Figure 24.2 shows the trajectory generated by the proposed algorithm. The robot uses the Lyapunov-based control scheme to move toward the destination. However, it comes to a halt just before the obstacle because the artificial potential field nears zero. The ACO–kinematic algorithm then generates points, and the robot moves to that point. The algorithm then switches back to Lyapunov-based control scheme, which successfully navigates the robot to the destination.

Figure 24.3 shows that the point-mass robot avoids multiple obstacles to reach its destination. The Lyapunov-based control scheme navigates the point-mass robot toward the last obstacle (line segment) but could not avoid that obstacle. Therefore, the ACO–kinematic algorithm generates points that navigate the robot around that obstacle. Once the artificial potential field is in the safe range, the algorithm switches back to Lyapunov-based control scheme.

24.8 Conclusion

A new hybrid algorithm presented in this paper is used to solve the collinear problem of the Lyapunov-based control scheme. The algorithm comprises the strategic formulation of the ACOR algorithm with kinematic equations and LbCS. The ant colony optimization algorithm will only be executed when the robot driven by LbCS is about to be trapped in local minima. The equation of motion will use the points generated by the ACOR algorithm to move the robot to a new location. Suppose the new location

24 A Solution to Collinear Problem in Lyapunov-Based Control Scheme

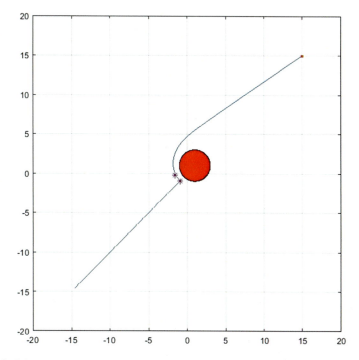

Fig. 24.2 Point-mass robot with initial position $(-15, -15)$ and target position $(15, 15)$

Fig. 24.3 Point-mass robot with initial position $(5, 5)$ and target position $(45, 45)$

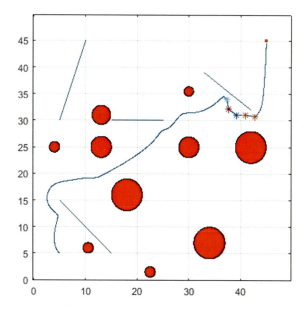

is still part of the local minima. The algorithm will continue executing ant colony optimization to generate new points to avoid the local minima. The ant colony optimization algorithm will only be executed when there is a chance of LbCS getting trapped in local minima. Otherwise, LbCS is responsible for avoiding obstacles and governing the motion of the robot.

The proposed algorithm was applied to two scenarios. The first scenario based on one robot and one obstacle showed that the algorithm was able to release the robot from its location to a new one. Likewise, scenario 2, consisting of one robot and multiple circular and line obstacles, showed that the robot was able to reach its destination despite the existence of the collinear problem. In other words, the robot would not have reached its destination if it had been governed solely by LbCS.

The current work is limited to only static obstacles with known locations and has been applied to medium environment complexity. In future work, the authors will use the proposed algorithm in an environment where obstacles are static and the locations are unknown to the robot. Also, the environmental complexity will be increased by adding more robots and including different types and sizes of obstacles. Due to the limited scope of this paper, the authors will conduct a mathematical analysis of the algorithm in the future.

References

1. Abbas, N.: Optimization of the path planning for the robot depending on the hybridization of artificial potential field and the firefly algorithm. pp. 040004. (2023)
2. Ajeil, F., Ibraheem, I., Azar, A., Humaidi, A.: Grid-based mobile robot path planning using aging-based ant colony optimization algorithm in static and dynamic environments. Sensors **20**(7), 1880 (2020)
3. Akka, K., Khaber, F.: Mobile robot path planning using an improved ant colony optimization. Int. J. Adv. Robot. Syst. **15**(3), 172988141877467 (2018)
4. Ali, S., Yonan, J., Alniemi, O., Ahmed, A.: Mobile robot path planning optimization based on integration of firefly algorithm and cubic polynomial equation. J. ICT Res. Appl. **16**(1), 1–22 (2022)
5. Bortoff, S.: Path planning for UAVs. In: Proceedings of the 2000 American Control Conference. ACC (IEEE Cat. No.00CH36334), vol.1. pp. 364–368 (2000)
6. Brand, M., Yu, X.-H.: Autonomous robot path optimization using firefly algorithm. In: 2013 International Conference on Machine Learning and Cybernetics, pp. 1028–1032. (2013)
7. Chand, R., Chand, R., Kumar, S.: Switch controllers of an n-link revolute manipulator with a prismatic end-effector for landmark navigation. PeerJ. Comput. Sci. **8**, e885 (2022)
8. Chand, R., Raghuwaiya, K., Vanualailai, J.: Leader-follower strategy of fixed-wing unmanned aerial vehicles via split rejoin maneuvers. pp. 231–245. (2023)
9. Chaudhary, K., Prasad, A., Chand, V., Sharma, B.: ACO-kinematic: a hybrid first off the starting block. PeerJ. Comput. Sci. **8**, e905 (2022)
10. Dewang, H., Mohanty, P., Kundu, S.: A robust path planning for mobile robot using smart particle swarm optimization. Proc. Comput. Sci. **133**, 290–297 (2018)
11. Dunbabin, M., Marques, L.: Robots for environmental monitoring: significant advancements and applications. IEEE Robot. Autom. Magazine **19**(1), 24–39 (2012)
12. Khurshid, J., Bing-Rong, H. (n.d.).: Military robots—a glimpse from today and tomorrow. In: ICARCV 2004 8th Control, Automation, Robotics and Vision Conference, pp. 771–777. (2004)

13. Kloetzer, M., Mahulea, C., Gonzalez, R.: Optimizing cell decomposition path planning for mobile robots using different metrics. In: 2015 19th International Conference on System Theory, Control and Computing (ICSTCC), pp. 565–570. (2015)
14. Kumar, S., Chand, R., Chand, R., Sharma, B.: Linear manipulator: motion control of an n-link robotic arm mounted on a mobile slider. Heliyon **9**(1), e12867 (2023)
15. Kumar, S., Vanualailai, J., Sharma, B.: Lyapunov functions for a planar swarm model with application to nonholonomic planar vehicles. In: 2015 IEEE Conference on Control Applications (CCA), pp. 1919–1924. (2015)
16. Kumar, S., Vanualailai, J., Sharma, B.: Lyapunov-based control for a swarm of planar nonholonomic vehicles. Mathem Comput Sci **9**(4), 461–475 (2015)
17. Kumar, S., Vanualailai, J., Sharma, B., Prasad, A.: Velocity controllers for a swarm of unmanned aerial vehicles. J. Indus. Inform. Integrat. **22**, 100198 (2021)
18. Lee, M.C., Park, M.G. (n.d.): Artificial potential field based path planning for mobile robots using a virtual obstacle concept. In: Proceedings 2003 IEEE/ASME International Conference on Advanced Intelligent Mechatronics (AIM 2003), pp. 735–740. (2003)
19. Murphy, R., Kravitz, J., Stover, S., Shoureshi, R.: Mobile robots in mine rescue and recovery. IEEE Robot. Autom. Magazine **16**(2), 91–103 (2009)
20. Patle, B., Pandey, A., Jagadeesh, A., Parhi, D.: Path planning in uncertain environment by using firefly algorithm. Defence Technol. **14**(6), 691–701 (2018)
21. Prasad, A., Sharma, B., Vanualailai, J.: Motion control of a pair of cylindrical manipulators in a constrained 3-dimensional workspace. In: 2017 4th Asia-Pacific World Congress on Computer Science and Engineering (APWC on CSE), pp. 75–81. (2017)
22. Prasad, A., Sharma, B., Vanualailai, J., Kumar, S.: Motion control of an articulated mobile manipulator in 3D using the Lyapunov-based control scheme. Int. J. Control **95**(9), 2581–2595 (2022)
23. Raj, J., Raghuwaiya, K., Vanualailai, J.: Collision avoidance of 3D rectangular planes by multiple cooperating autonomous agents. J. Adv. Transp. **2020**, 1–13 (2020)
24. Raj, J., Raghuwaiya, K., Sharma, B., Vanualailai, J.: Motion control of a flock of 1-trailer robots with swarm avoidance. Robotica **39**(11), 1926–1951 (2021)
25. Raj, J., Raghuwaiya, K., Vanualailai, J., Sharma, B.: Navigation of car-like robots in three-dimensional space. In: 2018 5th Asia-Pacific World Congress on Computer Science and Engineering (APWC on CSE), pp. 271–275. (2018)
26. Reshamwala, A., Vinchurkar, D.P.: Robot path planning using an ant colony optimization approach: a survey. Int. J. Adv. Res. Artif. Intell. **2**(3) (2013)
27. Sharma, B., Vanualailai, J., Singh, S.: Tunnel passing maneuvers of prescribed formations. Int. J. Robust and Nonlinear Control **24**(5), 876–901 (2014)
28. Socha, K., Dorigo, M.: Ant colony optimization for continuous domains. Europ. J. Operat. Res. **185**(3), 1155–1173 (2008)
29. Tuba, E., Strumberger, I., Zivkovic, D., Bacanin, N., Tuba, M.: Mobile robot path planning by improved brain storm optimization algorithm. In: 2018 IEEE Congress on Evolutionary Computation (CEC), pp. 1–8. (2018)
30. Wang, H., Zhou, Z.: A heuristic elastic particle swarm optimization algorithm for robot path planning. Information **10**(3), 99 (2019)
31. Zips, P., Böck, M., Kugi, A.: Fast optimization based motion planning and path-tracking control for car parking. IFAC Proc Volumes **46**(23), 86–91 (2013)

Chapter 25
A Detailed Review of Ant Colony Optimization for Improved Edge Detection

Anshu Mehta and Deepika Mehta

Abstract Due to rapid enhancement in image processing, there is need to design and implement an improved edge detection algorithm in order to analyzing the edges of an original image. Optimization mechanism based on ant colony optimization technique has been used in present work. Research work is focused on implementation of edge detection using ant colony optimization algorithm on MATLAB and to improve the drawbacks of that algorithm and comparing it with the new improved algorithm. Present research is focused on the performance parameters, namely RMSE and PSNR. Thus, edge detection process The edge detection process considers selection of the image as input, and image is saved in a 256 color bitmap format. Then edge pixel values and generated the edges in image is calculated to generate the results with improved quality edges. Finally, comparison of the results of both algorithms and represent those results are made graphically. Proposed research is supposed to play significant role in area of image processing and quality enhancement.

Keywords Image · Ant colony optimization · Edge detection · Shortest path · PSNR & RMSE

25.1 Introduction

Image processing, ED, and ACO are some of the topics that will be covered in this chapter. The most recent research looked at how ACO and photo processing may benefit from the identification and compression of edges in images. The method of image processing known as edge detection may be used in order to improve photographs in a straightforward manner suggested by Eason et al. [1]. Nonetheless, the ACO's primary mission is to identify the course of action that will result in the best possible resolution to the problem at hand. In order to get the best possible

A. Mehta (✉) · D. Mehta
Chandigarh University, Mohali, Punjab, India
e-mail: anshu.e13356@cumail.in

outcomes from the research that is going to be done, it has been advised by Maxwell [2] that an ACO be used with an edge detection approach and compression.

25.1.1 Image Edge Detection

It is possible to define an edge as a collection of linked pixels that lies along the border between two different areas. Rippel and Bourdev [3] suggested that a cluster of pixels that appear at an orthogonal step transition in gray level might be referred to as an edge. The technique of identifying locations in a picture where there are abrupt changes in intensity is referred to as edge detection. This procedure, which must be completed in order to comprehend the contents of a picture, has applications in image analysis as well as machine vision. Edge detection seeks to pinpoint the borders of objects in a picture and considerably decreases the quantity of data that has to be processed described by Ansari and Anand [4]. Figure 25.1 illustrates the image edge detection. It is of the utmost importance to reclaim information about the form, structure, and any other crucial aspects of the picture. The method that has been developed makes use of a number of ants that migrate across the picture as a result of the local fluctuation of the image's intensity values. These ants are used to build up a pheromone matrix that determines where the image's edge pixels are located.

Ghrare et al. [7] recommended that traditional approaches to edge detection include convolving the picture with an operator that is designed to be sensitive to big gradients in the image while returning values of zero in areas of the image that are uniform. There are a great many ED operators, each of which is meant to be accurate with regard to a certain category of edges. The computation performed by the operator resolves a distinctive direction in which it is more specific to edges. Reforming operators allows one to investigate if edges are horizontal, vertical, or diagonal used by Gholizadeh-Ansari et al. [8].

Fig. 25.1 Edge detection [5]

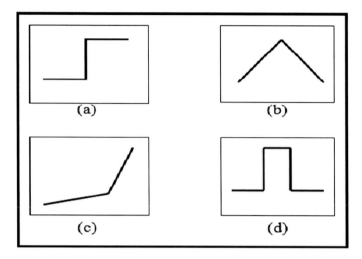

Fig. 25.2 Types of image edges [6]

25.1.2 Image Edge

A local notion, the edge, and contrasts with the global idea represented by the border. Daou et al. [9] advised that a collection of pixels that are placed at an orthogonal step transition in gray level is an example of an ideal edge. In addition to this, blurry edges may be caused by issues or defects that occur when the optics, sampling, and image collection systems are being used. Various types of edges are represented in Fig. 25.2. The edges that may be extracted from natural photos are almost never at all suitable for use as step edges described by Zhou et al. [10]. In most cases, however, they are generally influenced in some way by one or more of the following effects:

- Focal blur is the result of a finite depth-of-field in combination with a finite point spread function.
- Penumbral blur is brought by shadows cast by light sources with radii that are not zero.
- Shading at an item with a smooth surface.

25.1.3 Criteria for Edge Detection

Objectivity may be achieved while assessing the quality of the edge detection by using a variety of criteria. Some of the criteria are based on the needs for application and execution, while others might be presented in the form of mathematical measurements. In every one of the situations described above, it is necessary to utilize photos in which the real edges have been determined in order to conduct a quantitative analysis of performance.

- **Good detection**: The criterion should take into account a maximum signal noise ratio (SNR) and a minimal number of erroneous edges put forward by Zhu et al. [11]. After performing a threshold operation, the edges may be identified. A high threshold will assist to cut down on the number of edges that are fake, but it will also cut down on the number of edges that are really there presented by Yuan et al. [12].
- **Noise sensitivity**: In situations with tolerable noise, such as Gaussian noise, uniform noise, or impulsive noise, the use of a robust algorithm is necessary in order to identify edges. Annamalai and Lakshmikanthan [13] summarized that an edge detector is responsible for detecting edges while simultaneously amplifying background noise.
- **Good localizationx**: Accuracy in edge localization is required for the edge location, which means that it must be positioned on the proper spot.
- **Orientation sensitivity**: The operator is successful in accurately detecting both the edge magnitude and the edge orientation. Post-processing makes advantage of the orientation to link edge segments, eliminate noise, and trigger non-maximum edge magnitude.
- **Speed and efficiency**: The method has to have a quick execution time in order for it to be useful in an IP system. It is possible for an algorithm to be more efficient if it supports recursive implementation or allows for separate processing proposed by Chen et al. [14].

25.1.4 Jing Tian's Approach Based on ACO Technique

ACO is an algorithm that is influenced by the natural phenomenon that occurs when ants lay pheromone on the ground in order to indicate same path that should be followed by other ants in the colony. Chowdhary and Acharjya [15] recommended that this natural phenomenon occurs because ants want other ants in their colony to follow the same path. It is a kind of indirect communication in which the ants sought to make touch with one another despite their physical separation by responding and generating in response to their respective stimuli. While they are out searching for food, they do so by spreading pheromone, which is a chemical-like substance, on the ground. When other ants from the same colony pass along a certain route, it responds in a certain manner, which makes it simpler for the entire colony to look for food and saves them time represented by Lee et al. [16]. Steps for the edge detection for an image are represented in Fig. 25.3.

The perfect edge detector would provide a collection of linked curves that could be used to represent the borders of objects, the boundaries of marks, and any breaks in the surface's orientation. Therefore, adding an edge detection algorithm to an image has the potential to dramatically decrease the quantity of data that has to be processed and filter out information that is less relevant, all while maintaining the fundamental structural aspects of an image. Orujov et al. [17] suggested that

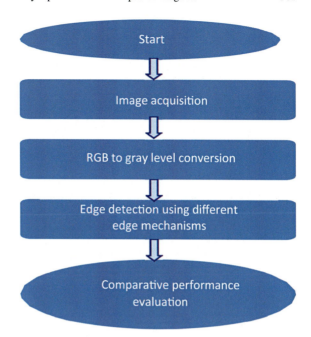

Fig. 25.3 Flowchart of edge detection

there are two primary ways for edge detection, and they are search-based and zero-crossing-based, respectively. Edge detection is the foundation for a wide variety of IA, and machine vision applications are recommended by Flores-Vidal et al. [18]. Its primary purpose is to pinpoint the locations of the edges of moving objects. Traditional methods of edge detection need a significant amount of processing power due to the fact that each set of operations is performed for each individual pixel are represented by Dhivya and Prakash [19]. With unconventional methods, the amount of time needed for computing rapidly grows in proportion to the size of the picture referred by Moustakidis and Karlsson [20].

25.1.5 Algorithm for Edge Detection

Step 1. Take a picture using color.
Step 2. Smoothing: Without wrecking genuine edges, destroy as adequate noise as accessible.
Step 3. Enhancement: By applying differentiation, the quality of edges can be enhanced.
Step 4. Threshold: The determination of edge pixels based on the use of the edge magnitude threshold, which helps assess which edge pixels need to be maintained and which ought to be dismissed as noise.

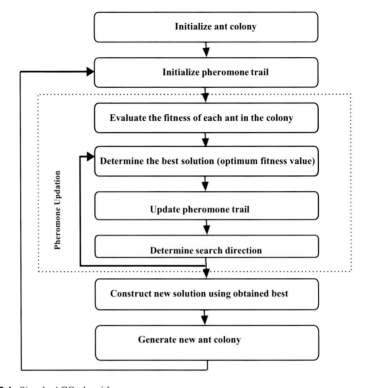

Fig. 25.4 Simple ACO algorithm

Step 5. Localization: Find the postulate edge bearings and record them. Evaluation using the fore mentioned algorithms.

Edge detection is totally based on ACO technique illustrated below in Fig. 25.4.

25.1.6 Application of ACO

Ant colony optimizations strategies for TSP:

TSP is first instance of problem description where the initial testing of the ACO algorithms was being done. This led to the development of ACO. This is due to a number of factors, including the following: The NP-hard optimization issue is particularly relevant for many different problems are suggested by Cococcioni et al. [21]. The ACO algorithms are straightforward to implement in TSP. It is assumed that the performance of all ACO algorithms has improved if there is a significant increase in the application's performance while running on the TSP. TSP was included into ACO by: The Ant system was first ACO algorithm to be presented by Liu et al., Marias [22, 23], and it was first used to illustrate how the TSP might be used. The

AS has served as a source of motivation for the creation of a number of different ACO algorithms, including the Elitist AS, the Rank- based AS, and the Max–Ong et al. [24] used the ant colony system, often known as ACS, is the most crucial algorithm of them all. The difference between ACS and the AS may be summarized using these three primary elements.

- First, when it comes to the more aggressive action rule, the ACS is going to be looked at as a far better usage than the AS is going to be.
- Second, on the arcs that are shown in the best- so-far tour, both pheromone deposition and pheromone evaporation are going place.
- Third, when an ant is traveling from point I to point J, it removes some pheromone from the arc it is going along in order to improve its ability to explore a new route within the search area introduced by Panda and Shemshad [25].

25.2 Literature Review

For instance, a piece of scientific research may be referred to as a literature review or a literature survey if it compiles the most current discoveries as well as theoretical and methodological contributions to a certain subject area. It covers a number of studies and research that have been done on the specific issue, as well as the conclusions that have been published in the past. There have been periodic appearances in a variety of journals and Web sites of a number of publications that are pertinent to our theme. The following is a brief summary of some of the articles that were discussed.

25.2.1 Review of Literature

Gandhi et al. [1] focused on image asymmetry priming for edge detection in order for computer vision to work; a single image has to be broken down into its component parts. The purpose of any segmentation approach was to stop segmentation as soon as it became meaningless. In this study, they will use preprocessing to remove superfluous noise and obtain just the needed edges. Song et al. [2] considered multi-task learning network for edge detection and stereo matching, or "Edge Stereo." Edge Stereo is a complex multi-task learning network that integrates disparity estimate with edge detection, allowing for accurate end-to-end predictions of the disparity and edge maps. Rippel et al. [3] suggested a pyramidal analytic autoencoder with an adjustable coding module and regularization of the projected code length. To generate visually pleasing reconstructions at low bitrates, we additionally use adversarial training optimized for compression. They showed that it is possible to compress lossy images using machine learning in real time and that the results are superior than those of any currently used codec. Several compression methods, their evolution, and their use in rapidly expanding medical fields including telemedicine and teleconsultation were examined and contrasted by Ansari et al. [4]. Even while the future of medical image compression is bright, it still has a way to go before it can satisfy the growing needs of the medical community.

Medical images should be of the highest possible quality since data loss is undesirable in many medical applications including disease detection and compression, as stated by Ghrare et al. [5]. Rather of relying on lossy compression methods that achieve high compression ratios, mathematicians have developed lossless compression methods. Yujing [6] considered VR multi-operator dynamic weight detection picture edge identification technique. Virtual reality (VR) was a computer generated three-dimensional, dynamic, and real-time representation of the world. The extraction of edge and contour properties from pictures has been the subject of much study in the fields of ID, processing, and analysis. Pashaei et al. [7] looked the extraction and classification of accident photos, and a convolution neural network is paired with a mix of extreme learning machines. Deep learning was utilized to extract characteristics from the accident photos, and an expert was employed to categorize the images. Zhao [8] discussed handwritten digits recognition using multiple classifier fusion and CNN. feature extraction on numerous datasets; some datasets need powerful classifiers, whereas others require weak classifiers. For example, the same classifier may perform differently on multiple tests sets due to the fact that image instances might be written differently by different people on the same digits due to their handwriting styles. Daou et al. [9] presented Iris tissue identification using GLDM features and an MLPNN- ICA hybrid classifier. Iris tissue identification was a reliable and accurate procedure. Their technique included segmentation, normalization, feature extraction, and matching. Using gray-level differences, the authors proposed new feature extraction and classification approaches.

Chowdhary et al. [10] reviewed MIS and Extraction: A Systematic Review The ability to see into the human body was made possible by medical imaging. Their method included creating data sets of "normal" and "abnormal" images and using them to make diagnoses about medical conditions. Visible-light and invisible-light medical imaging were two different medical imaging modalities. Lee et al. [11] discussed using basic spine X-ray images to estimate bone density in a Korean population. Data from previous health tests, such as X-rays of the spine and dual energy X-ray absorptiometry, was utilized in their study' (DXA). Series of individuals with varying levels of normal or abnormal bone mineral density were eventually selected. Deep convolution networks were used to extract visual attributes from X-ray images. We utilized machine learning to create prediction models for aberrant BMD using imaging data. Orujov et al. [12] considered method for detecting image edges using fuzzy logic, with application to retinal images. Using Mandeni fuzzy rules, they developed a contour detection method for finding blood veins in images of the retinal fundus. The recording quality was improved by using a median filter and CLAHE to lower the background noise. The edges of an image may be determined by using the gradient value in conjunction with the Mandeni fuzzy rules. Table 25.1 represents the comparison of previous studies done by various authors.

Table 25.1 Literature survey

S. No.	Author/Year	Objective	Methodology	Limitation
1	Moustakidis [20]	Use of Siamese convCNN to extract features for intrusion detection	Feature extraction, neural networks	Scope of this research is very less
2	Dhivya [19]	Fuzzy logic used for edge identification in a satellite picture	Edge detection	There is no implication in future
3	Flores-Vidal [19]	Fuzzy clustering as basis for novel global evaluation-based edge detection technique	Edge detection, clustering	There is no security in this system
4	Orujov [17]	Algorithm based on fuzzy pictures to locate blood vessels in retinal images	Edge detection	Lack of technical work
5	Lee [16]	Research on feature extraction and machine learning in the Korean Population to estimate bone density using basic spine X-ray pictures	Feature extraction, machine learning	Performance of this research is very low
6	Chowdhar [15]	Medical Imaging Segmentation and Feature Extraction: A Comprehensive Review	Feature extraction	There is no security in this system
7	Ahmadi (2020)	Recognition of iris tissue using a mixture of the MLPNN and ICA classifiers	Feature extraction	Lack of accuracy
8	Zhao [8]	For handwritten digit recognition, a combination of several classifiers and CNN feature extraction is used	CNN, feature extraction	Lack of security
9	Pashaei (2020)	Feature extraction and categorization from accident photographs using a CNN combined with a variety of extreme learning machines	CNN, feature extraction	Lack of accuracy
10	Chen [14]	A multi-operator dynamic weight detection technique for picture edge identification in a virtual reality situation	Detection, image edge, and virtual reality scenario	There is lack of performance

25.2.2 Research Gap

The process of edge detection is used in digital photography to pinpoint regions with dramatic tonal shifts. Several problems arise during edge detection, such as displaying a crisp and clear image. There is a break in the brightness of the images, making it difficult to see edges. Due to the presence of noise in the images, the detection process is very complicated. Edges are identified with the help of gradients, first and second derivatives, and so on. Further challenges include the problem of extended detection times and the difficulty in identifying and localizing edges. The proposed approach leverages ACO to improve the precision of edge detection and provide a more complete edge profile. In actual use, picking the best method for edge extraction from images is conditional. Feature extraction needs further work. During the edge detection phenomena, various issues might develop, such as the difficulty in showing an image with sufficient clarity and sharpness. Edge identification is challenging because of the discontinuity in picture intensity that occurs in the images. Furthermore, the identification procedure is already challenging due to the presence of some noisy material in the photos. Multiple techniques, including the first and second derivatives and gradients, may be used to locate edges. There are a number of issues that must be dealt with while attempting to detect edges, including the occurrence of misleading and erroneous edges (for a number of reasons), the identification and localization of edges, and the increased detection time.

By using ACO, my proposed work improves the accuracy of edge identification, leading to an image with a richer edge profile. The truth is that several methods of image edge extraction are used depending on the specifics of each case. There are pros and cons to using any given algorithm. The ant colony optimization edge detection technique combines the benefits of many methods to improve upon the shortcomings of individual techniques. MATLAB has been used to actualize the edge detection method. As a program, MATLAB facilitates both numerical calculation and visual representation. The matrix is the primary data structure. In MATLAB, an image is handled as a matrix. The programming language and development environment MATLAB is user-friendly. RMSE and PSNR are employed here as performance indicators. Jing Tian's algorithmic steps from the ACO technique were used to identify the edges in the image.

25.3 Problem Statement

While all valid edge detectors exist, issues with false edge detection, edge localization, excessive computing time, missing actual edges, noise, and so on plague even the best of them. The vast majority of edge detectors merely generate points where the edges of pictures are. A numerical value is returned for each pixel, as well as the image's orientation, as the typical results. It is possible that in the future we may be able to tweak the algorithm in order to account for lighting changes. There are several problems that may develop with edge detection, including incorrect edge identification, edge localization, lengthy computation times, missing true edges, and noise concerns. Typically, only points representing image boundaries are returned by edge detectors. Typically, a numerical value and directional label are produced for each pixel. It is possible that the algorithm will be modified so that it can operate in a wider range of illumination levels. Feature extraction also has to be refined.

25.4 Need of Research

Ant colony optimization (ACO) techniques are a type of metaheuristic that can be used to solve complex optimization problems. Due to the advancement in the field of image processing, image edge detection gained a lot of popularity. When compared to traditional approaches, ACO algorithms proved superior accuracy and efficiency for edge detection problems. This is due to their ability to incorporate problem-specific information into their optimization process, allowing for tailored solutions that are better suited for the task at hand. To enhance the performance and the scope of various applications of ACO, it is essential to conduct the research on their implementation for image edge detection.

25.5 Future Scope

In image edge detection, there are several applications of ACO that can be explored in future. One potential application is to use ACO algorithms to automatically detect edges in images that are not easily discernible by traditional edge detection techniques. This could be particularly useful in medical imaging, especially where a little variation in pixel intensity has a significant diagnostic value. Additionally, ACO algorithms can be used to track the evolution of edges over time, allowing for the detection of changes in images that may not be immediately visible. Current edge detection methods accuracy can be enhanced by using ACO with the integration of prior knowledge of the image. Finally, for more accurate and efficient edge detection method, ACO algorithms can be used potentially leading to more robust image analysis.

References

1. Eason, G., Noble, B., Sneddon, I.N.: On certain integrals of Lipschitz-Hankel type involving products of Bessel functions. Phil. Trans. Roy. Soc. London **A247**, 529–551 (1955)
2. Clerk Maxwell, J.: In: A Treatise on Electricity and Magnetism, 3rd ed. vol. 2. Oxford, Clarendon, pp. 68–73. (1892)
3. Rippel, O., Bourdev, L.: Real-time adaptive image compression. In: 34th International Conference Machine Learning ICML 2017, vol. 6, pp. 4457–4473. (2017)
4. Ansari, M.A., Anand, R.S.: Recent trends in image compression and its application in telemedicine and Teleconsuktation. In: National System Conference, pp. 59–64. (2008)
5. BogoToBogo Open CV3 Canny Edge Detection Homepage.: https://www.bogotobogo.com/python/OpenCV_Python/images/Canny/Canny_Edge_Detection.png. Last Accessed 24 Dec 2022
6. https://www.researchgate.net/profile/Vijayarani-Mohan/publication/339551773/figure/fig2/AS:863426553327619@1582868340615/Different-types-of-edges-a-Step-Edge-The-intensity-of-image-abruptly-varies-from-one.png

7. Ghrare, S.E., Ali, M.A.M., Jumari, K., Ismail, M.: An efficient low complexity lossless coding algorithm for medical images. Am. J. Appl. Sci. **6**(8), 1502–1508 (2009). https://doi.org/10.3844/ajassp.2009.1502.1508
8. Gholizadeh-Ansari, M., Alirezaie, J., Babyn, P.: Low-dose CT denoising using edge detection layer and perceptual loss. In: Annual International Conference of the IEEE Engineering in Medicine and Biology Society, vol. 2019, pp. 6247–6250. (2019). https://doi.org/10.1109/EMBC.2019.8857940
9. Abi Zeid Daou, R., El Samarani, F., Yaacoub, C., Moreau, X.: Fractional derivatives for edge detection: application to road obstacles. EAI/Springer Innov. Commun. Comput. 115–137 (2020). https://doi.org/10.1007/978-3-030-14718-1_6
10. Zhou, J., et al.: Optical edge detection based on high-efficiency dielectric metasurface. Proc. Natl. Acad. Sci. U.S.A. **166**(23), 11137–11140 (2019). https://doi.org/10.1073/pnas.1820636116
11. Zhu, T., et al.: Generalized spatial differentiation from the spin hall effect of light and its application in image processing of edge detection. Phys. Rev. Appl. **11**(3), 1 (2019). https://doi.org/10.1103/PhysRevApplied.11.034043
12. Yuan, J., Guo, D., Zhang, G., Paul, P., Zhu, M., Yang, Q.: A resolution-free parallel algorithm for image edge detection within the framework of enzymatic numerical P systems. Molecules **24**(7) (2019). https://doi.org/10.3390/molecules24071235
13. Annamalai, J., Lakshmikanthan, C.: In: An Optimized Computer Vision and Image Processing Algorithm For Unmarked Road Edge Detection. vol. 900. Springer, Singapore (2019)
14. Chen, Y., Wang, D., Bi, G.: An image edge recognition approach based on multi-operator dynamic weight detection in virtual reality scenario. Cluster Comput. **22**, 8069–8077 (2019). https://doi.org/10.1007/s10586-017-1604-y
15. Chowdhary, C.L., Acharjya, D.P.: Segmentation and feature extraction in medical imaging: a systematic review. Proc. Comput. Sci. **167**(2019), 26–36 (2020). https://doi.org/10.1016/j.procs.2020.03.179
16. Lee, S., Choe, E.K., Kang, H.Y., Yoon, J.W., Kim, H.S.: The exploration of feature extraction and machine learning for predicting bone density from simple spine X-ray images in a Korean population. Skeletal Radiol. **49**(4), 613–618 (2020). https://doi.org/10.1007/s00256-019-03342-6
17. Orujov, F., Maskeliūnas, R., Damaševičius, R., Wei, W.: Fuzzy based image edge detection algorithm for blood vessel detection in retinal images. Appl. Soft Comput. J. **94** (2020). https://doi.org/10.1016/j.asoc.2020.106452
18. Flores-Vidal, P.A., Olaso, P., Gómez, D., Guada, C.: A new edge detection method based on global evaluation using fuzzy clustering. Soft Comput. **23**(6), 1809–1821 (2019). https://doi.org/10.1007/s00500-018-3540-z
19. Dhivya, R., Prakash, R.: Edge detection of satellite image using fuzzy logic. Cluster Comput. **22**, 11891–11898 (2019). https://doi.org/10.1007/s10586-017-1508-x
20. Moustakidis, S., Karlsson, P.: A novel feature extraction methodology using Siamese convolutional neural networks for intrusion detection. Cybersecurity **3**(1) (2020). https://doi.org/10.1186/s42400-020-00056-4
21. Cococcioni, M., Rossi, F., Ruffaldi, E., Saponara, S.: Fast deep neural networks for image processing using posits and ARM scalable vector extension. J. Real-Time Image Process. **17**(3), 759–771 (2020). https://doi.org/10.1007/s11554-020-00984-x
22. Liu, Y. et al.: A 4e-–2e- cascaded pathway for highly efficient production of H2 and H2O2 from water photo-splitting at normal pressure. Appl. Catal. B Environ. **270**, 118875 (2020). https://doi.org/10.1016/j.apcatb.2020.118875
23. Marias, K.: The constantly evolving role of medical image processing in oncology: from traditional medical image processing to imaging biomarkers and radiomics. J. Imaging **7**(8) (2021). https://doi.org/10.3390/jimaging7080124

24. Ong, J.W., Chew, W.J., Phang, S.K.: The application of image processing for monitoring student's attention level during online class. J. Phys. Conf. Ser. **2120**(1) (2021). https://doi.org/10.1088/1742-6596/2120/1/012028
25. Panda, A., Shemshad, A.: Automated class student counting through image processing **1**(1), 24–29 (2021)

Chapter 26
Machine Learning-Based Sentiment Analysis of Twitter COVID-19 Vaccination Responses

Vishal Shrivastava and Satish Chandra Sudhanshu

Abstract The COVID-19 pandemic has caused significant fear, anxiety, and complex emotions or feelings in a large number of people. A global vaccination campaign to end the SARS-CoV-2 epidemic is now in progress. People's feelings have become more complex and varied since the introduction of vaccinations against coronavirus. The use of social media platforms such as Twitter enables users to communicate with one another and share information and perspectives on a wide variety of topics, spanning from local to international concerns, from global to personal. Twitter will prove to be a helpful source of information that can be tracked regarding views and sentiments regarding the SARS-CoV-2 vaccination. To better understand public views, concerns, and emotions that may influence the achievement of herd immunity targets and limit the pandemic's impact, this study uses deep learning to identify the themes and sentiments in the public about COVID-19 immunization on Twitter. Moreover, this paper consists of a detailed explanation of the sentiment analysis with their challenges, classification, approaches, applications, and VADER.

Keywords COVID-19 · Vaccination · Sentiment analysis · Opinion mining · Social media · Twitter · VADER · Machine-learning

26.1 Introduction

The current COVID-19 outbreak has had significant repercussions for the healthcare industry, and as a direct consequence, our fundamental understanding of safety has been disrupted. Isolation from others has the potential to halt or significantly delay the propagation of the coronavirus. At this time, it is necessary to take precautions

V. Shrivastava (✉) · S. C. Sudhanshu
Department of Computer Science and Engineering, Arya College of Engineering and IT, Jaipur, India
e-mail: vishal500371@yahoo.co.in

such as washing your hands frequently, using a mask, and avoiding close personal contact as much as possible. However, these can only reduce the risk of transmitting the coronavirus; they cannot eliminate the risk. In light of the current circumstances, vaccination has emerged as the only strategy capable of battling and possibly wiping out coronavirus. Pfizer's early research on mRNA vaccines involved more than 40,000 participants, while a more recent immunization study included 30,000 participants. Together, these numbers represent the number of persons who received the vaccine. In both studies, the efficiency of the vaccine was measured at 94% on average, and there was not a single fatality associated with either study. Another viral vector-based vaccine called Johnson & Johnsen is efficient against coronavirus which increases response of the immune system of those who get it. It has a success rate of 85%, and there are no noticeable bad effects associated with getting it [1, 2].

A significant number of analysts have made use of ML techniques in order to investigate how people talk about COVID-19 vaccinations online. The rise of social media may be one contributor to the decline in the number of people ready to be vaccinated. Analysing the messages that were communicated to the public in this context reveals the public's opinion on the matter of COVID-19 immunizations. At this time, if it has been determined that 90% of vaccinations are successful, then immunizations will be done starting in the U.K. from 8 December 2020. In this study, tweets pertaining to vaccines were analysed in order to have a better understanding of their effect. In this study, Twitter was used for increasing vaccination compliance, decreasing vaccination reluctance and resistance, and enhancing vaccination acceptability [3]. It is possible that the spread of misinformation about vaccinations may be mitigated if public health workers had a more nuanced awareness of the perspectives and attitudes around the topic. Authorities could use Twitter to actively promoting the use of vaccines among the general public while reducing vaccine hesitancy the general public. This would be a way of influencing people's attitudes around vaccination. People's views and beliefs regarding the matter were altered as a result of various attacks made against vaccines during the outbreak.

The development and approval of the COVID-19 immunization have given people renewed optimism that the pandemic can be put an end to and that normal life can be resumed. Unfortuitously, there is a substantial obstacle in the way of getting vaccination rates, and that obstacle is hesitation over-vaccination, which is occasionally motivated by misinformation. One of the most difficult components of machine learning is processing data in a way that detects emotions using methods that allow us to evaluate whether people have positive or negative perspectives on a topic. Although social media and microblogging sites are excellent sources of information, their primary function is to allow users to communicate thoughts and beliefs that are uniquely their own.

26.2 Background Study of COVID-19

The 2019 coronavirus disease (COVID-19) outbreak was initially discovered in Wuhan, China, in December of this year. It then rapidly grew to become a global epidemic that affected millions of people worldwide. The novel severe acute respiratory syndrome coronavirus-2 (SARS-CoV-2) has been attributed to COVID-19, and clinical manifestations of the virus have ranged from asymptomatic and mild symptoms related to flu to pneumonia, acute respiratory distress syndrome (ARDS), and in some cases, death. COVID-19 is associated with SARS-CoV-2.It is hoped that the spread of the virus could be contained by social isolation, the use of masks, the development of novel antiviral medications, and the production of an efficient vaccine. Establishing herd immunity through natural immunity through diseases is possible, but doing so could have catastrophic consequences, as was seen in Sweden, where authorities believed that infecting up to 60% of the population would be enough to protect the more vulnerable population through herd immunity. This plan backfired, however, as there are at least five times as many COVID-19-related fatalities per million people in Sweden than there are in Germany. As a result, the production of an efficient vaccine is of the utmost importance and is regarded as the only viable option for achieving herd immunity.

Researchers from all around the world are actively toiling away around the clock in an effort to produce a vaccine that is effective against COVID-19. There are now around 200 potential vaccines that are in the process of going through various stages of development. These vaccines include the AZD1222 vaccine developed by AstraZeneca and Oxford and the mRNA-1273 vaccine developed by Moderna. There are presently 30 vaccines being tested in clinical studies. Although it is possible that production capacity will not be sufficient for meeting the global demand for vaccines, it would be advantageous if a small selection of vaccines were available for emergency use for population sections that are more vulnerable. The goal is to achieve global vaccine distribution for stopping and limiting the impact of COVID-19.

In order to successfully create a vaccine that is both safe and efficient, it is essential that all stages of trials and testing be carried out with extreme caution to prevent serious adverse effects. Accelerating COVID-19 Therapeutic Interventions and Vaccinations (ACTIV), the Gavi alliance, the World Health Organization (WHO), and the Bill & Melinda Gates Foundation (BMGF) must collaborate to secure appropriate financing for vaccines and a coordinated response to the ongoing COVID-19 epidemic. This review provides a summary of the immune response and biology demonstrated by previous coronavirus infections and SARS-CoV-2, the impact of Twitter as an information-providing service, and the potential problems that may occur as a result of speeding up the production of vaccines. The development of vaccines could take around 15 years, but with advanced technologies and the urgent need for vaccines, the development time could be reduced to one and half years or less or less. This could potentially raise issues regarding safety and efficacy affecting public acceptance of vaccines.

26.2.1 Application Scenarios on COVID-19 Data

As a result of having to deal with the aftermath of COVID-19, many people are experiencing a wide range of mental health problems. During COVID-19, many researchers worked to analyse public opinion [4].

(1) *Mental Health Analysis of Students During the Lockdown*

The practice of social distance, which resulted in less encounters between people, was implemented in an effort to halt the progression of Covid. A number of nations went into lockdown, which included closing their airspace as well as educational and other institutions. As a result of the lockdown, people, particularly students, had to remain a significant distance from their houses, be confined within their dormitories, and cease their educational activities. This results in students experiencing anxiety and tension. During the lockdown, students expressed their feelings via social media platforms, and researchers attempted to investigate those feelings. Data from Twitter was analysed in order to gain a better understanding of those feelings.

(2) *Reopening After COVID-19*

As a result of the coronavirus, billions of individuals all over the world have been affected in some way. It has generated economic turmoil all around the world which is a roadblock to reopening. The permanent stagnation of the economy poses a risk to the continued existence of any nation. People are being forced to reopen companies and get back to living their usual lives as a result of these factors. The researchers focused their efforts on determining what kinds of businesses people are considering reopening after COVID-19.

(3) *Restaurant Reviews*

Customers have the ability to voice their opinions and provide feedback regarding the quality of the products or services provided by a variety of companies in today's digital world. These reviews are beneficial to other consumers who are about to utilize the service or purchase the goods since they assist them make judgments. The Internet reviews are tied to the overall star rating, which in turn influences the amount of money that the restaurant makes. People were especially concerned about the spread of Covid during Covid, so special SOPs were announced for eateries during Covid. As a result, numerous eateries received poor evaluations due to their chilly outdoor areas and their delayed service. Researchers studied the comments that individuals had made about restaurants, which assisted restaurant management in preserving the high quality of both the cuisine and the atmosphere.

(4) *Racial Sentiments and Vaccine Sentiments*

The production of a vaccine against Covid may prove useful in stopping the disease's further spread. As a result, a great number of industries are putting their efforts into developing various types of vaccines. However, the key necessity in order to reduce Covid with vaccines is for people to be willing to accept and take their vaccinations. In the event that people are unwilling to get themselves, there will be a significant obstacle in the way of the control of Covid. Researchers investigated the opinions of the general public regarding vaccinations. Additionally, Covid was responsible for feelings of discrimination across international borders, which led to an increase in people's racist attitudes.

26.3 Sentiment Analysis

Opinion mining and sentiment analysis are both terms that refer to the same thing: the automated computer study of people's attitudes, emotions, and expressions in relation to a certain objective. Any person, event, or subject matter could be taken as the target object. Analysis of sentiment and opinion mining are interchangeable phrases that can be used in the same context. However, according to the findings of some studies, these two terms refer to somewhat distinct mental images. Sentiment analysis involves identifying and analysing the emotion conveyed in a piece of text, while opinion mining involves obtaining and analysing people's opinions on an entity. Therefore, the purpose of sentiment analysis is to automate the process of discovering opinions, determining the emotions that those opinions represent, and then categorizing the polarity of those emotions. In many different spheres, the consideration of public opinion is an absolutely necessary step in the decision-making process. Before purchasing a certain item, a person who is interested in making a purchase could find it helpful to inquire about other people's experiences with the item in question. In the real world, companies and organizations often solicit client opinions on the quality of their goods and services. In recent years, applications of sentiment analysis have grown over a wide variety of fields, including ad placements, trend prediction, and recommendation systems as well as politics and health care. The past several years have seen a meteoric rise in the prevalence of social media on the Internet, such as online reviews, comments, forums, blogs, and comments on social networking sites. The majority of organizations are basing their choices on the contents of these reports. As a result of the vast amounts of data that are readily available to the public today, modern enterprises no longer need to rely on opinion polls, surveys, or focus groups. The necessity of checking each unique Website makes the work of mining opinions a challenging and difficult one. Finding the relevant Websites and gleaning the opinions contained within them can be an extremely challenging task for a human reader. As a result, automated sentiment analysis is something that is desperately needed. The majority of companies are relying on their own in-house research and analytic systems to learn what their customers think. Opinion mining and sentiment analysis are typically carried out utilizing one of two methods: (1) An approach based on machine learning (2) Derived from a lexicon. The strategy that is based on machine learning makes use of a number of different supervised and unsupervised learning algorithms in order to classify sentiment. For the purpose of sentiment classification, lexicon-based algorithms make use of a dictionary containing words connected to specific domains that convey a range of emotions. It is possible to learn whether a certain term is positively or negatively connoted by consulting a dictionary and detecting the polarity of the words by comparing the word in the sentence with the words in the dictionary.

The most important part of opinion mining is identifying the type of sentence. Sentences have to classified either subjective or objective. Researchers are using both supervised and unsupervised learning techniques for providing different methods of sentiment analysis. In general, sentiment analysis includes advanced processes. It has a totally different set of tasks: Subjective or objective analysis, opinion extraction, sentiment classification (supervised or unsupervised). Labelling any text document or sentence as subjective or objective can be done using subject-level analysis. Sentiment classification includes probing the sentiment polarity of the filtered sentences. All sentences or texts are divided into positive, negative, or neutral types depending on the emotions we get from the texts or sentences [5].

26.3.1 Valence Aware Dictionary for Sentiment Reasoning (VADER)

The NLTK module VADER (Valence Aware Dictionary for Sentiment Reasoning) generates sentiment scores from the words in a document. It is a sentiment analyser based on rules, where words are assigned positive or negative labels based on their semantic orientation. The VADER method is built on lexicons and makes use of gold-standard heuristics in addition to English-language sentiment lexicons. Lexicographies are subjected to human inspection and verification. They make use of qualitative methods in order to increase the effectiveness of the emotion analyser [6]. The VADER corpus is the result of pooling together several different datasets. The polarity of the emotions was provided by the initial corpus, in contrast to VADER, which contains an additional element that shows the intensity of that polarity score. Its corpus has over 7500 dictionaries, including slang and abbreviations. Scores might range from -4.0 to $+4.0$. These values indicate an attitude threshold, with scores below 4 suggesting negative sentiments and scores over $+4$ indicating good sentiments.

VADER uses grammatical and syntactic rules in addition to a sentiment vocabulary to indicate the severity and polarity of the sentiments being expressed. Utilizing a wide variety of language features, such as emoticons and acronyms, the VADER lexicon comprises over 7500 sentiment qualities. Since the emotional weight of a word is established by taking grammatical constraints into account, word sentiment scores might vary [7, 8].

26.4 Related Work

COVID-19 vaccine topic from Twitter was studied in [9]. It was found that people had different feelings about the Chinese vaccine compared to those about vaccines made in other countries, and the value of those feelings could be affected by the number of deaths and cases reported in the daily news as well as the nature of the most pressing problems in the communication network.

The positive appeals of recent news on the safety of the COVID-19 vaccination and the government's proactive risk communication were reflected in the finding that positive views outlasted negative feelings for 56 days (62.20%), were found in [10]. There was also a considerable correlation between positive vaccination attitudes and rising vaccination rates, which was statistically significant.

In this study, Qorib et al. [11] the reluctance over the COVID-19 vaccine is analysed using three different approaches of emotion computation: Azure Machine Learning, VADER, and TextBlob. Demonstrates that people's resistance to the COVID-19 vaccination lessens over time, which suggests that the general population may eventually feel more optimistic and happy about becoming vaccinated against COVID-19.

Yousefinaghani et al. [3] shown that there is a difference, albeit a little one, between the frequency of positive and negative emotions, with the former being the more common polarity and eliciting more responses. More time was spent talking about people's fears and reservations about vaccinations than actually learning about them, according to the study's findings. It found that some anti-vaccination accounts were run by Twitter bots or political activists, whereas pro-vaccination accounts tended to be associated with more authoritative figures or organizations.

In [12] when compared to states in other regions of the United States, states in the South showed a much higher incidence of negative tweets, but states with higher incomes reported a lower prevalence of negative tweets. Due to the fact that our data indicate the existence of negative vaccine attitudes as well as geographic variation in these opinions, it is necessary to customize our efforts to promote vaccinations, particularly in the southern portion of the United States.

Bokaee Nezhad and Deihimi [13] revealed a statistically significant difference (although a little one) in the amount of people in Iran who had a favourable view of domestic and imported vaccines, with the latter having the more prevalent positive polarity. The number of people who are worried about vaccines, both at home and abroad, has increased noticeably in recent months. Conclusions: There were no statistically significant differences in the percentage of Iranians who had a positive and unfavourable perspective about immunization.

By analysing the tone of 2,678,372 tweets on the COVID-19 vaccine posted between 1 November 2020 and 31 January 2021, researchers in [14] found that 42.8% were positive and 30.3% were negative. The public's mood and the number of tweets both spiked after Pfizer announced that the first COVID-19 immunization had attained 90% efficacy, and both continued to rise until the end of December, when they finally settled at a neutral emotion. Furthermore, people's perspectives varied depending on where they were located.

26.5 Proposed Methodology

This section provides an explanation of the processes that were utilized to examine the efficacy of the suggested machine learning strategy in the classification of the sentiments contained within COVID-19 tweets. In the present investigation, we created a model based on machine learning for analysing the sentiment of vaccination responses in reaction to an increase in the number of newly reported cases of COVID-19. We utilized an open-source dataset titled "All COVID-19 Vaccines Tweets," which contained each and every tweet ever sent out regarding the COVID-19 vaccine. The collection contains all COVID-19 vaccine-related tweets ever sent anywhere in the world. There was feeling behind every tweet, whether or not it was pro-vaccine. Our initial task was to determine the overall sentiment polarity of all the tweets. The positive, negative, or neutral nature of a tweet is indicated by the polarity of its expression. What the tweeter really wants to say is made clear here. With this data, we can assess the global effect of the COVID-19 vaccine. To determine the general tone of the tweets, we coded a Python tool. We have been using the Python Tweepy package for this purpose. Tweepy is a free and easy-to-use module for Python that acts as a gateway between your Python app and Twitter's API. This was used to collect the polarity values of the emotions. Now, the polarity value of emotion can be either zero, a positive number, or a negative number. We only considered tweets that expressed either good or negative emotions in our analysis. After carrying out the aforementioned steps, we had a dataset that could be used by our algorithms. When processing the data, we used the train-test split function, which is a key part of the machine-learning method. Train and test data were split 75%:25%. Reality's text data is a mess. Consequently, certain pre-processing operations have to be executed prior to feeding the dataset into the ML algorithms. Data pre-processing includes operations such as stemming, stop-word removal, tokenization, de-tokenization, URL removal, punctuation removal, URL removal, removal of incidents, removal of double spaces, and so on. All of these steps are essential components of the data preparation method. Then, analyse the results with VADER sentiment. Three different machine learning algorithms (NB, LR, and VC) were utilized to analyse the data.

All of the processes involved in the study approach are unmasked and discussed in this part very briefly:

i. **Data Gathering**

Collecting relevant data is the starting point for the planned work. The "COVID-19 All Vaccines Tweets" dataset was used throughout this study.

ii. Data Pre-processing

The quality of the data pre-processing used in constructing an ML model is directly related to the effectiveness of that model.

The Natural Language Toolkit (NLTK) Python package was used in this part to prepare the text. The text can be pre-processed in a number of different ways. Reduced text size, elimination of URLs, removal of punctuation, tokenization, stopword removal, and stemming are all pre-processing techniques.

iii. Sentiment Analysis

Opinions, attitudes, emotions, and views can be automatically mined from audio, text, tweets, and database sources via a process called sentiment analysis (SA), which employs natural language processing (NLP). In a SA, the text is parsed for positive, negative, and neutral sentiments. Other names for SA include opinion mining, assessment extraction, and evaluation extraction. Though they are often used interchangeably, "opinion," "sentiment," "view," and "belief" each have distinct meanings [13].

1. Valence Aware Dictionary for Sentiment Reasoning (VADER)

VADER is a lexicon-based method that makes use of heuristics considered to be the gold standard in the field, as well as lexicons of sentiment expressed in the English language. Lexicons are checked and verified by humans. To improve the efficacy of the emotion analyser, they employ qualitative methods [14]. According to, VADER can produce sentiment analysis results that are on par with those produced by human raters. The VADER corpus is the result of pooling together several different datasets. The polarity of the emotions was provided by the initial corpus, in contrast to VADER, which contains an additional element that shows the intensity of that polarity score. Its corpus has about 7500 lexicons, the sum of which includes slang and abbreviations. The possible range of scores is from 4.0 (being the lowest) to 4.0 (being the highest). Scores below 4 indicate negative sentiments, and scores above + 4 indicate positive sentiments; hence these values serve as a threshold for attitudes. The results from the VADER look like this (neg, neu, pos, and compound). In this situation, the compound score is calculated as the average lexical score of the entire text or a single sentence and can vary from 0 (no score) to 1 (perfect score).

iv. Dataset Splitting

After performing above procedures, the data is split in two sets for training and testing in the ratio of 75% and 25%, respectively.

v. Classification Technique

To classify unlabelled data, ML has produced a number of classification methods that make use of different approaches. In this research, we employed ML-based classification strategies as NB, LR, and voting classifier.

26.6 Results Analysis and Discussions

In this section, the results of the analysis that was done in the previous section are presented, and then an assessment of the performance is made. Python, a general-purpose, open-source programming language and interactive platform used for data analysis, scientific visualization, and scientific computation, was utilized for the simulation. Tensorflow, SciPy, NumPy, Pandas, Matplotlib, SciKit-learn, Keras, Pytorch, Scrappy, and Theano are only some of the excellent Python libraries for data science that were utilized.

a. **Performance Evaluation Metrics**

Measures of how well a classifier performs are controversial, with no one method currently dominating the field. Its effectiveness is measured with the confusion matrix, as well as precision, accuracy, and recall. We will then go on to explore the various metrics used to assess the classifier's performance.

b. **Confusion Matrix**

The capacity of a classifier to discriminate between tuple instances of different classes can be evaluated using a confusion matrix. It keeps track of both the actual and expected categories generated by a given method. Matrix data is frequently employed in evaluating the efficiency of such systems.

i. *Accuracy*

The ratio of accurate predictions to total predictions is a measure of accuracy. The accuracy of a classifier can be evaluated using the matrix that was shown earlier, as is shown down below:

$$\text{Accuracy} = TP + (TN/TP) + FP + FN + TN \tag{26.1}$$

ii. *Precision*

Precision is utilized to circumvent the constraint of accuracy. The precision indicates the proportion of positive predictions that were accurate.

$$\text{Precision} = TP/(TP + FP) \tag{26.2}$$

iii. *Recall*

Recall seeks to assess the proportion of actual positives that were wrongly detected.

$$\text{Recall} = TP/(TP + FN) \tag{26.3}$$

iv. *F1–Score*

The F1-score is calculated by finding the harmonic mean of the recall and precision scores.

$$F1\text{-score} = 2^*(\text{Recall}^*\text{Precision})/(\text{Recall} + \text{Precision}) \quad (26.4)$$

c. Simulation Results and Discussion

The results of our tests and an evaluation of our methodology are presented and analysed in this section. We begin by contrasting the efficacy of several approaches to sentiment analysis using information collected from the hashtag #vaccines on Twitter. In this part, we examine the accuracy of Naive Bayes, a voting classifier, and a logistic regression classifier on the COVID-19 vaccines Twitter dataset after the training phase is complete, using the tweets as examples (2). Here, we present the classification results achieved by various techniques.

d. Results of Naïve Bayes Classifier

These are the outcomes of running the Naive Bayes algorithm on the sample data:

Figure 26.1 is a report generated by NB machine's learning classifier while using the Covid vaccine dataset for classification purposes. Do the NBCs in three different areas of figure drawing? With a f1-score of 72%, 81%, and 76% for positive data, 86%, 89%, and 88% for neutral data, and 77%, 42%, and 54% for negative data, the overall NBC accuracy is 81%.

Figure 26.2 presents the NBC confusion matrix. In this graph, the actual label from the dataset is displayed along the x-axis, while the predicted label is displayed along the y-axis. The NBC model is 81% accurate. As an illustration, 1609 of the immunization records are considered positive, 16,650 are classified as neutral, and 7276 are considered negative.

	precision	recall	f1-score	support
0	0.72	0.81	0.76	8990
1	0.86	0.89	0.88	18627
2	0.77	0.42	0.54	3860
accuracy			0.81	31477
macro avg	0.78	0.71	0.73	31477
weighted avg	0.81	0.81	0.80	31477

Accuracy : 0.8114

Fig. 26.1 NBC classification report

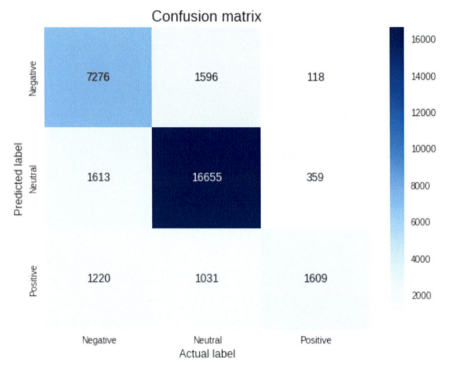

Fig. 26.2 NBC confusion matrix

The Naive Bayes machine learning classifier's multiclass ROC curve is depicted in Fig. 26.3. Methods exist, such as the receiver operating characteristics (ROC) graph, for organizing classifiers and demonstrating algorithmic efficiency on training data values of positive, neutral, and negative R-squared correlations. The largest ROC area indicates that the POSITIVE class performs better than the other classes in the dataset, so we can conclude that this dataset is positively biased. Improved performance and accuracy were also achieved while classifying weighted values for ROC area data determined using Naive Bayes.

v. *Results of Proposed Voting Classifier*

Results of sentiment analysis on the dataset may be predicted with 90% accuracy using a voting classification method.

Figure 26.4 is a report generated by voting machine's learning classifier on the Covid vaccine dataset. There are three different labels displayed for the findings. Precision, recall, and f1-score for the positive data, labelled as 0, are 94%, 86%, and 90%, respectively; for the neutral data, labelled as 1, they are 88%, 99%, and 93%; for the negative data, labelled as 2; they are 95%, 54%, and 69%, for an overall VC accuracy of 90%.

26 Machine Learning-Based Sentiment Analysis of Twitter COVID-19 ...   323

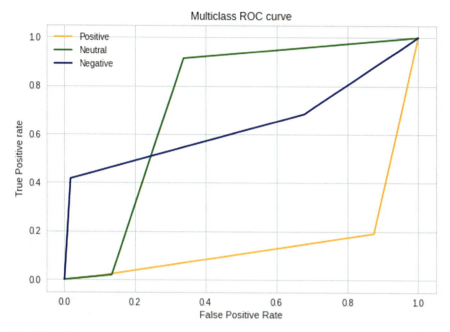

Fig. 26.3 NBC multiclass ROC curve

```
              precision    recall  f1-score   support

           0       0.94      0.86      0.90      8990
           1       0.88      0.99      0.93     18627
           2       0.95      0.54      0.69      3860

    accuracy                           0.90     31477
   macro avg       0.92      0.80      0.84     31477
weighted avg       0.90      0.90      0.89     31477

Accuracy  : 0.8977
```

Fig. 26.4 Voting classifier classification report

A representation of the confusion matrix for the voting classifier is shown in Fig. 26.5. The x-axis shows the actual label from the dataset, while the y-axis shows the predicted label. The image depicts 2077 successful vaccination records, 1847 neutral label records, and 7706 failure immunization instances with an accuracy of 90%.

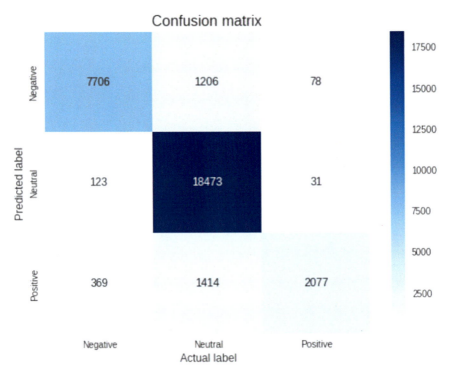

Fig. 26.5 Voting classifier confusion matrix

Multiclass ROC curve of voting machine learning classifier is shown in Fig. 26.6. On the x-axis of the graph is the percentage of false positives, and on the y-axis, the percentage of true positives in the dataset is displayed.

vi. **Results of Proposed Logistics Regression Classifier**

The dataset is subjected to a third analysis, this time using a logistic classifier. To conduct sentiment analysis on Twitter data, we apply this method to the dataset and find that it yields 94% accurate results. Positivity, negativity, and a neutral 0 are all valid values for expressing probabilities.

The classification produced by the LR machine learning classifier while employing the Covid vaccination dataset is shown in Fig. 26.7. Three different types of results are displayed. Precision, recall, and f1-score for the positive data (labelled as 0) are 95%, 92%, and 94%, respectively; for the neutral data (labelled as 1), they are 93%, 99%, and 96%; for the negative data (labelled as 2); they are 94%, 73%, and 82%, for an overall LR accuracy of 94%.

The LRC confusion matrix is shown in Fig. 26.8. In terms of making forecasts, the LRC model has a 94% accuracy. There are favourable tweets about vaccinations in 2810 records, no opinions in 18,462, and negative tweets about vaccinations in 8307.

26 Machine Learning-Based Sentiment Analysis of Twitter COVID-19 …

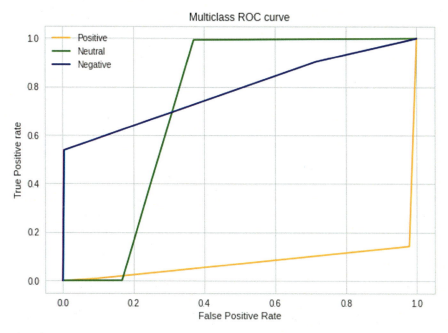

Fig. 26.6 Voting classifier multiclass ROC curve

Fig. 26.7 Logistic regression classification report

The LR machine learning classifier's multiclass ROC curve is displayed in Fig. 26.9. The x-axis of the graph shows the percentage of false positives, while the y-axis shows the percentage of real positives in the dataset.

Figure 26.10 displays the PR curve for the LRC model, which outperforms the other models. The correlation between precision and recall is depicted by a straight line called a precision–recall (PR) curve. It can also be written as: TP/(TP + FN) on

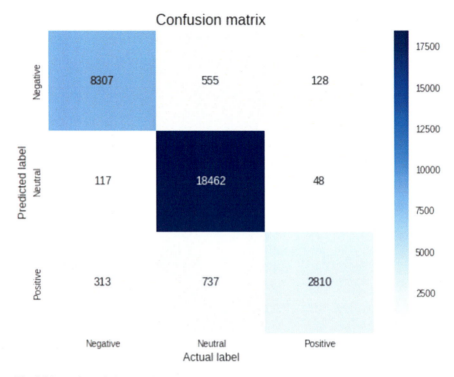

Fig. 26.8 LRC confusion matrix

the y-axis, TP/(TP + FP) on the x-axis, and so on for the entire PR curve. Besides its more common name, "positive predictive value," "precision" describes how likely something is to be correct (PPV). Figure 26.10 shows a PR curve of 0.97% for the positive class, 0.99% for the neutral class, 0.90% for the negative review class, and 0.97% for the over-average precision of the LR model on calls.

The accuracy of the categorization findings is displayed in Fig. 26.11. As shown in Table 26.1 and Fig. 26.11, the proposed Naive Bayes model achieves an accuracy of 81.14%, the voting classifier achieves an accuracy of 89.77%, and the third-best logistic regression model achieves an accuracy of 93.94%.

vii. *Comparative Results*

Figure 26.12 and Table 26.2 compare the original LSTM (89.30%) and Bi-LSTM (85.51%) model to the selected Naive Bayes (81.14%), VC (89.77%), and LR (93.97%) suggested models for evaluating the vaccine's sentiment analysis. In light of Twitter data, it seems that the former is the optimal approach. The logistic regression model excels at classification compared to other methods (93.97%).

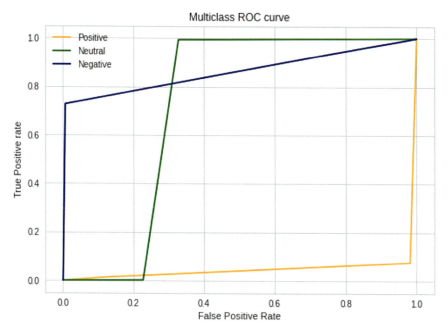

Fig. 26.9 LRC multiclass ROC curve

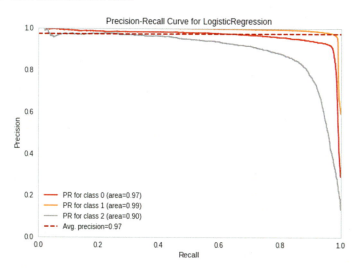

Fig. 26.10 LRC PR curve

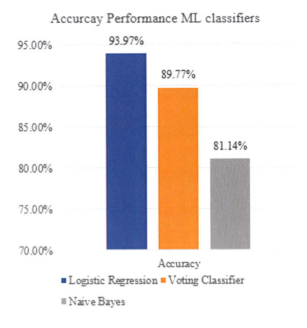

Fig. 26.11 Bar graph of proposed model's accuracy

Table 26.1 Machine learning classifier performance in terms of accuracy

Classifier	Accuracy (%)
Logistic regression	93.97
Voting classifier	89.77
Naive Bayes	81,14

26.7 Conclusion and Future Scope

The global SARS-CoV-2 coronavirus disease pandemic (COVID-19) poses a serious risk to public health. The pandemic has, without a doubt, changed the way we look at the world. A number of people who have received the COVID-19 vaccine have resorted to Twitter to discuss their experience. We offer a tool that can analyse Twitter data for sentiment, which can then be used in research.

To determine the user's perspective on ML, we analysed public tweets on COVID-19 vaccines using machine learning. Our study shows that machine learning techniques can be successfully used for sentiment analysis tasks. Simple natural language processing (NLP)-based sentiment analysis techniques were developed using the positive, negative, and neutral emotion polarities. Few machine learning (ML) algorithms exist now. Using a voting classifier, logistic regression, and naive Bayes, we evaluated the accuracy of our predictions and analyses. The results of the network visualization demonstrate that in order to combat the infodemic and increase vaccination rates, local-government health organizations and healthcare professionals need to be aware of the current state-of-the-art techniques in applying sentiment analysis.

Fig. 26.12 Bar graph of accuracy comparison between base and proposed models'

Table 26.2 Comparative analysis of base and proposed models

Parameter performance	Base models (DL)		proposed models (ml)		
	LSTM	BI-LSTM	NBC	VC	LRC
Accuracy (%)	89.30	85.51	81.14	89.77	93.77

All the models had excellent F-1 scores, confusion matrices, precision, and recall and precision; the logistics regression scored 93.97%, the voting classifier scored 89.77%, and the NB classifier scored 81.14%. Large numbers of people have decided to get vaccinated, but there are still many who are reluctant to do so because they are either unsure of the process, terrified of needles, or both.

Medical researchers will gain insight from this study as they learn more about the difficulties of the immunization process. A clear image of the vaccine's efficacy can be obtained by vaccine producers, health ministries, and governments across nations, and agencies such as World Health Organization. Those involved will have a clearer picture of what has to be done to restore faith in immunizations. We hope that our efforts, however small, can help frontline workers in the fight against this novel coronavirus and save lives.

References

1. Alam, K.N. et al.: Deep learning-based sentiment analysis of COVID-19 vaccination responses from Twitter data. Comput. Math. Methods Med. (2021). https://doi.org/10.1155/2021/4321131
2. Lincoln, T.M. et al.: Taking a machine learning approach to optimize prediction of vaccine hesitancy in high income countries. Sci. Rep. (2022). https://doi.org/10.1038/s41598-022-05915-3
3. Yousefinaghani, S., Dara, R., Mubareka, S., Papadopoulos, A., Sharif, S.: An analysis of COVID-19 vaccine sentiments and opinions on Twitter. Int. J. Infect. Dis. (2021). https://doi.org/10.1016/j.ijid.2021.05.059
4. Umair, A., Masciari, E.: A survey of sentimental analysis methods on COVID-19 research. CEUR Workshop Proc. **3194**, 167–174 (2022)
5. Mehta, P., Pandya, S.: A review on sentiment analysis methodologies, practices and applications. Int. J. Sci. Technol. Res. **9**(2), 601–609 (2020)
6. Amin, A., Hossain, I., Akther, A., Alam, K.M.: Bengali VADER: a sentiment analysis approach using modified VADER. (2019). https://doi.org/10.1109/ECACE.2019.8679144
7. Gouthami, S., Hegde, N.P.: A survey on challenges and techniques of sentiment analysis. Turkish J. Comput. Math. Educ. **12**(06), 4510–4515 (2021)
8. Alsaeedi, A., Khan, M.Z.: A study on sentiment analysis techniques of Twitter data. Int. J. Adv. Comput. Sci. Appl. (2019). https://doi.org/10.14569/ijacsa.2019.0100248
9. Xu, H., Liu, R., Luo, Z., Xu, M.: COVID-19 vaccine sensing: Sentiment analysis and subject distillation from twitter data. Telemat. Inform. Report. **8**, 100016 (2022). https://doi.org/10.1016/j.teler.2022.100016
10. Rahmanti, A.R. et al.: Social media sentiment analysis to monitor the performance of vaccination coverage during the early phase of the national COVID-19 vaccine rollout. Comput. Methods Programs Biomed. **221**, 106838 (2022). https://doi.org/10.1016/j.cmpb.2022.106838
11. Qorib, M., Oladunni, T., Denis, M., Ososanya, E., Cotae, P.: Covid-19 vaccine hesitancy: text mining, sentiment analysis and machine learning on COVID-19 vaccination Twitter dataset. Expert Syst. Appl. **212**, 118715 (2023). https://doi.org/10.1016/j.eswa.2022.118715
12. Sun, R., Budhwani, H.: Negative sentiments toward novel coronavirus (COVID-19) vaccines. Vaccine **40**(48), 6895–6899 (2022). https://doi.org/10.1016/j.vaccine.2022.10.037
13. Bokaee Nezhad, Z., Deihimi, M.A.: Twitter sentiment analysis from Iran about COVID 19 vaccine. Diabetes Metab. Syndr. Clin. Res. Rev. **16**(1), 102367 (2022). https://doi.org/10.1016/j.dsx.2021.102367
14. Liu, S., Liu, J.: Public attitudes toward COVID-19 vaccines on english-language Twitter: a sentiment analysis. Vaccine **39**(39), 5499–5505 (2021). https://doi.org/10.1016/j.vaccine.2021.08.058

Chapter 27
Exploring Sentiment in Tweets: An Ordinal Regression Analysis

Vishal Shrivastava and Dolly

Abstract The fundamental goal of sentiment analysis is to find and categorize any views or feelings that are communicated in a text. Nowadays, discussing thoughts and expressing feelings through social networking sites is widespread. Consequently, a vast amount of data is generated every day, which can be mined successfully to extract valuable information. Performing sentiment analysis on such data can be useful for producing an aggregated view of particular products. Due to the prevalence of slang and misspellings, sentiment analysis on Twitter is frequently a challenging undertaking. Additionally, we are constantly exposed to new terms, which makes it more difficult to assess and compute the sentiment compared to traditional sentiment analysis. Twitter limits a tweet's length to 140 characters. Consequently, obtaining important information from brief messages is another obstacle. Knowledge-based approaches and machine learning can significantly contribute to the sentiment analysis of tweets. The amount of data produced by people, i.e., users of a certain social site, is growing exponentially as a result of changing behavior of various types of networking sites like Snapchat, Instagram, Twitter, etc. The purpose of this paper is to determine the emotions underlying these posts. We have decided to use Twitter as our platform for this. In this study, we investigate the views expressed by Twitter users concerning certain companies. By computing a basic sentiment score and then categorizing them as positive or negative, the corporation would be provided with critical feedback about its products from individuals around the world. The proposed LSTM model has proved to be 93% efficient in comparison with previous models which were accurate up to 86%.

Keywords Twitter · Sentiment analysis · Machine learning · LSTM model

V. Shrivastava (✉) · Dolly
Department of Computer Science and Engineering, Arya College of Engineering and IT, Jaipur, India
e-mail: vishal500371@yahoo.co.in

27.1 Introduction

Twitter is one of the fastest-growing platforms for sending or receiving messages or tweets to a large number of people. Typically, users interact or express their thoughts and opinions regarding a topic via tweets. Blogging Web sites are becoming increasingly popular as a means of expressing one's viewpoint, which aids marketing campaigns in sharing consumer's perspectives on many themes pertaining to companies and products. Additionally, researchers utilize web data to do sentiment analysis on the public's perception of a product or issue.

Typically, [1], sentiment analysis helps identify and extracts subjective information from a text. Sentiment analysis is the most extensively utilized text classification method for analyzing concepts and determining whether the underlying sentiment is positive, negative, or neutral [1].

Challenges in Twitter sentiment analyses:

- Some tweets are often written in unpleasant language, whereas other brief messages lack emotional tone indicators.
- URLs, Hashtags, emojis, abbreviations, and acronyms are utilized often on Twitter.

The classification of sentiments in a text source is referred to as sentiment analysis. YouTube, Twitter, Facebook, and others play a key part in pandemic scenarios [2]. Twitter is a prominent, successful, and pervasive social media platform where millions of users share their ideas on a wide range of topics [2]. These systems include a vast amount of data. 90% of this information is textual or media-based. Analysis of text, reviews, and online comments can be conducted using a method called "sentiment analysis," which identifies the polarity of emotions such as sadness, rage, wrath, happiness, grief, and affection.

The method of determining the tone of a text in relation to a particular information source is referred to as opinion mining and is also called by its other name, sentiment analysis [3]. Due to numerous slang phrases, misspelled words, abbreviated forms, diverse characters, regional dialects, repeated characters, and incoming emojis, sentiment analysis is a complex and rapidly expanding study subject. Social media is one of the places where sentiment analysis is utilized effectively [3].

Nakov [4] In place of traditional two-or three-point scale, the ubiquitous 5 points "HIGHLY POSITIVE, NEUTRAL, NEGATIVE, POSITIVE, and HIGHLY NEGATIVE" scale is used anywhere human judgment is required in business sector, e.g., TripAdvisor, Amazon, and help, all utilize a five-point scale for rating sentiment toward products, hotels, and restaurants. Changing from a two- or three-point categorical scale to a five-point ordered scale is known as an "ordinal" classification change in world of machine learning (a.k.a. ordinal regression) [4].

27.1.1 Research Objective

- Examine the application of ordinal regression for LSTM-based sentiment prediction on Twitter.
- Check to see if this method can produce better outcomes than lexicon-based methods and conventional machine learning techniques.
- Solve difficulties with pattern recognition that can arise when predicting sentiment on an ordinal scale.
- Examine the possibility of enhancing sentiment analysis on Twitter data through the use of machine learning techniques.
- Provide advice and suggestions to those working as practitioners and scholars in the field of sentiment analysis.
- Participate in the continuing development of techniques for sentiment analysis on social media platforms that are more effective and efficient.

27.2 Literature Review

Several scholars have also contributed to lexicon-based sentiment analysis, which employs tokenization, stop-word removal, and stemming. Nevertheless, accuracy is lower when compared to previous work which used machine learning. Neutral, positive, and negative sentiments are classified using the following procedures.

This study [5], focused on LGBT sentiment analysis, which has become a prominent and polarizing topic of debate in current culture. Before determining the emotion of these Tweets from fifty states in United States of America, basic processing is performed. They evaluate five sentiment classification methods, including logistic regression, Naive Bayes, XGBoost, linear support vector machine, pattern analyzer, TextBlob, and on both unprocessed and preprocessed data. Finding that logistic regression without text preprocessing produces highest F1-score (70.87%) was the most significant finding. When they applied their sentiment classifier to U.S.-based tweets about the LGBT community, they found that the vast majority of messages fell into the "neutral" category [5].

The research [6] was undertaken to examine public's sentiment regarding this development, which was classified into 3 categories: positive, negative, and neutral. contra, neutral, and pro. Two distinct Doc2Vec models are used, namely the distributed model and distributed bag of words. As classifiers, this system utilizes logistic regression and support vector machines. Almost all of the models and classifiers have an accuracy rate of greater than 75%, and results show that they are opposed to development of Rinca Island [6].

In this study [7] proposed architecture, news content is encoded collectively. Taking into account the semantic data of news, text branch encodes semantic content information. At same time, information about visible aspects of a news image is extracted and encoded by visual branch. The next step is to construct a multimodal framework for joint sentiment categorization that makes use of inherent correlation between visual and textual data through use of a fusion layer [7]. This framework's goal is to classify the user's feelings. In last step, a decision-level fusion technique is applied to all three models in order to effectively merge cross-modal data for prediction of ultimate emotion. Based on in-house experiments conducted on an Assamese dataset, they find that multimodal features that are contextually integrated yield superior performance 89.3% compared to best unimodal features 85.6%.

The goal of this study [8] is to accurately analyze sentiment of trending tweets in data stream provided by Twitter API by combining a number of different algorithms in order to arrive at a consensus. Support vector machine, Naive Bayes, TextBlob, and lexicon approach were all put into action. They hope that by combining these approaches, they would get better results [8]. The results from an evaluation of their model employing labelled dataset indicate that combination of these four approaches resulted in a 68.29% overall accuracy, which is highest of any approach tested.

This study [9] focuses on developing a machine learning model that can analyze language patterns found in Twitter user data and determine whether or not a user is depressed. They developed diagnostic models using both random forest and support vector machines training and found that random forest performed better [9]. They believe that the findings of this study can be used to establish a new method for identifying depressed users on social media platforms.

The study [10] also constructed a broad metadata framework for the classification of hate speech on Twitter in order to address issues with Twitter's data streams. Compared to other techniques, the created generic metadata architecture demonstrated superior performance across all assessment criteria for hate speech identification, achieving for accuracy is 0.95, F1-score is 0.93, recall is 0.92, and precision is 0.93, respectively. Similarly, created generic metadata architecture for hate speech sentiment categorization outperformed comparable approaches with an F1-score of 91.5% [10].

In this study, [11] examine possibilities of employing hierarchical clustering for Twitter sentiment analysis, single linkage (SL), complete linkage (CL), and average linkage (AL) hierarchical clustering algorithms are explored. The concept of selecting optimal cluster for tweets is operationalized through majority voting, and this is accomplished by constructing a collaborative framework that is comprised of AL, SL, and CL. They compare hierarchical clustering methods with k-means and two other modern classifiers (SVM and Naive Bayes).Clustering and classification are evaluated based on their accuracy and effectiveness with respect to time [11]. The experimental findings show that cooperative clustering based on majority voting produces clusters of high quality, but at expense of time efficiency.

In this study, [12] Using Naive Bayes Algorithm in RapidMiner tools, they were able to acquire an accuracy of 86.43% in their testing, that is significantly higher than accuracy they achieved using Random Forest and Decision Tree (both of which yielded an accuracy of 82.91%) [12].

This work [13] describes various deep learning and machine learning models and how they were trained to utilize a dataset of tweets gathered from a GitHub repository. The suggested system was tested in a case study that predicted the outcome of elections in Punjab in February 2017 by analyzing public sentiment on Twitter toward various candidates [13]. The proposed SA system does sentiment analysis in real time, displaying analysis findings in sync with Twitter postings. To show users the outcomes, a dashboard has been developed. The display is updated each minute with real-time tweets and graphical representations of outcomes. A CSV file is also created to save tweets for future forecasting purposes.

Islam [14] studied the issue of sadness among Facebook users. They utilized KNN algorithm to identify sad moods in Facebook Data. On various matrices, their work yielded results between 60 and 70% [14].

In this study, [15] employ machine learning techniques that have suggested a connection between Bitcoin's price movement and mood of its users. They have laid out their implementation strategy, including their final analysis and factors they considered when setting prices. They applied an approach that was based on sentiment to challenge problem of predicting variations in price of bitcoin in order to establish the significance of public opinion in cryptocurrency business. In addition to that, the findings of this research reveal an innovative technique for utilizing the information obtained from social networking Web sites [15].

The goal of this [16] research is to find optimal combination of text transformations (including entity removal, stemming, and lemmatization), tokenizers (such as word n-grams), and token-weighting algorithms for a support vector machine classifier trained on two Spanish datasets. The methodology entails doing a comprehensive study of each and every conceivable combination of text alterations and the factors that are associated with them in order to identify the characteristics that are typical of the best-performing classifiers. In addition, a novel method fusing n-grams with words and q-grams with characters is presented. When applied to INEGI and TASS'15 datasets, this novel combination of words and characters produces a classifier that achieves 11.17 and 5.62 percentage point improvements, respectively, over standard word-based combinations [16].

In this work [17] following a description of Twitter data sentiment analysis, existing tools for sentiment analysis, related work methodologies, and a case study displaying effort, the paper moves on to its findings [17]. Researchers can see from the data that 50% of the responses were positive, 20% were negative, and 30% were neutral.

This study also [18] examines connected business insights within the communications services industry. PT XL Axiata Tbk, PT Telkomsel Tbk, and PT Indosat all received 32.3, 19.0, and 10.9, respectively, out of a possible 40 points on the NBR scale. Tbk, respectively, after taking into consideration an overall evaluation of these five products [18].

27.3 Methodology

This section discusses statement of problem, research technique, and sequential processes utilized in study's approach. In addition, a detailed flowchart of complete research process and an algorithm that has been developed with step-by-step instructions are provided in this section.

27.3.1 Problem Statement

There is no universal machine learning model applicable to all data mining issues. Experimentation is integral to data mining. As a result, they analyzed a large number of classification schemes for Tweets into certain ontologies in order to choose the most appropriate one, and then assigned a rating to every aspect contained within them. Their goal is to develop a method that accepts a phrase or sub-sentence as input and returns a sentiment score for this particular segment of speech. A prerequisite for this is the construction of a second mechanism that will take a text (such as a tweet) and divide it into as many subsentences as the number of ontologies it contains. This is already the subject of study being undertaken concurrently with this dissertation at our university.

27.3.2 Proposed Methodology

Twitter API is used to gather a dataset, and data are tagged as negative or positive tweets. The dataset can be accessed by general public via Natural Language Toolkit (NLTK) corpora repository, which is well-known and widely used in a wide range of different types of research. The corpus consists of 10,000 tweets, with 5000 negative tweets and 5000 positive tweets. Then we used some libraries such as numpy, pandas, tensor flow, sklearn, genism, and seaborn. The data has been preprocessed in which we have checked the null value and removed null value, all emojis, URLs from tweets, Twitter handles, punctuation, extra spaces, numbers and special characters. Then applied LSTM model and achieved performance matrix including precision, accuracy, recall, and F1-score.

In this section, several strategies and techniques are described and planned for implementation to achieve the objectives. Additionally, this section predicts results that these procedures will yield. Some of the strategies are explained in sections that follow:

a) **Data Collection**

In this work, we have utilized a dataset from Kaggle data repository that identifies Twitter sentiment analysis.

b) **Data Preprocessing**

Processing raw data into a form which can be read and analyzed by computers and machine learning is called data preprocessing, and it is an integral part of data mining and analysis process. Real-world information in the form of text, photos, video, etc., is chaotic. In addition to occasionally including errors and inconsistencies, it is frequently lacking in completeness and does not follow a regular or conventional layout. The analysis of information by machines is most successful when it is neat and ordered, and machines read data as a series of zeros and ones. As a result, calculating structured data, such as whole numbers and percentages, is a relatively simple task. However, textual and visual data that has not been preprocessed in a standard way must be cleaned and formatted before analysis can begin.

Feature Extraction

"Feature extraction" refers to the process of transforming unprocessed data into numerical features that can be controlled while maintaining the integrity of the original data set. Utilizing machine learning on raw data does not produce the same level of results.

Tokenizer: This technique breaks the given text into tokens (small parts) and removes any punctuation from textual data. This study employed nltk. Tokenization is performed through tokenize methods (a built-in function of nltk toolkit).

c) **Data Splitting**

Data splitting is when data is divided into two or more subsets. Typically, with a two-part split, one part is used to evaluate or test the data and the other to train the model. In this study, the set of data was divided into two parts for data split. 75% of training set and 25% of testing set were used.

d) **Classification**

A classification algorithm is a type of supervised learning algorithm that learns how to classify new observations based on existing ones. While we were sorting things out. A computer algorithm needs to be trained on a dataset of observations before it can reliably classify new data. Make use of machine learning for a classification analysis and ordinal regression on Twitter using LSTM model.

- LSTM Model

Long short-term memory (LSTM) models could unearth long-lasting dependencies. The gradient difficulties are resolved by the dynamic LSTM/LBU device. The LSTM model has fewer nodes but performs similarly to more complex networks with a certain layout. When one mechanism's mistakes are coupled with the other's mass, the former is known as a Constant Error Carousel (CEC). When you increase or reduce the period's interval width, your target times will advance or regress. The phasing in of CEC has two functions: First is the introduction of previous experience recall and identification of mental condition that causes it, and second is introduction of multiplicative units. Standard LSTMs have been demonstrated to be superior to

Fig. 27.1 LSTM model

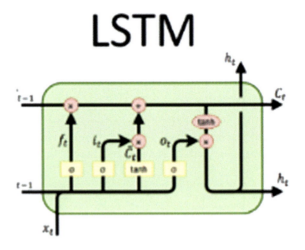

recurrent neural networks (LSTMs) in modeling long-term dependencies as show in Fig. 27.1.

A. *Proposed Algorithm*

    Input: NLTK dataset.
    Output: Get classified outcomes.
    Strategy:

Step 1: Start implementation process
Step 2: Import set of data (NLTK)
Step 3: Processing of raw data before it is used

- Check null value
- Remove punctuation
- Remove emoji
- Remove Twitter handles
- Remove newline character

Step 4: Apply feature extraction techniques

- Tokenizer

Step 5: Split dataset into training and testing set that divided into 75:25

- Training set (75%)
- Testing set (25%)

Step 6: A proposed machine learning model

- LSTM memory

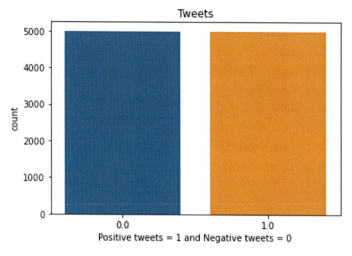

Fig. 27.2 Twitter sentiment analysis graph

Step 7: Evaluation parameters (recall, accuracy, precision, and F1-score)
Step 8: Get results.
Step 9: End

27.4 Results Illustrations

This section provides an overview of dataset, performance measures, and experimental outcomes. Python programming experiments have been conducted using Jupyter notebook in this suggested work.

27.4.1 Dataset Description

In this graph, we have divided tweets into two equal parts in which 5000 negative tweets and 5000 positive tweets are found as shown in Fig. 27.2.

27.4.2 Performance Metrics

(1) **Accuracy**

In context of sentiment analysis on Twitter data, accuracy is a commonly used metric to assess performance of a classification model. The formula for accuracy is the same as general classification problem:

$$\text{Accuracy} = \frac{|TN| + |TP|}{|TN| + |FN| + |TP| + |FP|} \qquad (27.1)$$

(2) **Precision**

Precision reflects the frequency with which the anticipated result of a classifier is accurate when it represents true. The preciseness formula is:

$$\text{Precision} = \frac{|TP|}{|FP| + |TP|} \qquad (27.2)$$

(3) **Recall**

A recall is a metric used to assess performance of a classification model, particularly in context of imbalanced datasets. It is the ratio of a total number of accurate occurrences in training data to fraction of correct predictions. The formula for recall in the context of sentiment analysis on Twitter data is:

$$\text{Recall} = \frac{|TP|}{|FN| + |TP|} \qquad (27.3)$$

(4) **F1-Score**

F1-score is the answer to the problem of inaccurate results on unbalanced data [19]. When data is unbalanced, we use the F1-score to help us out. The F1-score is a combination of the recall and accuracy scores [19]. The F1 rule of score is

$$F - \text{measure} = 2 \times \frac{\text{Recall} \times \text{Precision}}{\text{Recall} + \text{Precision}} \qquad (27.4)$$

27.4.3 Experimental Results

Figure 27.3 shows confusion matrix of proposed LSTM model utilizing NLTK dataset. In graph, axis shows the actual label, and y-axis shows predicated label each label shows. The number of successful and unsuccessful classifications made by algorithms is tabulated in a confusion matrix. "True Positive" means both projected and actual values are positive. The confusion matrix that we have here has a true positive value of 171, and a true negative value indicates that both the actual and anticipated values are negative. Here, in this confusion matrix, the true negative value is 29.

Table 27.1 compares the dataset performance of numerous classifiers. LSTMs, RFCs, DTs, and MLRs are being compared. Each classifier's recall, precision, accuracy, and F1-score are listed. A classifier's accuracy is its ability to accurately classify data items. LSTM has the best accuracy at 93%, followed by RFC at 86%, DT at

27 Exploring Sentiment in Tweets: An Ordinal Regression Analysis

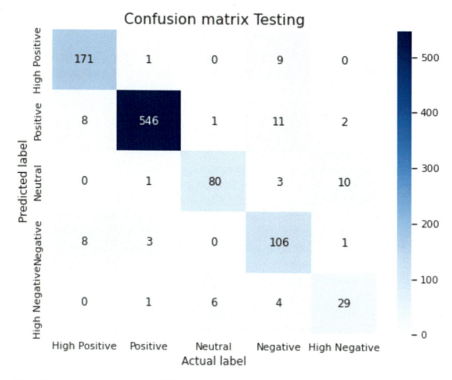

Fig. 27.3 Confusion matrix on LSTM

Table 27.1 Comparison of proposed and base models using NLTK dataset

Parameters	Proposed classifier	Base classifier		
	LSTM	RFC	DT	Multinomial linear regression
Accuracy (%)	0.93	86.00	84.17	80.71
Precision (%)	0.93	82.92	80.63	75.87
F1-score (%)	0.93	68.99	80.67	71.45
Recall (%)	0.93			

84.17%, and multinomial linear regression at 80.71%. Precision refers to percentage of "yes" data that was accurately identified. LSTM classifier has the highest precision at 93%, followed by the RFC, DT, and multinomial linear regression. F1-score indicates a classifier's precision-recall balance. In this table, the LSTM classifier has the greatest F1-score (93%), followed by the DT (80.67%), multinomial linear regression (71.45%), and RFC (68.99%). Recall, also known as sensitivity or true positive rate (TPR), is proportion of positive data items correctly detected by the classifier.

27.5 Conclusion

The goal of Twitter sentiment analysis is to ascertain general tone of a tweet by identifying and extracting subjective information utilizing NLP and ML methods (positive, negative, and neutral). This work has a lot of benefits, most notably time savings while solving equations, which will be valuable for many fields whose timetables include time-consuming equation solving and analysis. This study effort concludes the whole suggested study on Twitter spam streaming analysis and grouping of emotions. A feature extraction approach based on a blend of random forest, linear regression, and principal component analysis (PCA) is presented for extracting specific feature sets for boosting classification accuracy and using machine learning classifiers to expose spam tweets. When compared to other existing works, simulation outcomes suggest that proposed work has a higher detection ratio. When employing a vast quantity of data, the results achieved in this suggested study show a very big difference in terms of accuracy, recall, precision, and F1-score when compared to other classifiers. Furthermore, this hybrid technique is applied for sentiment classification of tweets with modest alterations in the suggested algorithm in terms of positive and negative tweets, resulting in good classification accuracy Sentiment analysis on Twitter is a subfield of text and opinion mining. It analyzes tweet emotions and trains a machine learning model to measure its accuracy so we can utilize it in future. Sentiment detection, text preprocessing, data collection, sentiment classification, model training, and testing are all a part of the process. Over past decade, advancements in this area of study have led to model efficiencies of 85–90%. Diverse information is still missing, however. It also has a lot of problems in practical use due to slang and abbreviations. Increasing the number of classes often results in poor performance for many analyzers. The model's applicability to areas other than the one under discussion has not yet been thoroughly examined. The proposed LSTM Model has proved to be 93% efficient in comparison with previous models which were accurate up to 86%. Thus, sentiment analysis offers tremendous potential for growth in the years to come.

References

1. Sentamilselvan, P.K.K.K., Aneri, D., Athithiya, A.C.: Twitter sentimental analysis using machine learning techniques. Int. J. Innov. Technol. Explor. Eng. **9**(3), 2249–8958 (2020). https://doi.org/10.35940/ijeat.c6281.029320
2. Santhosh Baboo, S., Amirthapriya, M.: Sentiment analysis and automatic emotion detection analysis of Twitter using machine learning classifiers. Int. J. Mech. Eng. **7**(2), 974–5823 (2022)
3. Gupta, B.V.: Comparison of sentiment analysis algorithms using Twitter and review dataset. Int. J. Res. Appl. Sci. Eng. Technol. **10**(4), 2299–2304 (2022). https://doi.org/10.22214/ijraset.2022.41785
4. Nakov, P., Ritter, A., Rosenthal, S., Sebastiani, F., Stoyanov, V.: SemEval-2016 task 4: sentiment analysis in Twitter. (2016). https://doi.org/10.18653/v1/s16-1001
5. Aldinata, Soesanto, A.M., Chandra, V.C., Suhartono, D.: Sentiments comparison on Twitter about LGBT. Proc. Comput. Sci. **216**, 765–773 (2023). https://doi.org/10.1016/j.procs.2022.12.194

6. Jaya Hidayat, T.H., Ruldeviyani, Y., Aditama, A.R., Madya, G.R., Nugraha, A.W., Adisaputra, M.W.: Sentiment analysis of twitter data related to Rinca Island development using Doc2Vec and SVM and logistic regression as classifier. Proc. Comput. Sci. **197**, 660–667 (2022). https://doi.org/10.1016/j.procs.2021.12.187
7. Das, R., Singh, T.D.; A multi-stage multimodal framework for sentiment analysis of Assamese in low resource setting. Expert Syst. Appl. **204**, 117575 (2022). https://doi.org/10.1016/j.eswa.2022.117575
8. Motz, A., Ranta, E., Calderon, A.S., Adam, Q., Alzhouri, F., Ebrahimi, D.: Live sentiment analysis using multiple machine learning and text processing algorithms. Proc. Comput. Sci. **203**, 165–172 (2022). https://doi.org/10.1016/j.procs.2022.07.023
9. Azam, F., Agro, M., Sami, M., Abro, M.H., Dewani, A.: Identifying depression among Twitter users using sentiment analysis (2021). https://doi.org/10.1109/ICAI52203.2021.9445271
10. Ayo, F.E., Folorunso, O., Ibharalu, F.T., Osinuga, I.A.: Machine learning techniques for hate speech classification of twitter data: state-of-the-art, future challenges and research directions. Comput. Sci. Rev. **38**, 100311 (2020) https://doi.org/10.1016/j.cosrev.2020.100311
11. Bibi, M., Aziz, W., Almaraashi, M., Khan, I.H., Nadeem, M.S.A., Habib, N.: A cooperative binary-clustering framework based on majority voting for twitter sentiment analysis. IEEE Access **8**, 68580–68592 (2020). https://doi.org/10.1109/ACCESS.2020.2983859
12. Fitri, V.A., Andreswari, R., Hasibuan, M.A.: Sentiment analysis of social media Twitter with case of anti-LGBT campaign in Indonesia using Naïve Bayes, decision tree, and random forest algorithm. Proc. Comput. Sci. **161**, 765–772 (2019). https://doi.org/10.1016/j.procs.2019.11.181
13. Gupta, Y., Kumar, P.: Real-time sentiment analysis of tweets: a case study of punjab elections. (2019). https://doi.org/10.1109/ICECCT.2019.8869203
14. Islam, M.R., Kamal, A.R.M., Sultana, N., Islam, R., Moni, M.A., Ulhaq, A.: Detecting depression using K-nearest neighbors (KNN) classification technique (2018). https://doi.org/10.1109/IC4ME2.2018.8465641
15. Rahman, S., Hemel, J.N., Junayed Ahmed Anta, S., Al Muhee, H., Uddin, J.: Sentiment analysis using r: an approach to correlate cryptocurrency price fluctuations with change in user sentiment using machine learning. In: 2018 Joint 7th International Conference on Informatics, Electronics and Vision (ICIEV) and 2018 2nd International Conference on Imaging, Vision and Pattern Recognition (icIVPR), pp. 492–497. (2018). https://doi.org/10.1109/ICIEV.2018.8641075
16. Tellez, E.S., Miranda-Jiménez, S., Graff, M., Moctezuma, D., Siordia, O.S., Villaseñor, E.A.: A case study of Spanish text transformations for twitter sentiment analysis. Expert Syst. Appl. **81**, 457–471 (2017). https://doi.org/10.1016/j.eswa.2017.03.071
17. Mishra, P., Rajnish, R., Kumar, P.N.: Sentiment analysis of Twitter data: case study on digital India (2017). https://doi.org/10.1109/INCITE.2016.7857607
18. Vidya, N.A., Fanany, M.I., Budi, I.: Twitter sentiment to analyze net brand reputation of mobile phone providers. Proc. Comput. Sci. **72**, 519–526 (2015). https://doi.org/10.1016/j.procs.2015.12.159
19. Barzenji, H.A.S.: Sentiment analysis of Twitter posts using machine learning algorithms. pp. 980–983. (2021)

Chapter 28
Automated Classification of Alzheimer's Disease Stages Using T1-Weighted sMRI Images and Machine Learning

Nand Kishore and Neelam Goel

Abstract Alzheimer's disease is the most common forms of dementia. Dementia is the general term for cognitive decline severe enough to impede with daily activities. Early diagnosis of Alzheimer's disease is important for slowing or stopping the disease's development, and experts can start preventive treatment right away. The experts must be capable of identifying Alzheimer's disease in its earliest and most challenging phases. The fundamental objective of this study is to create a machine learning model that can automatically diagnose disease using MRI, a widely used diagnostic tool. This research employed structural MRI to find out the difference between patients with Alzheimer's disease (AD), stable mild cognitive impairment (sMCI), progressing mild cognitive impairment (pMCI), and normal cognitive functioning (CN). In this research paper, machine learning models, namely SVM, RF, DT, and CNN, are used for multi-class classification. CNN obtained the highest testing accuracy of 88.84% among the four models, with a precision of 80.42%, a recall of 73.17%, and an F1-score of 76.62% for the CN versus sMCI versus pMCI versus AD multi-class classification.

Keywords Alzheimer's disease · GLCM · T1w-sMRI · Machine learning · Convolutional neural network

28.1 Introduction

Alzheimer's disease (AD), which causes cognitive impairment and behavioural difficulties, is the most common neurodegenerative disorder in older individuals, accounting for 60–80% of dementia cases. About 5–20% of the population, above the age of 65 suffers from mild cognitive impairment (MCI), which manifests as slight to moderate memory loss and other cognitive difficulties. MCI is a precursor

---

N. Kishore · N. Goel (✉)
University Institute of Engineering and Technology, Panjab University, Chandigarh 160014, India
e-mail: erneelam@pu.ac.in

to AD, with 10–15% of MCI patients developing AD each year [1]. Furthermore, over the next two to three decades, this number will rise steadily. Even though the aetiology of Alzheimer's disease has not been fully elucidated and no treatment has been developed, early diagnosis of the AD is critical for both prevention and treatment [2]. There are about 50 million people living with Alzheimer's disease (AD) across various regions of the world. It is predicted that the number of people diagnosed with Alzheimer's disease would be more than double by the year 2050 as a result of the ageing of the world's population [3]. There are several biomarkers for detection of Alzheimer's disease, such as those discovered in neuroimaging, CSF, genetic, blood, urine biomarkers, and others. Additional categories for neuroimaging biomarkers include structural imaging biomarkers, functional imaging biomarkers, and molecular imaging biomarkers, in which the brain's functions are determined by the brain's morphology and the molecules that are present. However, there is not yet sufficient evidence to determine which biomarker is the most accurate in detecting Alzheimer's disease [4]. The diagnostic and therapeutic roles of dementia biomarkers have the potential to change to life an AD patients [5]. In recent years, non-invasive brain imaging has become more and more common for diagnosing Alzheimer's disease. This imaging technique is completely safe for human brain tissues, making it a highly useful diagnostic tool. Non-invasive medical brain imaging is now one of the most preferred diagnostic tools by the physicians [6].

In recent years, machine learning algorithms for interpreting biological imaging and neuroimaging data have become increasingly popular. Deep learning, a machine learning framework based on "artificial neural networks," has emerged in recent years", due to the fact that it is capable of accurately predicting significant clinical outcomes. CNN models, which stand for "convolutional neural network," are potential deep learning methods for detecting and classifying objects. These models are used rather extensively in the field of medical imaging. CNN models are able to recognize patterns in visual data sets and accurately predict outcomes for fundamental tasks such as categorizing. This ability has been demonstrated through extensive testing of CNNs as specialized artificial neural networks that perform image analysis by convolving several filters into a single input, each of which is about the size of a small patch of the image, to look for similar spatial characteristics across the whole image [7]. This study helps doctors make better decisions about neurodegenerative diseases, especially Alzheimer's disease (AD).

The most important contributions that are made by this study are: To demonstrate an effective method for the multi-stage classification of AD. Texture feature extraction carried out by the application of the gray-level co-occurrence matrix (GLCM) for multi-stage categorization.

28.2 Related Work

This section provides a concise overview of the different structural MRI (sMRI) techniques used to diagnose Alzheimer's disease. There are three main types of such methods: whole brain, region of interest, and landmark-based. Brain mapping methods, such as voxel-based morphometry (VBM), tensor-based morphometry (TBM) and deformation-based morphometry (DBM) track morphological changes to determine brain atrophy. Voxel-based methods measure tissue density in the human brain voxel by voxel. Tensor-based methods examine the local Jacobians of deformation fields in order to determine structural issues. DBM is utilized to accommodate nonlinear deformation fields to a standard anatomical template [1]. SVM classifiers and grey matter voxels were used to distinguish AD from CN in [8]. Firstly, the hippocampus areas of several subjects were segmented, and then, their form was quantified using spherical harmonics features extraction [9]. Traditional approaches for diagnosing Alzheimer's disease often have poor performance because they rely on manually produced visual information that are either difficult or impossible to optimize. As a result, these methods are typically ineffective.

Deep learning algorithms aid in the early diagnosis of Alzheimer's disease (AD) symptoms, and the monitoring of the progression from mild cognitive impairment (MCI) to Alzheimer's disease is discussed in [1]. Existing methods, even though they are popular, use unrelated brain areas or less precise landmarks to diagnose AD. Densely connected neural networks were employed to extract multi-scale information from pre-processed images. In addition, a connection-wise attention mechanism was used to the process of diagnosing Alzheimer's disease [2]. A large number of studies make use of convolutional neural networks (CNNs) to record deep-level medical image data in an effort to diagnose Alzheimer's disease (AD) and forecast clinical scores [3]. 2D anatomical slices were used in the advanced diagnostic learning process for Alzheimer's disease (AD) [10]. Alzheimer's disease was diagnosed utilizing the lateral ventricles, periventricular white matter, and cortical grey matter [11]. Several statistical methods, such as support vector machines, logistic regression, decision trees, and random forests, were used in the process of AD diagnostic prediction [12]. A fuzzy hyperplane-based least square twin support vector machine for feature categorization and early AD detection using sagittal plane slices from 3D MRI imaging data are described in [13]. Researchers are trying to improve their ability to detect Alzheimer's disease by using subject-level, patch-based, and slice-based computational neuroimaging methods [4]. Clinical data and a hybrid texture, edge, color, and density feature extraction were utilized to diagnose AD [14]. The use of deep learning-based multiclass classification to distinguish between the stages of Alzheimer's disease for early diagnosis is discussed in [15]. Axial brain images from a three-dimensional MRI were fed into a convolutional neural network (CNN) for multiclass categorization [16]. Voxel-based morphometry was used to extract grey matter from the brain and then correlated clinical cognitive scores in MCI with grey matter in order to diagnose Alzheimer's disease [17]. Multi-modal medical imaging has the potential to improve Alzheimer's disease (AD) categorization and diagnosis [6]. Multimodal adaptive-similarity feature selection for AD diagnosis is discussed in [18].

28.3 Material and Methods

28.3.1 Data Collection

ADNI, which stands for the Alzheimer's disease neuroimaging initiative, data is used in this research. https://ida.loni.usc.edu. ADNI has different phases like ADNI-1, ADNI-GO, ADNI-2, and ADNI-3. The ADNI-1 baseline dataset used in this work comprises of 342 subjects including 89 CN, 100 sMCI, 110 pMCI, and 43 AD. Cognitive normal (CN) class includes healthy controls with no conversion into any other stage of Alzheimer's disease within 36 months of baseline follow up. Mild cognitively impaired individuals who were nevertheless capable of carrying out everyday tasks were kept in MCI class. Stable MCI (sMCI) subjects were those who had been diagnosed with MCI and had not changed or returned to any other class after at least 36 months from baseline. Progressive MCI (pMCI) class included subjects who were initially diagnosed with MCI but later developed AD within 36 month. Patients in the AD class were those who had been first diagnosed with Alzheimer's disease and had no signs of reversal within 24 months.

All the T1weighted-sMRI data is acquired with a 1.5 T T1-weighted 3D MP-RAGE sequence. The main reason for choosing structural MRI in this work is that these scans are more detailed and cheaper as compared to other imaging modalities like SPECT, PET, and DTI. sMRI is mainly used to capture the pathology and anatomy of brain. Other Imaging Protocol details include Acquisition Type: 3D; Field Strength: 1.5 T; Slice Thickness: 1.2 mm; Flip Angle: 8.0 degrees; X, Y, and Z each have 256.0 pixels; Visit—baseline and screening. In order to assess cognitive impairment, the 30-point MMSE questionnaire is often used in clinical and research contexts. MMSE can assist the clinicians to assess the level of dementia and also amenable to patients and their families. The subjects details are provided in Table 28.1.

The proposed methodology of AD classification is shown in Fig. 28.1. Detailed descriptions of the steps are provided in the subsections that follow.

Table 28.1 Subject's demographic information

Diagnosis	Number of subjects	Age (years)	MMSE
CN	89	65–85	28.11–30.11
sMCI	100	65–85	25.51–29.05
pMCI	110	65–85	24.87–28.29
AD	43	65–85	21.31–25.29

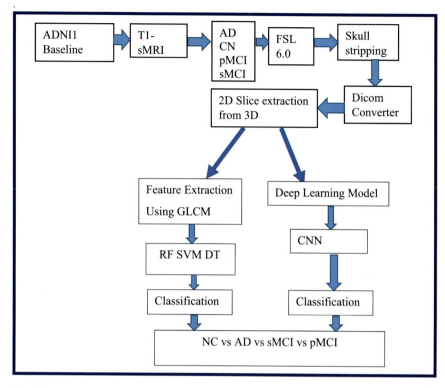

Fig. 28.1 Proposed methodology

28.3.2 MRI Image Pre-processing

To standardize the data into the appropriate shape and format, preprocessing is conducted for each brain structural MRI. It is a preparatory process that is intended to remove undesired non-brain tissues such as the skull, eyes, fat, and other things that are present in brain MRI images [19]. In this work, skull stripping is conducted by using brain extraction tool (BET) of FSL 6.0 from https://fsl.fmrib.ox.ac.uk/. The outcome of skull stripping is shown in Fig. 28.2. Once BET has reduced the extraneous material, the brain itself can be isolated from the whole brain scan. 50 middle slices of each subject that contain useful information about Alzheimer's disease are selected and remaining images are discarded.

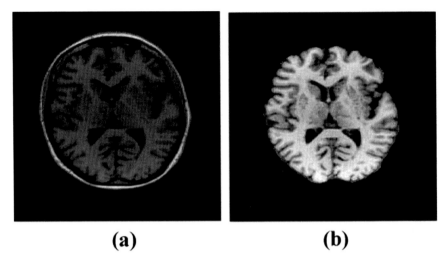

Fig. 28.2 Skull stripping **a** raw image **b** skull stripped image

28.3.3 Feature Extraction

As shown in Fig. 28.1, after preprocessing, 3D images are transformed to many groups of 2D images along sagittal direction. In the process of classification, the use of texture information reduced the amount of noise in the optical picture pixels. Gray-level co-occurrence matrix (GLCM), which is a common statistical method, was used for feature extraction for T1w-sMRI images. The features extraction from images described in Table 28.2.

Table 28.2 GLCM texture features

Features	Formula
Energy	$Ene = \sum_{i,j=1}^{M}(p(i,j))^2$ (1)
Contrast	$Con = \sum_{i,j=1}^{M}((i-j))^2 \, p(i,j)$ (2)
Correlation	$Corr = \sum_{i,j=1}^{M} \frac{(i-\mu i)(j-\mu i)p(i,j)}{\sigma i \sigma j}$ (3)
Homogeneity	$Hom = \sum_{i,j=1}^{M} \frac{p(i,j)}{1+|i-j|}$ (4)
Dissimilarity	$Dissim = \sum_{i,j=1}^{M}|i-j|p(i,j)$ (5)

28.3.4 Machine Learning Techniques

Even it is widely acknowledged that machine learning falls under the umbrella term of artificial intelligence, there are several points of view within the academic community that place more emphasis on its status as a subfield of computer science rather than AI. Machine learning can be thought of as both multidisciplinary and interdisciplinary field due to its usage of ideas and methods adopted from a wide range of disciplines. Machine learning classifies and diagnoses Alzheimer's disease severity. Three classifiers, namely SVM, RF, and DT are applied in this work.

Support Vector Machine: The support vector machine, more commonly known as SVM, is a well-known method of supervised learning that is applicable for classification and regression analysis. However, its most common use is in machine learning classification tasks. The SVM method figures out the best line or decision boundary to divide n-dimensional space into classes. This makes it easier to put future data points into the right category.

Decision Tree: Decision tree is a supervised learning technique. It looks like a tree, with fundamental nodes that show the features of a dataset, branches that show how to make decisions, and leaf nodes that show the final classes. It is possible to "learn" a tree by first dividing the source set into multiple subsets, each of which is evaluated using an attribute value test. Decision tree classifiers are excellent for knowledge discovery's exploratory phase as they do not require domain knowledge or parameters.

Random Forest: Feature randomization, often called feature bagging, provides a random set of characteristics to reduce decision tree correlation. This is one of the most significant differences between decision trees and random forests. While decision trees take into account all possible feature splits, random forests select only a fraction of the possible combinations.

28.3.5 Deep Learning

Machine learning has employed artificial neural networks and representation learning, a subset of which is deep learning. Deep learning applications in speech recognition, computer vision, machine translation, natural language processing, bioinformatics, medical imaging, climate research, board game programming, and material inspection have utilized architectures such as deep belief networks, deep neural networks, recurrent neural networks, deep reinforcement learning, convolutional neural networks, transformers, and others.

Convolutional neural networks are a component of deep neural networks. In one layer, these networks perform convolution instead of matrix multiplication.

CNNs are a type of deep learning architecture typically employed for image classification and recognition tasks. It has numerous layers, including convolutional, pooling, and fully linked layers. The convolutional layer uses image filtering to pull out features from the input image. The pooling layer reduces sample sizes, which in turn reduces processing time, and the fully connected layer offers the final forecast. Backpropagation and gradient descent are used to train the network to acquire the most efficient filters.

28.4 Results and Discussion

In this work, multiclass categorization of Alzheimer's disease is being accomplished by the application of machine learning and deep learning models. For this purpose, the total dataset is divided 80:10:10 between training, validation, and testing, respectively. All the experiments are carried out using python 3.5 and i5 processor with 8 GB of RAM. Keras, a deep learning package, is used in conjunction with Tensorflow as the backend to construct the neural network. The performance of the model is evaluated using the parameters, precision, recall, accuracy, and F1-score. The results of performing multiclass classification (AD vs. CN vs. pMCI vs. sMCI) using a machine learning and deep learning techniques are presented in Table 28.3.

Out of the four models, CNN (scratch) achieved the highest accuracy of 88.84%, precision of 80.42%, recall of 73.17%, and F1 score of 76.62%. It is clear from the above results that deep learning model performed better than traditional machine learning models. The bar graph of evaluation metrics of all four models is shown in Fig. 28.3.

It is clear from Fig. 28.3 that CNN outperformed machine learning techniques in terms of precision, recall and F1-score. The overall performance of CNN is better than machine learning models.

The results of the proposed method are compared with two state-of-art methods: (1) deep multi-task multi-channel learning (DM^{2}L) [20] (2) deep residual neural networks (ResNet) [21]. In the first method, a landmark-based approach is used to automatically generate discriminative image patches from MRI images which were further utilized for multi-class classification [20]. In the second method, a deep residual neural networks are applied for predicting progression from mild cognitive impairment (MCI) to Alzheimer's disease (AD) [21]. The comparison results are presented in Table 28.4.

Table 28.3 Result of various machine learning and deep learning models (Unit:%)

Networks	Modality	ACC
Random forest	T1weighted-sMRI	50.38
Support vector machine	T1weighted-sMRI	39.96
Decision tree	T1weighted-sMRI	39.84
CNN (scratch)	T1weighted-sMRI	**88.84**

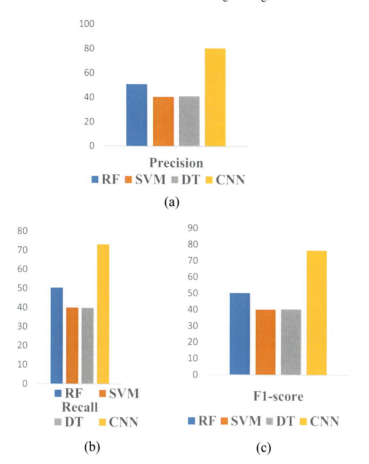

Fig. 28.3 Proposed models results **a** Precision **b** Recall **c** F1-score

Table 28.4 Comparison with state-of-the-art methods

Approach	Dataset	Modality	Classification task	Accuracy (%)
Liu et al. [20]	ADNI	T1w-sMRI	CN versus sMCI versus pMCI versus AD	51.8
Abrol et al. [21]	ADNI	sMRI	CN versus sMCI versus pMCI versus AD	83.01
Proposed method	ADNI	T1w-sMRI	CN versus sMCI versus pMCI versus AD	**88.84**

It is clear from the above results that deep learning techniques performed better than landmark-based approach. The proposed convolutional network outperformed deep residual network.

Machine learning techniques have one main drawback that they require manual feature extraction which makes these techniques time consuming. On the other side, deep learning algorithms have in-built potential for feature extraction and feature selection which makes them more efficient.

28.5 Conclusions

AD is a neurodegenerative disease that gets worse over time and affects people's health, society, and the economy all over the world. As there is no medication or drug to cure or reverse AD, many researchers in numerous fields have been focusing on early detection and prognosis using various approaches. In particular, neuroimaging data provides quantitative and qualitative measures on which computational models can be built to predict AD progression. In this paper, an attempt is made for multi-class categorization of Alzheimer's disease. The structural MRI scans are preferred over PET, DTI, and SPECT because they are clearer, more detailed, and less expensive. In this work, machine learning and deep learning models are proposed to perform multi-class classification of Alzheimer's disease using T1w-sMRI. The deep learning model performed better than machine learning techniques. However, deep learning techniques require large data and computation as compared to machine learning approaches. Also, the results confirmed that sMRI is a potential biomarker for early diagnosis of AD. In future, the model will be trained with more data. Other modalities like PET, CSF, and SPECT can be explored to obtain better results for prediction of Alzheimer's disease.

References

1. Chen, Y., Xia, Y.: Iterative sparse and deep learning for accurate diagnosis of Alzheimer's disease. Pattern Recognit. **116**, 107944 (2021). https://doi.org/10.1016/j.patcog.2021.107944
2. Zhang, J., Zheng, B., Gao, A., et al.: A 3D densely connected convolution neural network with connection-wise attention mechanism for Alzheimer's disease classification. Magn Reson Imaging **78**, 119–126 (2021). https://doi.org/10.1016/j.mri.2021.02.001
3. Lao, H., Zhang, X.: Regression and classification of Alzheimer's disease diagnosis using NMF-TDNet features from 3D brain MR Image. IEEE J Biomed. Heal. Inform. **26**, 1103–1115 (2022). https://doi.org/10.1109/JBHI.2021.3113668
4. Goenka, N., Tiwari, S.: AlzVNet: a volumetric convolutional neural network for multi-class classification of Alzheimer's disease through multiple neuroimaging computational approaches. Biomed. Signal Process Control **74**, 103500 (2022). https://doi.org/10.1016/j.bspc.2022.103500

5. Vishnu, V.Y., Modi, M., Garg, V.K., et al.: Role of inflammatory and hemostatic biomarkers in Alzheimer's and vascular dementia—a pilot study from a tertiary center in Northern India. Asian J Psychiatr **29**, 59–62 (2017). https://doi.org/10.1016/j.ajp.2017.04.015
6. Kong, Z., Zhang, M., Zhu, W., et al.: Multi-modal data Alzheimer's disease detection based on 3D convolution. Biomed. Signal Process Control **75**, 103565 (2022). https://doi.org/10.1016/j.bspc.2022.103565
7. Qiu, S., Chang, G.H., Panagia, M., et al.: Fusion of deep learning models of MRI scans, mini-mental state examination, and logical memory test enhances diagnosis of mild cognitive impairment. Alzheimer's Dement Diagnosis, Assess Dis. Monit. **10**, 737–749 (2018). https://doi.org/10.1016/j.dadm.2018.08.013
8. Klöppel, S., Stonnington, C.M., Chu, C., et al.: UKPMC funders group automatic classification of MR scans in Alzheimer's disease. Brain **131**, 681–689 (2008). https://doi.org/10.1093/brain/awm319.Automatic
9. Gerardin, E., Chételat, G., Chupin, M., Cuingnet, R., Desgranges, B., Kim, H.S., Niethammer, M., Dubois, B., Lehéricy, S., Garnero, L., Eustache, F.: Multidimensional classification of hippocampal shape features discriminates Alzheimer's disease and mild cognitive impairment from normal aging. Neuroimage **47**(4), 1476–1486 (2009). https://doi.org/10.1016/j.neuroimage.2009.05.036
10. Al-Khuzaie, F.E.K., Bayat, O., Duru, A.D.: Diagnosis of Alzheimer disease using 2D MRI slices by convolutional neural network. Appl. Bionics. Biomech. 6690539 (2021). https://doi.org/10.1155/2021/6690539
11. Ocasio, E., Duong, T.Q.: Deep learning prediction of mild cognitive impairment conversion to Alzheimer's disease at 3 years after diagnosis using longitudinal and wholebrain 3D MRI. PeerJ. Comput. Sci. **7**, 1–21 (2021). https://doi.org/10.7717/PEERJ-CS.560
12. Ghosh, M., Raihan, M.M.S., Raihan, M., et al.: A comparative analysis of machine learning algorithms to predict liver disease. Intell. Autom. Soft Comput. **30**, 917–928 (2021). https://doi.org/10.32604/iasc.2021.017989
13. Sharma, R., Goel, T., Tanveer, M., Murugan, R.: FDN-ADNet: fuzzy LS-TWSVM based deep learning network for prognosis of the Alzheimer's disease using the sagittal plane of MRI scans. Appl. Soft. Comput. **115**, 108099 (2022). https://doi.org/10.1016/j.asoc.2021.108099
14. Raghavaiah, P., Varadarajan, S.: A CAD system design for Alzheimer's disease diagnosis using temporally consistent clustering and hybrid deep learning models. Biomed. Signal Process Control **75**, 103571 (2022). https://doi.org/10.1016/j.bspc.2022.103571
15. Khan, R., Qaisar, Z.H., Mehmood, A., et al.: A practical multiclass classification network for the diagnosis of Alzheimer's disease. Appl. Sci. **12** (2022). https://doi.org/10.3390/app12136507
16. Lim, B.Y., Lai, K.W., Haiskin, K., et al.: Deep learning model for prediction of progressive mild cognitive impairment to Alzheimer's disease using structural MRI. Front Aging Neurosci. **14**, 1–10 (2022). https://doi.org/10.3389/fnagi.2022.876202
17. Huang, H., Zheng, S., Yang, Z., et al.: Voxel-based morphometry and a deep learning model for the diagnosis of early Alzheimer's disease based on cerebral gray matter changes. 1–10 (2022). https://doi.org/10.1093/cercor/bhac099
18. Shi, Y., Zu, C., Hong, M., et al.: ASMFS: adaptive-similarity-based multi-modality feature selection for classification of Alzheimer's disease. Pattern. Recognit. **126**, 108566 (2022). https://doi.org/10.1016/j.patcog.2022.108566
19. Wen, J., Thibeau-Sutre, E., Diaz-Melo, M., et al.: Convolutional neural networks for classification of Alzheimer's disease: overview and reproducible evaluation. Med. Image Anal. **63**, 101694 (2020). https://doi.org/10.1016/j.media.2020.101694
20. Liu, M., Zhang, J., Adeli, E., Di, S.: Joint classification and regression via deep multi-task multi-channel learning for Alzheimer's disease diagnosis. IEEE Trans. Biomed. Eng. **66**, 1195–1206 (2019). https://doi.org/10.1109/TBME.2018.2869989
21. Abrol, A., Bhattarai, M., Fedorov, A., et al.: Deep residual learning for neuroimaging: an application to predict progression to Alzheimer's disease. J. Neurosci. Methods **339**, 108701 (2020). https://doi.org/10.1016/j.jneumeth.2020.108701

Chapter 29
Employing Tuned VMD-Based Long Short-Term Memory Neural Network for Household Power Consumption Forecast

Sandra Petrovic, Vule Mizdrakovic, Maja Kljajic, Luka Jovanovic, Miodrag Zivkovic, and Nebojsa Bacanin

Abstract Estimating household power consumption energy usage patterns can assist households in planning and managing their power consumption. To address elaborate time-series data, long short-term memory artificial neural networks are a promising strategy. However, a decomposition-forecasting method called variation mode decomposition is necessary to handle challenging time series. The accuracy and effectiveness of machine learning models are influenced by their hyperparameter values. This paper suggests using a altered sine cosine algorithm to optimize the hyperparameters of the long short-term memory model. This algorithm enhances the accuracy and performance of household energy consumption forecasting. The proposed model is compared to other long short-term memory models that are optimized by advanced metaheuristics. Simulation results indicated the improved sine cosine algorithm surpassed other advanced approaches in terms of standard time-series forecasting metrics.

Keywords Long short-term memory · Variation mode decomposition · Swarm intelligence · Sine cosine algorithm · Energy forecasting

---

S. Petrovic · V. Mizdrakovic · M. Kljajic · L. Jovanovic · M. Zivkovic · N. Bacanin (✉)
Singidunum University, Danijelova 32, 11010 Belgrade, Serbia
e-mail: nbacanin@singidunum.ac.rs

S. Petrovic
e-mail: sandra.petrovic.211@singimail.rs

V. Mizdrakovic
e-mail: vmizdrakovic@singidunum.ac.rs

M. Kljajic
e-mail: mkljajic@singidunum.ac.rs

L. Jovanovic
e-mail: luka.jovanovic.191@singimail.rs

M. Zivkovic
e-mail: mnzivkovic@singidunum.ac.rs

29.1 Introduction

A key component of contemporary energy management, energy consumption forecasting (ECF) is essential to ensure effective and sustainable energy use. Accurate energy forecasting is increasingly crucial due to the adoption of the Internet of Things (IoT) technologies. IoT and sensors may collect in-the-moment data on energy-use trends, empowering individuals and organizations to make wise energy-use decisions. Households may manage energy expenses, cut waste, and implement more energy-efficient activities with the aid of accurate forecasts. As human behavior causes highly unpredictable time-series data, the ECF for households has received growing attention as part of the advanced metering infrastructure (AMI) program for smart grid building. Conventional methods, like physical model-based approaches, have proven challenging to predict. In response, technologies have been extensively utilized for ECF in individual households, in line with the rapidly expanding use of artificial intelligence (AI). The primary concern in ECF research is the lack of precision and consistency in predictions due to oscillations and rapid changes in sensor data, as well as the influence of mistakes in raw data. The main objectives and challenges of this research are noise elimination and volatility management to improve ECF predictions for individual households.

Various machine learning (ML) approaches can be used for time-series predictions. The best outcomes are achieved by using recurrent neural networks, particularly long short-term memory (LSTM) neural networks. However, each ML model is very reliant on its configuration. It is necessary to tune the model for each concrete task, by adapting the values of the hyperparameters, itself an NP-hard problem. Unfortunately, tackling this problem requires a lot of trial and error, and it is very time consuming. To tackle this issue, scientists have employed different metaheuristic algorithms to determine a set of hyperparameters' values and determine the model with the best possible level of performance. One of the most promising groups of algorithms is the swarm intelligence metaheuristics, which are commonly used to solve NP-hard problems in a reasonable time frame.

With all this in mind, this research uses LSTM neural network to forecast household energy consumption, with variational mode decomposition (VMD) to reduce data complexity. An enhanced variation of the sine cosine algorithm (SCA) is applied to tune the hyperparameters of LSTM. This improved SCA was devised in such a way as to overcome the known limitations of the elementary version. Therefore, the primary contributions of this paper are as follows:

- An enhanced variant of SCA is suggested that is capable to improve the already superior level of performance of the plain SCA.
- VMD decomposition approach has been utilized to tackle data complexity.
- LSTM neural network was tuned by the newly suggested enhanced SCA to attain supreme performance for predicting household energy consumption time series.

The manuscript is comprised of the following parts. Section 29.2 shows the survey of the neural networks and time-series predictions, LSTM, and metaheuristics optimization. Section 29.3 explains the basic and improved SCA algorithm. Experimental outcomes are disclosed in Sect. 29.4, while Sect. 29.5 provides final remarks and concludes this work.

29.2 Related Works and Background

29.2.1 LSTM Overview

Hochreiter and Schmidhuber [15] introduced the LSTM neural network as one of the modifications of recurrent neural network (RNN) architecture [19]. The LSTM neural network developed to handle flaws of RNNs, particularly the disappearing and ballooning gradients.

The unique composition of an LSTM neural network includes a designated layer used for inputting information, followed by a set of internal layers, with a final layer used to output the results. The neuron count in these layers is dependent on problem dimensionality. The distinguishing factor of an LSTM originates within a set of cells used to retain data in network layers. Each memory cell comprises gates, a forget f_t, input i_t, and output gate o_t, where t signifies the timestep. These gates have an important function in modifying the cell state S_t of the memory cell. Together, these gates fulfill the three primary functions of the memory cell. With every timestep t, network executes the a multiple phase procedure:

In the first phase, the network decides whether to discard information from the previous cell state, s_{t-1}. To do this, it applies a *sigmoid* activation function to the input sequence, transforming it into a range of 0 to 1. This determines whether to keep or remove information from the preceding cell state. The resulting outcome for the the forget gates, f_t, is calculated using the following formula:

$$f_t = \text{sigmoid}(W_x, x_t + W_h, h_t - 1 + b_f), \qquad (29.1)$$

where W_f, x and W_f, h are the weight matrices, x_t is the input vector at timestep $t, h - 1$ is the output at timestep $t - 1$, and bf is the bias vector of the forget gates.

During the second phase, network employs the *sigmoid* function to determine which information will be incorporated into the cell state S_t. Following, these successive inputs are scaled to a range $[-1, 1]$ via the *tanh* transformation. The procedure involves computing two values: the candidate number S_t as well as initialization number i_t used for input gates. These values are gained using the following method:

$$\tilde{S}_t = \tanh\left(W_{\tilde{s},x} x_t + W_{\tilde{s},h} x_t + b_{\tilde{s}}\right), \qquad (29.2)$$

$$i_t = \text{sigmoid}\left(W_{i,x} x_t + W_{i,h} x_t + b_i\right), \qquad (29.3)$$

where $W_{\tilde{s},x}$, $W_{\tilde{s},h}$, $W_{i,x}$ and $W_{i,h}$ are the weight matrices and $b_{\tilde{s}}$ and b_i represent+ bias vectors.

In phase 3: The novel status s_t is determined using:

$$s_t = f_t \circ s_{t-1} + i_t \circ \tilde{s}_t, \qquad (29.4)$$

where $\circ$ denotes the Hadamard product.

In phase 4: The output h_t will be computed using the sigmoid and *tanh* activation functions and the following two formulas:

$$o_t = \text{sigmoid}\left(W_{o,x}x_t + W_{o,h}h_{t-1} + b_o\right), \qquad (29.5)$$

$$h_t = o_t \circ \tanh(s_t), \qquad (29.6)$$

where $W_{o,x}$ and $W_{o,h}$ represent weight matrices, b_o signified output gate bias vector.

As a result of its unique three-gate structural design, LSTM represses the impact in the short-term, and even information from previous time steps may be represented as an contemporary cell value.

29.2.2 Variational Mode Decomposition

Data that represents complex time series is challenging for typical ML algorithms to handle. A relatively new approach with significant potential for handling complex sets of data is VMD [14]. Information decomposition techniques may be employed to identify, analyze, and determine trends in signal fluctuations by treating financial data as a signal and organizing it. Authors [26] have used in their paper a hybrid model for energy demand prediction by employing VMD-based LSTM. Namely, the original sequence was divided into stronger subsequences using the VMD approach, as part of the data pretreatment process. The maximum relevance minimum redundancy (mRMR) is employed to determine inputs, which involved examining the link between components and each feature while removing any overlap between the features. According to research, the introduced combined model may augment forecast precision.

29.2.3 Metaheuristics Optimization

Swarm intelligence algorithms, that belong to the subfield of AI, are inspired by nature to create systems that are able to address complex optimization challenges and are considered very powerful optimizers. Population-based algorithms are stochastic, and they start the execution by producing a random initial population of individuals. They consist of two distinctive phases: exploration and exploitation. During the exploration, the algorithm tries to cover large regions inside the provided search realm, aiming to locate the promising sub-domains where better solutions can be discovered. During the exploitation, the algorithm tries to narrow down the search to certain promising sub-domains, with the goal to discover the best solution within these interesting regions. Among many available algorithms, distinguished methods include the firefly algorithm (FA) [31], the whale optimization algorithm (WOA) [23], and artificial bee colony (ABC) [20]. Moreover, novel methods that have an admirable level of performance are the reptile search algorithm (RSA) [1], the chimp optimization algorithm (ChOA) [21], SCA (used in this paper as well) [22], and arithmetic optimization algorithm (AOA) [2], to name the few.

Population-based methods are more than capable to tackle very hard real-life challenges, belonging to the non-deterministic polynomial problems (also known shortly as NP-hard problems), making them substantially popular within academic circles. They were applied in numerous recent practical problems, including feature selection task [5, 37], crypto-currency forecasting [24, 27], oil price predictions [17], COVID-19 challenges [35, 38], workflow scheduling on cloud-edge systems [8], air pollution prediction [6, 18], computer-assisted medical diagnostics [10, 12, 34], tuning of a variety of artificial neural networks [3, 4], global numerical optimization [11], intrusion, security, and spam detection [9, 16, 39], energy and power supply predictions [7], and different wireless sensor networks problems [36].

29.2.4 Hybrid Artificial Intelligence-Based Systems in Time-Series Data Forecast

Time-series data prediction approaches may be broadly classified as model based or data driven [28]. Approaches that are based on data, such as convolutional neural networks (CNNs) and LSTM neural networks, are better suited for ECF.

There are two types of data-driven models: solitary models and hybrid models. Decision trees, random forests, support vector regression, multilayer perceptrons, convolutional neural networks, recurrent neural networks, and long and short-term memory neural networks are examples of singular models. In addition, hybrid models integrate unique models to improve prediction performance.

Wei et al. [29] introduced a combined model that hybridizes optimization algorithms and LSTM to forecast daily fossil gas consumption. The authors focused on the model's advantage across different climate zones rather than time periods. Additionally, it is widely acknowledged that data decomposition is crucial in enhancing forecasting accuracy. Yuan et al. [33] proposed using singular spectrum analysis (SSA) to anticipate building power usage. However, this paper does not investigate the diverse predicting performance over different time frames. On the other hand, SSA denoising has the ability to precisely identify noise in data. Additionally, SSA eliminates Gaussian noise. Neeraj et al. [13] applied SSA-LSTM to predict grid load.

During recombination stage, the non-singular matrix is merely using the decomposition feature for denoising instead of training, ignoring the crucial role of SSA for feature extraction.

29.3 Proposed Method

29.3.1 Elementary SCA

In 2016, the SCA algorithm was introduced as a metaheuristic solution to optimization problems that utilizes population-based techniques [22]. This approach leverages trigonometric sine and cosine expressions to adjust individuals' positions toward the optimal solution.

The SCA generates a range of potential solutions by allowing them to either deviate from or converge toward the optimal solution. This technique also incorporates numerous adjustable and random variables to enhance and diversify the population participating in the optimization. Studies have determined that the SCA may effectively consider multiple regions of the available space, avoiding local sub-optimals and advancing the population in the direction of the best option while exploring intriguing areas of the solution space during the optimization cycle.

While locating a promising area, the following equations are applied to determine agent positioning: [22]:

$$X_i^{t+1} = X_i^t + r_1 \cdot \cos(r_2) \cdot |r_3 P_i^t - X_i^t| \qquad (29.7)$$

$$X_i^{t+1} = X_i^t + r_1 \cdot \sin(r_2) \cdot |r_3 P_i^t - X_i^t| \qquad (29.8)$$

In reality, the preceding formulae are combined to generate the following formula:

$$X_i^{t+1} = \begin{cases} X_i^t + r_1 \cdot \sin(r_2) \cdot |r_3 P_i^t - X_i^t|, & r_4 < 0.5 \\ X_i^t + r_1 \cdot \cos(r_2) \cdot |r_3 P_i^t - X_i^t|, & r_4 \geq 0.5 \end{cases} \qquad (29.9)$$

wherein variables used to avoid getting trapped in a local best and to combine intensification and exploration methods include X_i^t, which represents the current candidate i at the $t-th$ iteration in the $d-th$ dimension, P_i^t, which represents the position of the best candidate at the $t-th$ iteration in the $d-th$ dimension, and randomly generated factors r_1, r_2, r_3, and r_4.

The factor r_1 determines whether a solution moves closer ($r_1 < 1$) or farther away ($r_1 > 1$), and it declines linearly from a constant (a) to 0 to facilitate blend intensification and diversification search behaviors [22]. The following is used to change r_1:

$$r_1 = a - t\frac{a}{T} \qquad (29.10)$$

wherein variable T represents the highest iterative count, while t represents the ongoing cycle, while a remains a fixed number.

The factor r_2 determines the severity of advances relative to the best agent and is expressed in the range $[0, 2\pi]$. The random factor r_3 generates an arbitrary weight to the goal, allowing the operator to either consider ($r_3 > 1$) or ignore ($r_3 < 1$) the target's influence on the location chances of all agents. r_3 is within the range of $[0, 2]$. Lastly, an arbitrary value in component r_4 ranges from $[0, 1]$ and acts as a variation in Eq. 29.9 to switch the utilized methods with either sine or cosine capabilities. The pseudo-code for the elementary SCA method can be found in [22].

29.3.2 Suggested Modified SCA

The SCA is one of the recent metaheuristics algorithms, and it is well known as a very versatile approach. Nevertheless, as with any other algorithm, it has notable weaknesses that can affect its performance. More precisely, in some runs, the basic SCA is hindered by slow convergence. Aiming to help SCA to converge, this paper proposes hybridization with bat algorithm (BA) [32]. BA is highly regarded as it exhibits powerful search. The observed individual bat is denoted by X_i^{t-1}, while the new, calibrated location in iteration t of the i-th bat is noted as X_i^t. This location is calculated in Eq. (29.11), where the bat's speed is represented by v_i^t.

$$X_i^t = X_i^{t-1} + v_i^t \tag{29.11}$$

The bat's velocity during round t is obtained by utilizing Eq. (29.12).

$$v_i^t = v_i^{t-1} + (X_i^{t-1} - X_*)f_i, \tag{29.12}$$

wherein the fresh global optimum location is depicted as x_*, while f_i denotes the frequency used by the i-th bat.

To enable switching between two search procedures, an additional parameter, search mode sm has been introduced. It is utilized to decide for each solution within the population if the search will be executed by performing the SCA or BA procedure. Each solution in every iteration produces a random number rnd inside limits $[0, 1]$, and, if $rnd > sm$, that solution will perform the SCA search, otherwise, it will continue by employing the BA search. The value of this parameter has been fixed to 0.3, and it was determined empirically.

Moreover, a supplementary mechanism that draws inspiration from the *trial* parameter described in ABC algorithm [20] has also been employed. The *trial* parameter is initialized to value 0 for all solutions, and if the particular solution has not been enhanced in the current iteration, the value of its *trial* is increased. If this value reaches *limit*, which is the predetermined threshold, that individual is rejected from the populace, and a quasi-reflective opposite individual x_{qrl} is introduced instead, generated by applying the quasi-reflection-based learning (QRL) process [25]. Recent publications suggested that the QRL procedure is efficient in producing the solution in the opposite part of the search region [3].

The hybridized metaheuristics have been simply named improved SCA-ISCA. The implementation of this approach described its functionality as given in Algorithm 1.

Algorithm 1 Pseudo-code of improved sine cosine algorithm - ISCA

Generate. A collection of N agents, referred to as X
Set iterative limit T.
Set $trial = 0$ for each solution.
while $t < T$ **do**
    for Each solution X from the population **do**
        Produce arbitrary number rnd.
        if $rnd > 0.3$ **then**
            Execute SCA search, according to the Eq. (29.9).
        else
            Execute BA search, according to the Eq. (29.12).
        end if
        if solution not improved **then**
            $trial = trial + 1$
            if $trial == limit$ **then**
                Replace current solution with quasi-reflective opposite solution x_{qrl}
                $trial = 0$
            end if
        end if
    end for
    Update predefined parameters.
    Update candidate solutions' positions.
end while
return the fittest agent.

29.4 Empirical Results, Comparison, and Discussion

This segment begins with a description of the utilized dataset and the VMD decomposition, as well as the basic experimental setup. It then continues with a comparison to other contemporary metaheuristics, which are adapted and evaluated under identical conditions as the proposed ISCA algorithm.

29.4.1 Employed Dataset

The dataset used in this research has been obtained from the London Data Store, which can be accessed on Kaggle at https://www.kaggle.com/code/rheajgurung/energy-consumption-forecast. The research aims to assess the capabilities of the proposed approach by analyzing electricity demand information for 5567 households in London. The dataset covers the period from November 2011 to February 2014.

29 Employing Tuned VMD-Based Long Short-Term Memory ...

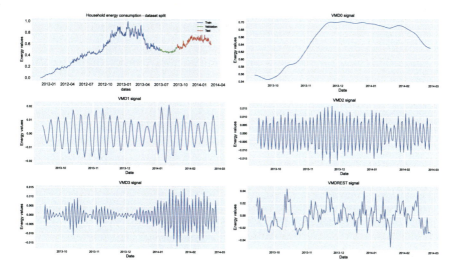

Fig. 29.1 Original data and signals generated by VMD after normalization for testing data

The data was gathered using smart power meters that recorded daily usage, resulting in a total of 19,752 data points. These individual household power consumption data points were combined to form 829 samples, which served as the target variable for univariate time-series forecasting. The collected data was divided into three parts: 70% was used for training, 10% for validation, and the remaining 20% was reserved for testing.

Furthermore, in order to enhance prediction accuracy, the VMD was used with $K = 4$ and the target variable is decomposed into three basic and residual signals, which transformed this challenge into multivariate time-series forecasting problem. The number of components for the original signal decomposition was determined through empirical methods. Before use in simulations, all features were normalized to fall within a specific range of [0, 1]. Figure 29.1 shows the original data series divided into three sets, as previously explained, as well as signals generated after applying VMD.

29.4.2 Basic Experimental Setup

Solutions in all metaheuristics used in the analysis represent a set of LSTM hyperparameters that were optimized of a length $D = 6$. Since it is an extremely resource-intensive experiment, the agent count for a population is limited to 8 ($N = 8$) which have been incrementally improved over the course of six iterations $T = 6$. Furthermore, because of stochastic mechanisms in metaheuristics, all approaches were evaluated over 20 self-contained executions ($R = 20$). The *limit* parameter for ISCA metaheuristics was set to 2, and this value empirically determined for this specific problem according to the expression: $roundup((2 \cdot T)/N)$, where the $roundup$ function rounds the number to the nearest integer of higher value.

The LSTMs have been given six steps of input information and are expected to predict the next three steps. The LSTM parameters that have been chosen for optimization and their corresponding ranges include the learning rate range of [0.0001, 0.01], the number of training epochs range of [300, 600], the dropout rate range of [0.05, 0.2], the number of hidden layers range of [1, 2], and the number of neurons in the hidden layers range of [50, 200]. These parameters have been selected due to empirical evidence showing they have the greatest impact on performance. Furthermore, since some of these parameters have a continuous range and others have a discrete range, this optimization is considered to be a mixed NP-hard challenge.

The overall, average metrics for all three steps for the objective function (mean square error—MSE) are captured and reported in terms of best, worst, mean, median, variance, and standard deviation over 20 runs. Furthermore, the coefficient of determination (R^2), mean absolute error (MAE), and root mean squared error ($RMSE$) are also reported for the best-performing solutions (LSTM structure).

Also, it should be noted that for this research, all proposed methods have been independently developed. The implementation was done using Python and commonly used ML libraries including Pandas, NumPy, scikit, and TensorFlow.

29.4.3 Comparision with Other Methods and Discussion

The evaluation process involved a comparative analysis of several modern metaheuristic algorithms. In addition to the proposed ISCA algorithm, several other algorithms were used to determine the satisfying hyperparameters of an LSTM network. These included the original SCA [22] algorithm, the widely recognized ABC [20], FA [31], and two relatively novel approaches: ChOA [21] and RSA [1].

Table 29.1 demonstrates the outcomes of the overall objective function (MSE) based on 20 independent runs in terms of best, worst, mean, median, standard deviation, and variance. Furthermore, the best-performing runs are given in detail in Table 29.2, while the best obtained LSTM hyperparameters values for all metaheuristics are presented in Table 29.3.

The findings indicate that the LSTM-ISCA method, as proposed, exhibits better performance when considering the optimal LSTM structure produced. From results presented in Table 29.2, it can be noticed that the LSTM-ISCA achieved the highest R^2 and lowest MSE (objective) score for all three steps in average (overall), as well as for the one and three steps ahead. The best results for the best-generated network two steps ahead were generated by the LSTM-FA approach.

Additionally, when comparing overall metrics for objective function (Table 29.1), the method introduced in this research clearly outperforms all other approaches when the best solution is considered; however, the LSTM-FA proved as the most robust method outscoring all others metaheuristics for mean and worst metrics. This is in

29 Employing Tuned VMD-Based Long Short-Term Memory ...

Table 29.1 Overall indicators for objective function (MSE)

Method	Best	Worst	Mean	Median	Std	Var
LSTM-ISCA	**0.0009890**	0.0010127	0.0010005	0.0010015	8.38E−06	7.01E−11
LSTM-SCA	0.0009944	0.0010065	0.0010003	0.0009995	4.83E−06	2.34E−11
LSTM-ABC	0.0009979	0.0010020	0.0009992	**0.0009979**	**1.68E−06**	**2.83E−12**
LSTM-FA	0.0009899	**0.0010013**	**0.0009972**	0.0009984	3.79E-06	1.43E−11
LSTM-ChOA	0.0009945	0.0010023	0.0009979	0.0009984	2.79E−06	7.78E−12
LSTM-RSA	0.0009963	0.0010042	0.0009991	0.0009980	2.94E−06	8.67E−12

Table 29.2 R^2, MAE, MSE, and $RMSE$ denormalized metrics for best-generated LSTM model per each step and overall

	Error indicator	LSTM-ISCA	LSTM-SCA	LSTM-ABC	LSTM-FA	LSTM-ChOA	LSTM-RSA
One-step ahead	R^2	**0.741599**	0.738957	0.740634	0.740924	0.740914	0.740873
	MAE	**0.024768**	0.024908	0.025055	0.024971	0.024917	0.024975
	MSE	**0.001000**	0.001010	0.001003	0.001002	0.001002	0.001002
	RMSE	**0.031617**	0.031778	0.031676	0.031658	0.031659	0.031661
Two-step ahead	R^2	0.744366	0.744349	0.741761	**0.744974**	0.743668	0.742866
	MAE	0.024707	**0.024627**	0.025125	0.024707	0.024784	0.024994
	MSE	0.000989	0.000989	0.000999	**0.000987**	0.000992	0.000995
	RMSE	0.031447	0.031448	0.031607	**0.031410**	0.031490	0.031539
Three-step ahead	R^2	**0.747042**	0.745549	0.743764	0.746410	0.744208	0.743667
	MAE	0.024531	**0.024521**	0.024911	0.024704	0.024883	0.024952
	MSE	**0.000979**	0.000984	0.000991	0.000981	0.000990	0.000992
	RMSE	**0.031282**	0.031374	0.031484	0.031321	0.031457	0.031490
Overall results	R^2	**0.744336**	0.742952	0.742053	0.744103	0.742930	0.742469
	MAE	**0.024669**	0.024685	0.025030	0.024794	0.024861	0.024974
	MSE	**0.000989**	0.000994	0.000998	0.000990	0.000994	0.000996
	RMSE	**0.031449**	0.031534	0.031589	0.031463	0.031535	0.031564

Table 29.3 Determined hyperparameters for best models generated by each method

Method	nn layer1	Learning rate	Epochs	Dropout rate	No. of layers	nn. layer 2
LSTM-ISCA	300	0.010000	600	0.200000	1	/
LSTM-SCA	100	0.005288	576	0.050000	2	300
LSTM-ABC	198	0.004094	559	0.081498	2	170
LSTM-FA	300	0.009532	600	0.118731	1	/
LSTM-ChOA	170	0.005746	530	0.126817	2	164
LSTM-RSA	184	0.009998	600	0.058570	1	/

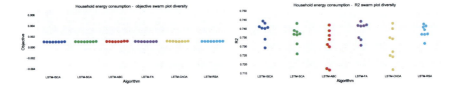

Fig. 29.2 Visualization of obtained results for all metaheuristics swarm plots

line with the "No Free Lunch" theorem of optimization [30], which states that no single approach is best in all cases. However, the proposed LSTM-ISCA manages to generate the best-performing LSTM structure with only one LSTM layer (Table 29.3).

Finally, to conclude the comparative analysis, results for all simulations are visualized in and shown. The population diversity of final iteration of the best run in the form of swarm plot diagrams is shown in Fig. 29.2. Afterward, Fig. 29.3 presents the following for all evaluated methods: the R^2 and objective (MSE) convergence speed diagrams for the best run, distribution of results over 20 runs in a form of a box plot, and violin plot diagrams for objective and R^2 indicators.

All metaheuristic algorithms demonstrated admirable diversity in for both objective and R^2 metrics.

As demonstrated, the introduced metaheuristic improved convergence rates of the original. The introduced approach also outpaced all other competing algorithms in convergence rates further emphasizing the introduced enhancements.

29.5 Conclusion

The global shift toward renewable resources highlights the demand for powerful prediction models. The ongoing power crisis further emphasizes the importance of forecasting power consumption on an individual household level. This demand moti-

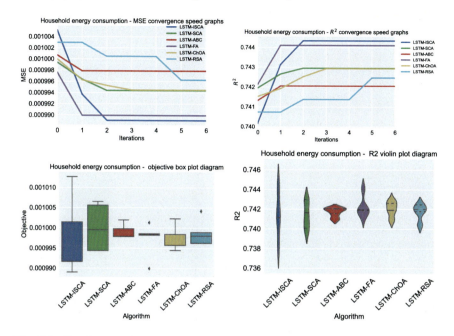

Fig. 29.3 Visualization of obtained results for all metaheuristics with comparisons

vated this research that presents a new approach for predicting power consumption optimized using a metaheuristics algorithm. First, the modified version of the SCA metaheuristics was developed, which copes with the known weaknesses of the basic variant. This was achieved by hybridizing the SCA with the BA algorithm, which increases the converging speed with its powerful search. Then, the proposed algorithm was employed to adjust the control parameters of LSTM models for the energy consumption time-series prediction task.

The proposed LSTM-ISCA model was tested on the dataset that contains the electricity consumption of London's households over three years. The proposed model was compared to LSTM tuned by other modern and powerful optimizers. The simulation outcomes display the supremacy of the suggested LSTM-ISCA method. Future studies in this field could be aimed at validation of the proposed model using additional energy consumption datasets and investigating its feasibility for solving other time-series forecasting problems.

Limitations regarding data availability and high computational demands of testing are just some of the challenges faced in this work. Future work will focus on overcoming these challenges by expanding testing data as new datasets become available as well as finding implementations of the introduced metaheuristic for optimization in improvements in other fields of application.

References

1. Abualigah, L., Abd Elaziz, M., Sumari, P., Geem, Z.W., Gandomi, A.H.: Reptile search algorithm (rsa): a nature-inspired meta-heuristic optimizer. Expert Syst. Appl. **191**, 116158 (2022)
2. Abualigah, L., Diabat, A., Mirjalili, S., Abd Elaziz, M., Gandomi, A.H.: The arithmetic optimization algorithm. Comput. Methods Appl. Mech. Eng. **376**, 113609 (2021)
3. Bacanin, N., Bezdan, T., Venkatachalam, K., Zivkovic, M., Strumberger, I., Abouhawwash, M., Ahmed, A.B.: Artificial neural networks hidden unit and weight connection optimization by quasi-refection-based learning artificial bee colony algorithm. IEEE Access **9**, 169135–169155 (2021)
4. Bacanin, N., Bezdan, T., Zivkovic, M., Chhabra, A.: Weight optimization in artificial neural network training by improved monarch butterfly algorithm. In: Mobile Computing and Sustainable Informatics, pp. 397–409. Springer (2022)
5. Bacanin, N., Petrovic, A., Zivkovic, M., Bezdan, T., Antonijevic, M.: Feature selection in machine learning by hybrid sine cosine metaheuristics. In: International Conference on Advances in Computing and Data Sciences. pp. 604–616. Springer (2021)
6. Bacanin, N., Sarac, M., Budimirovic, N., Zivkovic, M., AlZubi, A.A., Bashir, A.K.: Smart wireless health care system using graph lstm pollution prediction and dragonfly node localization. Susta. Comput.: Inf. Syst. **35**, 100711 (2022)
7. Bacanin, N., Stoean, C., Zivkovic, M., Rakic, M., Strulak-Wójcikiewicz, R., Stoean, R.: On the benefits of using metaheuristics in the hyperparameter tuning of deep learning models for energy load forecasting. Energies **16**(3), 1434 (2023)
8. Bacanin, N., Zivkovic, M., Bezdan, T., Venkatachalam, K., Abouhawwash, M.: Modified firefly algorithm for workflow scheduling in cloud-edge environment. Neural Comput. Appl. **34**(11), 9043–9068 (2022)
9. Bacanin, N., Zivkovic, M., Stoean, C., Antonijevic, M., Janicijevic, S., Sarac, M., Strumberger, I.: Application of natural language processing and machine learning boosted with swarm intelligence for spam email filtering. Mathematics **10**(22), 4173 (2022)
10. Basha, J., Bacanin, N., Vukobrat, N., Zivkovic, M., Venkatachalam, K., Hubálovskỳ, S., Trojovskỳ, P.: Chaotic harris hawks optimization with quasi-reflection-based learning: an application to enhance cnn design. Sensors **21**(19), 6654 (2021)
11. Bezdan, T., Petrovic, A., Zivkovic, M., Strumberger, I., Devi, V.K., Bacanin, N.: Current best opposition-based learning salp swarm algorithm for global numerical optimization. In: 2021 Zooming Innovation in Consumer Technologies Conference (ZINC), pp. 5–10. IEEE (2021)
12. Budimirovic, N., Prabhu, E., Antonijevic, M., Zivkovic, M., Bacanin, N., Strumberger, I., Venkatachalam, K.: Covid-19 severity prediction using enhanced whale with salp swarm feature classification, pp. 1685–1698. Computers, Materials and Continua (2022)
13. Choudhary, N., Mathew, J., Agarwal, M., Behera, R.: Long short-term memory-singular spectrum analysis-based model for electric load forecasting. Electr. Eng. **103**, 1–16 (04 2021). https://doi.org/10.1007/s00202-020-01135-y
14. Dragomiretskiy, K., Zosso, D.: Variational mode decomposition. IEEE Trans. Sig. Proc. **62**(3), 531–544 (2013)
15. Hochreiter, S., Schmidhuber, J.: Long short-term memory. Neural Comput. **9**(8), 1735–1780 (1997)
16. Jovanovic, D., Antonijevic, M., Stankovic, M., Zivkovic, M., Tanaskovic, M., Bacanin, N.: Tuning machine learning models using a group search firefly algorithm for credit card fraud detection. Mathematics **10**(13), 2272 (2022)
17. Jovanovic, L., Jovanovic, D., Bacanin, N., Jovancai Stakic, A., Antonijevic, M., Magd, H., Thirumalaisamy, R., Zivkovic, M.: Multi-step crude oil price prediction based on lstm approach tuned by salp swarm algorithm with disputation operator. Sustainability **14**(21), 14616 (2022)
18. Jovanovic, L., Jovanovic, G., Perisic, M., Alimpic, F., Stanisic, S., Bacanin, N., Zivkovic, M., Stojic, A.: The explainable potential of coupling metaheuristics-optimized-xgboost and shap in revealing vocs' environmental fate. Atmosphere **14**(1), 109 (2023)

19. Jörges, C., Berkenbrink, C., Stumpe, B.: Prediction and reconstruction of ocean wave heights based on bathymetric data using lstm neural networks. Ocean Eng. **232**, 109046 (2021)
20. Karaboga, D., Akay, B.: A comparative study of artificial bee colony algorithm. Appl. Math. Comput. **214**(1), 108–132 (2009)
21. Khishe, M., Mosavi, M.R.: Chimp optimization algorithm. Expert Syst. Appl. **149**, 113338 (2020)
22. Mirjalili, S.: Sca: A sine cosine algorithm for solving optimization problems. Knowl.-Based Syst. **96**, 120–133 (2016)
23. Mirjalili, S., Lewis, A.: The whale optimization algorithm. Adv. Eng. Softw. **95**, 51–67 (2016)
24. Petrovic, A., Jovanovic, L., Zivkovic, M., Bacanin, N., Budimirovic, N., Marjanovic, M.: Forecasting bitcoin price by tuned long short term memory model. In: 1st International Conference on Innovation in Information Technology and Business (ICIITB 2022), pp. 187–202. Atlantis Press (2023)
25. Rahnamayan, S., Tizhoosh, H.R., Salama, M.M.: Quasi-oppositional differential evolution. In: 2007 IEEE Congress on Evolutionary Computation, pp. 2229–2236. IEEE (2007)
26. Ruan, Y., Wang, G., Meng, H., Qian, F.: A hybrid model for power consumption forecasting using vmd-based the long short-term memory neural network. Front. Energy Res. **9**, 917 (2022)
27. Salb, M., Zivkovic, M., Bacanin, N., Chhabra, A., Suresh, M.: Support vector machine performance improvements for cryptocurrency value forecasting by enhanced sine cosine algorithm. In: Computer Vision and Robotics, pp. 527–536. Springer (2022)
28. Wang, Y., Wang, J., Zhao, G., Dong, Y.: Application of residual modification approach in seasonal Arima for electricity demand forecasting: a case study of china. Energy Pol. **48**, 284–294 (2012)
29. Wei, N., Li, C., Peng, X., Li, Y., Zeng, F.: Daily natural gas consumption forecasting via the application of a novel hybrid model. Appl. Energy **250**, 358–368 (2019)
30. Wolpert, D.H., Macready, W.G.: No free lunch theorems for optimization. IEEE Trans. Evolut. Comput. **1**(1), 67–82 (1997)
31. Yang, X.S.: Firefly algorithms for multimodal optimization. In: International Symposium on Stochastic Algorithms, pp. 169–178. Springer (2009)
32. Yang, X.S., Hossein Gandomi, A.: Bat algorithm: a novel approach for global engineering optimization. Eng. Comput. **29**(5), 464–483 (2012)
33. Yuan, Z., Wang, W., Wang, H., Mizzi, S.: Combination of cuckoo search and wavelet neural network for midterm building energy forecast. Energy **202**, 117728 (2020)
34. Zivkovic, M., Bacanin, N., Antonijevic, M., Nikolic, B., Kvascev, G., Marjanovic, M., Savanovic, N.: Hybrid cnn and xgboost model tuned by modified arithmetic optimization algorithm for covid-19 early diagnostics from x-ray images. Electronics **11**(22), 3798 (2022)
35. Zivkovic, M., Bacanin, N., Venkatachalam, K., Nayyar, A., Djordjevic, A., Strumberger, I., Al-Turjman, F.: Covid-19 cases prediction by using hybrid machine learning and beetle antennae search approach. Susta. Cities Soc. **66**, 102669 (2021)
36. Zivkovic, M., Bacanin, N., Zivkovic, T., Strumberger, I., Tuba, E., Tuba, M.: Enhanced grey wolf algorithm for energy efficient wireless sensor networks. In: 2020 Zooming Innovation in Consumer Technologies Conference (ZINC), pp. 87–92. IEEE (2020)
37. Zivkovic, M., Stoean, C., Chhabra, A., Budimirovic, N., Petrovic, A., Bacanin, N.: Novel improved salp swarm algorithm: an application for feature selection. Sensors **22**(5), 1711 (2022)
38. Zivkovic, M., Stoean, C., Petrovic, A., Bacanin, N., Strumberger, I., Zivkovic, T.: A novel method for covid-19 pandemic information fake news detection based on the arithmetic optimization algorithm. In: 2021 23rd International Symposium on Symbolic and Numeric Algorithms for Scientific Computing (SYNASC), pp. 259–266. IEEE (2021)
39. Zivkovic, M., Tair, M., Venkatachalam, K., Bacanin, N., Hubálovský, Š, Trojovský, P.: Novel hybrid firefly algorithm: an application to enhance xgboost tuning for intrusion detection classification. PeerJ Comput. Sci. **8**, e956 (2022)

Chapter 30
A 4-element Dual-Band MIMO Antenna for 5G Smartphone

Preeti Mishra and Kirti Vyas

Abstract A four-element planner, dual-band MIMO antenna for 5 generation (5G) smartphone application was proposed which can be implemented in wireless handsets. In this research, the design of an H-shaped monopole antenna was developed and simulated using computer simulation technology (CST) software. The antenna was fabricated on an FR4 substrate with the dimensions of 150 × 75 × 0.8 mm^3. The two 5G new radio bands, n79 band (4.4–5 GHz), and LTE band 46 (5.1–5.9 GHz) are covered by the antenna without using any additional decoupling structure. The performance characteristics of the antenna were analyzed, including the reflection coefficient, radiation pattern, envelope correlation coefficient, and efficiency. The antenna displayed excellent characteristics, such as good impedance matching (return loss > 10 dB), high isolation (> 18.8 dB), high efficiency (> 60%), and low envelope correlation coefficient (ECC, < 0.03) across the operating frequencies. The proposed MIMO antenna is simple and compact, leaving enough space inside handheld mobile terminals for the integration of other circuits.

Keywords Four element · 5G · MIMO antenna · Smartphone application

30.1 Introduction

The utilization of multiple input, multiple output (MIMO) technology is aimed at addressing the need for increased data rates by improving channel capacity and spectral efficiency [1]. However, to enable MIMO technology, mobile devices must have multiple antenna elements. Incompatibilities between the size of MIMO antennas and the available space within mobile devices limit element separation and bandwidth, resulting in lower transmission speeds and making MIMO antennas less desirable.

P. Mishra (✉) · K. Vyas
Department of ECE, Arya College of Engineering & IT, University in Kukas, Jaipur, Rajasthan, India
e-mail: preetimishra128@gmail.com

As a result, there has been increasing interest in recent years in how to build wideband MIMO antennas with acceptable isolation in mobile devices of limited size. Various methods have been used to enhance the performance of the antenna [2–30]. If in the mobile device MIMO antenna position is fixed, on the one hand, the simplest decoupling method is to expand the distance between the antennas, and reducing the antenna element's size is a way to increase physical distance within a given area if gain and bandwidth had to be compromised on the spot [2, 3]. Another approach is to decrease the density of antenna elements in an area and increase the spatial separation between them to maintain gain and bandwidth [4–10]. But it tolerates channel loss. However, other researchers were focused on improving element segregation by including decoupling structures [11–15], which are used for improving isolation. On the other hand, new portable gadgets are getting thinner and lighter, requiring greater computing power [16]. In fourth generation (4G) and 4G long-term evolution (LTE) technologies. The MIMO antenna system based on four elements is used for achieving high data rates.

Additionally, they are extensively utilized in modern cellular technologies [17–19]. Yang et al. [18] proposed a 4-element MIMO antenna design with folded box shape for mobile LTE devices. A decoupling structure consisting of ground slots and L-branches was employed to promote isolation between the elements. This architecture makes things more complicated and restricts the potential uses of emerging technology, like 5G, in current gadgets like tablets and smartphones. Choi et al. proposed a 4-element reconfigurable link loop antenna system for long-term evolution (LTE) technology [19].

Xu et al. [20] presents correspond to a hybrid arrangement with four elements printed on the sides and corners of the case with an ECC of no more than 0.3 between the two radiating elements. However, due to the complexity of the design, such hybrid structures have limited practical applications. Future mobile device antenna systems will be provided with a variety of designs, assemblies, and chassis thanks to the studies (mentioned above). In [21] or wideband applications, a dielectric resonator antenna in the shape of an H is described. Despite having a straightforward structure, the antenna's size prevents it from being used in chassis applications.

The main goal and impetus of this research are to investigate an antenna system that can meet various criteria for 5 generation technology such as data speed, low latency, and bandwidth [22–30]. The contributions of MIMO systems proposed to address these features are as follows:

- We have developed a simple monopole radiating four-element MIMO antenna for 5G technology which can operate in the frequency spectrum below 6 GHz (LTE band 43).
- Without the use of any decoupling structures or techniques, a low level of isolation between the radiating parts is accomplished, providing room for additional components and RF parts space for other components and RF parts.
- Additionally, the antenna system is easy to manufacture and integrate with other RF systems, components, and subsystems. In addition, the proposed improvement

30 A 4-element Dual-Band MIMO Antenna for 5G Smartphone

mandated that contiguous in-band carrier aggregation should be used to improve data throughput. Next, we describe the detailed antenna design of the proposed research.

This research suggests the most straightforward, low-profile, small H-shaped monopole antenna that can be employed for future smartphone applications over the 5G bands between 4.4–5 GHz and 5.1–5.9 GHz. The 4-port MIMO antenna system was created to achieve great isolation without any decoupling structures. The planned low-profile unit antenna elements make it simple to fabricate and integrate them with the other circuits in the handheld terminal. Additionally, the antennas were positioned on corner sides of the ground, providing ample room for the integration of other RF components.

30.2 Antenna Design

This section explains the functioning of a MIMO antenna by outlining the operation of a single radiating element. A monopole antenna with an H-shaped design was constructed on an FR4 substrate, which is easy to produce, cost-effective, and widely accessible. The substrate's loss tangent and dielectric constant are 4.4 and 0.02 respectively. One side of the PCB has all the radiating elements, feeding strip, and ground plan, and on the other side, the copper is etched. It provides space for other RF devices and prevents short circuits in the chassis. To achieve the size of the smartphone, a double-sided printed circuit board of $150 \times 75 \times 0.8$ mm^3 is used to create the proposed system. In Fig. 30.1, proposed antenna system is shown with all dimensions. It is significant to remember that the proposed antenna is built to modern smartphone commercial standards.

Figure 30.2 shows the surface current density of the antenna at resonating frequencies for further analysis of the structure's functioning. Therefore, the proposed work can be reliably assumed to be compatible with antenna systems used in modern commercial mobile phones. At the corner of the board, the radiating elements are

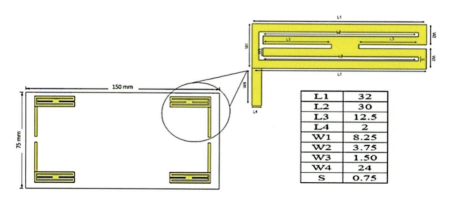

Fig. 30.1 MIMO antenna system

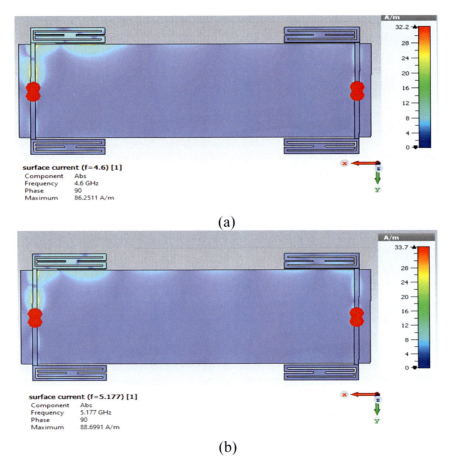

Fig. 30.2 Surface current density (**a**) at 4.6 GHz (**b**) 5.1 GHz, when antenna-1 is excited

situated and have a 50-Ω feed line and share the same ground plane. The proposed antenna has a dual-band response at the frequencies 4.6 and 5.1 GHz. Figure 30.2 shows the surface current density at resonating frequencies for further analysis of the structure's functioning.

30.2.1 Result and Discussion

A MIMO antenna system with four ports was built on the FR4 substrate. The front and back views of the MIMO antenna prototype are shown in Fig. 30.4. The following paragraphs discuss performance measures for return loss, gain, radiation pattern, and ECC.

30.2.2 Parametric Analysis

The impact of length L1 on the return loss of the antenna is illustrated in Fig. 30.3. With a 0.5 mm decrement, the length ranges from 30 to 32 mm. We find that the radiation frequency shifts to the right as the length decreases. This is because the frequency shifted down the band as the antenna length increased. Five distinct values are examined for the length parameter. Similar results may be derived for this investigation, and the best values for various design parameters are chosen based on these parametric analyses.

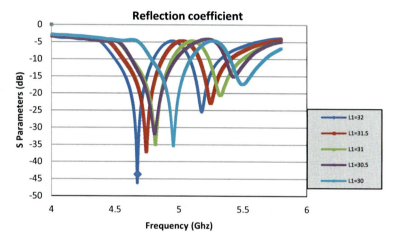

Fig. 30.3 Reflection coefficient for various values of L1

Fig. 30.4 Fabricated prototype

30.2.3 Fabrication and Measurement

Computer simulation technology (CST) software based on full wave electromagnetics is used for designing and simulation of the proposed antenna system, and A N5234A PNA-L network analyzer is used to measure the prototype. Figure 30.4 depicts the suggested MIMO antenna prototype.

30.2.4 S-Parameters and Radiation Pattern

The suggested antenna's S-parameters are described in this section. To measure the S-parameters of the suggested antenna, An Agilent Technologies Precision Network analyzer (PNA) N5234A was used. Figure 5a displays the reflection coefficients obtained from both simulated and measured data. It demonstrates that the antenna can resonate at frequencies between 4.4–5 GHz and 5.1–5.9 GHz. It is important to note that the system has a 400 MHz (4.4–4.8 GHz) and 500 MHz (5–5.5 GHz) impedance bandwidth of 6 dB and is resonating at 4.6 GHz and 5.1 GHz.

Figure 30.5b is depicted the isolation between the antenna elements. It should be observed that for both frequency ranges the adjacent antenna elements are isolated by more than 12 dB. The maximum isolation between Ant. 1 and Ant 4, for both the frequency band is > − 31.8 dB and > 33.4 dB, respectively. Figure 30.5a shows some discrepancy between the results (measured and simulated), which is caused by measurement setup losses and fabrication tolerances.

The radiation efficacy and overall performance of the MIMO antennas are shown in Fig. 30.6. Figure 30.6a demonstrates that the radiation efficiency for both bands is greater than 70%, with the 4.6 GHz frequency band's overall efficiency ranging between 50 and 72% and the 5.1 GHz frequency band's between 34 and 70% (see Fig. 30.6a).

The suggested system's 4.6 and 5.1 GHz radiation characteristics are shown in Fig. 30.7. Figure 30.7a illustrates the far-field patterns for the $\theta = 0$ and $\varphi = 90$ planes. It is significant to note that the radiating elements are positioned on the board to produce a wideband, quasi-isotropic overall pattern for the system. Wideband in this sense refers to the half-power beam width, and isotropic means that radiation is uniformly distributed around a sphere. The bulk of the power of the supplied antenna emits its power towards the main lobe with directivity, or a maximum directivity of 6.175 dBi is achieved.

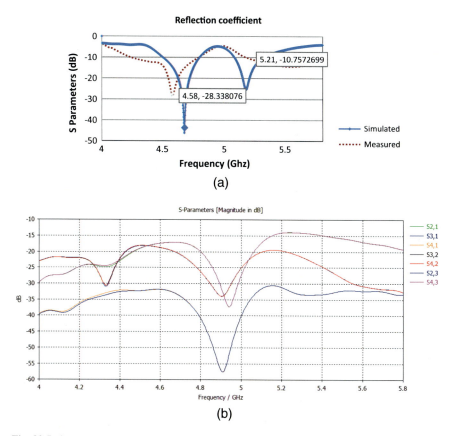

Fig. 30.5 Scattering parameters **a** reflection coefficients **b** transmission coefficients

30.2.5 MIMO Parameters

The envelope correlation coefficient (ECC) is becoming more significant in MIMO systems [22]. It specifies the degree to which the MIMO antenna components influence one another. In other words, there is minimal to no interference between the radiating components, and decreased ECC encourages greater MIMO system performance.

The 3D electric field patterns of radiating elements inside an array were used to compute the ECC. Remembering that a uniform incident wave environment is assumed, while computing ECC is essential [23]. The ECC result is shown in Fig. 30.8. All of the scenarios taken into consideration have an ECC that is considerably below 0.035, which is in line with the international specifications for 5G MIMO antenna systems (ECC 0.5).

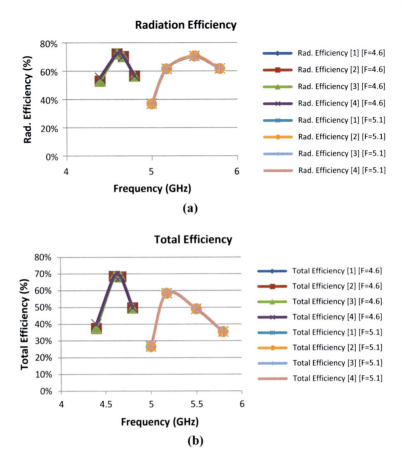

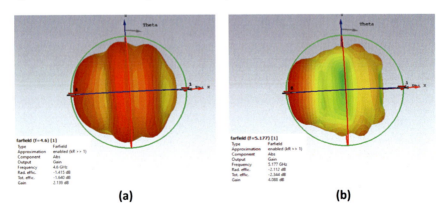

Fig. 30.6 **a** Radiation efficiency **b** total efficiency

Fig. 30.7 Radiation pattern **a** at 4.6 GHz, **b** at 5.1 GHz

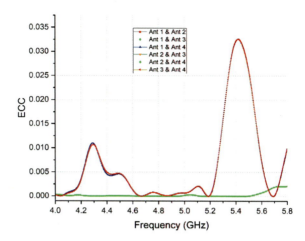

Fig. 30.8 Envelop correlation coefficient (ECC)

30.2.6 Comparative Analysis with Published Works

The differences between the 4- Port 5G MIMO antenna under consideration and the earlier studies highlight in Table 30.1. It makes it very clear that the suggested antenna covers the 5G new radio LTE band 46 and band n79 without using any additional decoupling techniques. The isolation between resonant elements is likewise quite good, which reduces design complexity. Indicating improved diversity performance, the envelope correlation coefficient value achieved is likewise quite low.

30.3 Conclusions

The main objective of this research is to suggest a straightforward antenna system that can meet the requirements of 5G technology in the sub-6 GHz frequency band. This antenna can operate 5G new radio bands: n79 (4.4–5 GHz) and LTE band 46 (5.1–5.9 GHz). Each antenna is mounted across a non-ground area measuring 32 mm by 8.25 mm. No additional decoupling approach is needed because the antennas are carefully positioned on the edges of the substrate. The radiating elements are positioned on one side of the board to allow for the integration of other RF components and devices. The performance of the suggested system is examined using a variety of important operational parameters. Mutual coupling is substantially below −10 dB in both bands. The Max. gain of 4.05 dB is achieved at 5.1 GHz, and greater than 2 dB gain is achieved for the other frequency. The MIMO antenna's ECC is far lower than the industry norm of 0.035. Additionally, a prototype is made, and it was discovered that the actual and calculated findings match very well. Additionally, the provided antenna's modest antenna footprint leaves room for other circuits that are present inside the mobile terminal. It is fair to conclude that the suggested MIMO system can be used with a variety of wireless technologies in the future.

Table 30.1 Comparison between the proposed antenna and previously published research

Reference	Bandwidth (GHz)	5G Bands	Elements Size	No. of antennas	ECC	Efficiency (%)
Proposed	4.4–5.9 (−10 dB)	n79, LTE Band 46	32 * 8.25 mm	4	< 0.03	70–72
[2]	3.4–3.6 (−10 dB)	Partially n77, n78	21.5 × 3	8	< 0.05	62–76
[15]	3.3–4.2 4.8–5.0	Partially n77, n78,79	49.6 * 7	8	< 0.12	62.6–79.1
[16]	3.4–3.6 (−10 dB)	Partially n77, n78	14.2 × 9.4	8	< 0.2	> 40
[21]	3.4–3.6 (−10 dB)	Partially n77, n78	8.5 × 3	6	< 0.15	50–60
[25]	3.4–3.6 (−10 dB)	Partially n77, n78	14 × 6	8	< 0.05	62–76
[26]	3.3–3.7 (−6 dB)	Partially n77, n78	4.6 × 5.6	8	< 0.1	50–70
[27]	3.4–3.6 (−6 dB)	Partially n77, n78	12.5–18.5	8	< 0.2	42–65
[28]	3.3–5 (−6 dB)	n77, n78, n79	10 * 8	4	< 0.018	–
[29]	3.4–3.6 (−17 dB)	Partially n77, n78	20 * 7	4	< 0.1	60–72
[30]	3.3–3.6 (−17 dB)	Partially n77, n78	25 * 7	4	< 0.05	56.2–64.7

References

1. Jensen, M.A., Wallace, J.W.: A review of antennas and propagation for MIMO wireless communications. IEEE Trans. Antennas Propag. **52**(11), 2810–2824 (2004). https://doi.org/10.1109/TAP.2004.835272
2. Li, Y., Sim, C.Y.D., Luo, Y., Yang, G.: High-Isolation 3.5 GHz eight-antenna MIMO array using balanced open-slot antenna element for 5G smartphones. IEEE Trans. Antennas Propag. **67**(6), 3820–3830 (2019). https://doi.org/10.1109/TAP.2019.2902751
3. Zhao, A., Ren, Z.: Size reduction of self-isolated MIMO antenna system for 5G mobile phone applications. IEEE Antennas Wirel. Propag. Lett. **18**(1), 152–156 (2019). https://doi.org/10.1109/LAWP.2018.2883428
4. Zhao, A., Ren, Z.: Wideband MIMO antenna systems based on coupled-loop antenna for 5G N77/N78/N79 applications in mobile terminals. IEEE Access **7**, 93761–93771 (2019). https://doi.org/10.1109/ACCESS.2019.2913466
5. Wang, H., Zhang, R., Luo, Y., Yang, G.: Compact eight-element antenna array for triple-band MIMO operation in 5G mobile terminals. IEEE Access **8**, 19433–19449 (2020). https://doi.org/10.1109/ACCESS.2020.2967651
6. Zhang, X., Li, Y., Wang, W., Shen, W.: ultra-wideband 8-Port MIMO antenna array for 5G metal-frame smartphones. IEEE Access **7**, 72273–72282 (2019). https://doi.org/10.1109/ACCESS.2019.2919622

7. Sim, C.Y.D., Liu, H.Y., Huang, C.J.: Wideband MIMO antenna array design for future mobile devices operating in the 5G NR frequency bands n77/n78/n79 and LTE band 46. IEEE Antennas Wirel. Propag. Lett. **19**(1), 74–78 (2020). https://doi.org/10.1109/LAWP.2019.2953334
8. Cai, Q., Li, Y., Zhang, X., Shen, W.: Wideband MIMO antenna array covering 3.3-7.1 GHz for 5G metal-rimmed smartphone applications. IEEE Access **7**, 142070–142084 (2019). https://doi.org/10.1109/ACCESS.2019.2944681
9. Chen, H.D., Tsai, Y.C., Sim, C.Y.D., Kuo, C.: Broadband eight-antenna array design for Sub-6 GHz 5G NR bands metal-frame smartphone applications. IEEE Antennas Wirel. Propag. Lett. **19**(7), 1078–1082 (2020). https://doi.org/10.1109/LAWP.2020.2988898
10. Singh, A., Saavedra, C.E.: Wide-bandwidth inverted-F stub fed hybrid loop antenna for 5G sub-6 GHz massive MIMO enabled handsets. IET Microwaves, Antennas Propag. **14**(7), 677–683 (2020). https://doi.org/10.1049/iet-map.2019.0980
11. Jiang, W., Liu, B., Cui, Y., Hu, W.: High-Isolation eight-element MIMO array for 5G smartphone applications. IEEE Access **7**, 34104–34112 (2019). https://doi.org/10.1109/ACCESS.2019.2904647
12. Guo, J., Cui, L., Li, C., Sun, B.: Side-Edge frame printed eight-port dual-band antenna array for 5G smartphone applications. IEEE Trans. Antennas Propag. **66**(12), 7412–7417 (2018). https://doi.org/10.1109/TAP.2018.2872130
13. Hu, W., et al.: Dual-band eight-element MIMO array using multi-slot decoupling technique for 5G terminals. IEEE Access **7**, 153910–153920 (2019). https://doi.org/10.1109/ACCESS.2019.2948639
14. Jiang, W., Cui, Y., Liu, B., Hu, W., Xi, Y.: A dual-band MIMO antenna with enhanced isolation for 5g smartphone applications. IEEE Access **7**, 112554–112563 (2019). https://doi.org/10.1109/ACCESS.2019.2934892
15. Cui, L., Guo, J., Liu, Y., Sim, C.Y.D.: An 8-Element dual-band MIMO antenna with decoupling stub for 5G smartphone applications. IEEE Antennas Wirel. Propag. Lett. **18**(10), 2095–2099 (2019). https://doi.org/10.1109/LAWP.2019.2937851
16. Han, C.Z., Xiao, L., Chen, Z., Yuan, T.: Co-Located self-neutralized handset antenna pairs with complementary radiation patterns for 5G MIMO applications. IEEE Access **8**, 73151–73163 (2020). https://doi.org/10.1109/ACCESS.2020.2988072
17. Ren, Z., Zhao, A., Wu, S.: MIMO antenna with compact decoupled antenna pairs for 5G mobile terminals. IEEE Antennas Wirel. Propag. Lett. **18**(7), 1367–1371 (2019). https://doi.org/10.1109/LAWP.2019.2916738
18. Chang, L., Yu, Y., Wei, K., Wang, H.: Polarization-orthogonal co-frequency dual antenna pair suitable for 5G MIMO smartphone with metallic bezels. IEEE Trans. Antennas Propag. **67**(8), 5212–5220 (2019). https://doi.org/10.1109/TAP.2019.2913738
19. Xu, H., Gao, S.S., Zhou, H., Wang, H., Cheng, Y.: A highly integrated MIMO antenna unit: differential/common mode design. IEEE Trans. Antennas Propag. **67**(11), 6724–6734 (2019). https://doi.org/10.1109/TAP.2019.2922763
20. Yang, L., Li, T.: Box-folded four-element MIMO antenna system for LTE handsets. Electron. Lett. **51**(6), 440–441 (2015). https://doi.org/10.1049/el.2014.3757
21. Choi, J., Hwang, W., You, C., Jung, B., Hong, W.: Four-Element reconfigurable coupled loop MIMO antenna featuring LTE full-band operation for metallic-rimmed smartphone. IEEE Trans. Antennas. Propag. **67**(1), 99–107 (2019). https://doi.org/10.1109/TAP.2018.2877299
22. Abdullah, M., Ban, Y.-L., Kang, K., Li, M.-Y., Amin, M.: Eight-Element antenna array at 3.5 GHz for MIMO wireless application (2017)
23. Liang, X.L., Denidni, T.A.: H-shaped dielectric resonator antenna for wideband applications. IEEE Antennas Wirel. Propag. Lett. **7**, 163–166 (2008). https://doi.org/10.1109/LAWP.2008.922051
24. Abdullah, M., Kiani, S.H., Iqbal, A.: Eight element multiple-input multiple-output (MIMO) antenna for 5G mobile applications. IEEE Access **7**, 134488–134495 (2019). https://doi.org/10.1109/ACCESS.2019.2941908
25. Kiani, S.H., et al.: Eight element side edged framed MIMO antenna array for future 5G smart phones. Micromachines (Basel) **11**(11), (2020). https://doi.org/10.3390/mi11110956

26. Abdullah, M., et al.: Future smartphone: MIMO antenna system for 5G mobile terminals. IEEE Access **9**, 91593–91603 (2021). https://doi.org/10.1109/ACCESS.2021.3091304
27. Kiani, S.H., et al.: Mimo antenna system for modern 5g handheld devices with healthcare and high rate delivery. Sensors **21**(21), (2021). https://doi.org/10.3390/s21217415
28. Biswas, A., Gupta, V.R.: Design and development of low profile MIMO antenna for 5G new radio smartphone applications. Wirel. Pers. Commun. **111**(3), 1695–1706 (2020). https://doi.org/10.1007/s11277-019-06949-z
29. Wong, K.L., Tsai, C.Y., Lu, J.Y.: Two asymmetrically mirrored gap-coupled loop antennas as a compact building block for eight-antenna MIMO array in the future smartphone. IEEE Trans. Antennas Propag. **65**(4), 1765–1778 (2017). https://doi.org/10.1109/TAP.2017.2670534
30. Wong, K.L., Lin, B.W., Lin, S.E.: High-isolation conjoined loop multi-input multi-output antennas for the fifth-generation tablet device. Microw Opt. Technol. Lett. **61**(1), 111–119 (2019). https://doi.org/10.1002/mop.31505

Chapter 31
Optimization of Controller Parameters for Load Frequency Control Problem of Two-Area Deregulated Power System Using Soft Computing Techniques

Dharmendra Jain, M. K. Bhaskar, and Manish Parihar

Abstract It is very difficult to obtain the optimal parameter of controllers for load frequency control (LFC) problem of a multiarea power system in deregulated environment. Deregulated power system contains multisources and multistakeholders; therefore, conventional LFC methods are not effective and competent. The primary goal of LFC in a deregulated system is to restore the frequency to its original value as soon as feasible while also minimizing uncontracted power flow in tie line between neighboring control regions and tracking load balancing contracts. Gains of PID controller are required to be optimized in order to fulfill the objectives of LFC. Here genetic algorithm as well as particle swarm optimization techniques are presented in this paper for optimization of controller parameters in order to achieve the purposes of LFC of two-area deregulated system taking suitable objective function that are to minimize the frequency deviations of both the areas and to maintain tie line power flow according to contractual conditions. System has been simulated under MATLAB/Simulink, and dynamic responses have been obtained for many contractual conditions between GENCOS and DISCOS. It is confirmed by the results that the soft computing–based PID controllers are capable of maintaining the frequency in the pre-specified range and keep the tie line power flow as per the contractual conditions. An analysis has been done by comparing the dynamic responses of the system with PSO-based controller and GA-based controller.

Keywords LFC · Deregulated · PID · GA · PSO

D. Jain (✉) · M. K. Bhaskar · M. Parihar
Department of Electrical Engineering, MBM University, Jodhpur, Rajasthan 342001, India
e-mail: dharmendrajainmbm@gmail.com

31.1 Introduction

LFC has been considered as one of the greatest noteworthy services in the interconnected power system. LFC has an important goal to restore the frequency to its operational value with maintaining the tie line power flow among the surrounding areas to the suitable values as explained by Donde et al. [1] and Dharmendra Jain et al. [10]. LFC has been developed as more noteworthy in recent time due to waste size and complicated structure of power system. To improve the power system operation, some major changes have been made in the structure of the power system by the process of restructuring the electrical power industry and making it available for competition. The main engineering features have been reformulated in deregulation process power system while critical ideas are the same.

In the deregulated power system, every area should first fulfill its own demand and aster that it should maintain its contracted power. Any discrepancy between the demanded and generated power will lead to a deviation in frequency of all the connected areas. Automatic generation control is used to for making balance between the generation and power demand.

The deregulated structure of power system comprises of many GENCOs and DISCOs; therefore, a DISCO has the liberty to choose and make a contract with any of the GENCO to fulfill if power demand. A DISCO can collaborate with a GENCO in any other control area. These types of contracts are termed as bilateral contracts. The transactions are monitored by an independent and unbiased entity defined as an independent system operator. Ancillary services are controlled by ISO. AGC is also one of the ancillary services. LFC is one of the most gainful ancillary services. The market player control both the generation and load demand with keeping the stability of whole power system under very competitive market and distributed environment. LFC plays a critical role and perform unending task in reformed power system.

A lot of studies have been conducted about various LFC issues in a deregulated power system to overcome these situations. Kothari et al. [2] explained the automatic generation control of deregulated power system. Optimum megawatt LFC was presented by Elgerd and Fosha [11] and [12]. Decentralized load frequency control in deregulated environment was presented by Tan et al. [4]. To solve LFC, many of the researchers used PID controllers because of its accuracy and high speed. The performance of PID controller directly depends on its parameters tuning as explained by Dharmendra Jain et al. [3]. Cohn [9] explained the concept of tie line bias control. IEEE report [13] represented operating problems associated with automatic generation control. Therefore, many researchers like Concordia and Kirchmayer [8] and Babahajiani et al. [7] used soft computing–based techniques like neural networks and fuzzy logic, honey bee algorithm is used by Abedinia et al. [14], and Sekhar et at. [5] and [6] used firefly algorithm or other methods for tuning of parameters in order to optimize the gain of controllers. Intelligent demand response contribution in frequency control of multiarea power systems was explained by Babahajiani et al. [7]. Sahoo [16] explained the application of soft computing neural network tools to line congestion study of electrical power systems. Reduced-order observer method was used by Rakhshani et al. [17]. Abd-Elazim and Ali [15] explained load frequency controller design of a two-area system composing of PV grid and thermal generator via firefly algorithm. In this paper, genetic algorithm optimization technique and particle swarm optimization techniques are used to tune the parameters of the controller for LFC of two-area interconnected power system in deregulated environment. The superiority of PSO-based proposed controller is shown by comparing the results with GA-based controller in deregulated power system.

31.2 Interconnected Power System in Deregulated Environment

Contracts are formed between corporations in a deregulated power market based on regulations as well as relationships in order to establish an equilibrium among GENCOs and DISCOs. These contracts might be bilateral or it might be Poolco, or mix mode of bilateral and Poolco, that is hybrid. According to the Poolco contract, each DISCO gets all of its electricity from generators in its own territory. However, under the bilateral contract, each DISCO can do business with any GENCO in any area. Here in this research work, two areas are considered in deregulated environment of power system. Area-1 and area-2 are made up of two thermal generation units in each area.

In the deregulated power system, generation companies (GENCOs) have freedom to participate or not to participate in AGC task, whereas distribution companies (DISCOs) can make contracts with any of the GENCOs in their own or other areas. Due to this freedom, DISCOs and GENCOs may have many possible combinations of contracts between them. The concept of distribution participation matrix (DPM) is used here to express possible contracts in the two-area deregulated model. DP matrix can be defined as a matrix with GENCO represents the rows, number of rows equal to the number of GENCOs and DISCO represents the column, the number of columns equal to the number of DISCOs in the system. Here a new term contract participation factor (cpf) is defined which can be given as entity in the DPM that represents the portion of a DISCO total contracted load demands being met by a GENCO as explained by Donde et al. [1]. This must be noted that the addition of all the entities of a column in DP matrix is always one. If a DISCO j demands power from a GENCO1, this will give the ijth entity of the DP matrix and will be represented by cpfij.

A two-area power system in deregulated environment has been considered here. Two GENCOs are there in each area and also each area has two DISCOs. Also, it is assumed that area-1 consists of GENCO1 and GENCO2 and DISCO1 and DISCO2 as shown in Fig. 31.1.

The DP matrix for this can be given as shown in Eq. 31.1.

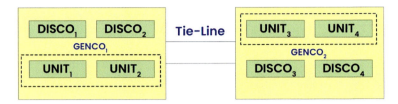

Fig. 31.1 Schematic of a two-area system in restructured environment

$$DPM = \begin{bmatrix} cpf11 & cpf12 & cpf13 & cpf14 \\ cpf21 & cpf22 & cpf23 & cpf24 \\ cpf31 & cpf32 & cpf33 & cpf34 \\ cpf41 & cpf42 & cpf43 & cpf44 \end{bmatrix} \quad (31.1)$$

If there is any change in load demand of a DISCO, it is represented as a local load ΔP in the same area to which DISCO belongs. Therefore, the local loads $\Delta PL1$ and $\Delta PL2$ for area-1 and area-2 should be replicated in the power system AGC block diagram. The area control error must be distributed in all GENCOs in the proportion of their participation as there are many GENCOs in each area. ACE participation factors (apfs) is defined as coefficients which distribute ACE to GENCOs according to participation. Sum of all apfs of a column is unity as shown in Eq. 31.2.

$$\text{Note that } \sum_{i=1}^{m} a_{ji} = 1 \quad (31.2)$$

Participation factor of i-th GENCO in j-th area is represented by the term a_{ji} and number of GENCOs in j-th area is represented by the term m.

Equations 31.3 and 31.4 represent scheduled tie line power flow.

$$\Delta P_{\text{tie1-2 scheduled}} = (\text{demand of DISCOs in area-2 from GENCOs in area-1 })$$
$$- (\text{demand of DISCOs in area-1 from GENCOs in area-2 }) \quad (31.3)$$

$$\Delta P_{\text{tie1-2 scheduled}} = \sum_{i=1}^{i=2} \sum_{j=3}^{j=4} CPF_{ij} \Delta PL_j - \sum_{i=3}^{i=4} \sum_{j=1}^{j=2} CPF_{ij} \Delta PL_j \quad (31.4)$$

The tie line power error $\Delta P_{\text{tie1-2}}$, error at any given time, is defined as in Eq. 31.5.

$$\Delta P_{\text{tie1-2,error}} = \Delta P_{\text{tie1-2 actual}} - \Delta P_{\text{tie1-2,scheduled}} \quad (31.5)$$

As the actual tie line power flow reaches the actual value, $\Delta P_{\text{tie1-2, error}}$ tends to zero in the steady state condition. Respective ACE signal is generated using this error signal which is same as in the traditional situation. ACE for area-1 and area-2 are shown in Eqs. 31.6 and 31.7, respectively.

$$ACE_1 = B_1\Delta f_1 + \Delta P_{\text{tie1-2,error}} \tag{31.6}$$

$$ACE_2 = B_2\Delta f_2 + \Delta P_{\text{tie2-1,error}} \tag{31.7}$$

where

$$\Delta P_{\text{tie1-2,error}} = -\left(\frac{P_{r1}}{P_{r2}}\right)\Delta P_{\text{tie1-2,error}} \tag{31.8}$$

And P_{r1}, P_{r2} are the rated powers of areas 1 and 2, respectively. Therefore,

$$ACE_2 = B_2\Delta f_2 + \alpha_{12}\,\Delta P_{\text{tie1-2,error}} \tag{31.9}$$

where

$$\alpha_{12} = -\left(\frac{P_{r1}}{P_{r2}}\right) \tag{31.10}$$

The Eqs. 31.11 and 31.12 represent contracted power supplied by i-th GENCO in a two-area system.

$$\Delta P_i = \sum_{j=1}^{n\text{ disco}=4} CPF_{ij}\Delta PL_j \tag{31.11}$$

For $i = 1$,

$$\Delta P_1 = CPF_{11}\Delta P_{L1} + CPF_{12}\Delta P_{L2} + CPF_{13}\Delta P_{L3} + CPF_{14}\Delta P_{L4} \tag{31.12}$$

ΔP_2, to ΔP_4 can be calculated easily in the similar manner.

Figure 31.2 shows the MATLAB/Simulink diagram for LFC issues in two-area deregulated power system. Structure of simulation diagram is based on the idea of Donde et al. [1] and Jain et al. [3]. Demand signals are shown by dashed lines. The $\Delta P_{1\text{LOC}}$ and $\Delta P_{2\text{LOC}}$ represent local loads in areas 1 and 2, respectively. Whereas ΔP_{uc1} and ΔP_{uc2} represent uncontracted power (if any).

Equations 31.13 and 31.14 show the local power of area-1 and area-2, respectively.

$$\Delta P_{1LOC} = \Delta P_{L1} + \Delta P_{L2} \tag{31.13}$$

$$\Delta P_{2LOC} = \Delta P_{L3} + \Delta P_{L4} \tag{31.14}$$

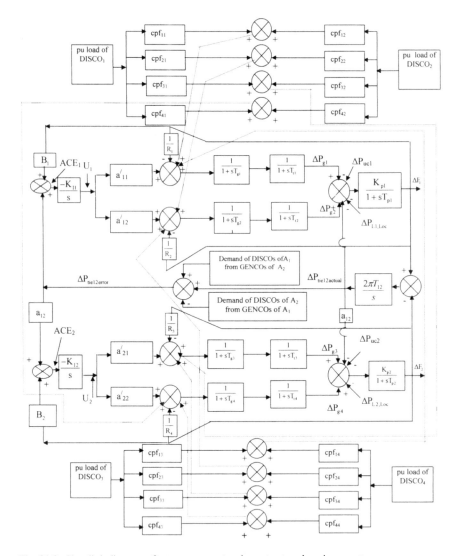

Fig. 31.2 Simulink diagram of a two-area system in restructured environment

31.3 PID Controller

Three separate controller named as proportional, integral and derivative controller when act together, the resultant controller is termed as PID controller. The controller output is called as manipulated variable. Equation 31.15 represents the output of the PID controller.

$$U(t) = MV(t) = Kpe(t) + Ki \int_0^t e(t)dt + Kd\frac{d}{dt}e(t) \qquad (31.15)$$

In order to obtain the desired response, the proper values of controller parameter required. The process of obtaining the proper values of PID controller parameter is called as tuning.

31.4 Tuning of PID Controller Parameters

Adjustment of controller gains for desired response of the system is called as tuning. The proper values of K_P, K_I, and K_D form a tuned PID controller. The imperfect tuning will lead to poor performance of the system. So tuning is very important as proper tuning improve the system response on the other hand improper tuning degrade the system performance.

There are numerous methods of tuning of PID controller parameters. Z-N method and IMC methods of tuning are used by Jain et al. [10]. But nowadays, soft computing methods are more popular due to several advantages. In this paper, GA as well as PSO methods are used for the optimization of controller parameters.

31.5 Genetic Algorithm

There are many soft computing techniques one of which is genetic algorithms. GAs are multipurpose search algorithms that leverage ideas derived from natural genetics to solve issues. They have been used effectively to a wide range of scientific and technical challenges, including optimization, machine learning, automated programming, transportation difficulties, adaptive control, and so on. Genetic algorithms are adaptable systems that strive to learn, adapt, and act like biological or natural entities. Figure 31.3 depicts a flowchart that describes the basic mechanism. GA has been implemented here to tune the PID controller with the objective to minimizing the time multiplied absolute error (ITAE).

The steps involved in tuning PID controller gains are given as below:

Step 1: Identify the transfer function of the plant or or identify the system model.
Step 2: Initialize controller gains K_P, K_I and K_D, and calculate ITAE.
Step 3: Obtain the value particle best value (pbest) and global best value (gbest).
Step 4: Apply mutation to Calculate new population.
Step 5: Obtain updated particle best (pbest1) and updated global best (gbest1) values.
Step 6: Compare previous and updated particle best values (pbest and pbest1).
Step 7: Compare previous and updated global best values (gbest and gbest1).

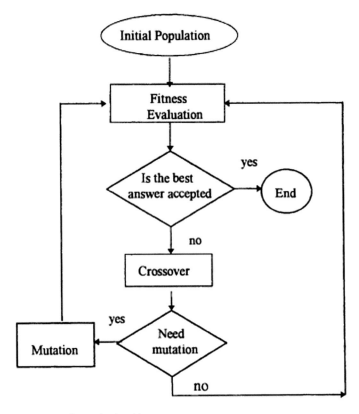

Fig. 31.3 Flow chart of genetic algorithm

Step 8: Obtain the updated values of controller parameters K_P, K_I & K_D, and find out the response for the system.

Step 9: If stopping criteria reached or maximum number of iterations reached: stop otherwise go to step 4.

31.6 Particle Swarm Optimization Technique

Eberhart and Kennedy developed a optimization technique which is popularly known as particle swarm optimization (PSO) technique. It is inspired by the communal conduct of schooling fish or swarming birds. In this study, it is used to explore the search space of a given issue in order to determine the best gain values of controller parameters necessary to meet the LFC objectives. PSO starts with a set of random particles (solutions) and then updates the solutions to get the best gains. Each particle is represented by two vectors, i.e., position 'xi' and velocity 'vi'.

The position of every single particle at any given time is seen as a solution to the challenge at hand. The particles then fly about the search region, changing their speed and position to locate the optimal place at each time. All of the particles have fitness values that are evaluated by the fitness function to be optimized, as well as velocities that direct the particles' flight. Following the current optimal particles, the particles fly through the problem space. The position and velocity of the ith particle in a physical d-dimensional search space are represented by the vectors shown in Eqs. 31.16 and 31.17, respectively

$$X_i = [X_{i1}, X_{i2}, \ldots X_{id}] \quad (31.16)$$

$$V_i = [V_{i1}, V_{i2}, \ldots V_{id}] \quad (31.17)$$

The particles are updated to pbest and gbest, two "best" values that each particle has so far attained. The best value that has been attained far by every particle in the population, gbest. Pbest is the optimal location that produces the ith particle's best fitness value, and gbesti is the optimal position over the whole swarm population. Best values of ith particle are represented by Eqs. 31.18 and 31.19, respectively.

$$\text{pbest} = [\text{pbest}_i^1, \text{pbest}_i^2 \ldots \text{pbest}_i^d] \quad (31.18)$$

$$\text{gbest} = [\text{gbest}_i^1, \text{gbest}_i^2, \ldots \text{gbest}_i^d] \quad (31.19)$$

The PSO algorithm updates its velocity and position using the following equation. The velocity updating equation is given as in Eq. 31.20.

$$v_i^d(j+1) = w(j)v_i^d(j) + c_1r_1\left[\text{pbest}_i^d(j) - x_i^d(j)\right] + c_2r_2\left[gbest_i^d(j) - x_i^d\right] \quad (31.20)$$

The velocity at jth iteration can be given by $V^d{}_i(j)$ which is velocity of 'i'th particle in 'd'th dimension.

Equation 31.21 is used to update the particle position, once the velocity for each particle has been calculated. Each particle's position will be updated by applying the new velocity to the particle's previous position.

$$x_i^d(j+1) = x_i^d(j) + v_i^d(j+1) \quad (31.21)$$

The goal of the optimization is to reduce the integral of time multiplied absolute error while maintaining the PID gains KP, KI, and KD of the controllers within the specified maximum and minimum ranges. This is done by minimizing the fitness function. PSO flow chart is shown in Fig. 31.4. The PSO algorithm consists of just few steps as shown in flow chart. Steps are repeated till the desired response obtained

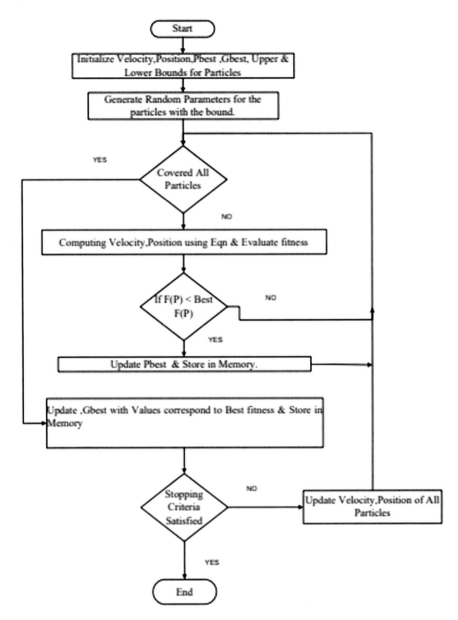

Fig. 31.4 PSO flow chart

Fig. 31.5 Area-1 frequency change

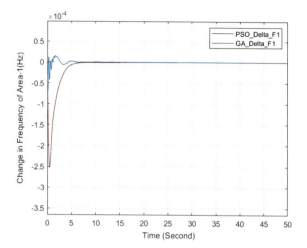

or some stopping criteria has been reached. Using this, optimized parameters of the controllers are obtained.

31.7 Simulation and Result

The diagram shown in the Fig. 31.2 has been used for simulation purpose. Controller based on PSO and GA optimization techniques are used, and dynamic responses have been obtained for various contractual conditions.

31.7.1 Case-I

It is the base case. Total load demand of all the DISCOs is 0.005 pu MW. Comparative responses using GA optimized controller and PSO-based controllers are given in Figs. 31.5, 31.6, 31.7, 31.8 and 31.9.

31.7.2 Case-II

Additional load demand of 0.0025 pu-MW is raised by area-1 at $t = 25$ s. and it is supplied by only GENCO1 of area-1. It is a contract violation case. Comparative responses using GA-based controller and PSO-based controller are shown in Figs. 31.10, 31.11, 31.12, 31.13 and 31.14.

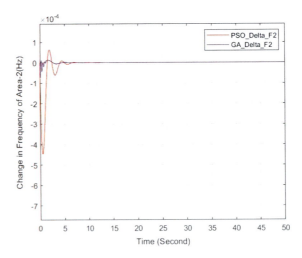

Fig. 31.6 Area-2 frequency change

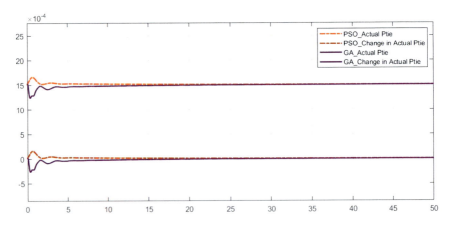

Fig. 31.7 Tie line flow of power: Actual and change

31.8 Conclusion

For the best operation of power system, contracted power and frequency has to stay almost constant or should be close to the scheduled values in interconnected deregulated power system. The efficiency, reliability, and operation of a power system have influence of frequency variations. Therefore, a proper control strategy is required. Soft computing techniques like PSO and GA optimization techniques have been used for the optimization of the controller parameters. In order to apply the controller and check its responses, simulation model of a two-area interconnected deregulated power system developed in MATLAB–Simulink. Comparative responses using various control strategies have been obtained and shown. Comparative analysis shows

Fig. 31.8 GENCO responses of area-1

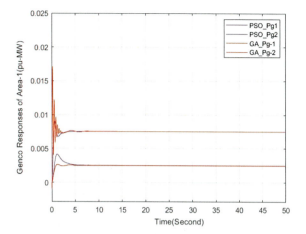

Fig. 31.9 GENCO responses of area-2

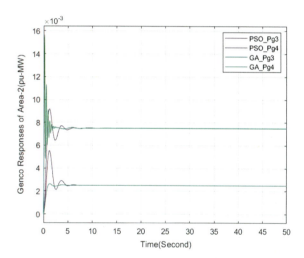

that PSO-based controller provides the best response for two-area interconnected and restructured power system as compared to other controllers used in this work.

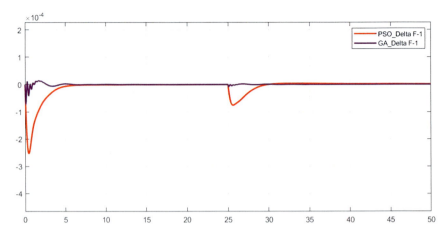

Fig. 31.10 Area-1 frequency change

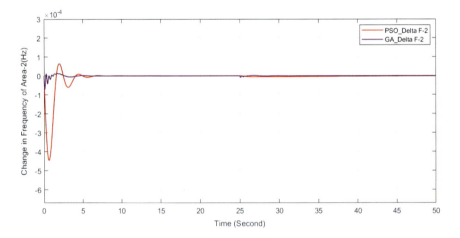

Fig. 31.11 Area-2 frequency change

31 Optimization of Controller Parameters for Load Frequency Control … 399

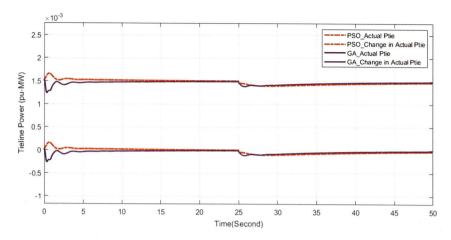

Fig. 31.12 Tie line flow of power: actual and change

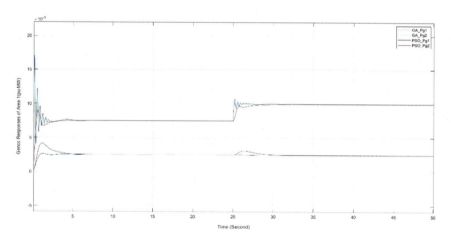

Fig. 31.13 GENCO responses of area-1

Fig. 31.14 GENCO responses of area-2

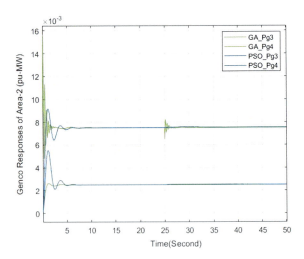

References

1. Donde, V., Pai, M.A., Hiskens, I.A.: Simulation and optimization in an AGC system after deregulation. IEEE Trans. Power Syst. **16**(3), 481–9 (2001)
2. Kothari, M.L., Sinha, N., Rafi, M.: Automatic generation control of an interconnected power system under deregulated environment. Proc. IEEE **6**, 95–102 (1998)
3. Jain, D. et al.: Analysis of load frequency control problem for interconnected power system using PID controller. IJETAE. **4**(11), (2014). ISSN 2250–2459, ISO 9001
4. Tan, W., Zhang, H., Yu, M.: Decentralized load frequency control in deregulated environments. Int. J. Electr. Power Energy Syst. **41**(1), 16–26 (2012)
5. Sekhar, G.C., Sahu, R.K., Baliarsingh, A., Panda, S.: Load frequency control of power system under deregulated environment using optimal firefly algorithm. Int. J. Electr. Power Energy Syst. **74**, 195–211 (2016)
6. Sood, Y.R.: Evolutionary programming based optimal power flow and its validation for deregulated power system analysis. Int. J. Electr. Power Energy Syst. **29**(1), 65–75 (2007)
7. Babahajiani, P., Shafiee, Q., Bevrani, H.: Intelligent demand response contribution in frequency control of multiarea power systems. IEEE Trans. Smart Grid **9**, 1282–1291 (2018)
8. Concordia, C., Kirchmayer, L.K.: Tie line power and frequency control of electric power systems. Amer. Inst. Elect. Eng. Trans. Pt. II. **72**, 562 572 (1953)
9. Cohn, N.: Some aspects of tie-line bias control on interconnected power systems. Amer. Inst. Elect. Eng. Trans. **75**, 1415–1436 (1957)
10. Jain, D., et al.: Comparative analysis of different methods of tuning the PID controller parameters for load frequency control problem. IJAREEIE **3**(11), (2014)
11. Elgerd, O.I., Fosha, C.: Optimum megawatt frequency control of multiarea electric energy systems. IEEE Trans. Power App. Syst. **PAS-89**(4), 556–563 (1970)
12. Fosha, C.E., Elgerd, O.I.: The megawatt -frequency control problem: A new approach via optimal control theory. IEEE Trans. Power App. Syst. **PAS-89**(4), 563–567 (1970)
13. IEEE PES Committee Report: Current operating problems associated with automatic generation control. IEEE Trans. Power App. Syst. **PAS-98** (1979)
14. Abedinia, O., Naderi, M.S., Ghasemi, A.: Robust LFC in deregulated environment:fuzzy PID using HBMO. Proc IEEE **1**, 1–4 (2011)
15. Abd-Elazim, S., Ali, E.: Load frequency controller design of a two-area system composing of PV grid and thermal generator via firefly algorithm. Neural Comput. Appl. **30**, 607–616 (2018)

16. Sahoo, P.K.: Application of soft computing neural network tools to line congestion study of electrical power systems. Int. J. Inf. Commun. Technol. **13**(2), (2018)
17. Rakhshani, E., Sadeh, J.: Simulation of two-area AGC system in a competitive environment using reduced-order observer method. Proc IEEE **1**, 1–6 (2008)

Chapter 32
Quantization Effects on a Convolutional Layer of a Deep Neural Network

Swati, Dheeraj Verma, Jigna Prajapati, and Pinalkumar Engineer

Abstract Over the last few years, we have witnessed a relentless improvement in the field of computer vision and deep neural networks. In a deep neural network, convolution operation is the load bearer as it performs feature extraction and dimensionality reduction on a large scale. As the models continue to go deeper and bulkier for better efficiency and accuracy there is a rapid increment in storage requirements too. The problem arises when performing computation with efficient numerical representations for embedded devices. Transitioning from floating-point representation to fixed-point could potentially reduce computation time, storage requirements, and latency with some accuracy loss. In this paper, an analysis of the effects of quantization of the first convolutional layer on the accuracy, and memory storage requirement with varying bit-width for fixed-point integer values of network parameters has been carried out. The approach adopted is post-training quantization with a mixed-precision format to avoid model re-training and minimize accuracy loss by using root-mean-square-error (RMSE) as a performance metric. Various combination has been analyzed and compared to find the optimal precision to implement on a resource-constraint device. Based on the analysis, the suggested bit-width of I/O data for this implementation is selected as <10,5> and mid-data be <20,10> instead of <16,8> and <32,16> respectively. This combination of bit-widths has reduced memory consumption such as BRAM by 10%, DSPs by 98.6% and FFs by 40.27% with some accuracy loss.

Keywords Quantization · Convolution · Fixed-point precision

Swati (✉) · D. Verma · J. Prajapati · P. Engineer
Department of Electronics Engineering, SVNIT, Surat, India
e-mail: d20eced012@svnit.ac.in

P. Engineer
e-mail: pje@eced.svnit.ac.in

© The Author(s), under exclusive license to Springer Nature Singapore Pte Ltd. 2024
P. K. Jha et al. (eds.), *Proceedings of Congress on Control, Robotics, and Mechatronics*,
Smart Innovation, Systems and Technologies 364,
https://doi.org/10.1007/978-981-99-5180-2_32

32.1 Introduction

During recent years, there has been a huge breakthrough in the field of artificial intelligence, and specifically, with computer vision problems like image classification, object detection, etc. Ever since the introduction of AlexNet [1] by Krizhevsky et al., there has been an explosion in deep neural networks and it continues to evolve. The computer vision tasks like object classification, detection, and even object tracking became possible to achieve using DNN. However, the model size has also increased with parameters ranging in millions. Network training is usually carried out with GPUs with high memory bandwidth and high computation capabilities. When dealing with inference for an end application, it gets exhausting to perform computation on a resource-constrained device like an embedded device or an FPGA. This creates a problem when realizing real-time inference where there is a necessity for lighter models with smaller memory footprints and lower energy consumption.

There are quite a few ways to reduce the model complexity and size in order to fit in a limited resource environment. Network pruning is one method where redundant connections are dropped while little to no effects on accuracy [2]. Model compression follows a pipeline of pruning, weight quantization, and Huffman coding, which collectively reduce the storage requirement by $35\times$ to $49\times$ by not affecting their accuracy [3]. Using mixed-precision for different storage requirements could reduce model complexity and storage [4]. There are few co-design approaches where the network is re-trained with reduced precision for inferencing purposes [5]. The re-training method ensures that there is minimal loss of accuracy, but this is a separate task of analysis and training. We also can achieve reduced network size by applying quantization to the trained network just for inference.

This enables us to avoid the pain of re-training and perform co-design of the model with fixed precision with some loss in accuracy. Here in this paper, we analyze the variable bit-widths for mixed-precision analysis with respect to the accuracy and storage requirement for a deep neural network. Typically, deep learning models deal with the floating-point data type. Each and every data being calculated, manipulated, and stored at every layer are floating-point data types. During the inference phase, when implementing on resource-constrained environments, if we move from floating-point representation to fixed-point representation, it would presumably reduce memory footprint and computation efficiency. This conversion mechanism is attained by following the quantization scheme. There are two techniques for performing quantization, and one way is to quantize a pre-trained model by directly converting the floating-point parameters to fixed-point parameters, which might result in some loss of accuracy. The other way is re-training the network with quantized parameters in order to maintain the accuracy and convergence point. These two methods are known as post-training quantization (PTQ) and quantization-aware training (QAT), respectively [6]. In this work, we are focusing on the first discussed method, as re-training the network requires a lot of work and resources. A thorough analysis of quantization with variable bit-widths and their effect on accuracy and resource allocation has been performed here to find an efficient combination of fixed-point representation for different parameters.

32.2 Related Work

In the recent few years, there has been a lot of work targeting fixed-point implementation of DNNs. In [7], authors have proposed a quantization scheme regarding a co-design of training procedure to preserve the model accuracy post-training with integer-only arithmetic inference. In their work, they have suggested using of mixed-point precision for different parameters to ensure the model performance. The work done in [8], proposes a 4-bit quantization for a pre-trained model and compares mean-square-error (MSE) for both weights and activations without re-training of the model.

Wang et al. [9] has proposed a fixed-point factorized network (FFN) to convert the weights into ternary values, i.e., [−1, 0, +1] with no re-training required, and the resultant network achieved low-bit fixed-point arithmetic operations. Yuan et al. [10] have implemented a variable bit-width quantization strategy on weights and activation by grouping layer and channel separately to compare the performance, respectively. Wu et al. [11] has proposed a differentiable neural architecture search (DNAS) framework to find an optimal neural network architecture from the network search space. They have implemented a mixed-precision quantization for different layers, but here re-training is also required. Wu et al. [12] have proposed partial quantization where the precision of sensitive layers is left as *float32* to minimize accuracy loss. They have also suggested quantization-aware training (QAT) to recover accuracy loss. This work is intended to find a workaround to avoid re-training and leverages the mixed-precision quantization concept without the need for re-training.

32.3 Post-training Quantization Scheme

Quantization could be understood as the mapping of input values from a large set to output values of a smaller set.

32.3.1 Different Types of Quantization

There are three different types of quantization: (1) Uniform Quantization, (2) Non-Uniform Quantization, and (3) Hybrid Quantization.

In uniform quantization, the quantized values are uniformly spaced, and in non-uniform quantization they might vary, as can be observed from Fig. 32.1 as well. The recovery of real values from quantized values is possible and known as de-quantization, but there is a possibility of some truncation/quantization error. Here Eq. (32.1) represents quantization formula where r is real-valued input, S is the scaling factor, and Z is integer zero point and the de-quantization formula is mentioned as shown in Eq. (32.2).

Fig. 32.1 Uniform quantization and non-uniform quantization [13]

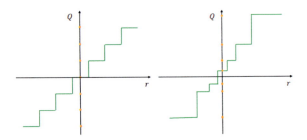

$$Q(r) = \text{Int}(r/S) - Z \tag{32.1}$$

$$r' = S(Q(r) + Z) \tag{32.2}$$

In uniform quantization, scaling factor S distributes over the range of real values into sections that can be calculated as shown in Eq. (32.3)

$$S = \frac{\beta - \alpha}{2^b - a} \tag{32.3}$$

where $[\alpha, \beta]$ refers to the range that we are restricted out quantized values, b is the quantization bit-width in order to calculate the scaling factor, we need to choose the range, which generally is the maximum and minimum real values. Figure 32.2 depicts the difference between symmetric and asymmetric quantization. In case of asymmetric quantization, $\alpha = r_{\max}$ and $\beta = r_{\min}$, and in case of symmetric quantization, $-\alpha = \beta = \max(|r_{\max}|, |r_{\min}|)$.

Hybrid Quantization Another method of performing approximation for the models is layer-wise quantization. In this method, we can have different ranges and scaling factors for individual layers and analyze the performance. In this method, we have the flexibility to be selective when performing approximation to accuracy-sensitive layers, as we can not perform the same aggressive approximation to all the layers. We also have an option of performing mixed-precision where we decide different precision bit-width for individual layers evaluating the effects on accuracy and other parameters.

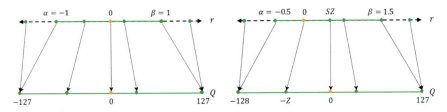

Fig. 32.2 Symmetric quantization and asymmetric quantization [13]

32.4 Experiments

32.4.1 Numerical Representation

In the numerical representation of fixed-point notation, there is a decimal-point separating integer values and binary values. In a few cases, there is a sign bit reserved for storing the sign and a partition between integer and fraction values as shown in Fig. 32.3. In computer architecture, we have a combination [Total bits (T), Integer portion (I)]. When there is a need for low precision representation, the decimal point is moved and fixed to some different location with respect to the bit-width as shown in Fig. 32.4.

We can refer to the example in Fig. 32.5 if we have a 32-bit representation, then each integer and fraction value is stored in 16 bits. We can analyze the range and can come to a conclusion regarding how many minimum bits are required for storing the parameters. Here, in the case of a CNN implementation, we first normalize the values to fit in the range where we can easily convert the bit representation. In an 8-bit image, the maximum pixel value goes about 255, so we normalize the range using the maximum value. We need to keep the factor in mind when we need to compare the accuracy.

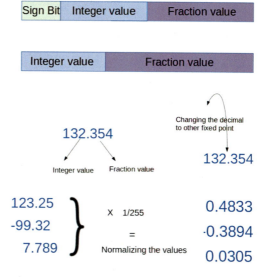

Fig. 32.3 Numerical representation in a fixed point

Fig. 32.4 A visualization of fixing the decimal in a fixed-point representation

Fig. 32.5 Normalizing the values of a 8-bit image

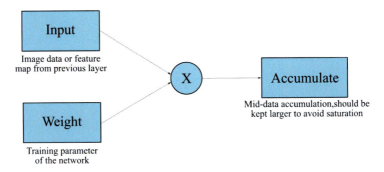

Fig. 32.6 A typical convolution operation in a CNN

32.4.2 Convolution Operation

In a CNN model, most of the operations are done by a convolution operation. The first layer has the maximum computation as it reduces the input image size to a large extent as an output feature map. In this work, the first layer of a CNN is targeted, intending to find the optimal bit-width to get the best results with the least accuracy and memory consumption.

Figure 32.6 shows a typical convolution operation, where input data is the input image fed to the block or an output feature map from the previous layer used as input for consecutive layers. The other input parameters are weight which is pre-trained parameter and is fetched from the model for individual layers, respectively. If biases are available they are also added after multiplying weights and input data and the resultant data is accumulated together for the next layer using the Eq. (32.4).

$$y = f\left(\sum_{i=1}^{n} w_{ij} \times x_i + b\right) \quad (32.4)$$

Here in this work, the quantification of the first convolution layer has been focused on. The same methodology could work for all the other convolutional layers accordingly with a similar analysis.

32.4.3 Work Done

Here, a pre-trained model is targeted, and its weights and other parameters are extracted from software implementation in ***Tensorflow***. The software implementation of the network deals with the full precision 32-bit floating-point data type. The optimization was intended to get an implementation for resource-constrained devices like FPGA boards. In order to fit the entire network, a thorough layer-wise analysis was gathered.

Table 32.1 Architecture of YOLO v3-tiny with all layers and parameters information

Layer	Type	Filters	Size/strides	Input	Output
0	Convolutional	16	$3 \times 3/1$	$416 \times 416 \times 3$	$416 \times 416 \times 16$
1	Maxpool		$2 \times 2/2$	$416 \times 416 \times 16$	$208 \times 208 \times 16$
2	Convolutional	32	$3 \times 3/1$	$208 \times 208 \times 32$	$208 \times 208 \times 32$
3	Maxpool		$2 \times 2/2$	$208 \times 208 \times 32$	$104 \times 104 \times 32$
4	Convolutional	64	$3 \times 3/1$	$104 \times 104 \times 64$	$104 \times 104 \times 64$
5	Maxpool		$2 \times 2/2$	$104 \times 104 \times 64$	$52 \times 52 \times 64$
6	Convolutional	128	$3 \times 3/1$	$52 \times 52 \times 64$	$52 \times 52 \times 128$
7	Maxpool		$2 \times 2/2$	$52 \times 52 \times 64$	$26 \times 26 \times 64$
8	Convolutional	256	$3 \times 3/1$	$26 \times 26 \times 64$	$26 \times 26 \times 256$
9	Maxpool		$2 \times 2/2$	$26 \times 26 \times 256$	$13 \times 13 \times 256$
10	Convolutional	512	$3 \times 3/1$	$13 \times 13 \times 256$	$13 \times 13 \times 512$
11	Maxpool		$2 \times 2/1$	$13 \times 13 \times 512$	$13 \times 13 \times 512$
12	Convolutional	1024	$3 \times 3/1$	$13 \times 13 \times 512$	$13 \times 13 \times 1024$
13	Convolutional	256	$1 \times 1/1$	$13 \times 13 \times 1024$	$13 \times 13 \times 256$
14	Convolutional	512	$3 \times 3/1$	$13 \times 13 \times 256$	$13 \times 13 \times 512$
15	Convolutional	256	$1 \times 1/1$	$13 \times 13 \times 512$	$13 \times 13 \times 256$
16	YOLO				
17	Route 13				
18	Convolutional	128	$1 \times 1/1$	$13 \times 13 \times 256$	$13 \times 13 \times 128$
19	Up-sampling		$2 \times 2/1$	$13 \times 13 \times 128$	$26 \times 26 \times 128$
20	Route 19,8				
21	Convolutional	256	$3 \times 3/1$	$13 \times 13 \times 384$	$13 \times 13 \times 256$
22	Convolutional	256	$1 \times 1/1$	$13 \times 13 \times 256$	$13 \times 13 \times 256$
23	YOLO				

The implementation of YOLO v3-tiny, an object detection model, was targeted here for hardware implementation. Table 32.1 illustrates the architecture of YOLO v3-tiny with all the layers and parameters size. The first layer has input image size $416 \times 416 \times 3$ is convolved with 16 filters of size 3×3, and stride 1 gives an output feature map of size $416 \times 416 \times 16$. With respect to quantization, input data, and weights are quantized to 16-bit values and stored as data files which are to be used in hardware-level programming for the architecture. For hardware-level programming, Vivado HLS [15] is used with C-level coding. In order to calculate the accuracy deviation, root-mean-square-error (RMSE) has been calculated at every stage to find the best-suited bit-width for the implementation.

32.5 Results and Discussion

We have implemented the first layer of convolution with variable bit-width for input data, mid-data, and result has been observed by comparing with golden output data. The variable bit-width selected here ranges depending on the use case. However, it should be kept in mind that the mid-data type is actually used to store accumulated value, which is usually followed after multiplication and addition. Therefore, a thumb rule is to double the size of input data.

32.5.1 Input-Output Width Varying with Mid-Data Width Constant

Here, the first condition under study is the analysis of varying input/output bit-width and keeping mid-data bit-width constant at <32,16>. The input/output width is varying from <16,8> to <8,4>. The observations have been tabulated and plotted in graph for visualization. As shown in Table 32.2, RMSE has drastically increased for lower bit-width data as the bit-width of I/O data is reduced. However, memory consumptions (BRAM, DSPs, and FFs) have reduced, but LUTs have increased after decreasing the I/O bit-width, respectively. Here, there is no effect in latency as the datapath of the design is not modified, only the input memory size. In Fig. 32.7, the graph is plotted with different Y-axis dimensions to incorporate all deviations in a single graph. For BRAM, DSPs, and RMSE, the graph is to be visualized with the right-hand and for FFs and LUTs with the left-hand Y-axis, respectively.

32.5.2 Input-Output Width Constant and Mid-Data Bit-Width Varies

The second condition for the analysis of varying mid-data bit-width and keeping input-output bit-width constant at <16,8>. The mid-data width is varied from <32,16> to <16,8>. The observations have been tabulated and plotted in the graph

Table 32.2 Table illustrating RMSE and resource consumption for I/O bit-width varying with mid-data bit-width constant table

I/O data bit-width	Mid data width	BRAM	DSPs	FFs	LUTs	Latency (ms)	RMSE
<16,8>	<32,16>	120	146	41620	28429	14.550	0.0469
<14,7>	<32,16>	120	146	36732	31757	14.550	17.472
<12,6>	<32,16>	120	146	31852	31581	14.550	25.194
<10,5>	<32,16>	108	2	27376	40662	14.550	45.194
<8,4>	<32,16>	96	2	18091	36966	14.550	1119.868

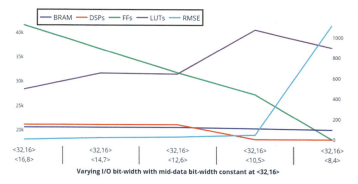

Fig. 32.7 Graph showing resource consumption and RMSE for input-output width varying with constant mid-data width respectively

Table 32.3 Table illustrating RMSE and resource consumption for constant input-output width and varying mid-data width

I/O data bit-width	Mid data width	BRAM	DSPs	FFs	LUTs	Latency (ms)	RMSE
<16,8>	<32,16>	120	146	41620	28429	14.550	0.0469
<16,8>	<28,14>	120	146	49557	33203	14.550	0.0465
<16,8>	<24,12>	120	146	48681	33035	14.550	0.0447
<16,8>	<20,10>	120	146	46997	32623	14.550	0.0368
<16,8>	<16,8>	120	146	46042	28987	14.550	0.0

for visualization. In Table 32.3, we can see as we reduce the bit width of mid-data, there are no changes in BRAM and DSPs however, some fluctuations in the FFs and LUTs are observed. The major observation is the RMSE value. We can observe here that it continues to maintain its value and for the lowest bit-width, it has a negligible error. From this analysis, we can conclude here that there is no harm in fine-tuning the bit-width combinations given some deviations in memory consumption. Here also, there is no effect in latency as the datapath of the design is not modified, only the input memory size. In Fig. 32.8, the graph is plotted with different Y-axis dimensions to incorporate all deviations in a single graph. For BRAM, DSPs, and RMSE the graph is to be visualized with the right-hand and for FFs and LUTs with the left-hand Y-axis respectively.

32.5.3 Both Data Type Width Varies

This section deals with the analysis of varying both I/O bit width and mid-data type bit width. The I/O data width is varied from <16,8> to <8,4> and the mid-data width is varied from <32,16> to <16,8>. Here, we have kept the mid-data width double of I/O width to avoid any saturation loss. The observations have been tabulated and

Fig. 32.8 Graph showing resource consumption and RMSE for constant input-output width and varying mid-data width respectively

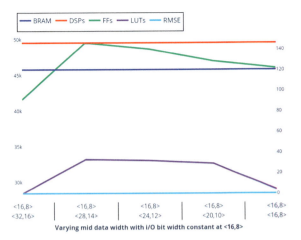

Table 32.4 Table illustrating RMSE and resource consumption for proportionally varying both input/output and mid-data and bit-width

I/O data bit-width	Mid data width	BRAM	DSPs	FFs	LUTs	Latency (ms)	RMSE
<16,8>	<32,16>	120	146	41620	28429	14.550	0.0469
<14,7>	<28,14>	120	146	36728	28637	14.550	17.472
<12,6>	<24,12>	120	146	31849	28053	14.550	25.194
<10,5>	<20,10>	108	2	24859	31257	14.550	47.313
<8,4>	<16,8>	96	2	20444	25735	14.550	1167.70

plotted in graph for visualization. In Table 32.4, we can see as we reduce the bit-width memory consumption also reduces; however, LUTs have spiked in between. But RMSE has constantly been increasing. Here again, there is no effect in latency as the datapath of the design is not modified, only the input memory size.

In Fig. 32.9, the graph is plotted with different Y-axis dimensions to incorporate all deviations in a single graph. For BRAM, DSPs, and RMSE, the graph is to be visualized with the right-hand and for FFs and LUTs with the left-hand Y-axis, respectively.

Based on the carried work, we can comprise the findings as it is safer to maintain the ratio of mid-data type and I/O data type in order to avoid any quantization loss or saturation loss. After comparing all the observations, we selected the optimum bit-widths combination as <12,6> and <24,12> for I/O and mid-data type respectively, as it has all the favorable parameter at the desired range.

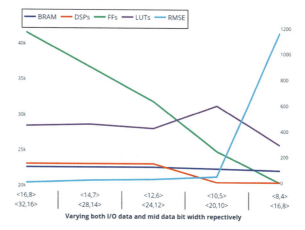

Fig. 32.9 Graph showing resource consumption and RMSE for varying both input/output and mid-data and bit-width respectively

32.6 Conclusion

Fixed-point-based implementation is required when end applications are intended to run on a resource-constrained device. It reduces the storage requirement, however with a comparable impact on the accuracy of the network. In this work, it has been observed that using mixed-precision bit-width has reduced the memory consumption for combination for I/O data and mid-data bit-width <10,5> and <20,10> has reduced the memory consumption for BRAM, DSPs, and FFs by 10%, 98.6%, and 40.27% respectively however the LUTs have increased by 9.9% with comparable accuracy loss. The other best mixed-precision combination <12,6> and <24,12> have reduced memory consumption for FFs and LUTs by 23.4% and 1.3% respectively and a lesser accuracy loss. Based on the above metrics and data, we can conclude that fine-tuning the data storage requirements could be achieved by restricting the data size in the design, and it has be an informed decision in terms of accuracy analysis and other resource utilization metrics.

In CNN, every individual layer performs a different level of feature extraction. In order to perform the efficient implementation of deeper networks onto resource-constraint devices, similar layer-wise analysis could be performed to find the most fitted precision for all layers respectively.

References

1. Krizhevsky, A., Sutskever, I., Hinton, G.E.: ImageNet classification with deep convolutional neural networks. In: Proceedings of the 25th International Conference on Neural Information Processing Systems, vol. 1 (NIPS12)pp. 1097–1105. Curran Associates Inc., Red Hook, NY, USA (2012)
2. Han, S., Pool, J., Tran, J., Dally, W.: Learning both weights and connections for efficient neural network. In: Advances in Neural Information Processing Systems, pp. 1135–1143 (2015)

3. Han, S., Mao, H., Dally, W.J.: Deep compression: compressing deep neural network with pruning, trained quantization and Huffman coding. Comput. Vis. Pattern Recogn. (2015)
4. Dally, W.: High-performance hardware for machine learning. In: Tutorial in Advances in Neural Information Processing Systems (2015)
5. Jacob, B., Kligys, S., Chen, B., Zhu, M., Tang, M., Howard, A., Adam, H., Kalenichenko, D.: Quantization and training of neural networks for efficient integer-arithmetic-only inference (2017)
6. Nagel, M., Fournarakis, M., Amjad, R.A., Bondarenko, Y., van Baalen, M., Blankevoort, T.: A white paper on neural network quantization. arXiv preprint arXiv:2106.08295 (2021)
7. Jacob, B., et al.: Quantization and training of neural networks for efficient integer-arithmetic-only inference. In: 2018 IEEE/CVF Conference on Computer Vision and Pattern Recognition, pp. 2704–2713. Salt Lake City, UT, USA (2018). https://doi.org/10.1109/CVPR.2018.00286
8. Choukroun, Y., et al.: Low-bit quantization of neural networks for efficient inference. In: 2019 IEEE/CVF International Conference on Computer Vision Workshop (ICCVW), pp. 3009–3018 (2019)
9. Wang, P., et al.: Unsupervised network quantization via fixed-point factorization. IEEE Trans. Neural Networks Learn. Syst. **32**, 2706–2720 (2020)
10. Yuan, Y., Chen, C., Hu, X., Peng, S.: Towards low-bit quantization of deep neural networks with limited data. In: 2020 25th International Conference on Pattern Recognition (ICPR), pp. 4377–4384 (2021)
11. Wu, B., Wang, Y., Zhang, P., Tian, Y.-D., Vajda, P., Keutzer, K.: Mixed precision quantization of convnets via differentiable neural architecture search. arXiv preprint arXiv:1812.00090 (2018)
12. Wu, H., Judd, P., Zhang, X., Isaev, M., Micikevicius, P.: Integer quantization for deep learning inference: principles and empirical evaluation. arXiv preprint arXiv:2004.09602 (2020)
13. Gholami, A., Kim, S., Zhen, D., Yao, Z., Mahoney, M., Keutzer, K.: A survey of quantization methods for efficient neural network inference (2022).https://doi.org/10.1201/9781003162810-13
14. He, Y., Huang, C.-W., Wei, X., Li, Z., Guo, B.: TF-YOLO: an improved incremental network for real-time object detection. Appl. Sci. **9**, 3225 (2019). https://doi.org/10.3390/app9163225
15. Vivado 2020.1—High-level synthesis (C based)—Xilinx. https://www.xilinx.com/support/documentation-navigation/design-hubs/dh0012-vivado-high-level-synthesis-hub.html

Chapter 33
Non-linear Fractional Order Fuzzy PD Plus I Controller for Trajectory Optimization of 6-DOF Modified Puma-560 Robotic Arm

Himanshu Varshney, Jyoti Yadav, and Himanshu Chhabra

Abstract The purpose of this research is to employ non-integer order calculus to enhance the control action of the non-linear fractional order fuzzy PD plus I (FOFPD + I) controller. To operate a non-linear 6-DOF Puma 560 robotic arm, a FOFPD + I controller is developed and implemented. Fractional order fuzzy proportional derivative (FOFPD) and fractional order fuzzy integral (FOFI) controllers are used to create the proposed controller. Because of the non-linear gains, the proposed control approach preserves the linear structure of the fractional order proportional derivative plus integral (FOPD + I) controller while still providing adaptive capabilities. Further, the PID controller is also derived to compare with the FOFPD + I controller. Both, FOFPD + I and PID controller's parameters are optimized using non-dominated sorting genetic algorithm-II (NSGA-II). The performance and effectiveness of the presented controller are examined in terms of trajectory tracking, tracking error, and robotic arm control efforts. Simulation shows excellent performance of FOFPD + I controller over traditional PID and fuzzy logic controllers. To be precise, on the current optimal solution, suggested controlling strategy is more than 80% efficient than other strategies.

Keywords FOFPD + I · Puma 560 · Robot dynamics · PID · NSGA-II

H. Varshney (✉)
Department of Mechanical Engineering, Indian Institute of Technology (Indian School of Mines), Dhanbad, Jharkhand, India
e-mail: 21dr0061@mech.iitism.ac.in

J. Yadav
Department of Instrumentation and Control Engineering, Netaji Subhas University of Technology, New Delhi, India

H. Chhabra
Department of Mechatronics, Parul Institute of Technology, Parul University, Vadodara, Gujrat, India

33.1 Introduction

Robotic manipulators are electronic mechanisms that are composed of numerous segments, which interact with their surroundings to carry out tasks. Due to their advanced programming, they can perform repetitive tasks at high speeds and with a level of precision that surpasses human capabilities [1]. Nowadays, robotic systems are vastly used in many industries such as the medical industry, automobile industry, space research organizations, power plants, and so on, for accurate position tracking, pick and place, palletizing, handling, painting and assembly, etc. [2]. Automating with robotic manipulators can enhance manufacturing processes by improving efficiency, reliability, and productivity. Therefore, a significant amount of attention has been directed toward creating models for robotic manipulators and developing practical controllers that are simple to implement and produce optimal performance [3–5]. Only with tight position control, the effective motion in the robotic arm can be achieved. PID controllers are broadly employed because of its simplicity, economical, and easy implementation. But PID controllers are not capable in controlling of non-linear systems. To rectify these drawbacks, various types of adaptive PID controllers with auto-tuning capability are available in [6–8].

In contemporary control research, the concept of fractional order has garnered considerable attention. Fractional order modeling and control, which employs fractional order derivatives and integrals, has been acknowledged as a viable approach to effectively tackle numerous robust control issues [9, 10]. Das et al. showed a comparative analysis of PID, fuzzy PID (FPID), fractional order PID (FOPID), fractional order fuzzy PID (FOFPID), and controller for a non-linear unstable process with delay in [11]. A study showed that the FOFPID controller is more robust and effective. Chhabra et al. designed a FOFPD + I controller for a two-link planar robotic arm [12]. The parameters of controllers are optimized by NSGA-II. Results showed the effectiveness of the proposed controllers over the PID controller. Furthermore, fractional order operators provide additional parameters to the controller, increasing the controller's design complexity. High control performance of these complex controllers can only be achieved by properly optimizing these controllers' parameters.

Most of the controllers are not self-tunable and are based on specialists' knowledge rather than precise mathematics. Hence, in this paper, a FOFPD + I controller is built, employing accurate mathematical equations while also keeping in mind the linear structure property of the PID controller. Controller gains depend on error and error rate and are also non-linear in nature, which makes the controller self-tunable. The NSGA-II is utilized to optimize FOFPD + I and PID controller's parameters by reducing the Integral Absolute Error (IAE) of the considered objective functions. The designed controller's control performance is analyzed for trajectory tracking and control efforts. This paper is divided into the following sections: After a brief introduction and literature review in Sects. 33.1 and 33.2 delves deeper into the formulation and mathematical modeling of the robotic arm. The mathematical analysis and implementation of the FOFPD + I controller are covered in Sect. 33.3.

The NSGA-II optimization technique is described in Sect. 33.4. Trajectory tracking performance has been analyzed in Sect. 33.5. Section 33.6 show the conclusion of this research and the future scope for the related research, respectively. In the end, references are mentioned.

33.2 Mathematical Modeling of Robotic Arm

Figure 33.1 shows the schematic diagram of the 6-DOF Puma 560 robotic arm. The Puma 560 robotic arm consists of six links ($i = 1, 2, \ldots, 6$) having mass m_i with their center of mass located at $r_i = (r_{i_x}, r_{i_y}, r_{i_z})$ and inertia tensor $I_i = (I_{i_{xx}}, I_{i_{yy}}, I_{i_{zz}}, I_{i_{xy}}, I_{i_{yz}}, I_{i_{zx}})$ respectively, as suggested in [13]. The kinematic modeling of a robotic arm is done based on modified D-H (Denavit-Hartenberg) parameters which include joint length (a), joint twist (α), joint offset (d), and joint variable (θ). These D-H parameters and the model's physical parameters are available in [14].

33.2.1 Kinematic Modeling

Kinematic modeling is grounded by the D-H parameters of links to define the position of links and the end effector. By using the general transformation matrix, we can easily calculate the actual transformation matrices ($^{i-1}T_i$) according to D-H method for ($i = 1, 2, \ldots, 6$). By using these transformation matrices, we can easily formulate the dynamic model of the Puma 560 robotic arm.

33.2.2 Dynamic Modeling

Dynamic modeling is done by using the Lagrange–Euler formulation. It is based on the principle of conservation of energy. For a robotic manipulator, the generalized Lagrangian operator (L) is expressed by,

$$L = \frac{1}{2} \sum_{i=1}^{n} \sum_{j=1}^{i} \sum_{k=1}^{i} \text{Tr}\left[(^0T_{j-1} Q_j^{j-1} T_i) I_i (^0T_{k-1} Q_k^{k-1} T_i)^T \right] \dot{q}_j \dot{q}_k + \sum_{k=1}^{i} m_i g\, ^0T_i^i \bar{r}_i \tag{33.1}$$

where Tr operator gives trace of the matrix, $\dot{q}$ is the joint velocity.

For the revolute joint,

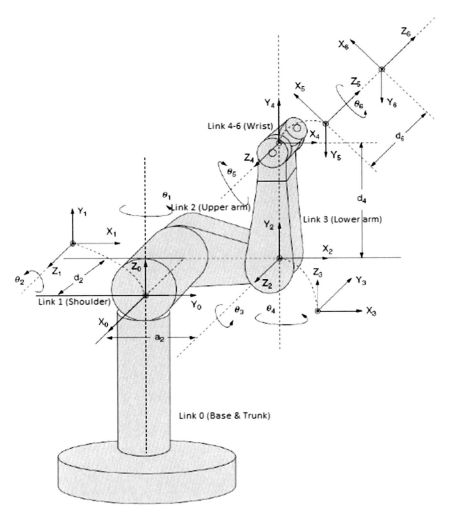

Fig. 33.1 Schematic diagram of 6-DOF Puma 560 robotic arm [15]

$$Q_j = \begin{bmatrix} 0 & -1 & 0 & 0 \\ 1 & 0 & 0 & 0 \\ 0 & 0 & 0 & 0 \\ 0 & 0 & 0 & 0 \end{bmatrix} \qquad (33.2)$$

So, according to the Lagrange–Euler formulation,

$$\tau_i = \frac{\mathrm{d}}{\mathrm{d}t}\left(\frac{\partial L}{\partial \dot{q}_i}\right) - \frac{\partial L}{\partial q_i} \qquad (33.3)$$

This leads to a general dynamic equation of a system, which is given by

$$\tau = M(\theta)_{n \times n} \ddot{\theta} + C(\theta\dot{\theta})_{n \times n} \dot{\theta} + G(\theta)_{n \times 1} \qquad (33.4)$$

where $M(\theta)$ is the inertia matrix, $C(\theta\dot{\theta})$ is the velocity coupling matrix which consists of both centripetal and coriolis components, and $G(\theta)$ is the gravity loading matrix, $q = \theta$ is used because all joints are revolute joints.

In the case of a robotic manipulator, (33.4) can be further expanded to a more generalized version, which is given by

$$\tau_i = \sum_{j=1}^{n} M_{ij}(q) \ddot{q}_j + \sum_{j=1}^{n}\sum_{k=1}^{n} h_{ijk} \dot{q}_j \dot{q}_k + G_i(q_i) \qquad (33.5)$$

for $i = 1, 2, 3, \ldots, 6$ for 6-DOF Puma 560 where

$$M_{ij} = \sum_{p=\max(i,j)}^{n} \mathrm{Tr}\left[d_{pj} I_p d_{pi}^T \right] \qquad (33.6)$$

$$h_{ijk} = \sum_{p=\max(i,j,k)}^{n} \mathrm{Tr}\left[\frac{\partial d_{pj}}{\partial q_p} I_p d_{pi}^T \right] \qquad (33.7)$$

$$G_i = -\sum_{p=i}^{n} m_p g d_{pi} \, {}^p \bar{r}_p \qquad (33.8)$$

$$d_{ij} = \begin{cases} {}^0T_{j-1} Q_j {}^{j-1} T_i & \text{if } j \leq i \\ 0 & \text{if } j > i \end{cases} \qquad (33.9)$$

$$\frac{\partial d_{ij}}{\partial q_k} = \begin{cases} {}^0T_{j-1} Q_j {}^{j-1} T_{k-1} Q_k {}^{k-1} T_i & \text{if } i \geq k \geq j \\ {}^0T_{k-1} Q_k {}^{k-1} T_{j-1} Q_j {}^{j-1} T_i & \text{if } i \geq j \geq k \\ 0 & \text{if } i < k \end{cases} \qquad (33.10)$$

After solving (33.5) with the help of (33.2), (33.6)–(33.10), the dynamics of 6-DOF Puma 560 can be described as (33.4). A complete description of all the matrices can be found by following the given algorithm as in [16].

33.3 Controller Design

Figure 33.2 depicts the $PI^\lambda D^\lambda$ controller, from which the FOFPD + I controller is derived. FOFPD + I controller is a combination of a FOFPD and a FOFI controller, both of which are derived from conventional PD^λ and I^λ controllers.

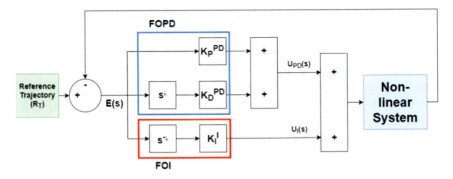

Fig. 33.2 FOPD + I controller block diagram [12]

33.3.1 FOFPD + I Controller

The control law of the FOFPD + I controller can be obtained by adding the FOFPD and FOFI control laws algebraically.

The FOFPD controller's control law can be expressed as,

$$\grave{u}_{PD}(n) = K_P e(n) + K_D r(n) \tag{33.11}$$

where $K_P = K_P^{PD} T^\lambda$, $K_D = K_D^{PD} T^\lambda$, $r(n) = D^\lambda e(n)$, $n = nT$ and

$$u_{PD}(n) = T^{-\lambda} \grave{u}_{PD}(n) = K_{uPD} \grave{u}_{PD}(n) \tag{33.12}$$

The FOFI controller's control law can be expressed as,

$$\Delta u_I(n) = K_I e(n) + K r(n) \tag{33.13}$$

where $K_I = K_I^I T^\lambda$, $r(n) = D^\lambda e(n)$, $n = nT$ and

$$u_I(n) = T^{-\lambda} D^{-\lambda} \Delta u_I(n) = K_{uI} D^{-\lambda} \Delta u_I(n) \tag{33.14}$$

The control law of the FOFPD + I controller will be

$$u_{\text{FOFPD}+I}(n) = u_{PD}(n) + u_I(n) = K_{uPD} \grave{u}_{PD}(n) + K_{uI} D^{-\lambda} \Delta u_I(n) \tag{33.15}$$

The FOFPD + I controller's action is expressed by (33.15). The fundamental implementation of the FOFPD + I controller is depicted in Fig. 33.3 as a block diagram.

33 Non-linear Fractional Order Fuzzy PD Plus I Controller for Trajectory ...

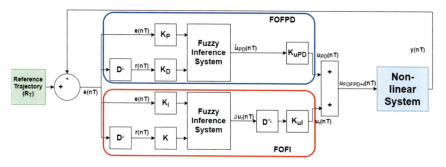

Fig. 33.3 Typical layout of FOFPD + I controller for non-linear system [12]

33.3.2 Fuzzy Inference System (FIS)

The basic procedure for designing a FOFPD + I controller consists of three steps: (i) fuzzification, (ii) design of the control rule base, and (iii) defuzzification.

Fuzzification. There are two phases in the fuzzification of the FOFPD + I controller. First, the FOFPD and FOFI controller components are fuzzified separately, and the two components are then merged to create the required control action. An error signal $e(n)$ and a fractional rate of error signal $r(n)$ are both inputs to the FOFPD + I controller, and an output $u_{FOFPD+I}(n)$. Figure 33.4 shows the input and output membership functions of both FOFPD and FOFI controllers, where the parameter L is greater than zero.

Control rule base design. The FOFPD controller is subject to the following control rules.

(R1): If $e(nT) = e.p$ AND $r(nT) = r.n$ THEN FOFPD-output $= o.p$	(R2): If $e(nT) = e.n$ AND $r(nT) = r.p$ THEN FOFPD-output $= o.n$
(R3): If $e(nT) = e.n$ AND $r(nT) = r.n$ THEN FOFPD-output $= o.z$	(R4): If $e(nT) = e.p$ AND $r(nT) = r.p$ THEN FOFPD-output $= o.z$

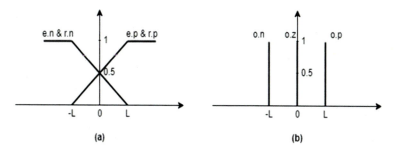

Fig. 33.4 a Membership functions for both inputs and **b** membership functions for output for both FOFPD and FOFI controller

where $e(n) = (R_T - y)$ denotes error, $r(n)$ denotes fractional error rate of order λ, "FOFPD-output" denotes FOFPD output $\dot{u}_{PD}(n)$, "$e.p$" denotes "positive error," "$r.n$" denotes "negative fractional error rate," and "$o.z$" denotes "zero output."

The following rules, on the other hand, are outlined for FOFI controllers.

(R5): If $e(nT) = e.p$ AND $r(nT) = r.n$ THEN FOFI-output $= o.z$	(R6): If $e(nT) = e.n$ AND $r(nT) = r.p$ THEN FOFI-output $= o.z$
(R7): If $e(nT) = e.n$ AND $r(nT) = r.n$ THEN FOFI-output $= o.n$	(R8): If $e(nT) = e.p$ AND $r(nT) = r.p$ THEN FOFI-output $= o.p$

Here, FOFI output $\Delta u_I(n)$ is referred by "FOFI-output."

Rule R1 will be triggered along with R5, rule R2 will be triggered along with R6, and so on, because both FOFPD and FOFI controllers have the same inputs. As a result, the controller must continue to increase the system output. The upward movement of the FOFPD controller can be maintained in rule R1 by maintaining positive output. The rest of the rules are obtained in the same way.

Defuzzification. We employed the "center of mass" technique for the defuzzification of output from FOFPD and FOFI controllers.

$$\Delta u(n) = \frac{\sum (\text{input membership value} * \text{output membership value})}{\sum \text{input membership value}} \quad (33.16)$$

The inputs of the proposed controller's FOFPD and FOFI components, error, and error rate are separated into 20 adjacent input combination (IC) regions. Figure 33.5 shows the regions in which the input error signal is displayed against the input error rate. The suitable control laws FOFPD are evaluated using rules (R1)–(R4) and IC regions.

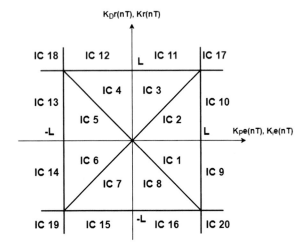

Fig. 33.5 IC regions for both FOFPD and FOFI controller

In IC3 region, both the inputs, i.e., error and error rate, lie in the range [0, L], and hence, according to Fig. 4a, condition $e.p > 0.5 > r.n$ is satisfied. Now, Zadeh's logical "&" is used to get the minimum [17].

R1:{"error = $e.p$ & rate = $r.n$"} = min {$e.p, r.n$} = $r.n$, yields.

(R1)	membership value at input – $r.n$ membership value at output – $o.p$	(R2)	membership value at input – $e.n$ membership value at output – $o.n$
(R3)	membership value at input – $r.n$ membership value at output – $o.z$	(R4)	membership value at input – $e.p$ membership value at output – $o.z$

Using the above rules and geometry in Fig. 33.5, (33.16) can be used to write defuzzified FOFPD controller output in region IC3 as

$$\grave{u}_{PD}(n) = \frac{e.p * o.z + r.n * o.p + r.n * o.z + e.n * o.n}{e.p + r.n + r.n + e.n} \quad (33.17)$$

where the value of $o.p = L$, $o.z = 0$, and $o.n = -L$.

The controller parameters are adaptive and self-tunable in nature which is a function of the signal $|r(n)|$. The FOFPD controller's control law for all IC regions is derived as

$$\grave{u}_{PD}(n) = \begin{cases} \frac{L(K_P e(n) + K_D r(n))}{2(2L - K_P |e(n)|)} & \text{for IC 1,2,5,6} \\ \frac{L(K_P e(n) + K_D r(n))}{2(2L - K_D |r(n)|)} & \text{for IC 3,4,7,8} \\ \frac{1}{2}(-K_D r(n) + L) & \text{for IC 9,10} \\ \frac{1}{2}(K_P e(n) - L) & \text{for IC 11,12} \\ \frac{1}{2}(-K_D r(n) - L) & \text{for IC 13,14} \\ \frac{1}{2}(K_P e(n) + L) & \text{for IC 15,16} \\ 0 & \text{for IC 17,19} \\ -L & \text{for IC 18} \\ L & \text{for IC 20} \end{cases} \quad (33.18)$$

Similarly, for the FOFI controller, the control law for all IC regions is derived as,

$$\Delta u_1(n) = \begin{cases} \frac{L(K_I e(n) + K r(n))}{2(2L - K_I |e(n)|)} & \text{for IC 1,2,5,6} \\ \frac{L(K_I e(n) + K r(n))}{2(2L - K |r(n)|)} & \text{for IC 3,4,7,8} \\ \frac{1}{2}(K r(n) + L) & \text{for IC 9,10} \\ \frac{1}{2}(K_I e(n) + L) & \text{for IC 11,12} \\ \frac{1}{2}(K r(n) - L) & \text{for IC 13,14} \\ \frac{1}{2}(K_I e(n) + L) & \text{for IC 15,16} \\ 0 & \text{for IC 17,19} \\ -L & \text{for IC 18} \\ L & \text{for IC 20} \end{cases} \quad (33.19)$$

33.4 NSGA-II Optimization Technique

NSGA-II is a multi-objective optimization (MOO) technique that can achieve a reasonable trade-off between several objectives that are often in conflict with one another. Pareto-optimal front, which indicates the optimum solutions for trade-offs between the conflicting objective functions, are used to explain the solution to a multi-objective problem.

$$w_1 = \sum |e_1(n)| + \sum |e_2(n)| + \sum |e_3(n)| + \sum |e_4(n)| \\ + \sum |e_5(n)| + \sum |e_6(n)| \tag{33.20}$$

$$w_2 = \sum |\tau_1(n+1) - \tau_1(n)| + \sum |\tau_2(n+1) - \tau_2(n)| \\ + \sum |\tau_3(n+1) - \tau_3(n)| + \sum |\tau_4(n+1) - \tau_4(n)| \\ + \sum |\tau_5(n+1) - \tau_5(n)| + \sum |\tau_6(n+1) - \tau_6(n)| \tag{33.21}$$

where $e_i(n)$ and $\tau_i(n)$ are the error signal and control effort, respectively, of the ith link. The goal is to find the optimum controller parameters while minimizing the conflicting objectives w_1 and w_2. At the true Pareto-optimal front, NSGA-II gives a wider range of solutions and also has greater convergence capability. NSGA-II overcomes all of NSGA's flaws, such as the highly complex computation of non-dominated sorting, the need of defining the sharing parameter, and the lack of elitism [18]. Hence, in this research, NSGA-II is employed for all optimization tasks. The algorithm is statistically compared, and for a fair comparison, the algorithm is run for 5 independent runs with 25,000 objective function evaluations. The number of populations considered is 50 and the number of generations is 500.

33.5 Results and Discussion

Non-linearity, vagueness, and other factors increase the complexity of a robotic arm in various working environments. As a result, the controller should be capable of providing tight position control under both favorable and unfavorable conditions. A thorough simulation analysis is carried out to evaluate the robustness of FOFPD + I and PID controllers in trajectory tracking. NSGA-II is used to find the optimum controller parameters. Did all the simulations in MATLAB and Simulink on an Intel core i3-4th generation 1.7 GHz processor backed with 8 Gigabytes of RAM. 4th order Runge–Kutta algebraic solver with 1 m-s sample time is utilized in simulations. The variation in joint torque is constrained within $[-70, 70]$ Nm, and the reference trajectory considered is a quintic polynomial for all the joints as shown below

33 Non-linear Fractional Order Fuzzy PD Plus I Controller for Trajectory ...

$$R_{T_i}(t) = a_{0_i} + a_{1_i}t + a_{2_i}t^2 + a_{3_i}t^3 + a_{4_i}t^4 + a_{5_i}t^5 \tag{33.22}$$

with a boundary as,

$$[R_{T_i}|_{t=0\,s}, \dot{R}_{T_i}|_{t=0\,s}, R_{T_i}|_{t=1\,s}, \dot{R}_{T_i}|_{t=1\,s}] \tag{33.23}$$

where $i = 1, 2, \ldots, 6$ for all the links and R_{T_i} denotes the current desired position of the ith link. Reference trajectories have the following boundary conditions for all the links,

Link 1: [0, 0, $pi/6$, 0]	Link 2: $\left[0, 0, \frac{-\pi}{4}, 0\right]$	Link 3: $[0, 0, \pi, 0]$
Link 4: $\left[0, 0, \frac{\pi}{3}, 0\right]$	Link 5: $\left[0, 0, \frac{\pi}{4}, 0\right]$	Link 6: $\left[0, 0, \frac{\pi}{2}, 0\right]$

The optimum controller parameters are found in the range [0 500] for K_{P_i}, K_{I_i}, K_{D_i}, K_i, K_{uPD_i} and K_{uI_i} and [0 1] for the fractional operator λ_i. For all FIS, the value of L is taken as 1. Fuzzy input and output gains will do all the required scaling. Figure 33.6 shows FOFPD + I controller implementation on 6-DOF Puma 560 robotic arm in which the FOFPD + I subsystem is implemented according to Fig. 33.3. Figure 33.7 shows the Pareto fronts of the FOFPD + I and PID controller which gives us the optimum controller parameter values. Figures 33.8, 33.9, and 33.10 show the suggested controller in action on the 6-DOF Puma 560 robotic arm. Figure 33.11 shows the locus of end effector X, Y, and Z coordinates in the 3D workspace. Table 33.1 shows the optimum parameters found by NSGA-II for the FOFPD + I and PID controllers. Table 33.2 shows the individual IAE of all 6 joints of the 6-DOF Puma 560 robotic arm in the case of FOFPD + I and PID controllers.

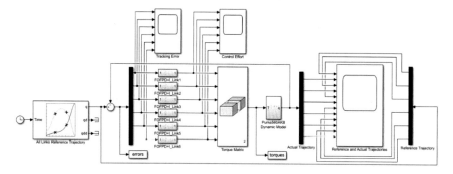

Fig. 33.6 Simulink implementation of FOFPD + I controller on modified 560 robotic arm

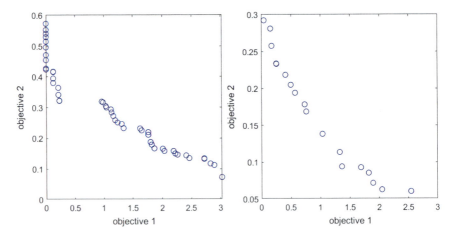

Fig. 33.7 FOFPD + I (left) and PID (right) controller Pareto front

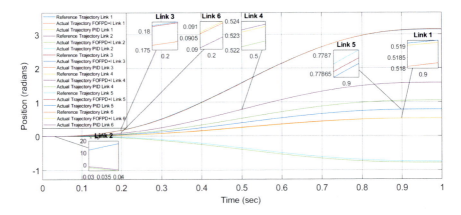

Fig. 33.8 Tracking performance (Trajectories)

The FOFPD + I and PID controllers for the Puma 560 robotic arm were designed in this research work. NSGA-II is used to optimize controller parameters. The objective functions are the sum of the IAE of tracking error and the sum of the IAE of change in controller output for all 6 links, and the controller's performance is analyzed for the tracking task. According to the literature, the maximum error can be seen as more than 0.08 radians in time 0 to 1 s which is for joint 2. While in this research, the maximum error came out to be less than 0.05 radians in the case of the PID controller and less than 0.002 radians in the case of the FOFPD + I controller in time 0 to 1 s. The result shows that the FOFPD + I controller outperforms other controllers.

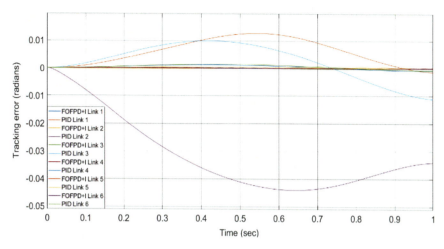

Fig. 33.9 Tracking error

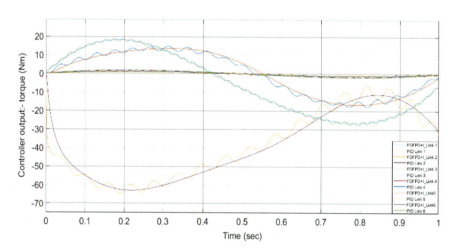

Fig. 33.10 Control effort

33.6 Conclusion and Future Scope

The present work proposes a non-linear FOFPD + I controller for 6-DOF modified Puma 560 robotic arm. The major goal of this work is to provide precise position control for a robotic manipulator while minimizing IAE and control effort. The NSGA-II multi-objective optimization technique is used to optimize the parameters of the controller. Robust analysis of the designed controller is carried out for trajectory tracking, change in the boundary condition, uncertainty in model parameters, and disturbance rejection. Simulation results demonstrate that the IAE of tracking error in

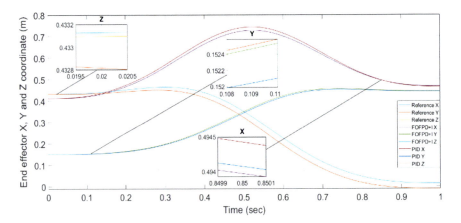

Fig. 33.11 Locus of end effector X, Y, and Z coordinates

Table 33.1 Optimum parameters of designed controllers

Controller parameters	FOFPD + I	PID
K_{P_1}	500	40.5519
K_{D_1}	0.01	387.637
K_{I_1}	291.21	89.3734
K_1	219.443	–
K_{uPD_1}	454.457	–
K_{uI_1}	8.69639	–
λ_1	0.77679	–
K_{P_2}	500	500
K_{D_2}	2.17476	496.202
K_{I_2}	207.467	498.709
K_2	236.796	–
K_{uPD_2}	414.381	–
K_{uI_2}	78.5723	–
λ_2	0.01	–
K_{P_3}	500	56.4563
K_{D_3}	0.01	492.733
K_{I_3}	216.418	32.2016
K_3	6.6071	–
K_{uPD_3}	500	–
K_{uI_3}	23.2991	–
λ_3	0.878869	–

(continued)

Table 33.1 (continued)

Controller parameters	FOFPD + I	PID
K_{P_4}	382.569	360.818
K_{D_4}	0.01	234.646
K_{I_4}	100.509	94.7102
K_4	361.699	–
K_{uPD_4}	71.0559	–
K_{uI_4}	15.4754	–
λ_4	0.94862	–
K_{P_5}	316.777	44.359
K_{D_5}	0.01	500
K_{I_5}	0.01	109.903
K_5	0.228051	–
K_{uPD_5}	270.635	–
K_{uI_5}	0.01	–
λ_5	0.182211	–
K_{P_6}	236.227	318.727
K_{D_6}	0.01	354.341
K_{I_6}	72.9148	67.1595
K_6	188.189	–
K_{uPD_6}	336.879	–
K_{uI_6}	0.01	–
λ_6	0.105482	–
IAE of error	0.000671684	0.045133

Table 33.2 Individual IAE of all the 6 joints of the 6-DOF Puma 560 robotic arm

Joint	FOFPD + I (10^{-3})	PID (10^{-3})
1	0.1216	6.1776
2	0.2048	31.3486
3	0.1384	5.9213
4	0.1423	0.6622
5	0.0190	0.2806
6	0.0455	0.7427

the case of the FOFPD + I controller is very less than that of the PID controller. This shows the excellent performance of the FOFPD + I controller over traditional PID and fuzzy logic controllers. To be precise, on the current optimal solution, the FOFPD + I controller is more than 80% efficient than PID and fuzzy logic controllers [19]. The simulation result demonstrates the robustness and effectiveness of the designed controller outperforming other controllers for a robotic manipulator.

Further in future research, different fuzzification strategies for the fuzzy inference system can be used. Along with this, the performance of the suggested controller may be further enhanced by increasing input and output membership functions. Besides NSGA-II, other optimization techniques like, the Spider Monkey optimization technique, etc., can be used.

References

1. Wu, X., Huang, Y.: Adaptive fractional-order non-singular terminal sliding mode control based on fuzzy wavelet neural networks for omnidirectional mobile robot manipulator. ISA Trans. **121**, 258–267 (2022)
2. Xie, Y., Zhang, X., Meng, W., Zheng, S., Jiang, L., Meng, J., Wang, S.: Coupled fractional-order sliding mode control and obstacle avoidance of a four-wheeled steerable mobile robot. ISA Trans. **108**, 282–294 (2021)
3. Wang, Y., Yan, F., Jiang, S., Chen, B.: Time delay control of cable-driven manipulators with adaptive fractional-order nonsingular terminal sliding mode. Adv. Eng. Softw. **121**, 13–25 (2018)
4. Sharma, R., Gaur, P., Mittal, A.P.: Performance analysis of two-degree of freedom fractional order PID controllers for robotic manipulator with payload. ISA Trans. **58**, 279–291 (2015)
5. Sharma, R., Rana, K.P.S., Kumar, V.: Performance analysis of fractional order fuzzy PID controllers applied to a robotic manipulator. Expert Syst. Appl. **41**(9), 4274–4289 (2014)
6. Åström, K.J., Hang, C.C., Persson, P., Ho, W.K.: Towards intelligent PID control. Automatica (Oxf.) **28**(1), 1–9 (1992)
7. Åström, K.J., Hägglund, T.: The future of PID control. Control Eng. Pract. **9**(11), 1163–1175 (2001)
8. Åström, K.J., Hägglund, T.: Advanced PID Control. ISA-The Instrumentation Systems, and Automation Society, Research Triangle Park (2006)
9. Shah, P., Agashe, S.: Review of fractional PID controller. Mechatronics **38**, 29–41 (2016)
10. Bingi, K., Ibrahim, R., Karsiti, M.N., Hassan, S.M., Harindran, V.R.: Fractional-Order Systems and PID Controllers, vol. 264 (2020)
11. Das, S., Pan, I., Das, S., Gupta, A.: A novel fractional order fuzzy PID controller and its optimal time domain tuning based on integral performance indices. Eng. Appl. Artif. Intell. **25**(2), 430–442 (2012)
12. Chhabra, H., Mohan, V., Rani, A., Singh, V.: Robust nonlinear fractional order fuzzy PD plus fuzzy I controller applied to robotic manipulator. Neural Comput. Appl. **32**(7), 2055–2079 (2020)
13. Craig, J.J.: Introduction to Robotics: Mechanics and Control. Pearson, Upper Saddle River, NJ (2005)
14. Corke, P.I., Armstrong-Helouvry, B.: A search for consensus among model parameters reported for the PUMA 560 robot. In: Proceedings of the 1994 IEEE International Conference on Robotics and Automation, pp. 1608–1613. IEEE Computer Society Press (1994)
15. Piltan, F., Emamzadeh, S., Hivand, Z., Shahriyari, F., Mirzaei, M., Apr, D., Ave, T.: PUMA-560 robot manipulator position sliding mode control methods using MATLAB/SIMULINK and their integration into graduate/undergraduate nonlinear. researchgate.net. **6**, 107 (2012)
16. Armstrong, B., Khatib, O., Burdick, J.: The explicit dynamic model and inertial parameters of the PUMA 560 arm. In: Proceedings. 1986 IEEE International Conference on Robotics and Automation, pp. 510–518. Institute of Electrical and Electronics Engineers (1986)
17. Zadeh, L.A.: Fuzzy sets. Inf. Control **8**(3), 338–353 (1965)

18. Deb, K., Pratap, A., Agarwal, S., Meyarivan, T.: A fast and elitist multiobjective genetic algorithm: NSGA-II. IEEE Trans. Evol. Comput. **6**(2), 182–197 (2002)
19. Abdel-Salam, A.-A.S., Jleta, I.N.: Fuzzy logic controller design for PUMA 560 robot manipulator. IAES Int. J. Robot. Autom. (IJRA) **9**(2), 73 (2020)

Chapter 34
Solving the Capacitated Vehicle Routing Problem (CVRP) Using Clustering and Meta-heuristic Algorithm

Mohit Kumar Kakkar, Gourav Gupta, Neha Garg, and Jajji Singla

Abstract The capacitated vehicle routing problem (CVRP) is a significant topic in distribution networks. CVRP is essentially a subset of the vehicle routing problem (VRP). As CVRP has many applications in transport, logistics, and telecommunications, exact methods are not properly suitable for finding the optimal solution to large-scale CVRP problems, so most of the researchers are focusing on meta-heuristics like genetic algorithms and ant colony algorithms. As the CVRP is an NP-hard problem, it means that an efficient algorithm for solving the problem optimally is unavailable. In this paper, approximate optimal solutions to the CVRP are generated using a two-phase method with the modified genetic algorithm. Phase one includes clustering, and the second phase is based on the genetic algorithm, which is a meta-heuristic algorithm used for finding the best solution. The proposed method is tested on a set of benchmark instances from the literature. We report computational results with this meta-heuristic algorithm on some instances taken from the literature.

Keywords K-means clustering · Genetic algorithm · Capacitated vehicle routing problem (CVRP)

M. K. Kakkar · G. Gupta · N. Garg (✉) · J. Singla
Chitkara University Institute of Engineering and Technology, Rajpura, Punjab, India
e-mail: nehagarg7883@gmail.com

M. K. Kakkar
e-mail: Mohit.kakkar@chitkara.edu.in

G. Gupta
e-mail: gourav.gupta@chitkara.edu.in

J. Singla
e-mail: jajjisingla.js@gmail.com

© The Author(s), under exclusive license to Springer Nature Singapore Pte Ltd. 2024
P. K. Jha et al. (eds.), *Proceedings of Congress on Control, Robotics, and Mechatronics*,
Smart Innovation, Systems and Technologies 364,
https://doi.org/10.1007/978-981-99-5180-2_34

34.1 Introduction

VRP stands for "vehicle routing problem." It is a type of optimization problem in which a set of vehicles with limited capacity needs to visit a set of locations to perform a certain task while minimizing the total cost or time required to complete the task. The aim of VRP is to find an optimal or near-optimal routing plan that assigns each vehicle to a set of locations and determines the order in which they should be visited. This problem becomes more complex as the number of vehicles and locations increases and as additional constraints, such as time windows and vehicle load balancing, are added. Various algorithms and techniques have been developed to solve VRP, including heuristics, meta-heuristics, and exact methods. CVRP, on the other hand, stands for capacitated vehicle routing problem. It is a specific type of VRP where each vehicle has a fixed capacity and the total demand of the locations visited by each vehicle cannot exceed that capacity. This work presents an alternative procedure to solve the capacitated vehicle routing problem. As it is a non-polynomial type problem (NP-Hard), the method of a genetic algorithm is used as a meta-heuristic to solve it. The range of uses that CVRP offers in fields including transportation, networking, logistics, railways, marine, and the military, among others, makes it a topic of tremendous interest. The vehicle routing problem (VRP), by Reong et al. [1], is defined as an optimization problem in the field of operational research and logistics. The capacitated vehicle routing problem (CVRP) is an NP-hard problem which examines the solutions in the form of routes for various vehicles starting from one initial depot to a set of customers and returning back to same depot under capacity constraints of each vehicle. This problem includes minimization of total cost of the combined routes for a fleet of vehicles. Since cost is closely related to distance, the objective is to minimize the distance traveled by a fleet of vehicles with varying constraints. The cost of CVRP is determined by the distance between any two points, where the depot-customer locations, the volume of requests, and the vehicle capacity are all known in advance.

This paper proposes a novel approach, clustering-based novel genetic algorithm (CNGA), for solving the capacitated vehicle routing problem (CVRP) by combining clustering and meta-heuristic techniques. The proposed method first clusters the customers based on their geographical proximity and demand and then applies a meta-heuristic algorithm to find the optimal routes for the vehicles. It is a two-phase method. First phase is based on the implementation of the angle-based clustering, and then, the modified genetic algorithm is used to provide a solution to the optimal route of each vehicle within each cluster.

34.2 Literature Review

In the literature, three different approaches to the solution can be distinguished for CVRP: the conventional, heuristic, and meta-heuristic approaches. Standard operational research techniques such as Branch and Bound or Branch and Cut can be used to solve CVRP problems with a small number of nodes, proposed by Naddef et al. [2] and Hosoda et al. [3]. The heuristic approach involves developing algorithms that are designed to quickly generate high-quality solutions, but without guaranteeing that the solution is optimal. Heuristics exploit the problem structure to find near-optimal solutions, often by iteratively improving a solution as suggested by Otoo et al. [4]. Heuristic algorithms are faster than conventional methods and can handle larger instances of the problem, but they may not always find the best solution. As proposed by Garg et al. [5] and Rani et al. [6], meta-heuristic algorithms include techniques such as tabu search, simulated annealing, genetic algorithms, particle swarm algorithm, and ant colony optimization.

In CVRP, a fleet of vehicles with uniform capacities must meet a group of customers' demands through a series of routes that start and end at a shared warehouse or depot. In recent years, however, so-called "meta-heuristic" approaches have emerged that exploits global search strategies for the exploration of the search space. The term "meta-heuristic" is defined as "A meta-heuristic is a general problem-solving strategy or framework that provides a set of guidelines or principles for developing heuristic optimization algorithms." As discussed by Nanda et al. [7], meta-heuristics are characterized by their ability to explore large search spaces and escape from local optima to find better solutions. Unlike traditional optimization algorithms, which are often tailored to a specific problem domain, meta-heuristics can be applied to a wide range of problems. They are typically designed to be flexible and adaptable, allowing them to be customized and modified for different types of optimization problems. Singla et al. [8] presented a novel approach to finding an initial basic feasible solution to transportation problems in an uncertain environment. Meta-heuristic approaches generally incorporate a stochastic component and are generally nature-inspired algorithms, i.e., they are designed as an imitation of animal behavior or evolutionary processes. Meta-heuristics have been applied to various types of vehicle routing problems (VRP), including the capacitated vehicle routing problem (CVRP), the vehicle routing problem with time windows (VRPTW), and the multi-objective vehicle routing problem (MOVRP) as in the work by Gendreau et al. [9] and Silvia et al. [10]. A performance evaluation of genetic algorithms and simulated annealing for the solution of tsp with profit using Python is presented in the work by Garg et al. [11].

CVRP is used in a variety of contexts, from logistics company delivery schedules to unmanned aerial vehicle (UAV) deliveries as presented in their work by Wang et al. [12], Song et al. [13], and Wang et al. [14]. Prior to employing mixed-integer programming (MIP) with valid inequalities to solve the TSP for each vehicle, employed k-means clustering to assign clients to a heterogeneous fleet of cars as proposed by Mostafa and Amr [15]. Fuzzy C-means (FCM) clustering was utilized by Shalaby et al. [16] to group customers into groups. They execute their method for up to 15 min and then provide the results for cases A-n32-k5, A-n33-k6, A-n36-k5, A-n33-k5, and A-n39-k6. In the article Alesiani et al. [17], authors demonstrated an efficient version of clustering that considers the constraints of the original problem to transform it into a more tractable version. To determine the optimum CVRP solution, Zhu et al. [18] presented an updated evolutionary algorithm with fuzzy method-based clustering.

34.3 Problem Description and Methodology

The capacitated vehicle routing problem (CVRP) is a combinatorial optimization problem that involves finding a set of minimum-cost routes to serve a set of customers from a central depot using a fleet of vehicles with a limited capacity. The CVRP is NP-hard because it belongs to a class of problems that are computationally intractable, and no known algorithm can solve them in polynomial time with respect to the problem size.

The aim is to minimize the total distance traveled by the vehicles while ensuring that each customer is served by a single vehicle and that the total demand of the customers served by each vehicle does not exceed its capacity. The problem is typically formulated as a mixed-integer programming (MIP) problem, which can be solved using optimization software such as CPLEX or Gurobi. However, exact methods can become computationally intractable for large instances of CVRP. To overcome this challenge, various heuristic and meta-heuristic algorithms have been developed to find high-quality solutions for the CVRP. The capacitated vehicle routing problem's main objective is to optimize a set of vehicle routes that satisfy each customer's demand while ensuring that the total number of deliveries along each route dos not exceed the truck or vehicle's capacity and that the overall distance traveled is kept to a minimum. As we know that hard constraints are constraints in a problem, it must be satisfied for a solution to be considered valid or feasible. Violating a hard constraint makes the solution invalid and unusable. The following three constraints are hard in our case:

- Each vehicle departs from the depot and returns there after servicing a portion of the customers.
- Only one specific vehicle visits each customer at a time.
- No route's total load may exceed the vehicle's carrying capability.

Each vehicle has capacity C and is required to satisfy the n customers, each with a positive demand. The objective function including constraints defined as follows, which will remain same as the fitness function:

$$\text{Minimize} \sum_{i=0}^{n} \sum_{j=0}^{n} \sum_{v=1}^{k} X_{ij}^{v} C_{ij} \qquad (34.1)$$

under the following constraints

$$\sum_{i=0}^{n}\sum_{v=1}^{k} X_{ij}^{v} = 1 \forall j = 1,\ldots n \quad (34.2)$$

$$\sum_{j=1}^{n} X_{0j}^{v} = 1 \forall v = 1,\ldots k \quad (34.3)$$

$$\sum_{i=1}^{n} X_{i0}^{v} = 1 \forall v = 1,\ldots k \quad (34.4)$$

$$\sum_{i=0}^{n} X_{ip}^{v} - \sum_{j=0}^{n} X_{pj}^{v} = 0 \forall p = 0,\ldots n \text{ and } v = 1,\ldots k \quad (34.5)$$

$$Y_{iv} = d_i \sum X_{ji}^{v} \forall i = 1,\ldots n \text{ and } v = 1,\ldots k \quad (34.6)$$

Equation (34.1) represents the objective function to be minimized. Equation (34.2) ensures that each customer must be visited only by one vehicle. Equation (34.3) ensures that each vehicle starts only once from the depot, and Eq. (34.4) ensures that each vehicle should return to the depot. Equation (34.5) ensures that each vehicle reached at the location of the customer, and Eq. (34.6) ensures that the quantity delivered to customer i by vehicle v is fully satisfied.

The solution methodology implemented for the development of the current work consists of the following two phases, i.e., generating the vehicle-affected clusters and then solving each as a travel agent problem using the novel genetic algorithm (GA).

Phase 1: Creation of the clusters for each one of the vehicle, depending on their capacity and location of customers with respect to the depot.

Phase 2: A solution is given to each cluster established in the previous phase, implementing the GA adapted to a problem of optimization of the symmetrical travel agent problem.

34.3.1 Phase 1: Clustering

Clustering in the capacitated vehicle routing problem (CVRP) is a technique used to group customers into clusters based on their proximity to each other. This approach is often used to simplify the problem by reducing the number of locations that the vehicles need to visit and, thus, improving the computational efficiency of the solution.

The clustering process involves partitioning the customers into a set of clusters such that customers in the same cluster are geographically close to each other, and customers in different clusters are far apart. Once the customers are clustered, the CVRP problem can be solved by treating each cluster as a single location and applying the standard routing algorithm to determine the optimal routes for the vehicles to visit the clusters. This approach can significantly reduce the complexity of the problem, as the number of locations that the vehicles need to visit is reduced, and the solution time can be improved.

Although clustering may seem like an obsolete solution in a scenario dominated by evolutionary and memetic algorithms, in the real case, where there are clear benefits in terms of computational efficiency and response timeliness, it remains a viable option for simplifying the CVRP problem. For the designed system, we are taking the k-means algorithm as a basis but making substantial changes to the clustering model:

Modify the distance function so as to create clusters of the desired shape; balancing of the nodes within the clusters in order to respect the constraint on vehicle capacity. As proposed by Jain et al. [19], the k-means is one of the most famous clustering models. The algorithm follows a base iterative procedure with the aim of minimizing the variance within the clusters created: Initially, k partitions (number of clusters) are created randomly or with certain heuristics; for each partition, a centroid is calculated, and all nodes are assigned to the partition with the closest centroid; every iteration, the centroids are therefore recalculated and the nodes reassigned, until the algorithm converges, i.e., the centroids do not change compared to the previous iteration. The Euclidean distance measure is used to calculate the distance between two nodes. However, a simple k-means is not suitable for the problem since it tends to form globular clusters; ideally, we would like long-limbed clusters, arranged in a radial pattern around the warehouse. To do this, we have to operate on the distance function used by the algorithm inspired by the work done by Cakir et al. [20], and we define the measure of distance between two nodes i and j as

$$\delta_{ij} = \alpha \theta_{ij} + \beta \rho_{ij} \tag{34.7}$$

where θ_{ij} is the angular distance between the two nodes, i.e., the angle they form with the depot, considered as the origin; ρ_{ij} is the radial distance between the two nodes; α and β are the respective weights of the two distances, fixed so that

$$\alpha + \beta = 1 \tag{34.8}$$

In this way, a higher value of α will make the clusters more circular around the warehouse or depot, while a high value of β indicates that customer and depot are far away.

34.3.2 Phase 2: GA Implementation

In this second phase, the route of each vehicle is optimized by a genetic algorithm in every cluster separately. For this, we have used the benchmark "An32k5." This benchmark problem ("An32k5") is solved by using a genetic algorithm. Genetic algorithms are more adaptable and can deal with a wider range of problems than heuristic methods. In order to solve search and optimization issues, genetic algorithms are adaptable techniques. They are based on how living things reproduce genetically. Darwin proposed that populations change through generations in accordance with the concepts of natural selection and survival of the fittest. Genetic algorithms were developed independently by several researchers in the 1960s and 1970s. However, John Henry Holland is widely credited as the founder of genetic algorithms. Genetic algorithms can solve problems in the real world by reproducing this process. Genetic algorithm is a stochastic optimization technique. This means that it involves a degree of randomness or probability in its operation. According to Zhang et al. [21], the genetic algorithm operates by simulating the process of natural selection, where the fittest individuals in a population are selected to produce the next generation. The randomness comes into play during the selection process, where the fitness of an individual is evaluated probabilistically based on some fitness function or objective function. Furthermore, genetic algorithm also involves other stochastic processes such as mutation and crossover. Mutation involves randomly changing the value of some parameters in an individual, while crossover involves randomly combining two individuals to produce a new individual. The use of stochastic processes in genetic algorithm helps to ensure diversity in the population, prevent premature convergence to local optima, and increase the chances of finding the global optimum solution. As presented by Garg et al. [22], the genetic algorithm takes two sets of chromosomes, called parents, and produces two offspring by exchanging parts of the parents' chromosomes during the crossover phase. The crossover phase accelerates the process of achieving better solutions. In the mutation phase, a single offspring is produced from each parent, aiming to maintain diversity in the population and avoid confinement to a local optimum. The new generation is created by selecting the best solutions, either offspring or parents, according to their quality (fitness) or suitability. A genetic algorithm terminates when the maximum number of iterations is reached, the solution stops improving, or an acceptable solution has been found. Figure 34.1 shows how the genetic algorithm works in the current work.

The role of the fitness function is to quantify how well each candidate solution solves the problem at hand, and it provides the basis for selecting which individuals will be used to generate the next generation of solutions. The fitness function takes as input a candidate solution represented by a chromosome and outputs a score that represents the solution's quality or fitness. The genetic algorithm uses the fitness function to guide the evolution of the population. In each generation, the algorithm selects the fittest individuals based on their fitness scores and uses them as parents to create offspring for the next generation. During the reproductive phase, individuals from the population are selected to interbreed and produce offspring, which, once

Fig. 34.1 Flowchart of genetic algorithm

mutated, will constitute the next generation of individuals. In a genetic algorithm, selection is the process of selecting individuals from a population to be used in the creation of the next generation. The individuals that are selected are typically those that have higher fitness scores or are well-suited to solving the problem at hand. In our work, we have used Roulette wheel selection method. In this method, each individual in the population is assigned a probability of being selected based on its fitness score. Individuals with higher fitness scores are more likely to be selected. After selection of two parents, the chromosomes of the two are often joined using the crossover and mutation operators. The crossover operator divides the chromosomal strings of two chosen parents into two substrings by randomly cutting the threads. It combines the genetic material of the selected individuals to create new offspring. The crossover operator exchanges genetic material between two parents to create new offspring. At each iteration, after recombination, a simple mutation process adds a random element to the entire population of the current solution. This mutation is an exchange of positions or swap of two random elements of a chromosome. This is used as a mechanism of disturbance in order to better explore the space of solutions and escape from local optima. Mutation is an important mechanism in genetic algorithms for introducing new genetic material and increasing the diversity of the population, which helps to prevent premature convergence and improve the quality of the solutions found.

34.4 Results

We have solved benchmark problem "An32k5" for the capacitated vehicle routing problem (CVRP) by using proposed method (CNGA), where a fleet of vehicles is used to serve a set of customers with known demand and location. The "An32k5" instance is a well-known problem in the CVRP literature and is often used as a benchmark to evaluate the performance of optimization algorithms for CVRP. The "An32k5" instance is particularly challenging because it involves a large number of customers and a limited number of vehicles, which requires effective routing and scheduling to minimize the total distance traveled while satisfying the capacity constraint. The problem has been extensively studied in the literature, and several approaches have been proposed to solve it, including exact algorithms, heuristic algorithms, and meta-heuristic algorithms such as genetic algorithms and ant colony optimization. But in this paper, we have used the modified clustering algorithm and the genetic algorithm to get the best solution for the benchmark problem "An32k5." For this, we have implemented this work on Python by using Intel(R) Core(TM) i3-7100 CPU @ 3.90 GHz on Acer-i3-893 with 4 GB RAM Windows 10 operating system. This problem has been solved by using the following parameters and specifications for the algorithm:

Number of clients: 31
Number of vehicles: 5
Vehicle capacity: 100
Optimal solution: 784
Number of generations: 2500
Population size: 800
Percentage of individuals that will inherit by crossing: 80%
Probability of mutation: 0.07

After 150 iterations of proposed cluster-based genetic algorithm (CNGA) for problem, An32k5 gives the following results as given in Table 34.1. We can say that the method provides satisfactory results in reasonable computing times for some instances like "An32k5."

In addition, we compare the outcomes of our strategy CNGA with those of given by Shalaby et al. [16], who created a cluster-first route-second method (CFRS) strategy in which consumers are first separated into clusters, and each cluster is then individually solved as a TSP. Fuzzy C-means (FM) clustering was utilized by Shalaby et al. [16] to group customers into groups. They execute their method for up to 15 min and then provide the results for some benchmark problems like An32k5, An33k6, An36k5, An33k5, and An39k6. The results of our method's (CNGA) solutions and their optimality gaps in comparison with the best-known solutions are presented in Table 34.2 that follows in our numerical experiments. Here, we can see that our method is showing good performance especially in the instance "An32k5," and in the other instances, it is performing almost close to the best solutions.

Table 34.1 Summary of the performed algorithm (CNGA)

Characteristics	Values
Number of iterations	150
Best solution	803
Worst solution	915.08
Average of the solutions	850.06
Best execution time (s)	26.569
Worst execution time (s)	31.689

Table 34.2 Comparison of our CNGA solutions and the FM solutions [16]

Instances	Best-known solution	CNGA	FM [18]
An32k5	784	803	840
An33k6	742	771	769
An36k5	799	860	857
An33k5	661	690	695

34.5 Conclusion

In CVRP, a set of customers with known demand for goods or services needs to be visited by a fleet of vehicles starting and ending at a central depot. The problem is to find an optimal set of routes for the vehicles such that the total distance traveled by the vehicles is minimized while respecting the capacity constraints of the vehicles. In this article, a two-phase solution methodology was developed for the CVRP. Based on the implementation of the constructive K-means for the generation of the clusters and the GA provided a solution to the optimal route of each vehicle within each cluster, the enhanced algorithm developed in the current work presents satisfactory results in computing times reasonable for some benchmark instances. Taking the analyzed meta-heuristic into account, we see that the results obtained were slightly lower than the optimum, but not by much and in a reasonable amount of time. The method (CNGA) in this paper has presented a satisfactory results in reasonable computing times for some instances like An32k5, An36k5, and An33k5. It can be seen that the best solution is close to the minimum margin of error of the optimal. So for the meta-heuristic algorithms, clustering-based genetic algorithm seems to be a very good tool for estimating capacitated vehicle routing problem as it provides us optimized solutions on time. Overall, the results obtained based on the two-phase meta-heuristic method (CNGA) for the vehicle routing problem with capacity constraints allow us to conclude that the application of meta-heuristic procedures can present reliable behavior in the instances of CVRP.

The proposed approach has the potential to be applied in real-world transportation and logistics applications to optimize the routing of vehicles and reduce transportation costs.

References

1. Reong, S., Wee, H.-M., Hsiao, Y.-L.: 20 years of particle swarm optimization strategies for the vehicle routing problem: a bibliometric analysis. Mathematics **10**(19), 3669 (2022)
2. Naddef, D., Rinaldi, G.: Branch-and-cut algorithms for the capacitated VRP. In: The Vehicle Routing Problem, pp. 53–84. Society for Industrial and Applied Mathematics (2002)
3. Hosoda, J., Maher, S.J., Shinano, Y., Villumsen, J.C.: A parallel branch-and-bound heuristic for the integrated long-haul and local vehicle routing problem on an adaptive transportation network. Available at SSRN 4341530 (2023)
4. Otoo, D., Amponsah, S.K., Sebil, C.: Capacitated clustering and collection of solid waste in kwadaso estate, Kumasi. J. Asian Sci. Res. **4**(8), 460–472 (2014)
5. Garg, N., Kakkar, M.K., Gupta, G., Singla, J.: Meta heuristic algorithm for vehicle routing problem with uncertainty in customer demand. ECS Trans **107**(1):6407 (2022)
6. Rani, S., Babbar, H., Kaur, P., Alshehri, M.D., Shah, S.H.A.: An optimized approach of dynamic target nodes in wireless sensor network using bio inspired algorithms for maritime rescue. IEEE Trans Intell Transp Syst (2022)
7. Nanda, M., Kumar, A.: Meta-heuristic algorithms for resource allocation in cloud. J. Phys. Conf. Ser. **1969**(1):012047 (2021)
8. Singla, J., Gupta, G., Kakkar, M.K., Garg, N.: A novel approach to find initial basic feasible solution of transportation problems under uncertain environment. AIP Conf. Proc. **2357**(1), 110007 (2022)
9. Gendreau, M., Laporte, G., Séguin, R.: A tabu search heuristic for the vehicle routing problem with stochastic demands and customers. Oper. Res. **44**(3), 469–477 (1996)
10. Mazzeo, S., Loiseau, I.: An ant colony algorithm for the capacitated vehicle routing. Electron Notes Discrete Math. **18**, 181–186 (2004)
11. Garg, N., Kakkar, M.K., Gupta, G., Singla, J.: A performance evaluation of genetic algorithm and simulated annealing for the solution of TSP with profit using python. In: Emerging technologies in data mining and information security: proceedings of IEMIS 2022, vol. 3, pp. 13–26. Springer Nature Singapore, Singapore (2022)
12. Wang, Z., Zhao, J., Deng, H., Xing, C., Gao, T.: Heterogeneous multi-UAV multi-task reallocation problem using mixed-integer linear programing. In: Proceedings of 2021 International Conference on Autonomous Unmanned Systems (ICAUS 2021), pp. 3453–3459. Springer Singapore, Singapore (2022)
13. Song, B.D., Park, K., Kim, J.: Persistent UAV delivery logistics: MILP formulation and efficient heuristic. Comput. Ind. Eng. **120**, 418–428 (2018)
14. Wang, K., Shao, Y., Zhou, W.: Matheuristic for a two-echelon capacitated vehicle routing problem with environmental considerations in city logistics service. Transp. Res. Part D: Transp. Environ. **57**, 262–276 (2017)
15. Mostafa, N., Eltawil, A.: Solving the heterogeneous capacitated vehicle routing problem using K-means clustering and valid inequalities. In: Proceedings of the International Conference on Industrial Engineering and Operations Management (2017)
16. Shalaby, M.A.W., Mohammed, A.R., Kassem, S.S.: Supervised fuzzy c-means techniques to solve the capacitated vehicle routing problem. Int. Arab J. Inf. Technol. **18**(3A), 452–463 (2021)
17. Alesiani, F., Ermis, G., Gkiotsalitis, K.: Constrained clustering for the capacitated vehicle routing problem (CC-CVRP). Appl. Artif. Intell. **36**(1), 1995658 (2022)

18. Zhu, J.: Solving capacitated vehicle routing problem by an improved genetic algorithm with fuzzy C-means clustering. Sci. Program. **2022** (2022)
19. Jain, A.K.: Data clustering: 50 years beyond K-means. Pattern Recogn. Lett. **31**(8), 651–666 (2010)
20. Cakir, F., Nick Street, W., Thomas, B.W.: Revisiting cluster first, route second for the vehicle routing problem (2015)
21. Zhang, P., Wang, J., Tian, Z., Sun, S., Li, J., Yang, J.: A genetic algorithm with jumping gene and heuristic operators for traveling salesman problem. Appl. Soft Comput. **127**, 109339 (2022)
22. Garg, N., Kakkar, M.K., Gupta, G., Singla, J.: Impact of genetic operators on the performance of genetic algorithm (GA) for travelling salesman problem (TSP). AIP Conf. Proc. **2357**(1), 100020 (2022)

Chapter 35
Drone Watch: A Novel Dataset for Violent Action Recognition from Aerial Videos

Nitish Mahajan, Amita Chauhan, Harish Kumar, Sakshi Kaushal, and Sarbjeet Singh

Abstract In recent developments, a lot has been done for computer vision applied to human action recognition and violence detection. Although various datasets are available for action and violence recognition, there is a clear lack of datasets that include non-violent and violent activities simultaneously from an aerial view. A new aerial video dataset for concurrent human action recognition, including violence detection, is presented in this study. It consists of 60 min of fully annotated data with two action classes, namely violent and normal (non-violent). The current dataset addresses various factors that are not considered in the existing datasets, like changes in the altitude of the drone, changes in the angle at which the video is being captured, video captured during motion, changes in frame rates, videos from different cameras with different configurations, multiple labels for every subject, and labels for violent activities. The resulting dataset is a multifaceted representation of the real-world scenarios, which addresses various shortfalls in the existing datasets. The current dataset will push forward computer vision applications for action recognition, particularly automated violence detection in real-time video streams from an aerial view. Furthermore, the curated dataset is validated for violence detection using machine and deep learning algorithms, namely Support Vector Machine (SVM), Long Short-Term Memory (LSTM), Bi-Directional LSTM (Bi-LSTM) and Adaptive Boosting (AdaBoost).

Keywords Computer vision · Human action recognition · Violence detection · Drone · Machine learning · Deep learning

---

N. Mahajan (✉) · A. Chauhan · H. Kumar · S. Kaushal · S. Singh
University Institute of Engineering and Technology, Panjab University, Chandigarh, India
e-mail: nitish7mahajan@gmail.com

35.1 Introduction

In the recent past, vision-based human activity recognition (HAR) has gained popularity for numerous applications that require detecting and tracking humans. HAR from UAV videos has attracted researchers in the field because of the potential of UAVs to tackle the problem of accessing and covering difficult areas like streets in slums or narrow pathways in a colony. Violent activity detection in videos captured from a bird eye view is a challenging application in developing automated video-based surveillance systems. Detecting violent activities in public areas is a crucial task so as to send automated alerts to public authorities and enable them to take appropriate actions. With the emergence of machine and deep learning techniques for human action recognition, various approaches have become prevalent in modeling violence identification systems. However, in a real-world scenario, the actual data and application requirements have become more and more diverse. Thus, violence detection systems also need to be more accurate and faster. Various problems arise as the distance between the camera and the objects under surveillance increases. Human appearance gets degraded due to serious occlusions, and identifying discriminating features becomes difficult. Many other issues exist in real life such as abrupt camera movements, changes in the angle of the camera, and changes in the height at which the drone is flying. Although multiple datasets exist in the art that are crafted using video data captured through drones, they have certain limitations as they do not take into consideration the problem of constraints associated with dynamic data acquisition sources, platforms, and environments. Also, most of the existing datasets focus on normal human activities like walking, running, sitting, jogging, exercising, etc. In this paper, a novel aerial video dataset that emphasizes recognizing concurrent human activities, including violent actions, is proposed. The dataset comprises: 60 min of fully annotated video-sequence data and contains two classes, namely violent and normal (non-violent). To determine the violent acts of humans in the videos, actions like punching, pushing, kicking, and throwing are used. The presented dataset addresses various drawbacks of currently available datasets and considers several factors including changes in the altitude of the drone, changes in the angle at which the video is being captured, video captured during motion, changes in frame rates, videos from different cameras with different configurations, multiple labels for every subject and most importantly violent activities. Furthermore, the performance of the violent action recognition system is evaluated on the developed dataset by deploying SVM, LSTM, Bi-LSTM, and AdaBoost classification algorithms. The following are the major contributions of the proposed research:

1. A novel dataset for violent activity recognition is proposed. The dataset consists of video-sequence data with actions for both violent and non-violent human behaviors.
2. The dataset meets the various dynamic requirements crucial for developing video surveillance systems. It addresses the limitations that the existing datasets have in terms of UAV and camera configurations, dynamic frame rates, multiple labels assigned to single object, and other challenges that are present in the real world.

3. The study proposes machine and deep learning models to classify human actions as violent or non-violent. The model is validated using the developed dataset as well as with several open datasets, and provides promising results in terms of accuracy and precision.

Remainder of the paper is arranged as follows. Section 35.2 provides a literature review of various existing datasets and approaches to human action recognition. Section 35.3 describes the configurational details of the UAV. Section 35.4 explains, in detail, the various characteristics and advantages of the developed aerial video dataset. Details of the architecture of the proposed drone-based video surveillance system and deep learning-based violence detection model are provided in Sect. 35.5. Section 35.6 presents the experimental results. Section 35.7 concludes the paper.

35.2 Related Work

In this section, we discuss various state-of-the-art datasets and techniques used for human activity recognition.

35.2.1 Datasets for Human Action Recognition

Soomro et al. [1] introduced a dataset called UCF101 for human activity recognition. The dataset contains 101 classes for different actions and is composed from video data of 27 h (over 13,000 clips). The videos used for the database are chosen to include various factors like poor lighting, camera motion, cluttered background, etc., to provide diversity to the dataset. A methodology for generating synthetic dataset to evaluate HAR systems is proposed by [2] that uses surveys for humans to capture their activities and behaviors, and utilizes simulator tools to model and generate synthetic sensor-based dataset with labels and time stamps. Authors claimed that there is a significant similarity between the developed dataset and a dataset generated by using actual sensors in real-world scenario. Considering the limitations that existing RGB + D-based action recognition datasets have, Shahroudy et al. [3] introduced a large-scale dataset for identifying human activities in RGB + D-based HAR systems. The dataset is collected from 40 different subjects and has about 56 thousand videos providing 4 million frames. The 60 distinct action classes are included in the dataset, and a novel RNN architecture is proposed by the authors for human action classification. Barekatain et al. [4] presented a novel aerial video dataset called Okutama-Action for the purpose of concurrent HAR. The dataset comprises of 12 action classes and addresses many challenges of real-life scenarios such as significant changes in aspect ratio, abrupt movement of camera, dynamic transition of actions, and multiple actions performed concurrently by a person. Authors evaluated their dataset by training a model for action detection [5] and took into consideration that performance of human activity detection degrades as objects captured through UAVs look tilted or distorted. Therefore, to enhance the performance of classification and detection models, authors re-trained the CNNs used in YOLOv2 with their own manually labeled aerial image dataset. A post-processing technique is proposed by the authors for classifying human activities as normal or abnormal. A survey of existing HAR approaches is presented by [6]. Authors included both traditional handcrafted and well-known deep learning-based action representation techniques for the study, and provided their comparisons and analyses. Furthermore, authors discussed about public datasets, namely KTH [7], UCF Sports [8, 9], Hollywood2 [10], ActivityNet [11], YouTube [12], UCF-101 [1], IXMAS [13], Weizmann [14], and HDMB-51 [15] that are available for HAR applications. Focusing on the importance of multi-modal datasets for HAR in indoor applications such as healthcare assistant robots, Moncks et al. [16] presented a novel dataset captured by four types of sensors that are often used in autonomous vehicles and indoor applications. The dataset incorporates nine activities performed by 16 volunteers. Authors also proposed a novel algorithm for data preprocessing to facilitate dynamic context-dependent feature extraction from the dataset that are then used by various machine learning models for posture identification. Wijekoon et al. [17] created a dataset called MEx: Multi-Modal Exercises Dataset by collecting data of seven exercises using four sensors, namely a depth camera, a pressure mat, and two accelerometers. Authors implemented this dataset to fulfill the requirement for identifying and evaluating quality of exercises performed by the patients with musculoskeletal disorders. The modalities captured in the dataset are video data, time-series data, and pressure sensor data. The dataset was evaluated by the authors on several standard classification algorithms to recognize human activities. In a survey presented by [18], a description and classification of existing HAR datasets is provided for researchers in the field enabling them to choose the befitting benchmarks according to their domain of study. A novel Event Recognition in Aerial Video dataset (ERA) is proposed by [19] for recognizing events in unconstrained and diverse aerial videos for remote sensing applications. A total of 2866 videos were collected comprising a dataset with 25 different class labels. Deep learning models were trained to validate ERA dataset. Mmereki et al. [20] modified existing pretrained CNNs in YOLOv3 [21] through transfer learning by re-training them with their own aerial action dataset. Authors customized YOLOv3 for detection and localization of human activities from aerial view. Considering the hassle in training deep neural networks with a large number of human action videos captured by drones, Sultani and Shah [22] explored two alternative sources to collect data for HAR. Authors aimed to improve the performance of aerial action classification models with few training instances. For that purpose, they collected aerial game action videos from video games, and generated aerial features from ground videos using GANs.

35.2.2 Approaches for Human Action Recognition

A deep learning-based drone surveillance system is proposed by [23] to identify violent humans in a scene. The system in the first phase identifies humans from aerial images using feature pyramid network (FPN). Once the human objects are detected, their pose estimation is performed by ScatterNet Hybrid Deep Learning (SHDL) network model. After the pose estimation phase, SVM model classifies human actions as violent or non-violent by utilizing the knowledge of orientations among the limbs of the estimated poses. A dataset to train deep learning-based models on aerial images to detect human actions is also introduced by the authors [24], and in their work, proposed an approach for HAR from UAV-captured videos that work in two phases. The first phase is the offline phase in which pretrained CNNs were applied to generate human/non-human classification and human activity models. The second phase is the inference phase to utilize the generated models to identify human objects and classify the activities performed by them. For the purpose of detecting potential regions of motions in the videos, scene stabilization is applied within the two said phases. The authors evaluated and dated their approach through various experiments performed on UCF-ARG dataset and compared the performance of the proposed models with existing techniques. Aviles-Cruz et al. [25] proposed a novel CNN-based approach for the classification of human activities. Three CNNs, namely coarse, medium, and fine CNN, are used parallelly to extract local features. The extracted features are integrated into the final classification stage. The proposed approach has been applied to recognize angle use movement using a tri-axial gyroscope and a tri-axial accelerometer installed in a smartphone. The model successfully classified six human activities and performed better than certain competitive techniques present in the art. A weakly-supervised approach to recognize complex human activities from video sequences is proposed by [26] that does not require explicit object detection and laborious data annotation. A multi-level contextual model is introduced to fuse low-level, mid-level, and high-level features simultaneously, and the recognition model is trained using only activity labels for each video. The proposed model was evaluated on realistic datasets for identifying human-to-human and human-to-object interactions. Ramzan et al. [27] presented an exhaustive review of various techniques present in the art for detecting violence from videos. Authors categorized the detection methods based on the classification model used: using conventional machine learning, using SVM, and using deep learning. Video features that are important for better classification results are also provided in the paper. Aktı et al. [28] proposed a deep network for vision-based violence detection using surveillance cameras. The authors trained a CNN model on fight scenes and integrated the trained model with Bi-LSTM for fight detection. An attention layer is also added after Bi-LSTM for improving the performance of the classification model. A new dataset for fight detection from surveillance cameras is also presented in the paper. In a survey by [29], various CNN models used for violence detection from videos are presented along with their advantages and disadvantages. The authors also discussed the characteristics of several public datasets available for violence detection. Challa et al. [30] proposed a hybrid network architecture that uses CNN and Bi-LSTM to classify human activities recognized using wearable sensors' data. The approach presented by the authors requires minimal preprocessing and works directly on raw sensor data to capture various local features and long-term dependencies in time-series data. The model was evaluated and validated on three open datasets and was observed to have outperformed other deep learning approaches for HAR present in the literature. A deep learning-based unsupervised approach for detecting anomalies in crowd surveillance videos is proposed by [31]. The model has two stages, namely a CNN auto-encoder and a sequence-to-sequence LSTM auto-encoder, and follows a one-class classification approach. The authors evaluated their model on multiple benchmark datasets for anomaly detection and reported a significant performance. Srivastava et al. [32] created a new dataset for violent activity recognition from drone surveillance videos. Videos for the dataset are captured by flying the drones at different heights in an unconstrained environment and is annotated with eight action classes. Authors proposed an approach for violence detection that first extracts key-points of detected human objects and then feeds those key-points as features to SVM and RF models for human action classification.

35.2.3 Research Gaps

According to the literature survey, a few research gaps have been identified and are considered while developing the proposed dataset and machine and deep learning models for violence detection, and are given as follows.

1. Although there are datasets available in the art that considers multiple factors related to the changes in the physical environment such as lighting, camera motion, and object movements, they are not specifically created for the purpose of violent activity identification.
2. On the contrary, the datasets that are accessible for violence detection lack the knowledge of various aspects of dynamicity of the real world.
3. Furthermore, the existing violent action classification models are not precise and well matured. Also, they do not track and address how human behaviors and postures change over a period of time.

Therefore, there is a need for a comprehensive dataset that takes into account major challenges of data collection, data annotation, and violence recognition. Also, a system to detect violent actions from aerial videos is required that is more accurate, fast, and reliable so as to enable it to be better utilized in real time.

35.3 Configurational Details of the UAV

This section provides the details of the major components of the UAV configured for capturing and transmitting the data. The UAV is equipped with a mobile computing device with an onboard processing capability and is powered by a backbone of high-performance cloud-based servers. High-resolution photo-sensors are mounted over the UAV to capture targets on the ground from a wider field of view. The captured data is transmitted to Cloud servers in real time over a secure high-speed 4G-LTE network for further processing. Following are the key hardware, software, and networking components of the proposed UAV-based system:

1. Raspberry Pi: It is a computer-on-a-chip that makes the communication between the ground station and the UAV. As the flight controllers work on pulse position modulation (PPM) signals, Raspberry Pi converts the digital signals sent from the ground station to PPM signals. Also, to transmit the captured data to Cloud servers, connection between the UAV and Cloud servers is established using Raspberry Pi.
2. XB Station: It is the software that is used to provide a secure and reliable 4G-LTE connection between the ground station and UAV. It provides the facility of low latency video streaming from UAV to the ground station using GStreamer [33]. XB Station needs to be installed on both the companion computer (Raspberry Pi) on UAV and the ground station.
3. Photo-sensors/Cameras: The Raspberry Pi camera module is used for capturing aerial images and videos. The module needs to be connected to and compatible with the Raspberry Pi. Other cameras with special features can also be deployed as per the desired requirements.
4. Cloud Servers: The Cloud servers provide services like storage, machine and deep learning-based violence detection models, and various analytical tools. Data from UAV is processed and analyzed in real-time using these services and can be streamed to desired authorities and stakeholders for taking necessary action.

35.4 Proposed Drone Surveillance System

This section presents the architecture of the proposed drone control and media transfer system in detail. The system can monitor public places, capture videos, and send them to the Cloud platform to detect the violence in the videos. This can discover violence in public places in time and give early warning to prevent criminal incidents. For the detection of fight incidents and violence, the AI models of our system identify human objects and label them violent or normal according to their postures and interactions among themselves. Once a fight scene is identified (i.e., one or more humans in the picture or video are labeled violent), alerts are generated and sent to the stakeholders such as Police or other authorities. Figure 35.1 depicts the architecture of the drone-based surveillance system for detecting violent activities.

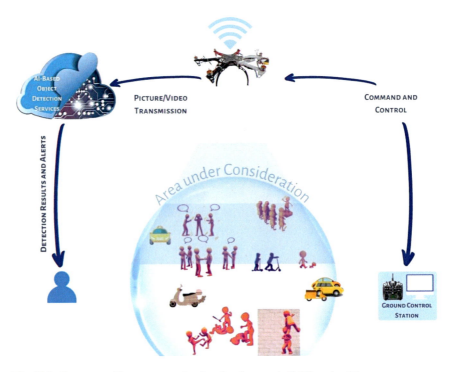

Fig. 35.1 Drone surveillance system showing the placement of different entities

35.4.1 *Framework of the Violence Detection Model*

Figure 35.2 demonstrates the structure of the proposed violent action recognition model. To detect violent activities in a video sequence, frames are first fed to YOLOv5 object detection model that identifies human objects in the frame and generates proposal frame with bounding boxes around the detected humans. Once the proposal is generated and objects are localized, a multi-person 2D pose estimation technique is applied to extract human key-points. The 17 key-points are extracted to generate 2D postures for each individual present in the frame, and x, y coordinates of all the extracted key-points along with the x, y coordinates of the bounding boxes are saved in a CSV file. After saving the key-points, pose normalization is performed by changing the numeric values of the x, y coordinates to a common scale. Also, the problem of missing key-points arises if the pose estimation algorithm is unable to detect key-points due to overlapped or occluded human objects, and this can affect the accuracy of the detection models. To mitigate this and create a stable dense vector, different techniques have been applied for different models. For SVM and AdaBoost, the missing key-points were substituted with the corresponding values in the previous frame to provide an approximation. In case of LSTM and Bi-LSTM models, an embedding layer is deployed that generates similar key-points for similar

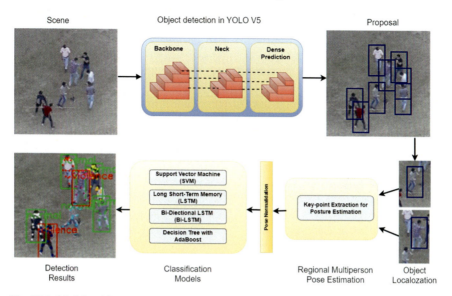

Fig. 35.2 Model architecture

human postures. After performing the data preprocessing steps, a dataset of size 16,400 × 38 is obtained for training the classification models. Following section explains the process of model training.

35.4.2 Model Training

The classification models are trained for the detection of violence in aerial videos on the dataset obtained after preprocessing. The data is split into train and test sets with size 12, 300 × 38 and 4, 100 × 38, respectively. The models were trained on Nvidia Tesla P100-PCIE-16 GB GPU. The first model trained is an SVM with L2 norm regularization called L2-SVM that optimizes the sum of squared errors. L2-SVM is chosen as it is differentiable and imposes a bigger loss for points which violate the margin. The model has a dense layer with 64 outputs and ReLU activation. The 500 epochs are run to train the model with a learning rate of 0.01 and a batch size of 10. The loss function used is hinge loss. To train the LSTM model for the identification of violent human behaviors, the vector from the embedding layer (as given in Sect. 35.4-A) is passed to the LSTM layer with 128 units. The number of units deployed has been decided after running multiple experiments. The vector from the LSTM layer is fed to a densely connected layer having sigmoid activation on the output. The 100 epochs are run for 100 batches with 0.1 learning rate. The loss function used in this model is binary cross-entropy. Next model is a Bi-Directional LSTM that trains two LSTMs on the input sequence: the first on the input sequence as it is, and the second on an inverted copy of the input sequence. This provides additional context to the network and results in faster and even fuller learning on the problem. A 64 unit forward and 64 unit backward Bi-LSTM is used to learn the transition of key-points over 30 units of time. The output of the LSTM layer is then fed to the attention layer that takes in a 3D input and outputs a 2D vector.

Attention mechanism is applied to the network so as to make it capable of remembering and differentiating human actions over a period of time. The output from the attention layer is then passed through a dense layer with sigmoid activation and a 0.5 dropout ratio. The 100 epochs for 100 batches are run with a learning rate of 0.1 and binary cross-entropy loss function. Lastly, AdaBoost classifier that has a Decision Tree with max-depth 1 as the base estimator is trained. The number of estimators used for training is 6, a learning rate of 0.5 is applied, and SAMME.R algorithm is adopted for additive model updating. The final output of the system is a video in which the humans that are performing violent actions are labeled as "Violence" with red colored bounding boxes and those that are performing non-violent actions are labeled as "Normal" with green colored bounding boxes, as shown in Fig. 35.2.

35.5 Details of Drone Watch

This section provides the major details of Drone Watch for violent action recognition. Various actions chosen for the purpose of data collection and annotation are explained, and the approaches for data collection and annotation are described.

35.5.1 Action Selection for Violent Activities

Replicating violence for dataset building poses two challenges: (i) selecting the actions to be performed and (ii) executing the actions based on a script. The action selection for violence detection is done by analyzing already existing videos of the prior art for violence detection that are available on the Internet. The videos consisted of various sources ranging from fights in hockey matches, low altitude UAV videos of fights [23], CCTV camera captured violence, and media reports. After analyzing the videos, the four most common violent actions were identified, namely pushing, punching, kicking, and throwing. Furthermore, the actions selected above were executed according to a script with different settings. The script was crafted to represent the real-world scenarios. An action scene consists of various objects, from moving vehicles to stationary objects like benches. In addition to violent actions, non-violent actions like running, standing, walking, and laying are also a part of the captured scenes. In addition to this, the fight scenes have varied scenarios in terms of the number of persons fighting.

35.5.2 Data Collection and Annotation

The data collection strategies have been designed by keeping in view that the recorded data represents real-world data. In order to assure real-world correspondence of the data, existing UAV-captured videos have been thoroughly studied. The angles of inclination, speeds, and movements of drones have been studied to determine the most widely used parameters. It is observed that the drone videos in the real world have different inclinations of capture, the drone's speed is low, and the drone movements are smooth while capturing the videos. The UAV motion is a very crucial aspect of the data collection process. Thus, the data videos must have hovering drone shots, shots with the drone in motion, including forward, sideways, pitch, and yaw movements. To include all of the properties mentioned earlier, the drone pilots were given different flight plans.

In addition to this, the interactions between people in the videos play a crucial role. After studying the videos, a script was developed where the actors go in and out of the frames, so there are partial occlusions. Also, there are people in the frames doing activities other than violence such as walking, laying, sitting, and running. Video sequences of 60 min are recorded by the drone from a height of 20–40 m at 30 fps. Various atomic and group activities are captured depicting both violence and non-violence. After the aerial videos have been captured, the next step is to annotate the data for classifying violent and normal activities. Annotation is performed on the CSV data that is obtained after key-point generation and pose normalization as explained in Sect. 35.4. The approach used for annotating the data is frame-by-frame annotation. The human object IDs that are indulged in violent activities such as punching, kicking, pushing, and throwing are labeled as violent, whereas those that are carrying out non-violent activities are labeled as normal.

35.6 Experimental Results and Analysis

In this section, the details of experiments and the performance of the classification models utilized for detecting violent actions on Drone Watch are presented. The proposed drone surveillance system first uses YOLOv5 object detection algorithm to identify humans in aerial videos and then passes the localized human objects to the regional multi-person pose estimator network to generate their estimated postures. These poses and orientations are then used as feature vectors for identifying violent human activities. Following subsections discuss the metrics used for evaluating the performance of the violent action classification models and the results of the experiments performed. Figure 35.3 shows a few shots from the m the Drone Watch dataset.

Fig. 35.3 Shots from the drone watch dataset

35.6.1 Evaluation Metrics

There are several metrics available for assessing and comparing the performances of machine and deep learning models such as confusion matrix, F1-score, root mean squared error, and log loss. Confusion matrix is the most widely used metric that calculates various parameters to assess the functioning of machine learning models like accuracy, precision, recall or sensitivity, specificity, etc. In this work, accuracy, precision, and recall metrics are calculated for each of the models in order to compare their violence detection.

35.6.2 Results and Discussion

This section presents the results of the trained models on test set. SVM model provides an accuracy of 84.94% with a loss of 34.16% in classifying human actions as violent and normal. The precision, recall, and F1-score obtained for SVM are 85.20%, 85.25%, and 85.22%, respectively. A classification accuracy of 87.21% is attained using LSTM model with 30.07% loss. The respective values obtained for precision, recall, and F1-score are 87.89%, 87.80%, and 87.84%. For Bi-LSTM model, an accuracy of 99.85% is achieved with a loss of 0.50%. The precision, recall and F1-score values for Bi-LSTM are 99.90%, 99.80%, and 99.95%, respectively. An accuracy of 98.24% is achieved with precision, recall, and F1-score as 98.52%, 98.52%, and 98.52%, respectively. Clearly, Bi-LSTM classifier outperforms other classification models in detecting violence from aerial videos by a decent margin.

For complex datasets in which violence features change rapidly and frequently, it is crucial to understand the context of each frame in respect of the whole video, i.e., both the past and future paths. In videos with greater dynamicity and heterogeneity, the same series of frames could follow one of several trajectories in the future. Bi-LSTM potentially acquires a better interpretation of the entire video as it can obtain long range context in both forward and backward directions of the time sequence of the input, and hence, performs better than other models presented in the study. AdaBoost with Decision Tree as base estimator also performs better than SVM and LSTM models on the proposed dataset. Each Decision Tree in the model learns and collects different information on the features (human actions) such that the final decision is obtained from a combined knowledge of the estimators. AdaBoost is adaptive as the subsequent learners are adjusted in favor of the instances that are misclassified by previous classifiers. Thus, it provides an optimized performance with better efficiency and accuracy.

35.7 Conclusion

A multifaceted dataset for concurrent human action recognition and violence detection from aerial videos is presented in the paper. The dataset comprises of 60 min of video data captured using UAVs and is fully annotated with two classes, namely violence and normal. Activities chosen to represent violence are punching, pushing, kicking, and throwing, while walking, standing, sitting, and laying are normal non-violent human actions. The proposed dataset addresses various shortcomings of the existing datasets and takes into consideration the factors such as abrupt camera motion, changes in the altitude at which the drone is flying, changes in the angle of the camera, changes in frame rates, multiple labels assigned to a single subject, and the like. The proposed dataset is validated by deploying SVM, LSTM, Bi-Directional LSTM, and AdaBoost classifier models for detecting violent human behaviors.

References

1. Soomro, K., Zamir, A.R., Shah, M.: Ucf101: a dataset of 101 human actions classes from videos in the wild. arXiv preprint arXiv:1212.0402 (2012)
2. Azkune, G., Almeida, A., Lopez-de Ipi´na, D., Chen, L.: Combining users' activity survey and simulators to evaluate human activity recognition systems. Sensors 15(4), 8192–8213 (2015)
3. Shahroudy, A., Liu, J., Ng, T.-T., Wang, G.: Ntu rgb+ d: a large scale dataset for 3d human activity analysis. In: Proceedings of the IEEE Conference on Computer Vision and Pattern Recognition, pp. 1010–1019 (2016)
4. Barekatain, M., Mart´ı, M., Shih, H.-F., Murray, S., Nakayama, K., Matsuo, Y., Prendinger, H.: Okutama-action: an aerial view video dataset for concurrent human action detection. In: Proceedings of the IEEE Conference on Computer Vision and Pattern Recognition Workshops, pp. 28–35 (2017)

5. Wang, H.-Y., Chang, Y.-C., Hsieh, Y.-Y., Chen, H.-T., Chuang, J.-H.: Deep learning-based human activity analysis for aerial images. In: 2017 International Symposium on Intelligent Signal Processing and Communication Systems (ISPACS), pp. 713–718. IEEE (2017)
6. Sargano, A.B., Angelov, P., Habib, Z.: A comprehensive review on handcrafted and learning-based action representation approaches for human activity recognition. Appl. Sci. **7**(1), 110 (2017)
7. Schuldt, C., Laptev, I., Caputo, B.: Recognizing human actions: a local SVM approach. In: Proceedings of the 17th International Conference on Pattern Recognition, 2004. ICPR 2004, vol. 3, pp. 32–36. IEEE (2004)
8. Rodriguez, M.: Spatio-temporal maximum average correlation height templates in action recognition and video summarization (2010)
9. Soomro, K., Zamir, A.R.: Action recognition in realistic sports videos. In: Computer Vision in Sports, pp. 181–208. Springer (2014)
10. Marszalek, M., Laptev, I., Schmid, C.: Actions in context. In: 2009 IEEE Conference on Computer Vision and Pattern Recognition, pp. 2929–2936. IEEE (2009)
11. Heilbron, F.C., Escorcia, V., Ghanem, B., Niebles, J.C.: Activitynet: a large-scale video benchmark for human activity understanding. In: Proceedings of the IEEE Conference on Computer Vision and Pattern Recognition, pp. 961–970 (2015)
12. Liu, J., Luo, J., Shah, M.: Recognizing realistic actions from videos "in the wild". In: 2009 IEEE Conference on Computer Vision and Pattern Recognition, pp. 1996–2003. IEEE (2009)
13. Weinland, D., Boyer, E., Ronfard, R.: Action recognition from arbitrary views using 3d exemplars. In: 2007 IEEE 11th International Conference on Computer Vision, pp. 1–7. IEEE (2007)
14. Gorelick, L., Blank, M., Shechtman, E., Irani, M., Basri, R.: Actions as space-time shapes. IEEE Trans. Pattern Anal. Mach. Intell. **29**(12), 2247–2253 (2007)
15. Kuehne, H., Jhuang, H., Garrote, E., Poggio, T., Serre, T.: Hmdb: a large video database for human motion recognition. In: 2011 International Conference on Computer Vision, pp. 2556–2563. IEEE (2011)
16. Moencks, M., De Silva, V., Roche, J., Kondoz, A.: Adaptive feature processing for robust human activity recognition on a novel multi-modal dataset. arXiv preprint arXiv:1901.02858 (2019)
17. Wijekoon, A., Wiratunga, N., Cooper, K.: MEx: multimodal exercises dataset for human activity recognition. arXiv preprint arXiv:1908.08992 (2019)
18. Singh, R., Sonawane, A., Srivastava, R.: Recent evolution of modern datasets for human activity recognition: a deep survey. Multimedia Syst. **26**(2), 83–106 (2020)
19. Mou, L., Hua, Y., Jin, P., Zhu, X.X.: Event and activity recognition in aerial videos using deep neural networks and a new dataset. In: IGARSS 2020—2020 IEEE International Geoscience and Remote Sensing Symposium, pp. 952–955. IEEE (2020)
20. Mmereki, W., Jamisola, R.S., Mpoeleng, D., Petso, T.: Yolov3-based human activity recognition as viewed from a moving high-altitude aerial camera. In: 2021 7th International Conference on Automation, Robotics and Applications (ICARA), pp. 241–246. IEEE (2021)
21. Farhadi, A., Redmon, J.: Yolov3: an incremental improvement. Comput. Vis. Pattern Recogn. **1804** (2018)
22. Sultani, W., Shah, M.: Human action recognition in drone videos using a few aerial training examples. Comput. Vis. Image Underst. **206**, 103186 (2021)
23. Singh, A., Patil, D., Omkar, S.N.: Eye in the sky: real-time drone surveillance system (DSS) for violent individuals identification using scatternet hybrid deep learning network. In: Proceedings of the IEEE Conference on Computer Vision and Pattern Recognition Workshops, pp. 1629–1637 (2018)
24. Mliki, H., Bouhlel, F., Hammami, M.: Human activity recognition from UAV-captured video sequences. Pattern Recogn. **100**, 107140 (2020)
25. Aviles-Cruz, C., Ferreyra-Ramírez, A., Zuñiga-López, A., Villegas-Cortez, J.: Coarse-fine convolutional deep-learning strategy for human activity recognition. Sensors **19**(7), 1556 (2019)

26. Ajmal, M., Ahmad, F., Naseer, M., Jamjoom, M.: Recognizing human activities from video using weakly supervised contextual features. IEEE Access **7**, 98420–98435 (2019)
27. Ramzan, M., Abid, A., Khan, H.U., Awan, S.M., Ismail, A., Ahmed, M., Ilyas, M., Mahmood, A.: A review on state-of-the-art violence detection techniques. IEEE Access **7**, 107560–107575 (2019)
28. Aktı, S., Tataroğlu, G.A., Ekenel, H.K.: Vision-based fight detection from surveillance cameras. In: 2019 Ninth International Conference on Image Processing Theory, Tools and Applications (IPTA), pp. 1–6. IEEE (2019)
29. Jain, A., Vishwakarma, D.K.: State-of-the-arts violence detection using convnets. In: 2020 International Conference on Communication and Signal Processing (ICCSP), pp. 0813–0817. IEEE (2020)
30. Challa, S.K., Kumar, A., Semwal, V.B.: A multibranch CNN-BiLSTM model for human activity recognition using wearable sensor data. Vis. Comput. 1–15 (2021)
31. Pawar, K., Attar, V.: Application of deep learning for crowd anomaly detection from surveillance videos. In: 2021 11th International Conference on Cloud Computing, Data Science and Engineering (Confluence), pp 506–511. IEEE (2021)
32. Srivastava, A., Badal, T., Garg, A., Vidyarthi, A., Singh, R.: Recognizing human violent action using drone surveillance within real-time proximity. J. Real-Time Image Process. 1–13 (2021)
33. GStreamer. https://gstreamer.freedesktop.org/documentation/?gilanguage=c. Last accessed Dec 2022

Chapter 36
Performance Analysis of Different Controller Schemes of Interval Type-2 Fuzzy Logic in Controlling of Mean Arterial Pressure During Infusion of Sodium Nitroprusside in Patients

Ayushi Mallick, Jyoti Yadav, Himanshu Chhabra, and Shivangi Agarwal

Abstract This research work is on designing an optimum and robust controller to improve the infusion of drug in patients during post-operative conditions in order to automatically control the mean arterial pressure. The control of mean arterial pressure of critically-ill patients is a prominent research field due to its ability to lessen recovery time, lower medical costs, and better medical staff management by lowering their burden. However, due to uncertainty in patient's sensitivity toward the drug infused and external disturbance/noise, this is a complicated task. In this work, two efficient controller schemes, interval type-2-fuzzy logic controller-proportional integral derivative and fractional order-interval type-2-fuzzy logic controller-proportional derivative + proportional integral controller are presented. For effective optimization of these control schemes, one of the most effective nature inspired optimization, cuckoo search algorithm is implemented for finding their optimal values of parameters. In the work, simulation results and performance index i.e., settling time, overshoot IAE values shows superior performance of fractional order-interval type-2-fuzzy logic controller-proportional derivative + proportional integral controller over IT2-FLC-PID controller.

Keywords Mean arterial pressure (MAP) · Proportional integral derivative (PID) · Fuzzy logic controller (FLC) · Interval type-2 fuzzy logic controller (IT2-FLC) ·

A. Mallick (✉) · J. Yadav
Department of Instrumentation and Control Engineering, NSUT, Sector-3, Delhi, India
e-mail: ayushimallick22@gmail.com

H. Chhabra
Department of Electrical and Communication Engineering, M. L. V. Textile and Engineering College, Bhilwara, Rajasthan, India

S. Agarwal
Department of Electronics Engineering, Ramrao Adik Institute of Technology, Nerul, Navi Mumbai, India

© The Author(s), under exclusive license to Springer Nature Singapore Pte Ltd. 2024
P. K. Jha et al. (eds.), *Proceedings of Congress on Control, Robotics, and Mechatronics*, Smart Innovation, Systems and Technologies 364,
https://doi.org/10.1007/978-981-99-5180-2_36

Interval type-2-fuzzy logic controller-proportional integral derivative (IT2-FLC-PID) · Fractional order–interval type-2-fuzzy logic controller-proportional derivative plus proportional integral (FO-IT2-FLC-PD + PD) · Sodium nitroprusside (SNP) · Cuckoo search algorithm (CSA)

36.1 Introduction

The control of MAP within a certain range by infusion of vasoactive drugs such as SNP in patients is required in numerous medical circumstances when patient's autoregulation of MAP is insufficient to maintain a certain physiological range of MAP [1]. However, the manual regulation of the infusion rates of these drugs in patients is a laborious job for clinical personnel, and therefore prone to human error [1, 2]. Hence, there is a need for closed-loop drug administration for patients so as to improve health care and reduce medical costs [1, 2]. One of the challenges is the lack of mathematical modeling that explains the relationship between the change in MAP due to SNP infusion in patients. This is by virtue of nonlinear behavior in patients due to external disturbance and differences in sensitivity for the same drug from patient to patient [1].

36.2 Related Work

Malagutti et al. proposed a control design using RMMAC showing capability to avoid transient instability for both variation in time-varying parameters and non-negative mean disturbances [1]. Sondhi et al. suggested a controller using fractional order PI as the control design, which showed better noise rejection [2]. Soltesz et al. presented the control design for auto-supervision of MAP of a brain-dead patient within a range of [60, 70] mmHg using a robustly tuned PID controller that showed satisfactory performance [3]. Su et al. presented a comparative study between the MPC-PSO controller and a PID controller, where the MPC-PSO controller handled the time delay of the control system better [4]. Silva et al. suggested an embedded version of a simple adaptive PI controller for auto-supervision of MAP for a known range of patients showing satisfactory performance but with minute overshoot of 4–6 mmHg [5]. Tasoujian et al. suggested the MAP regulation via a closed-loop control of loop sharing control and an IMC-PID controller showing lower lethargic closed-loop performance and preserves the bandwidth [6]. Sharma et al. proposed an optimal IT2-FLC for auto-supervision of infusion of vasoactive drugs to maintain MAP for post-surgical patients. The proposed controller was tested for robustness over time-varying parameter and adding Gaussian noise at the output of the system [7]. Chhabra et al. utilize the fractional order differentiation and integration in designing a vigorous nonlinear fractional-order (FO) fuzzy-PD + fuzzy-I controller design, which showed great performance in tracking of reference provided to the system and external noise rejection in a robotic arm [8]. Nagarsheth et al. proposed different types of PSO-tuned fractional-order PID controllers for a MIMO model controlling MAP and cardiac output (CO), showing better performance than a PSO-tuned integer-PI controller [9]. Kumar et al. presented a comparative study between a fractional-order two-layer FLC (FO-TLFLC) and an integer-order two-layer FLC controller tuned using gray wolf optimization, showing that FO-TLFLC improves the overshoot, settling time, and IAE of MAP [10].

36.3 Mathematical Modeling

The model relating the mean arterial pressure (MAP) to the rate of infusion of drugs, i.e., sodium nitroprusside (SNP), in patients is given by [1]:

$$\sum(s) = \frac{\Delta P_{MAP}(s)}{I(s)} = \frac{k(\tau_3 s + 1)e^{-st}}{(\tau_1 s + 1)[(\tau_2 s + 1)(\tau_3 s + 1) - \alpha]} \quad (36.1)$$

where the transfer function of the model is the ratio of the Laplace transform of the change in mean arterial pressure (MAP) $\Delta P_{MAP}(s)$ to the Laplace transform of the rate of infusion of SNP $I(s)$. This plant model has dependence over time-varying parameters K, T, and α, where K is the sensitivity of a patient's MAP response to infusion of SNP, which differs between patients as well as within a patient over a period of time, T is a time-delay parameter, and α is the fraction of drug being recirculated in the system. These parameters have a range of values within which they should be adjusted according to the patient [1]. Richa et al. have taken the value to K as -0.25 mmHg/ml/h [7]. Nicolo et al. have explained this plant model as a third-order, stable transfer function which consists of three first-order linear systems where each linear system represents pulmonary, systemic circulation, and drug action [1]. The values of the time constants $\tau_1 = 50$ s, $\tau_2 = 10$ s, and $\tau_3 = 30$ s are known and fixed [1, 2]. Malagutti has taken a wide range of $K \in [-0.25, 9.5]$ mmHg/ml/h so as to be able to describe the response of MAP of a wide range of patients varying in conditions [1]. Time-delay parameter is taken as $T \in (10, 50)$ and drug fraction parameter $\alpha \in (0.25, 0.75)$ [1, 7].

36.4 Controller Design

The primary objective of this controller design is to maintain the mean arterial pressure (MAP) at a value of 100 mmHg when an initial input of 150 mmHg, settling to 100 mmHg in 1 s, is given to the plant [7]. The other desired performance specifications of the designed controller are a faster rise time, a minimum settling time, the lowest possible steady state error, and no overshoot so as to improve the patient's post-operative recovery in some applications of the closed-loop drug administration

[7]. The limitations of the parameters of the plant model taken into account for our existing problem so as to design the controller accordingly are shown as follows [7]:

1. The mean arterial pressure of the patient should never be below 70 mmHg.
2. The settling time should be ≤ 10 min and strictly < 15 min.
3. The MAP response after settling must be within ± 5 mmHg of the final value of 100 mmHg.
4. The response of the proposed system should never be oscillatory at any point.
5. After settling, the value of the MABP should be within the desired limits of [70, 120] mmHg.

36.4.1 Interval Type-2 FLC-PID Controller

Fuzzy PID controllers are frequently suggested as a substitute to conventional PID controllers due to their similar nature of input–output relationships [11]. The output equation of the PID controller can be expressed as [7]:

$$U_{\text{pid}}(t) = K_p e_p(t) + K_i \int e_p(t) + K_d \frac{de_p(t)}{dt} \quad (36.2)$$

where $U_{\text{pid}}(t)$ is the output of conventional PID controller and $e_p(t)$ is the error between the desired output and the measured output of the controller. The gains K_p, K_d, and K_i are the proportional, derivative, and integral gains of the conventional PID controller. The primary design layout of proposed closed-loop system for administering infusion of drug is as shown in Fig. 36.1 [7]:

The error $e_p(t)$ for this controller design can be expressed as:

$$e_p(t) = D_{\text{MAP}_{\text{patient}}}(t) - \text{MAP}_{\text{patient}}(t) \quad (36.3)$$

where $D_{\text{MAP}_{\text{patient}}}(t)$ gives the value of desired MAP and $\text{MAP}_{\text{patient}}(t)$ is the measured MAP of the patient. In this controller design, an IT2-FLC is placed before the PID controller, so that it acts as a pre-compensator for the PID controller. This makes the transient performance of the overall control scheme better as IT2-FLC helps to

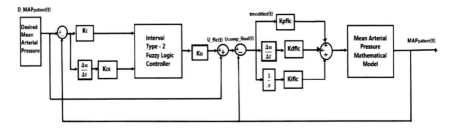

Fig. 36.1 Design scheme of IT2-FLC-PID controller

remove the overshoots and undershoots even in the presence of external noise and time-varying parameters [7, 11]. The tuning parameters are the scaling gains used to normalize the inputs to the controller so as to bound them in the range of $[-1, 1]$. The K_{DE} and K_E are the scaling gains to the input $e_p(t)$ and $\frac{de_p(t)}{dt}$. K_U is the scaling gain for the output of the IT2-FLC controller. The output of the IT2-FLC controller can be represented as [7]:

$$U_{\text{FLC}}(t) = K_U * f_{\text{IT2-FLC}}\left(K_E e_p(t), K_{DE}\frac{de_p(t)}{dt}\right) \quad (36.4)$$

where $U_{\text{FLC}}(t)$ is the output of the IT2-FLC controller with $f_{\text{IT2-FLC}}$ being a nonlinear mapping function of $e_p(t)$ and $\frac{de_p(t)}{dt}$. Nonetheless, the eventual input provided to PID control system is the summation of output of the IT2-FLC controller $U_{\text{FLC}}(t)$ and the desired MAP signal. The equation is expressed as [7]:

$$U_{\text{comp}_{\text{final}}}(t) = U_{\text{FLC}}(t) + D_{\text{MAP}_{\text{patient}}}(t) \quad (36.5)$$

where $U_{\text{comp}_{\text{final}}}(t)$ is the final compensated input provided to the PID controller. Using this input signal, the error signal to the PID controller is modified as follows [7]:

$$e_{\text{modified}}(t) = U_{\text{FLC}}(t) + U_{\text{comp}_{\text{final}}}(t) \quad (36.6)$$

The PID controller's output is now modified as follows [7]:

$$U_{\text{pid}}(t) = K_{P_{\text{flc}}} e_{\text{modified}}(t) + K_{I_{\text{flc}}} \int e_{\text{modified}}(t) + K_{D_{\text{flc}}} \frac{de_{\text{modified}}(t)}{dt} \quad (36.7)$$

where $K_{P_{\text{flc}}}$, $K_{D_{\text{flc}}}$, and $K_{I_{\text{flc}}}$ are the modified proportional, derivative, and integral gains of the PID controller.

36.4.2 Fractional Order-IT2-FLC-Proportional Derivative + Proportional Integral (FO-IT2-FLC-PD + PI) Controller

The blueprint of the proposed controller has two IT2-FLC in two paths is shown in Fig. 36.2.

The above controller scheme maintains the linear characteristics of the standard PID controller and incorporates the ability of interval type-2-FLC controller to deal with uncertain time-varying parameters as well as external disturbance present within the patient model [8]. The interval type-2 FLC works with two inputs error $e_p(t)$ and $\frac{d^\lambda e_p(t)}{dt}$, i.e., the fractional order derivative of error [8, 9]. The scaling gains K_p, K_d, K_i, and K_z are the tuning parameters whose main purpose is to limit the

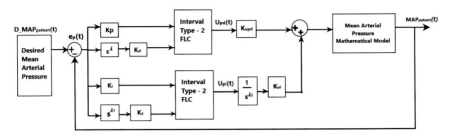

Fig. 36.2 Block diagram of (FO-IT2-FLC-PD + PI) controller

range of the input to the IT2-FLC block in the range of $[-1, 1]$ [8, 9]. These additional gains K_{upd} and K_{ul} are the scaling gains present at the output of each of the IT2-FLC dependent of input signals [8–10]. These gains have a self-tuning capability in set-tracking performance, making the controller to have a faster response time and less overshoot.

Fractional Order-IT2-FLC-Proportional Derivative

Fractional order PD controller can be expressed as in s-domain [8]:

$$U_{pd}(t) = K_p E(s) + s^\lambda K_d E(s) \tag{36.8}$$

where $K_p, K_d, E(s)$, and λ are the proportional gain, derivative gain, Laplace transform of error signal, and fractional order of derivative, respectively. This Eq. (36.8) can be modified using backward transformation $s = \frac{1-z^{-1}}{T}$ and substituting $T^\lambda U_{pd}(z) = \widehat{U_{pd}}(z)$ where T being the sampling period which in frequency domain as follows [8]:

$$\widehat{U_{pd}}(z) = (K_p T^\lambda + (1 - z^{-1}) K_d) E(z) \tag{36.9}$$

The binomial expansion is done as on Eq. (36.9) due to fractional number λ, i.e.,

$$\widehat{U_{pd}}(z) = K_p T^\lambda + K_d \sum_{k=0}^{k=\infty} (-1)^k b_k (z^{-1})^k E(s) \tag{36.10}$$

where $b_k = \frac{\lambda(\lambda-1)(\lambda-2)\ldots(\lambda-k+1)}{k!}$ are the binomial coefficients [8]. For sampled time-domain equation, inverse z-transform is applied on Eq. (36.10), which is as follows:

$$U_{pd}(nT) = K_p T^\lambda e(nT) + K_d \sum_{k=0}^{k=\infty} (-1)^k b_k (z^{-1})^k e((n-k)T) \tag{36.11}$$

The computation of Eq. (36.11) is impossible as there are infinite coefficients present along with delay terms [8]. As shown in Eq. (36.11), the following term of the equation illustrates Lubich formulation for fractional order of integration and/or derivation of order λ. The expression of Lubich formula for short memory size M, in this case $M = 100$, for any function $f(nT)$ is as follows [8]:

$$D^{\pm\alpha} f(nT) = T^{\pm\alpha} \sum_{k=0}^{M} \frac{(-1)^k \pm \alpha(\pm\alpha - 1)\ldots(\pm\alpha - k + 1)}{k!} f((n-k)T) \quad (36.12)$$

Using Eq. (36.12), Eq. (36.11) can be rewritten as:

$$U_{pd}(t) = K_{upd} * f_{\text{IT2-FLC}}\big(K_p T^\lambda e_p(nT), K_d T^\lambda D^\lambda e_p(nT)\big) \quad (36.13)$$

where $f_{\text{IT2-FLC}}$ is the same nonlinear mapping function as used before in IT2-FLC-PID controller.

Fractional Order-IT2-FLC-Proportional Integral

The output of the IT2-FLC for the second path for proportional gain and fractional order of λ_1 of derivative can be expressed as:

$$U_{pi}(t) = f_{\text{IT2-FLC}}\big(K_p T^{\lambda_1} e_p(nT), K_d T^{\lambda_1} D^{\lambda_1} e_p(nT)\big) \quad (36.14)$$

The fractional integral of the controller output $U_{pi}(t)$ in s-domain is as follows [8–10]:

$$U_{pi_{\text{int}}}(s) = \frac{1}{s^{\lambda_2}} U_{pd}(s) \quad (36.15)$$

where $U_{pd}(z)$ is the IT2-FLC controller's output in the second path of proposed controller design with λ_2 as the fractional order of integral using backward transformation of equation $s = \frac{1-z^{-1}}{T}$ and by substituting the $U_{pi_{\text{int}}}(z)(1 - z^{-1})^{\lambda_2} = \Delta U_{pi_{\text{int}}}(z)$ we can rewrite the equation as [8]:

$$\Delta U_{pi_{\text{int}}}(z) = T^{\lambda_2} U_{pi}(z) \quad (36.16)$$

By taking inverse z-transform, we can express the above Eq. (36.16), first 100 terms of binomial expansion of $(1 - z^{-1})^{\lambda_2}$ in the beginning according to Lubich formulation as shown in Eq. (36.16) [8]:

$$U_{pi_{\text{int}}}(nT) = T^{-\lambda_2} D^{-\lambda_2} \Delta U_{pi_{\text{int}}}(nT) \quad (36.17)$$

Replacing $T^{-\lambda_2}$ by K_{uI},

$$U_{pi_{\text{int}}}(nT) = K_{uI} D^{-\lambda_2} \Delta U_{pi_{\text{int}}}(nT) \quad (36.18)$$

The combined output of the controller which is applied as input to the patient model is the sum of Eqs. (36.13) and (36.18) which is expressed as:

$$U_{\text{controller\_output}}(nT) = U_{pd}(nT) + U_{pd_{\text{int}}}(nT) \tag{36.19}$$

36.4.3 Fuzzification, Control Rule Base, and Defuzzification of IT2-FLC

The fundamental elements of IT2-FLC comprise of a fuzzifier, control rule base, inference mechanism, and an output processing unit which has type-reducer and defuzzifier [7, 12]. The block diagram shown in Fig. 36.3 conveys the basic working of the IT2-FLC.

The only difference between the IT2-FLC and IT1-FLC is in its output processing stage where IT2-FLC has an additional type-reducer block which converts the type-2 fuzzy sets acquired from the inference mechanism in reduced type-1 fuzzy sets. These reduced type-1 fuzzy set are then converted into crisp output by the defuzzifier [7, 9].

Fuzzification

Type-2 fuzzy sets consist of two type-1 MFs, upper membership function (UMF), and lower membership function (LMF) which confides a footprint of uncertainty (FOU) between them [7]. The FOU represents the uncertainties in the input/output variables. For the presented controller design of IT2-FLC, for two input variables

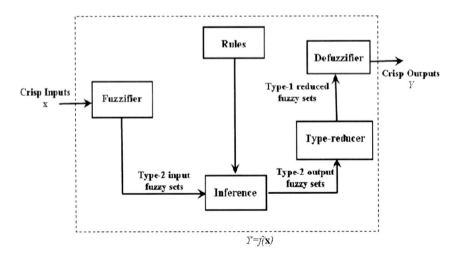

Fig. 36.3 Block diagram of interval type-2 fuzzy logic controller

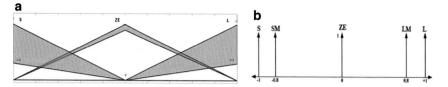

Fig. 36.4 **a** Input variable of IT2-FLC, **b** output variable of IT2-FLC

$e_p(t)$ and $\frac{de_p(t)}{dt}$, three triangular membership functions, namely small (S), zero (ZE), and large (L) are used as shown in Fig. 36.4a [7, 9]. For one output variable, five singleton membership functions are taken, namely small (S), small medium (SM), zero (ZE), large medium (LM), and large (L) as shown in Fig. 36.4b [7, 9].

The antecedent and consequent membership function can be represented as $\tilde{A}_{flc}$ and $\tilde{B}_{flc}$, with $(\overline{\mu}_{A_{flc}}, \overline{\mu}_{B_{flc}})$ and $(\underline{\mu}_{A_{flc}}, \underline{\mu}_{B_{flc}})$ as the functions for UMF and LMF, respectively. Consequently, the overall firing strength of nth rule is expressed as [7]:

$$\widetilde{fs_m} = \left[\underline{fs^m}, \overline{fs}^m\right] \quad (36.20)$$

where $\widetilde{fs_m}$ is the overall firing strength which can be expressed as [7]:

$$\underline{fs^m} = (\underline{\mu}_{A_{flc}} * \underline{\mu}_{B_{flc}}) \quad (36.21)$$

$$\overline{fs}^m = (\overline{\mu}_{A_{flc}} * \overline{\mu}_{B_{flc}}) \quad (36.22)$$

where * is the product implication between the two type-2 fuzzy sets.

Control Rule Base

With two input and one output variable for the controller, total nine rules are formulated. The rule base employed for IT2-FLC controller is given in Table 36.1 [7].

Table 36.1 Rule base of IT2-FLC

Error $e_p(t)$	Rate of change of error $\frac{de_p(t)}{dt}$		
	S	ZE	L
S	S	SM	ZE
ZE	SM	ZE	LM
L	ZE	LM	L

Defuzzification

In this paper, center of sets type-reduction technique is applied due to its appropriate computational complexity. It can be expressed as [7]:

$$\mu_{\text{flc}} = \frac{\mu_{lp} + \mu_{rp}}{2} \quad (36.23)$$

where μ_{lp} and μ_{rp} are the left and right end points for calculated reduced type-1 fuzzy sets using the following formula [7]:

$$\mu_{lp} = \frac{\sum_{m=1}^{lp} \overline{fs}^m G_m + \sum_{m=lp+1}^{M} \underline{fs}^m G_m}{\sum_{m=1}^{lp} \overline{fs}^m + \sum_{lp+1}^{M} \underline{fs}^m} \quad (36.24)$$

$$\mu_{rp} = \frac{\sum_{m=1}^{rp} \overline{fs}^m G_m + \sum_{m=rp+1}^{M} \underline{fs}^m G_m}{\sum_{m=1}^{rp} \overline{fs}^m + \sum_{rp+1}^{M} \underline{fs}^m} \quad (36.25)$$

where G_m is the consequent interval set and M is the total number of rules of the controller. The Karnik–Mendel type reduction technique is used for obtaining the switching points lp and rp [7, 11]. The surface plot of the interval type-2 fuzzy logic design (IT2-FLC) is shown in Fig. 36.5.

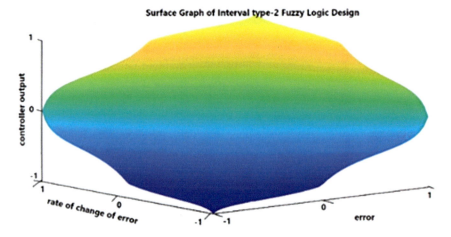

Fig. 36.5 Surface plot of IT2-FLC-PID control scheme

36.5 Cuckoo Search Optimization

Cuckoo search optimization technique is inspired by the brood parasitism of some of the species of cuckoo birds by laying their eggs in other host bird's nests [12]. CSA appears as a dormant technique among the other nature-inspired state-of-the-art algorithm because of its fewer initial parameters, independence of the convergence rate on parameters and longer step size during iteration [7, 12]. For yielding a new nest, Lévy flight law is used which is expressed as:

$$x_j(t_p + 1) = x_j(t_p) + \alpha \oplus \text{Levy}(\lambda) \quad (36.26)$$

where $\alpha > 0$ is the step size of the Levy flight law, $x_j(t_p + 1)$ is the new solution, and $x_j(t_p)$ is the current solution [7, 12]. In both the control scheme presented, parameters of the controllers need to be tuned to their optimal values for a robust and efficient controller design. To achieve that, a fitness function needs to be defined which needs to be minimized in order to get the optimal value of parameters. In the proposed work, the weighted sum of integral absolute error (IAE) and integral absolute change in output of the controller as expressed in equation [7]:

$$f_{\text{fitness\_value}} = w_1 * \int |e_p(t)| dt + w_2 * \int |\Delta I_{\text{patient}}| dt \quad (36.27)$$

where w_1 and w_2 are the weights assigned to each individual performance index taken into consideration equal to 0.5 [7]. The algorithm of the cuckoo search algorithm is as follows [7, 12]:

Step 1: Initialize an initial population of 25 bird nests. Initialize number of generations as 100. Initialize an iteration counter $I = 100$:
Step 2: While ($I < 100$):
  Step 3: Generate a cuckoo randomly using Levy flight. Evaluate its fitness value f_{new}.
  Step 4: Select a nest among the initial population of bird nests and evaluate its fitness value f_{old}.
    Step 5: If ($f_{\text{new}} < f_{\text{old}}$): Replace nest with f_{old} with new solution of f_{new}.
    Else: Retain the nest with f_{old} as new solution.
  Step 6: Remove a fraction of $p_a = 0.25$ of poor fitness value and replace them with new nest using Levy flight law. Increment $I = I + 1$.
Step 7: Keep the final best solution.

36.6 Result and Discussion

Simulation results of IT2-FLC-PID controller and FO-IT2-FLC-PD + PI controller for output response of MAP are presented for fixed $K_{patient}$ sensitivity $= -0.25$ mmHg/m/h, fluctuations in $K_{patient}$ sensitivity in the range of $(-0.25, -0.3)$ mmHg/m/h, and an external noise disturbance on the output of the mathematical model. The open-source type-2 fuzzy toolbox is used for implementation of IT2-FLC controller [7]. For obtaining optimal parameters and simulation results, MATLAB/Simulink platform is used. CSA is used for optimization of both controller designs presented in this research work. The parameters for IT2-FLC-PID controller using CSA are referred from [7]. For finding the optimal value of the tuning parameters of FO-IT2-FLC-PD + PI controller, a standard reference step input from 150 to 100 mmHg at a step time of 1 s is applied to the controller [8]. The optimal values of the parameters of FO-IT2-FLC-PD + PI for fixed $K_{patient}$ patient sensitivity is obtained and given in Table 36.2.

The convergence graph for FO-IT2-FLC-PD + PI controller is shown in Fig. 36.6 showing to settling after 50 iterations.

The MAP and error graph are shown in Fig. 36.7 for fixed $K_{patient}$ patient sensitivity $= -0.25$ mmHg/(mh^{-1}). From the graphs, we can observe that FO-IT2-FLC-PD + PI controller has a faster rise time while maintaining the MAP within $\pm$ 5mmHg.

For robust analysis of both proposed controllers, firstly $K_{patient}$ patient sensitivity is given sudden fluctuations within a range of $[-0.25, -0.95]$ mmHg/(mh^{-1}) to both the controller schemes. The fluctuations given to the patient model is shown in Fig. 36.8.

The MAP and error graph are shown in Fig. 36.9 for fluctuations on $K_{patient}$ patient sensitivity $= -0.25$ mmHg/(mh^{-1}). From the graphs, we can observe that FO-IT2-FLC-PD + PI controller has a faster rise while maintaining the MAP within $\pm$ 5mmHg as we have seen first case.

From these graphs, it is clear that FO-IT2-FLC-PD + PI controller settles faster than IT2-FLC-PID controller. The FO-IT2-FLC-PD + PI controller is well-tuned and performs satisfactorily according to the conditions of controller design. For second section of robust analysis, a white noise of 1 mmHg variance is introduced to the mean arterial pressure, i.e., output of patient model to both controller schemes. The noise given to the model is shown in Fig. 36.10.

Table 36.2 Optimal parameters of FO-IT2-FLC-PD + PI using cuckoo search algorithm

K_p	K_d	K_i
2.7593	0.6765	0.0546
K_z	K_{upd}	K_{uI}
0.8988	4.9748	4.5471
λ	λ_1	λ_2
0.0343	0.00137	0.0112

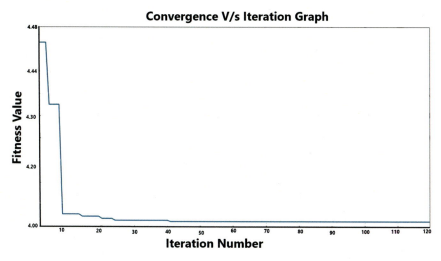

Fig. 36.6 Convergence versus iteration graph of optimization of cuckoo search algorithm for FO-IT2-FLC-PD + PI

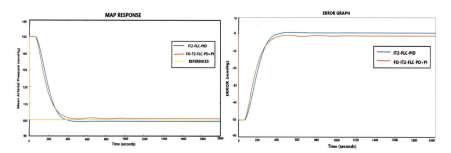

Fig. 36.7 MAP response and error graph of FO-IT2-FLC-PD + PI and IT2-FLC-PID controller for $K_{patient} = -0.25$ mmHg/(mh^{-1})

The MAP and error graph are shown in Fig. 36.11 for noise introduced at the output of the patient model.

There is no overshoot present in any case of both the controller scheme in both cases. For performance comparison of both the controllers, $Y_{settling\_time}$ and IAE values are presented in Table 36.3 where FO-IT2-FLC-PD + PI shows superior performance over the other controller in fixed patient gain as well as robust analysis.

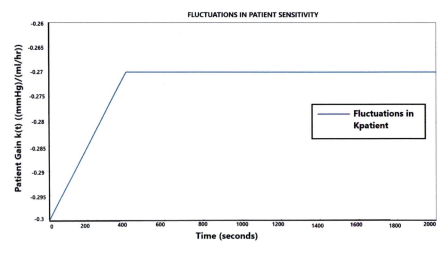

Fig. 36.8 Fluctuations in K_{patient} sensitivity in mmHg(mh^{-1})$^{-1}$

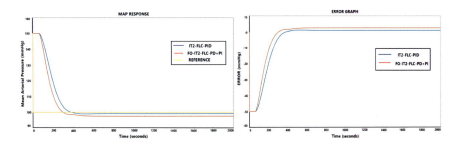

Fig. 36.9 MAP response and error graph of FO-IT2-FLC-PD + PI and IT2-FLC-PID controller for fluctuations in K_{patient} patient sensitivity

From Table 36.3, we can observe that FO-IT2-FLC-PD + PI controller shows better performance than IT2-FLC-PID controller under fixed patient sensitivity $K_{\text{patient}} = -0.25$ mmHg/(mh^{-1}) with better setting time and lower IAE value. In case of fluctuations in patient's sensitivity and in presence of external noise at the output introduced in the mathematical model, the simulation results of FO-IT2-FLC-PD + PI controller show better performance. The proposed controller FO-IT2-FLC-PD + PI also showed better disturbance/noise rejection than proposed controller due to the use of IT2-FLC in control design.

36 Performance Analysis of Different Controller Schemes of Interval ... 475

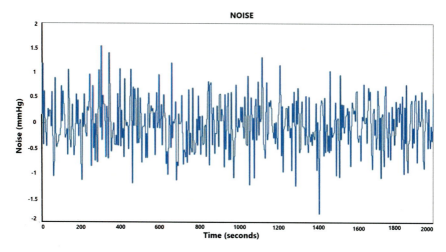

Fig. 36.10 Noise introduced to patient model

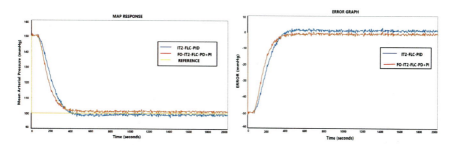

Fig. 36.11 MAP response and error graph of FO-IT2-FLC-PD + PI and IT2-FLC-PID controller in presence of noise

Table 36.3 Summary of performance specifications of two control schemes

Controller	Cases	$Y_{\text{settling\_time}}$ (s)	IAE
FO-IT2-FLC-PD + PI	Fixed K_{patient} patient sensitivity	332	1.014×10^4
	Fluctuation in K_{patient} patient sensitivity		1.051×10^4
	Noise rejection at output		1.09×10^4
IT2-FLC-PID [7]	Fixed K_{patient} patient sensitivity	350	1.155×10^4
	Fluctuation in K_{patient} patient sensitivity		1.076×10^4
	Noise rejection at output		1.16×10^4

36.7 Conclusion

In this paper, the designing of an optimal and robust controller for the auto-supervision of infusion of drugs for controlling MAP in critically-ill patients during surgery and in post-operative conditions is studied. The IT2-FLC is chosen as the base for designing the controller scheme for the above-mentioned problem due to its ability to cope with uncertainties in patient's sensitivity and presence of disturbance in the surroundings. Two efficient controller schemes, IT2-FLC-PID controller, and FO-IT2-FLC-PD + PI controller are presented. For effective tuning of parameters in both the controller schemes, cuckoo search algorithm is applied. The simulation results and the performance index, i.e., settling time, overshoot, and IAE values, show the superior performance of FO-IT2-FLC-PD + PI controller over IT2-FLC-PID controller.

References

1. Malagutti, N., Dehghani, A., Kennedy, R.A.: Robust control design for automatic regulation of blood pressure. IET Control Theory Appl. **7**(3):387–396 (2013). https://doi.org/10.1049/iet-cta.2012.0254
2. Sondhi, S., Hote, Y.: Fractional order PI controller with specific gain-phase margin for MABP control. IETE J. Res. **61** (2015). https://doi.org/10.1080/03772063.2015.1009395
3. Soltesz, K., Sjöberg, T., Jansson, T., Johansson, R., Robertsson, A., Paskevicius, A., Liao, Q., Qin, G., Steen, S.: Closed-loop regulation of arterial pressure after acute brain death. J. Clin. Monitor. Comput. **32**(3), 429–437 (2018)
4. Su, T.-J., Wang, S.-M., Vu, H.-Q., Jou, J.-J., Sun, C.-K.: Mean arterial pressure control system using model predictive control and particle swarm optimization. Microsyst. Technol. **24**, 147–153 (2018)
5. da Silva, S.J., Scardovelli, T.A., da Silva Boschi, S.R.M., Rodrigues, S.C.M., da Silva, A.P.: Simple adaptive PI controller development and evaluation for mean arterial pressure regulation. Res. Biomed. Eng. **35**, 157–165 (2019)
6. Tasoujian, S., Salavati, S., Franchek, M., Grigoriadis, K.: Robust IMC-PID and parameter-varying control strategies for automated blood pressure regulation. Int. J. Control Autom. Syst. **17**(7), 1803–1813 (2019)
7. Sharma, R., Deepak, K.K., Gaur, P., Joshi, D.: An optimal interval type-2 fuzzy logic control based closed-loop drug administration to regulate the mean arterial blood pressure. Comput. Meth. Program. Biomed. **185**, 105167 (2020)
8. Chhabra, H., Mohan, V., Rani, A., et al.: Robust nonlinear fractional order fuzzy PD plus fuzzy I controller applied to robotic manipulator. Neural Comput. Appl. **32**, 2055–2079 (2020). https://doi.org/10.1007/s00521-019-04074-3
9. Nagarsheth, S.H., Sharma, S.N.: The impact of fractional-order control on blood pressure regulation. Int. J. E-Health Med. Commun. (IJEHMC) **12**(3), 38–54 (2021). https://doi.org/10.4018/IJEHMC.20210501.oa3
10. Kumar, A., Raj, R.: Design of a fractional order two layer fuzzy logic controller for drug delivery to regulate blood pressure. Biomed. Signal Process. Control **78**, 104024 (2022) ISSN 1746-8094. https://doi.org/10.1016/j.bspc.2022.104024
11. Mendel, J.M.: Uncertain Rule-Based Fuzzy Systems: Introduction and New Directions, 2nd edn, pp. 229–234, 600–608. Springer (2017)
12. Joshi, A.S., Kulkarni, O., Kakandikar, G.M., Nandedkar, V.M.: Cuckoo search optimization—a review. Mater. Today Proc. **4**(8), 7262–7269 (2017). ISSN 2214-7853

Chapter 37
Early Detection of Alzheimer's Disease Using Advanced Machine Learning Techniques: A Comprehensive Review

Subhag Sharma, Tushar Taggar, and Manoj Kumar Gupta

Abstract Alzheimer's disease (AD) is a slow-paced irreversible brain disease and a neurodegenerative disorder that accounts for approximately 70% of the dementia cases estimated worldwide, the number of which totals to more than 46 million. AD affects the brain's thinking capacities along with significant memory loss. People with onset of aging are found to be more prone, with greater memory loss, cognitive difficulties, etc. Currently, there exists no fixed cure for AD, but early detection and characterization are proven to be helpful. Methodologies like electroencephalograms (EEG), magnetic resonance imaging (MRI), computed tomography, positron emission tomography (PET) scan, etc., are helpful in providing information regarding the persisting conditions of the brain cells. Computer-aided diagnosis (CAD) along with biomedical data processing when applied to machine learning and deep learning methodologies has vastly helped sophisticated techniques like CNNs, SVMs, etc., to evolve and achieve promising prediction accuracies. This paper provides a review and critical evaluation of recent research on early detection of AD using ML techniques which employ a variety of complex optimization and statistical techniques to obtain a better accuracy score. Along with advancement in computational capabilities, other factors such as preprocessing and feature extraction along with class imbalance have distinctively helped improve the prediction score which has overall helped produce better prediction with respect to earlier detection of AD.

Keywords CAD · SVM · CNN · MMSE · Feature selection · Alzheimer's disease · Mild cognitive impairment

---

S. Sharma · T. Taggar (✉) · M. K. Gupta
School of Computer Science and Engineering, Shri Mata Vaishno Devi University (SMVDU), Katra, India
e-mail: tushartaggar2804@gmail.com

S. Sharma
e-mail: mr.subhagsharma@gmail.com

M. K. Gupta
e-mail: manoj.cst@gmail.com

© The Author(s), under exclusive license to Springer Nature Singapore Pte Ltd. 2024
P. K. Jha et al. (eds.), *Proceedings of Congress on Control, Robotics, and Mechatronics*, Smart Innovation, Systems and Technologies 364,
https://doi.org/10.1007/978-981-99-5180-2_37

37.1 Introduction

Alzheimer's disease (AD) is a neurodegenerative disease paced by aging factors with most of the patients having observable symptoms after the age of 65 years with the onset of shrinking of the hippo-campus area of the brain marked by significant loss of memory. An exact cure for AD has not yet been found, and anti-AD drugs can only slow disease progression. Therefore, early diagnosis of AD is a key component in treating this disease [1]. MRI scans, PET scans, SPET scans, electroencephalogram (EEG), etc., are few of the non-invasive techniques which help in early detection and prognosis of AD. The preliminary stage of AD is defined as mild cognitive impairment (MCI) in which a significant downfall in cognitive abilities such as speaking, thinking, and remembering is observed [1]. It is considered an intermediate state between the normal cognitive and AD or any other forms of dementia. It is often contemplated as a 'preclinical' early stage of AD [2]. It was found that few MCI patients (between 6 and 25%) later on developed AD. The first stage is described as normal control including patients having no clear symptoms and are mostly non-dementated. The second, third, and the final stages are known as mild AD, moderate AD, and severe AD, respectively, and involve full caregiver dependence [1, 3]. Thus, prognosis and early detection in case of AD plays an important role. Machine learning techniques of both unsupervised and supervised have played an important role in computer-aided diagnosis (CAD) along with various techniques like advanced feature selection methodologies [4]. Various performance measuring probabilistic approaches have helped achieve better accuracy scores. The limitations of human brain, instinct, and standard measurements do not quite suffice the requirements of the current times, and hence, machine learning methodologies are gaining interests of various academicians and professionals in helping diagnose disease at an earlier stage that too with better accuracy scores. In general, the overview of various ML techniques is well directed toward achieving higher accuracy, specificity, and sensitivity. Various similar techniques with different feature selection methods provided different efficiency scores. Various forms of data are incorporated either in the form of 3D or 2D MRI scans [5–8], PET scans [9], EEG signals [2], computed tomography, and SPET scans [10]. The data is filtered and used according to the various parameters set for various experimental setups as described under various articles. However, the general presumption during the review of certain articles hinted toward the issue that convolutional neural networks (CNN) do not tend to be the ideal choice for many of the researchers. The problem with CNNs is common, especially in medical applications, where tagged data is often insufficient for training a CNN's filter bank. A CNN cannot train parameter tuning and learn highly distinguishing visual features if there is a lack of labeled datasets available for training. For this reason, few studies have considered the application of CNNs to medical applications [11]. However, the most common techniques that provided with high accuracy rates included SVM [10] as most common technique followed by CNN [11], etc. The paper is split into four sections, of which Sect. 37.2 describes various literature reviews.

Sections 37.3 and 37.4 describe the analysis and discussion. Finally, Conclusions are drawn in Sect. 37.5.

37.2 Literature Review

This section of the paper gives a detailed study on various techniques proposed by researchers for early diagnosis of possible patients of AD with help of various medical tools.

37.2.1 Data Acquisition

Data for most of the experiments was collected with conscious beforehand divisions made according to three groups defined in the introduction. Public datasets are used widely, but certain researchers tend to use various thresholds set for sampling like the ADNI dataset used extensively for the purpose with MRI and PET scans, uses mini mental state examination (MMSE) scores, clinical dementia rating (CDR) for classification of its subjects according to their MCI status and divided groups according to their respective scores [12]: (1) Healthy subjects (non-MCI), i.e., normal control (NC): MMSE scores between 24–30, a (CDR) of 0. (2) MCI subjects: MMSE scores between 24 and 30, education-adjusted Wechsler memory scale logical memory II scores, and a CDR of 0.5 indicate objective memory loss. (3) Mild AD: MMSE scores in the range 20–26 and a CDR of 0.5 or 1.0, and satisfying other criteria for probable advancement to AD [12]. Other datasets used widely include Kaggle Alzheimer's classification dataset (KCAD), recognition of Alzheimer's disease dataset (ROAD) [6], and OASIS [9].

37.2.2 Data Preprocessing and Feature Selection

MRI scans from the ADNI database were preprocessed, co-registered using statistical parametric mapping (SPM) software [13, 14] and voxel-based morphometric measurement 8 (VBM8) toolbox [8], and segmented using a variety of data preprocessing methods. For EEG signals, the discrete-wavelet-transform (DWT) is almost always used. This is to classify the wide time frames for dividing the low frequency windows and the narrow time windows for the high frequencies [15].

$$\psi a, b(t) = \frac{1}{\sqrt{a}} \psi \frac{(t-b)}{a}; \quad a > 0, -\infty < b < \infty \quad (37.1)$$

where a and b specify the wavelet scale and transform [9]. Other techniques widely include spatial normalization, noise filter, intensity normalization, and activation estimation [9] for PET scans to improve accuracy and score. Certain researchers have adopted various means of feature selection by both mathematical means like using Hjorth parameter which consist of three parameters: activity, mobility, and complexity defined as [9]:

$$\text{Activity} = \text{var}(y(t)) \tag{37.2}$$

$$\text{Mobility} = \sqrt{\left(\frac{\text{var}\left(\frac{dy(t)}{dt}\right)}{\text{var}(y(t))}\right)} \tag{37.3}$$

$$\text{Complexity} = \frac{\text{Mobility}\left(\frac{dy(t)}{dt}\right)}{\text{Mobility}(y(t))} \tag{37.4}$$

where $y(t)$ is the EEG signal. The activity denotes the signal power, mobility signifies the mean frequency, and complexity gives the standard deviation of the power spectrum [9] or by selection of various biomarkers like division between white matter area and gray matter area. It is evident that group-wise distribution during the sampling provided better results as in case of [8] when ICA feature extraction was applied to ADNI dataset and data was processed using SVM with nonlinear kernels. Similarly, targeting specific areas of the brain as biomarkers also showed variety in accuracies reported with some close to 99% [1]. Savio and Graña [17] uses deformation-based morphometry (DBM) [17] which selected features to boost accuracy of the proposed SVM. On the basis of selection of characters, methodologies may be divided into single modal [15] or multimodal approaches [18], both methods having respective advantages.

37.2.3 Techniques Used

Various methodologies when used gave various accuracy scores as presented in Table 37.1 for the detection of AD using various kinds of data either in the form MRI scans, PET scan, and EEG signal analyses; from the analysis of various techniques, it is very easy to analyze that both supervised and unsupervised also when tweaked with respect to various features and data processing techniques excellent accuracy was achieved. Linear support vector machine (LSVM) [2] implemented with sequential backward feature selection (SBFS) [2] gave outstanding results of 99.4 ± 1.8% [2]. Along with features selection, cross-validation methods like Leave One Subject Out (LOSO) [16] and tenfold cross-validation [10] helped overcome the problem of high biased and high variance making models for more flexible in terms of prediction

[2]. It was also observed in DL-based algorithms like correlation-based K-means and singular value decomposition (CLC-KSVD) which gave better results when data was collected from distinct zones of brains touching approx. 89% with spatial temporal zones [1]. As discussed previously that groups were divided on the basis of MMSE scores, it was observed that highest accuracy was achieved when group-wise data was divided reaching 97% for AD versus MCI [8, 11].

Some papers did include various methodology applied over a single dataset and reported best of accuracy. In advanced frameworks like ADGNET which is form of

Table 37.1 Comparison of various ML techniques, applied over various datasets

Ref. No.	Modality	Dataset details	Technique	Accuracy (%)
[9]	EEG	EEG recordings of mild cognitive impairment (MCI) patients of 53 patients	LDA	78.33
[21]	MRI	ADNI	Decision tree	85
[20]	MRI	3D MP-RAGE dataset	Artificial neuron network	86
[8]	MRI-2D subspaces	ADNI	SVM	86.4 for MCI versus AD
[1]	EEG	61 patients sampled	KSVD	88.9
[9]	PET	AR-mining for feature extraction[a]	Classification using feature extraction	91.33
[17]	MRI	OASIS	SVM	92
[11]	MRI	ADNI	PCANet + K-means clustering	92.5
[18]	MRI	(Dataset-66, dataset-160, and dataset-255) by Harvard Medical School	TSVM (Twin-SVM)	93.05
[10]	SPECT	3D SPECT brain images	SVM	96.91
[16]	EEG	86 patients sampled	KNN with DWT	97.64
[6]	MRI, 3D	ROAD	ADGNET	98.71
[19]	MRI	ADNI	SVM (supervised)	99.10
[2]	EEG	Dataset by [1]	LSVM with SBFS	99.4
[19]	MRI	ADNI	Logistic regression (unsupervised)	99.43
[6]	MRI, 2D	KACD	ADGNET	99.61

[a]Dataset undisclosed

CNN with single-input multiple-output (SIMO) architecture based on WSL framework on the backend and applying the classification subnetwork CSN and the reconstructed subnetwork RSN, we obtained an excellent accuracy of 99.61% in identifying and classifying AD using multimodal brain imaging data. The model is proven to be superior to the two methods based on SOTA WSL [6].

37.2.4 Critical Evaluation

Reviewing models, it is evident that advanced features extraction methods like variance, kurtosis, Shannon entropy, and Hjorth parameter in case of EEG signal helped classify and improve specificity, sensitivity, and accuracy scores [14]. Although in general, CNNs were scarcely used, but with advanced architecture shown in Fig. 37.1 [6], efficiency scores improved drastically reaching approx. 99% although both supervised and unsupervised learning tend to work efficiently; but it is denoted that various improvements can be made while performing various experiments. Dataset could be age matched instead of generalizing over a range of ages based on MMSE and CDR scores. It could be further helpful to separate amnestic and non-amnestic MCI cases. It would better prove accuracy scores [1]. It was also observed that various imaging techniques that provided multiple view scans when analyzed separately on based views performed better in case of one than another [22]. Liu et al. [23] discuss about patch-based deep multimodal learning (PDMML) to learn the loss of spatial information caused due to loss of flattening, help capture multi-view brain disease [25] and make the use of regions of interest (ROI) along with CNN methodologies. No doubt using standardized scores like MMSE and CDR have helped differentiate between probable and non-probable cases of AD, but, however, liberty may be at the helm of the researcher to choose from or before the data is collected to make models better adaptation of features and achieve a better clarity as to what marks the onset of AD. Advancement in deep learning is a huge benefit as it helps better predict the AD much before its onset and as a result of which the period of ailment can be elongated. Carcagni et al. in [26] describe the use of various pretrained models like ResNet and DensNet along with visual transformers and mask auto-encoders (MAE) to improve accuracy up to 7% same discussion for pretrained models which is carried out in [27]. Various researchers which describe the methods have failed to give the precise accuracy of their proposed models. As described previously, more of the permutations and combinations of features, methods, and data acquisition can be checked to achieve better performance.

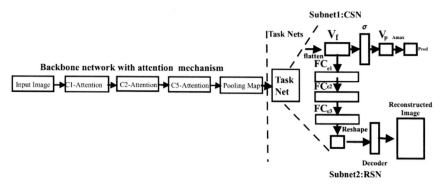

Fig. 37.1 Proposed architecture of CNN called ADGNET in [6] with two sub-networks (SN), classification SN (CSN), and the reconstruction SN (RSN) [6]

37.3 Analysis

After reviewing many of the techniques used for prognosis of AD, it is much clear that raw datasets when used provided lower efficiency as compared to that of when specific features were targeted. It can be further drawn from the comparison of various techniques used for the purpose as discussed in Table 37.1 that variations in datasets can also affect the results. In [22] using PCANet, the sagittal view provided better efficiency than the top view. The PCANet [22] employs a two-stages convolution learn by the principal component analysis for block-wise histograms, and for the feature mapping, outputting feature performed better reaching 97.01% AD versus MCI group, 92.6% for MCI versus NC, and 91.25% for AD versus MC 99.15% AD versus NC [11]. Along with henceforth mentioned techniques, separation of data from biomarkers such as gray matter (GM) mass and white matter (WM) mass has been considered promising [8]. Khedhe et al. [8] uses a combination of GM + WM and compares it with individual features separately. The outputs show clear effects of doing so with NC versus AD accuracy reaching 87.2%. Henceforth, it is noteworthy that advanced data collection techniques when applied to above-mentioned methods are bound to help improve the prediction rate of AD. Specific feature targeting as in case of EEG signals has also shown to improve efficacy drastically. However, it may be noted from various studies that using multiple features is always not helpful [24]. What really matters is to use appropriate methods with appropriate features to get better results, and applying all possible combinations of various methods at the helm of researchers is a good starting point. Datasets have also been a major influence over the score as in Table 37.1, it can be clearly seen that ADNI dataset is widely used and, hence, is consecutively updated and is one of the most preferred datasets; however, various techniques [1, 6, 8–10, 14–18] tend to use other datasets depending on modalities used and type of data along with other factors like conditions prevailing during tests and prerequisites for test. Tests like CDR and MMSE are considered to be ideal for selecting patients for sampling, and this demarcation has proved to be efficient, and we can thus conclude from [8, 11] that this method has provided with a better understanding to differentiate between MCI and non-MCI (both AD and Non-AD) and help draw better lines in terms of prognosis as far as AD is concerned.

37.4 Discussion

Although multiple fields of statistical and probabilistic approaches have already been explored, there still is a lack of unified approach both in terms of data collection and practical applications of various advanced methods to get feedback from real-world use cases. Multiple advancements have been made, but still many more are yet to be made to incorporate latest technologies and advancements in the medical research domain like, for example, a unified approach to collect medical data could be one of many steps. Using various advanced techniques and allowing researchers to assess higher quality datasets is obviously going to help the case. After reviewing many articles, a simple conclusion can be drawn that with advancement in the field of deep learning, it is necessary to provide appropriate data to work more effectively. Of all the articles reviewed for the sake of writing this paper, the beat of the accuracy was achieved in ADGNET using 2D gray scale images of MRI scans and the best accuracy for the supervised methods used of all techniques was achieved on using SVM on EEG dataset created by [1] research developments including the use of transfer learning methodologies which are proving to give interesting results and advancement on large public datasets is ought to improve efficacy [26, 27]. It would also be very interesting to know what early life habits cause AD in later stages of life as to avoid and prevent individuals and allow them to lead a healthy life in later stages of their lives. Implementation of various algorithms along with computer vision is still an unexplored area. Integration of the methodologies with newer instruments is ought to help the medical fraternity process information with a faster pace and which in turn is a boon for patients.

37.5 Conclusion

This study focuses on comparisons and evaluations of previous studies on the prognosis and prediction of AD utilizing supervised and unsupervised machine learning techniques. It provides a clear picture of recent trends in machine learning, including the types of data used in predicting early stages of Alzheimer's disease and the performance of machine learning methods. Using supervised method of machine learning, we came across the various methods like SVM [8, 10, 16] which perform astoundingly good and unsupervised learning method like ADGNET [6] was almost near to perfection, not only these basic techniques like linear discriminant analysis (LDA) provided not so bad results as well. The gap between human-level intelligence can be coped up using defined methods of data collection given in the previous section, and the data can be processed with any of the above-discussed methods to provide desired and better results. Open-source datasets need to be improved over time like that of ADNI, concurrent studies and experimentation need to be performed in the defined path to achieve better accuracy rates. The changes for sampling as defined require funding and research at a greater level. Also, incorporation of wider research techniques from the medical domain may help provide a better sense of the disease and provide insights on how the development takes place in various regions of the brain. Lifestyle changes may be adhered to so as for avoidance of related issues and complications in future.

References

1. Kashefpoor, M., Rabbani, H., Barekatain, M.: Supervised dictionary learning of EEG signals for mild cognitive impairment diagnosis. Biomed. Signal Process. Control **53**, 101559 (2019)
2. Movahed, R., Rezaeian, M.: Automatic diagnosis of mild cognitive impairment based on spectral, functional connectivity, and nonlinear EEG-based features. Comput. Math. Methods Med. **2022**, 1–17 (2022)
3. Alzheimer's Association: 2013 Alzheimer's disease facts and figures. Alzheimers Dement. **9**(2), 208–245 (2013)
4. Jie, B., Wee, C., Shen, D., Zhang, D.: Hyper-connectivity of functional networks for brain disease diagnosis. Med. Image Anal. **32**, 84–100 (2016)
5. Sathiyamoorthi, V., Ilavarasi, A., Murugeswari, K., Thouheed Ahmed, S., Aruna Devi, B., Kalipindi, M.: A deep convolutional neural network based computer aided diagnosis system for the prediction of Alzheimer's disease in MRI images. Measurement **171**, 108838 (2021)
6. Liang, S., Gu, Y.: Computer-aided diagnosis of Alzheimer's disease through weak supervision deep learning framework with attention mechanism. Sensors **21**(1), 220 (2020)
7. Ayadi, W., Elhamzi, W., Charfi, I., Atri, M.: A hybrid feature extraction approach for brain MRI classification based on bag-of-words. Biomed. Signal Process. Control **48**, 144–152 (2019)
8. Khedher, L., Illán, I., Górriz, J., Ramírez, J., Brahim, A., Meyer-Baese, A.: Independent component analysis-support vector machine-based computer-aided diagnosis system for Alzheimer's with visual support. Int. J. Neural Syst. **27**(03), 1650050 (2017)
9. Veeramuthu, A., Meenakshi, S., Manjusha, P.S.: A new approach for Alzheimer's disease diagnosis by using association rule over PET images. Int. J. Comput. Appl. **91**(9), 9–14 (2014)
10. Illán, I., et al.: Computer aided diagnosis of Alzheimer's disease using component based SVM. Appl. Soft Comput. **11**(2), 2376–2382 (2011)
11. Bi, X., Li, S., Xiao, B., Li, Y., Wang, G., Ma, X.: Computer aided Alzheimer's disease diagnosis by an unsupervised deep learning technology. Neurocomputing **392**, 296–304 (2020)
12. Zhang, D., Wang, Y., Zhou, L., Yuan, H., Shen, D.: Multimodal classification of Alzheimer's disease and mild cognitive impairment. Neuroimage **55**(3), 856–867 (2011)
13. Friston, K.J., Frith, C.D., Dolan, R.J., Price, C.J., Zeki, S., Ashburner, J.T., Penny, W.D.: Human Brain Function, 2nd edn. Academic Press, London (2003)
14. Friston, K.J., Ashburner, J., Kiebel, S.J., Nichols, T.E., Penny, W.D.: Statistical Parametric Mapping: The Analysis of Functional Brain Images. Academic Press, London (2007)
15. Amin, H.U., et al.: Feature extraction and classification for EEG signals using wavelet transform and machine learning techniques. Australas. Phys. Eng. Sci. Med. **38**(1), 139–149 (2015)
16. Safi, M., Safi, S.: Early detection of Alzheimer's disease from EEG signals using Hjorth parameters. Biomed. Signal Process. Control **65**, 102338 (2021)
17. Savio, A., Graña, M.: Deformation based feature selection for computer aided diagnosis of Alzheimer's disease. Expert Syst. Appl. **40**(5), 1619–1628 (2013)

18. Sharma, S., Mandal, P.: A comprehensive report on machine learning-based early detection of Alzheimer's disease using multi-modal neuroimaging data. ACM Comput. Surv. **55**(2), 1–44 (2022)
19. Alroobaea, R., et al.: Alzheimer's Disease Early Detection Using Machine Learning Techniques
20. Aguilar, C., et al.: Different multivariate techniques for automated classification of MRI data in Alzheimer's disease and mild cognitive impairment. Psychiatry Res. Neuroimaging **212**(2), 89–98 (2013)
21. Querbes, O., et al.: Early diagnosis of Alzheimer's disease using cortical thickness: impact of cognitive reserve. RöFo—Fortschritte auf dem Gebiet der Röntgenstrahlen und der bildgebenden Verfahren **182**(12) (2010).
22. Chan, T.H., Jia, K., Gao, S., Lu, J., Zeng, Z., Ma, Y.: PCANet: a simple deep learning baseline for image classification? IEEE Trans. Image Process. **24**, 5017–5032 (2015)
23. Liu, F., Yuan, S., Li, W., Xu, Q., Sheng, B.: Patch-based deep multi-modal learning framework for alzheimer's disease diagnosis using multi-view neuroimaging. Biomed. Signal Process. Control **80**, 104400 (2023)
24. Gallego-Jutglà, E., Solé-Casals, J., Vialatte, F., Elgendi, M., Cichocki, A., Dauwels, J.: A hybrid feature selection approach for the early diagnosis of Alzheimer's disease. J. Neural Eng. **12**(1), 016018 (2015)
25. Gao, S., Lima, D.: A review of the application of deep learning in the detection of Alzheimer's disease. Int. J. Cogn. Comput. Eng. **3**, 1–8 (2022)
26. Carcagnì, P., Leo, M., Del Coco, M., Distante, C., De Salve, A.: Convolution neural networks and self-attention learners for Alzheimer dementia diagnosis from brain MRI. Sensors **23**(3), 1694 (2023). https://doi.org/10.3390/s23031694
27. Sethi, M., Ahuja, S.: Alzheimer disease classification using MRI images based on transfer learning. In: Innovations in Computational and Computer Techniques: ICACCT-2021 (2022)

Chapter 38
Navigation of a Compartmentalized Robot Fixed in Globally Rigid Formation

Riteshni Devi

Abstract The subject of this research is navigation of a compartmentalized robot made up of several structures which are car-like with a global rigid formation. The system must move safely to a predetermined target in an already-known workspace that is filled with obstacles. A target convergence and obstacle avoidance method is put forth that is effective for any quantity of obstacles. The smallest distance between an obstacle and the point closest to it on each line segment that forms the rectangular protected zone of the car-like units can be calculated analytically using the minimum distance technique, as proposed in this paper. The Lyapunov-based control scheme (LbCS) is used to construct continuous acceleration-based controls. The computer simulations are presented to validate the proposed controllers, and the system demonstrates the rigorous upkeep of a rigid formation. The successful presentation of controllers opens further research in developing and applying controllers by considering various compartmentalized robots in swarming models, splitting and rejoining units, and applying rotational leadership concepts.

Keywords Compartmentalized robot · Navigation · Artificial potential fields

38.1 Introduction

From the beginning of the modern era, humanity has dreamed of a technologically advanced future in which humans and machines coexist and do similar tasks [1]. Multiple homogeneous or heterogeneous robots make up a multiple robot system. The challenge of motion planning and control problems in robotics is to produce continuous robot motion from one configuration to the next in a configuration space without colliding with obstacles. Since mobile robots must, by definition, move in the actual world to carry out their jobs, motion planning is especially important for these machines. Heuristic algorithms and classical algorithms are the two primary methods

R. Devi (✉)
University of the South Pacific, Suva, Fiji
e-mail: riteshnikaran@gmail.com

used in motion planning and control problems for numerous robots [2, 3]. There are several examples of classical algorithms that have been extremely successful, making them a crucial part of technological advancement. Among them are the novel artificial potential field (APF) technique, the creation of thorough roadmaps, the dismantling of cells, the use of neural networks and neuro-fuzzy systems, and the implementation of reactive strategies. [1, 4]. The Lyapunov-based control scheme (LbCS), which is just another application of the APF approach, is taken into consideration for the navigation problem in this paper. The control of the formation or decision-making in multiple robot systems can be centralized or decentralized, depending on the formation. In a centralized architecture, there is a central control agent, known as a leader robot, who has access to comprehensive environmental knowledge. The leader can communicate with other robots and has access to all of their information. An individual robot in the form of a computer might serve as the central control agent. The benefit of centralized control is that it allows for the creation of globally ideal plans since the leader has a comprehensive understanding of the whole universe. Additionally, a centralized control planner is used in this research. For big teams of robots, this control mechanism is ineffective and usually only works with a limited number of robots. One of the difficulties in keeping the formation rigid is retaining the patterns that the multiple car-like robots have formed. In contrast to globally rigid formations, where rigid maintenance of the formation is required, the split/rejoin formations allow for unlimited modifications to the pattern. When a locally rigid formation encounters obstacles and constraints, the pattern of the formation becomes momentarily deformed. Few current instances of formation control are [5, 6]

38.1.1 Contributions

This research proposes a novel method of formation control using a compartmentalized robot system in a confined environment. It integrates the Lyapunov-based control scheme (LbCS) and the centralized leader–follower design. Mathematical functions pertaining to constraints, inequalities, and other mechanical limitations associated with the robotic system are easier to design and implement in the controllers. As a result, the LbCS is typically preferred to other motion control approaches [5–11]. The motion planning and control problem have been given a new dimension due to introducing of a compartmentalized robot capable of doing many tasks simultaneously. The CR is divided into smaller units in this situation and maneuvers with the aid of a predetermined sub-unit that serves as the leader, in parallel, without colliding while preserving the rigid formation. The following is a list of the contributions that the paper has made:

1. Designing a set of new LbCS-compliant time-invariant nonlinear acceleration-based controllers for compartmentalized robots collision-free motions.
2. A new centralized leader to guide the motion of a compartmentalized robot.
3. Designing and controlling a globally rigid formation.

38.1.2 Related Works

In addition to microscopic approaches, macroscopic, temporal, and spatial compartmentalization have been developed due to the evolution of compartmentalized systems. Ongoing research is being conducted in materials science, synthetic biology, medical engineering, waste collection and transportation, protein engineering and agricultural crop treatments [12, 13]. Scientists created robots that team up to make bigger ones. The research objective, which was directed by Marco Dorigo of the IRIDIA AI intelligence lab at the University Libre de Bruxelles, was to give robots the ability to behave independently but to collaborate with one another when necessary. One main robot controls the merged group of robots. The primary robot is then in charge of the larger, newly produced robot [14]. The robots on exhibit can separate into individual bodies with separate controls, combine to form larger bodies with a single central controller, and self-heal by removing or replacing damaged sections. [14]. Several publications investigating motion planning and control have focused on various robotic systems in formation, including mobile manipulators, articulated robotic limbs, and wheeled platforms. This study introduces a hitherto unexplored notion known as the compartmentalized robot, a creative variation on the globally rigid formation. The work of [15, 16] served as inspiration for the idea. The compartmentalized robot used in this study has four units that are adjoint and are mathematically bound as a rigid formation of intelligently automated machines that are managed using a Lyapunov-based control strategy. The benefits include the successful execution of several activities that are impossible to accomplish with straightforward cooperative agents, great energy efficiency, and a reduction in total expenses [15, 16].

38.2 System Modeling

In this section, a kinematic model for compartmentalized robots is devised. As depicted in Fig. 38.1, it is composed of four car-like homogeneous units joined together in the Euclidean plane.

Definition 38.1 A compartmentalized robot C_0 is a mobile robot that has been separated into four car-like robots called units. Consider C_0 to be a rectangle with length $(2L + 4\epsilon_1)$ and width $(2l + 4\epsilon_2)$ and an orientation angle of θ_1 in the $z_1 z_2$ plane.

The term "car like unit" defined as follows is taken from [5].

Definition 38.2 The nth car-like unit C_n is disk-shaped with a radius of r_n and is centered at (x_n, y_n). Precisely, the nth unit is the set

$$C_n = \{(z_1, z_2) \in \mathbf{R}^2 : (z_1 - x_n)^2 + (z_2 - y_n)^2 \leq r_n\} \tag{38.1}$$

for $n \in \{1, 2, 3, 4\}$.

Fig. 38.1 A schematic diagram of a compartmentalized robot C_0

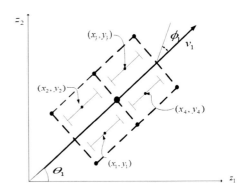

The equations for coordinates of the center of each unit are given by:

$$x_n = x_1 + \sum_{n=1}^{4}\left(2\left[\lceil\tfrac{n}{2}\rceil - \lceil\tfrac{n}{4}\rceil\right]\left(\tfrac{L}{2}+\epsilon_1\right)\cos\theta_n + \left[(-1)^{\lfloor\tfrac{n}{2}\rfloor}-1\right]\left(\tfrac{l}{2}+\epsilon_2\right)\sin\theta_n\right)$$

$$y_n = y_1 + \sum_{n=1}^{4}\left(2\left[\lceil\tfrac{n}{2}\rceil - \lceil\tfrac{n}{4}\rceil\right]\left(\tfrac{L}{2}+\epsilon_1\right)\sin\theta_n - \left[(-1)^{\lfloor\tfrac{n}{2}\rfloor}-1\right]\left(\tfrac{l}{2}+\epsilon_2\right)\cos\theta_n\right)$$

(38.2)

The system's kinematic model, which automatically accounts for nonholonomic restrictions, is explained as follows, using information taken from [5].

$$\dot{\theta}_n = \tfrac{v_n}{L}\tan\phi := \omega_n,$$
$$\dot{v}_n = \sigma_{n1},\ \dot{\omega}_n = \sigma_{n2},$$
$$\dot{x}_n = \dot{x}_1 - (2\left[\lceil\tfrac{n}{2}\rceil - \lceil\tfrac{n}{4}\rceil\right]\left(\tfrac{L}{2}+\epsilon_2\right)\dot{\theta}_n\sin\theta_n + \left[(-1)^{\lfloor\tfrac{n}{2}\rfloor}-1\right]\left(\tfrac{l}{2}+\epsilon_1\right)\dot{\theta}_n\cos\theta_n))$$
$$= v_1\cos\theta_1 - \tfrac{L}{2}\omega_1\sin\theta_1$$
$$- \sum_{n=1}^{4}(2\left[\lceil\tfrac{n}{2}\rceil - \lceil\tfrac{n}{4}\rceil\right]\left(\tfrac{L}{2}+\epsilon_2\right)\omega_n\sin\theta_n + \left[(-1)^{\lfloor\tfrac{n}{2}\rfloor}-1\right]\left(\tfrac{l}{2}+\epsilon_1\right)\omega_n\cos\theta_n)$$
$$\dot{y}_n = \dot{y}_1 + (2\left[\lceil\tfrac{n}{2}\rceil - \lceil\tfrac{n}{4}\rceil\right]\left(\tfrac{L}{2}+\epsilon_2\right)\dot{\theta}_n\cos\theta_n + (-1)^n\left[(-1)^{\lfloor\tfrac{n}{2}\rfloor}-1\right]$$
$$\left(\tfrac{l}{2}+\epsilon_1\right)\dot{\theta}_n\sin\theta_n)) = v_1\sin\theta_1 + \tfrac{L}{2}\omega_1\cos\theta_1$$
$$+ \sum_{n=1}^{4}(2\left[\lceil\tfrac{n}{2}\rceil - \lceil\tfrac{n}{4}\rceil\right]\left(\tfrac{L}{2}+\epsilon_2\right)\omega_n\cos\theta_n - \left[(-1)^{\lfloor\tfrac{n}{2}\rfloor}-1\right]\left(\tfrac{l}{2}+\epsilon_1\right)\omega_n\sin\theta_n)$$

(38.3)

The rotational and translational velocities are ω_n and v_n, respectively, and θ_n provides the orientation of C_n with the z_1-axis. σ_{n1} and σ_{n2} are the acceleration controllers of C_0. The rectangular protective regions of the C_n sub-units are built using the nomenclature of [16]. The vertices of the rectangular protection zone for C_n are first calculated with regard to the lead unit's center of mass $((x_1, y_1))$.

$$x_{ni} = x_1 + \left[(-1)^{\lceil \frac{n}{2} \rceil} + (-1)^{\lceil \frac{i}{2} \rceil} + 1\right]\left(\frac{L}{2} + \epsilon_1\right)\cos\theta_n$$
$$+ \left[(-1)^{\lfloor \frac{n}{2} \rfloor} + (-1)^{\lfloor \frac{i}{2} \rfloor} - 1\right]\left(\frac{l}{2} + \epsilon_2\right)\sin\theta_n,$$
$$y_{ni} = y_1 + \left[(-1)^{\lceil \frac{n}{2} \rceil} + (-1)^{\lceil \frac{i}{2} \rceil} + 1\right]\left(\frac{L}{2} + \epsilon_1\right)\sin\theta_n$$
$$- \left[(-1)^{\lfloor \frac{n}{2} \rfloor} + (-1)^{\lfloor \frac{i}{2} \rfloor} - 1\right]\left(\frac{l}{2} + \epsilon_2\right)\cos\theta_n.$$
(38.4)

Rectangular protective regions have the benefit of providing comparatively more open space when compared to circular ones.

Definition 38.3 The protective region's nth unit's ith boundary line is a line segment in the $z_1 z_2$-plane that runs from the points (x_{ni}, y_{ni}) to (x_{mi}, y_{mi}) where $m = i + 1 - 4\lfloor i/4 \rfloor$. The ith border line is specifically the set:
$Bl_{ni} = \{(z_1, z_2) \in \mathbf{R}^2 : (z_1 - X_{ni})^2 + (z_2 - Y_{ni})^2 = 0)\}$

The following are the parametric equations describing the peripheral line segments of the rectangular protective region: Let $S = \sin\theta_n$ and $C = \cos\theta_n$, then

$$X_{ni} = x_1 + \left[(-1)^{\lceil \frac{n}{2} \rceil} + (-1)^{\lceil \frac{i}{2} \rceil} + 1\right]\left(\frac{L}{2} + \epsilon_2\right)C$$
$$+ \left[(-1)^{\lfloor \frac{n}{2} \rfloor} + (-1)^{\lfloor \frac{i}{2} \rfloor} - 1\right]\left(\frac{l}{2} + \epsilon_1\right)S$$
$$+ (-1)^{\lfloor \frac{i}{2} \rfloor}\left[2(1 - (-1)^i)\left(\frac{L}{2} + \epsilon_2\right)C - 2(1 + (-1)^i)\left(\frac{l}{2} + \epsilon_1\right)S\right]\lambda_{ni},$$
$$Y_{ni} = y_1 + \left[(-1)^{\lceil \frac{n}{2} \rceil} + (-1)^{\lceil \frac{i}{2} \rceil} + 1\right]\left(\frac{L}{2} + \epsilon_2\right)$$
$$- \left[(-1)^{\lfloor \frac{n}{2} \rfloor} + (-1)^{\lfloor \frac{i}{2} \rfloor} - 1\right]\left(\frac{l}{2} + \epsilon_1\right)C$$
$$+ (-1)^{\lfloor \frac{i}{2} \rfloor}\left[2(1 - (-1)^i)\left(\frac{L}{2} + \epsilon_2\right)S + 2(1 + (-1)^i)\left(\frac{l}{2} + \epsilon_1\right)C\right]\lambda_{in}.$$
(38.5)

λ_{in} is a non-negative scalar, restricted to the interval $[0, 1]$ for all $i, n \in \{1, 2, 3, 4\}$.

38.3 Research Objective

A compartmentalized robot C_0 led by a sub-unit C_1 is an efficient variant of a globally rigid formation. There is no requirement for a virtual leader, as has been customary for globally rigid formations [17]. Throughout the motion, the spacing in between the units is maintained in the formation. In this study, C_0 is given the objective of avoiding obstacles under the leadership of C_1. The primary goal of the research is to develop controllers for the leader C_1, which will drive C_0 toward the target in an obstacle-filled environment. The LbCS is utilized to create potential field functions.

38.3.1 Car-Like Units Potential Field Functions

The units C_n for $n \in 2, 3, 4$ keep a fixed distance from the leader C_1 and from each other in the compartmentalized robot C_0. The other sub-units C_n do not necessitate distinct targets since the distance remains constant during the motion. Together with their leader C_1, they move rigidly toward a shared goal.

Definition 38.4 The target set for C_0 is a disk centered at (P_{11}, P_{12}) and radius r_T, described as the set

$$T = \{(z_1, z_2) \in \mathbf{R} : (z_1 - P_{11})^2 + (z_2 - P_{12})^2 \leq r_T\}. \tag{38.6}$$

The Euclidian distance between the point (x_1, y_1) of the car 1 and the target must be measured in order to calculate target attraction. Therefore,

$$M(x) = \frac{1}{2}\left[(x_1 - P_{11})^2 + (y_1 - P_{12})^2 + v_1^2 + \omega_1^2\right]. \tag{38.7}$$

An auxiliary function ensures that the compartmentalized robot converges to its destination and that the controllers vanish upon its arrival at the target.

$$B(x) = \frac{1}{2}[(x_1 - P_{11})^2 + (y_1 - P_{12})^2]. \tag{38.8}$$

38.3.2 The Kinematic Constraints

Consider a workspace filled with $k \in \mathbb{N}$ fixed obstacles. The C_0 must steer clear of these obstacles. The avoidance is accomplished by using equations for the units C_n.

Definition 38.5 The kth fixed obstacle is an elliptic disk with center (o_{1k}, o_{2k}). Precisely the disk-shaped obstacle is a set

$$FO_k = (z_1, z_2) \in \mathbb{N}^2 : \left(\frac{z_1 - o_{1k}}{a_k}\right)^2 + \left(\frac{z_2 - o_{2k}}{b_k}\right)^2 \leq 1, k \in \mathbb{N}. \tag{38.9}$$

For minimum distance technique, the distance function $D_{ni} = \sqrt{(X_{ni} - o_{1k})^2 + (Y_{ni} - o_{2k})^2}$ for each points on the outside 8 line segments of (X_{ni}, Y_{ni}), described in (5), and the fixed obstacle is determined. Note that the derivation given is for the special case where $a_k = b_k$. Differentiating D_{ni} with respect to λ_{ni} to obtain λ_{ni} where D_{ni} is the minimum results in:

$$\begin{aligned}\lambda_{ni} = \Big\{ &o_{1k} - [x_1 + [(-1)^{\lceil\frac{n}{2}\rceil} + (-1)^{\lceil\frac{i}{2}\rceil} + 1]\left(\tfrac{L}{2} + \epsilon_1\right)\cos\theta_n \\ &+ [(-1)^{\lfloor\frac{n}{2}\rfloor} + (-1)^{\lfloor\frac{i}{2}\rfloor} - 1]\left(\tfrac{l}{2} + \epsilon_2\right)\sin\theta_n] \\ &+(-1)^{\lfloor\frac{i}{2}\rfloor}\left[4\kappa\left(\tfrac{L}{2} + \epsilon_1\right)\cos\theta_n - 4\varphi\left(\tfrac{l}{2} + \epsilon_2\right)\sin\theta_n\right]\Big\}d_{ni} \\ + \Big\{ &o_{2k} - [y_1 + [(-1)^{\lceil\frac{n}{2}\rceil} + (-1)^{\lceil\frac{i}{2}\rceil} + 1]\left(\tfrac{L}{2} + \epsilon_1\right)\sin\theta_n \\ &- [(-1)^{\lfloor\frac{n}{2}\rfloor} + (-1)^{\lfloor\frac{i}{2}\rfloor} - 1]\left(\tfrac{l}{2} + \epsilon_2\right)\cos\theta_n] \\ &+(-1)^{\lfloor\frac{i}{2}\rfloor}\left[4\kappa\left(\tfrac{L}{2} + \epsilon_1\right)\sin\theta_n + 4\varphi\left(\tfrac{l}{2} + \epsilon_2\right)\cos\theta_n\right]\Big\}r_{ni}\end{aligned} \quad (38.10)$$

Let $C = \cos\theta_n$ and $S = \sin\theta_n$, then

$$d_{ni} = \frac{4\kappa\left(\tfrac{L}{2} + \epsilon_1\right)C - 4\varphi\left(\tfrac{l}{2} + \epsilon_2\right)S}{\left[4\kappa\left(\tfrac{L}{2} + \epsilon_1\right)C - 4\varphi\left(\tfrac{l}{2} + \epsilon_2\right)S\right]^2 + \left[4\kappa\left(\tfrac{L}{2} + \epsilon_1\right)S + 4\varphi\left(\tfrac{l}{2} + \epsilon_2\right)C\right]^2}$$

and

$$r_{ni} = \frac{\left[4\kappa\left(\tfrac{L}{2} + \epsilon_1\right)S + 4\varphi\left(\tfrac{l}{2} + \epsilon_2\right)C\right]^2}{\left[4\kappa\left(\tfrac{L}{2} + \epsilon_1\right)C - 4\varphi\left(\tfrac{l}{2} + \epsilon_2\right)S\right]^2 + \left[4\kappa\left(\tfrac{L}{2} + \epsilon_1\right)S + 4\varphi\left(\tfrac{l}{2} + \epsilon_2\right)C\right]^2}$$

where $\kappa = (1 - (-1)^i)/2$ and $\varphi = (1 + (-1)^i)/2$. The following equation is designed for obstacle avoidance.

$$O_{nk}(x) = \frac{1}{2}\left(\left(\frac{X_{in} - o_{1k}}{a_k}\right)^2 + \left(\frac{Y_{in} - o_{2k}}{b_k}\right)^2 - 1\right) \quad (38.11)$$

for $k \in \mathbb{N}$ and $n \in \{1, 2, 3, 4\}$.

38.3.3 Modulus Bounds on Velocities of C_0

The compartmentalized robot's translational speed and steering angle must be constrained for safety and to account for its built-in limits, which can be seen through its units. The values of the rotational and translational velocities of C_n are necessary to maintain the formation of C_0. Since C_1 is the leader, the velocity of $v_1 = v_n$. From [5, 6], the following are the further dynamical restrictions put on the rotational and translational velocities of C_1 (and hence the other units):

1. $|v_1| \leq v_{max}$ where v_{max} is C_1's maximal achievable speed.

2. $|\omega_1| \leq |v_1|/|\rho_{min}| \leq |v_{max}|/|\rho_{min}|$. This is due to the fact that $v_1^2 \geq \rho_{min}^2 \omega_1^2$ where ρ_{min} is the minimum turning radius and is given as $\rho_{min} = L/\tan\phi_{max}$.

To ensure that C_1 operates within these limitations, the following artificial obstacles are constructed in conformance with the LbCS.

$$AO_{11} = v_1 \in \mathbb{R} : v_1 \leq v_{max} \text{ or } v_1 \geq v_{max},$$
$$AO_{12} = \omega_1 \in \mathbb{R} : \omega_1 \leq v_{max}/|\rho_{min}| \text{ or } \omega_1 \geq v_{max}/|\rho_{min}|.$$

Avoidance functions from the work of [15, 16] are implemented to avoid artificial obstacles.

$$W_1(x) = \frac{1}{2}\left(v_{max}^2 - v_1^2\right), \tag{38.12}$$

$$W_2(x) = \frac{1}{2}\left(\frac{v_{max}}{|\rho_{min}|}^2 - \omega_1^2\right). \tag{38.13}$$

38.4 Nonlinear Controller Design

This section defines the Lyapunov function, from which the controllers are derived. Prior to that, the following tuning and control parameters are introduced:

1. $\alpha > 0$, for C_1 to converge to the target,
2. $\beta_1, \beta_2 > 0$, to avoid the artificial obstacles posed by dynamic constraints,
3. $\gamma_{nk} > 0, n \in \{1, 2, 3, 4\}$ and $k \in \mathbb{N}$, to avoid the kth disk-shaped obstacle.

Table 38.1 shows the numerical values of these parameters utilized in the simulations. The Lyapunov function for system (3) is generated using the control parameters:

$$V(\mathbf{x}) = M(\mathbf{x}) + B(\mathbf{x})\left[\frac{\beta_1}{W_1(\mathbf{x})} + \frac{\beta_2}{W_2(\mathbf{x})} + \sum_{k=1}^{p}\sum_{n=1}^{4}\frac{\gamma_{nk}}{O_{nk}(\mathbf{x})}\right]. \tag{38.14}$$

From which the LbCS control laws are formed.

Theorem 38.1 *Consider the motion of C_0 and its sub-units C_n, $n \in \{1, 2, 3, 4\}$, which is regulated by the ODEs defined by system(3). The goal is to designate C_1 as the leader and induce convergence to the target while preserving a specified rigid formation in a collision-free environment. The following acceleration-based controllers for C_0, which the leader C_1 regulates, are based on attractive and repulsive field functions.*

$$\sigma_{11} = -(\delta_1 v_1 + \frac{\partial V(\mathbf{x})}{\partial x_1} \cos\theta_1 + \frac{\partial V(\mathbf{x})}{\partial y_1} \sin\theta_1)/(1 + B\frac{\beta_1}{W_1^2}),$$
$$\sigma_{12} = -(\delta_2 \omega_1 - \frac{\partial V(\mathbf{x})}{\partial x_1}\frac{L}{2} \sin\theta_1 + \frac{\partial V(\mathbf{x})}{\partial y_1}\frac{l}{2} \cos\theta_1 + \frac{\partial V(\mathbf{x})}{\partial \theta_1})/(1 + B\frac{\beta_2}{W_2^2}). \quad (38.15)$$

38.5 Stability Analysis

Theorem 38.2 *A fixed point* $x_1^* = (P_{11}, P_{12}, \theta_1, v_1, \omega_1) \in \mathbb{R}^5$ *is the equilibrium point of* C_0, *regulated by the leader* C_1 *then* $x_e = x_1^* \in D(V(\mathbf{x}))$ *is the point of stability for system* (3).

Proof : The Lyapunov function $V(\mathbf{x})$ of system (3) is a continuous, defined and positive on the domain $D(V(\mathbf{x})) = \{x \in \mathbb{R}^{5n} : W_r(\mathbf{x}) > 0, r \in \{1, 2\}, O_{nk}(\mathbf{x}) > 0, \forall n = 1, 2, 3, 4, k \in \mathbb{N}\}$, it therefore verifies that $V(\mathbf{x})$ fulfills the following:

1. $V(\mathbf{x})$ continuous, defined and positive over the domain $D(V(\mathbf{x}))$ in the neighborhood of the point $\mathbf{x}_e$ of system (3),
2. $V(\mathbf{x}_e) = 0$ since $M(\mathbf{x}_e) = 0$ and $B(\mathbf{x}_e) = 0$,
3. $V(\mathbf{x}_e) > 0$ for all $\mathbf{x} \in D(V(\mathbf{x}))/\mathbf{x}_e$,
4. Given the convergence parameters $\delta_1, \delta_2 > 0$, then

$$\dot{V}(\mathbf{x}) = \left[-\delta_1 v_1^2 - \delta_2 \omega_1^2\right] \leq 0.$$

$V(\mathbf{x})$ is therefore classified as the system 3's Lyapunov function, and $\mathbf{x}_e$ is a stable equilibrium point.

38.6 Simulation Results

This section shows simulation results for a compartmentalized robot. The stability analysis derived from the Lyapunov function is quantitatively validated.

The first trajectory planning scenario captures a straightforward circumstance to show how well the control laws work in practice. As shown in Fig. 38.2, the entire unit goes from an initial condition to a final one. The values of different parameters, constraints, and other variables used in the simulation are shown in Table 38.1. The second case as illustrated in Fig. 38.3 shows how well the control laws work when there are arbitrary circular obstacles of varying sizes present. It could involve the use of autonomous robots that are intended to convey supply of products, conduct search and rescue operations after natural disasters, and launch enemy attacks. The compartmentalized robot C_0 in Fig. 38.3 navigates from an initial point to a target while avoiding obstacles of random sizes placed throughout the workspace. The robot was snapped at various points in its trajectory. The efficiency of minimum distance technique is demonstrated here by the fact that the obstacle and the closest point on the outer line segment of the rectangular protective zone avoid one another

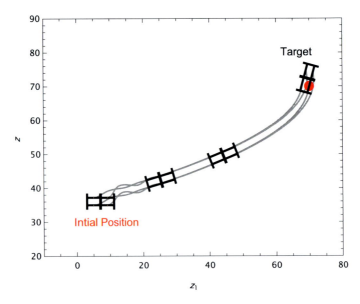

Fig. 38.2 *Scenario 1*. Integrated snapshot of C_0 with no obstacles in the workspace

as they approach each other. The controllers made sure that the equilibrium state and the system state would eventually converge. The leader units, C_1 translational and rotational velocities, are displayed in Fig. 38.4. The nonlinear controllers are shown in Fig. 38.5 . The plots' peaks and spikes demonstrate the compartmentalized robot's maximal obstacle avoidance and energy dissipation.

38.7 Conclusion

This study addresses the compartmentalization idea derived from cells, the basic unit of life, as an MPC problem for a car-like vehicle using the LbCS, a novel artificial potential field approach. To help a compartmentalized robot navigate a cluttered, obstacle-filled environment, a system of time-invariant, continuous nonlinear acceleration-based controllers is developed. Simulation results demonstrate the efficacy of the control laws and durability of the systems navigation in an environment flooded with obstacles.

In this study, the compartmentalized robot is arranged in a rigid structure. A single sub-unit, the leader, primarily facilitates the tasks carried out by the compartmentalized robot. The other units do not perform any task that may be the subject of future investigation.

All of the compartmentalized robot's units moved toward the target at the same speed as the leader unit in order to preserve the formation. It is evident that the units retained their distance relative to the leader from the starting state to the end state.

Table 38.1 Table of numerical values of initial state, constraints, and control and convergence parameters

Initial position	$(x_0, y_0) = (5, 35)$ m
Initial orientation	$(\theta_0) = 0$ rad
Initial translational velocity	$(v_0) = 0$ m/s
Initial rotational velocity	$(\omega_0) = 0$ rad/s
Length of car-like unit	$L = 4m$
Width of car-like unit	$l = 2m$
Radius of circular protection region	$r_v = 2.37$ m
Target center	$(P_{11}, P_{12}) = (75, 80)$ m
Radius of target	$r_T = 0.4$ m
Target center	$(o_{11}, o_{12}) = (50, 58)$ m
Radius of obstacle	$r_o = 6$ m
Workspace dimensions	$0 \leq z_1 \leq 100, 0 \leq z_2 \leq 100$
Number of CR	$n = 1$
Number of car-like units	$i = 4$
Number of obstacles	Scenario 1 : $k = 0$
	Scenario 2 : $k = 5$
Clearance parameters	$\epsilon_1 = 0.1m, \epsilon_2 = 0.1$ m
Control parameters	$\beta_1 = 0.001, \beta_2 = 0.002$.
	Simulation 1 : $\alpha = 0.00001$.
	Simulation 2 : $\alpha = 0.0005$.
Convergence parameters	Scenario 1 : $\delta_1 = 10, \delta_2 = 10$.
	Scenario 2 : $\delta_1 = 1, \delta_2 = 1$.

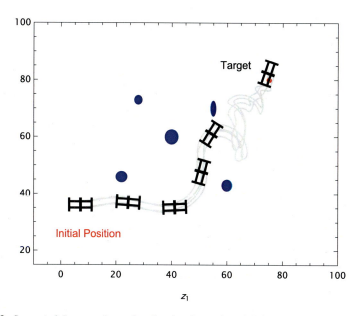

Fig. 38.3 *Scenario 2*. Integrated snapshot showing the motion of C_0 in obstacle ridden environment

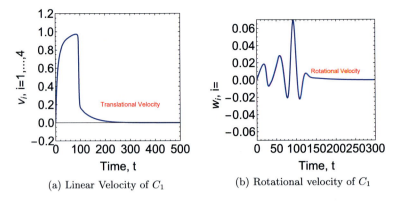

Fig. 38.4 Graphs linear and rotational velocity of C_1

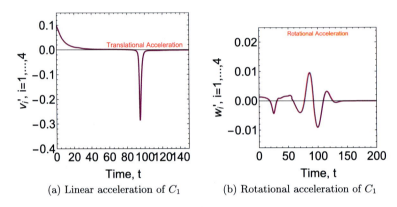

Fig. 38.5 Graphs of linear and translational acceleration of C_1

This illustrates quantitatively the strict maintenance of the rigid formation and, consequently, the globally rigid formation. The stability of the system is also guaranteed by the control laws suggested in this study. This has been demonstrated through the use of Lyapunov's direct method and numerical verification using computer simulations. The idea behind such a system is that robots will not be created and constructed for a single purpose anymore. Robots with segmented designs will be able to adjust independently to both shifting work demands and their own capabilities. This study is a theoretical exposition of the applicability of a new artificial potential field method, namely the Lyapunov-based control scheme, with a primary emphasis on demonstrating the effectiveness of the control laws. Future research in this field will utilize multiple units as a swarm model navigation of a compartmentalized robot.

References

1. Fan, X., Guo, Y., Liu, H., Wei, B., Lyu, W.: Improved artificial potential field method applied for autonomous underwater vehicle path planning. Math. Prob. Eng. (2020)
2. Masehian, Ellips, Sedighizadeh, Davoud: Classic and heuristic approaches in robot motion planning-a chronological review. World Acad. Sci. Eng. Tech. **23**(5), 101–106 (2007)
3. Chaudhary, K., Lal, G., Prasad, A., Chand, V., Sharma, S., Lal, A.: Obstacle avoidance of a point-mass robot using feedforward neural network. In: 2021 3rd Novel Intelligent and Leading Emerging Sciences Conference (NILES), pp. 210–215 (2021)
4. Khatib, O.: Real-time obstacle avoidance for manipulators and mobile robots. In: Autonomous robot vehicles, pp. 396–404. Springer (1986)
5. Sharma, B.N., Raj, J., Vanualailai, J.: Navigation of carlike robots in an extended dynamic environment with swarm avoidance. Int. J. Rob. Nonlinear Cont. **28**(2), 678–698 (2018)
6. Raghuwaiya, K., Sharma, B., Vanualailai, J.: Leader-follower based locally rigid formation control. J. Adv. Transp. (2018)
7. Prasad, A., Sharma, B., Vanualailai, J., Kumar, S.: Motion control of an articulated mobile manipulator in 3D using the Lyapunov-based control scheme. Int. J. Cont. pp. 1–15 (2021)
8. Kumar, S.A., Vanualailai, J., Sharma, B.: Lyapunov-based control for a swarm of planar non-holonomic vehicles. Math. Comput. Sci. **9**(4), 461–475 (2015). October
9. Chand, R.P., Kumar, S.A., Chand, R., Tamath, R.: Lyapunov-based controllers of an n-link prismatic robot arm. In: 2021 IEEE Asia-Pacific Conference on Computer Science and Data Engineering (CSDE), pp 1–5 (2021)
10. Chand, R., Kumar, S.A., Chand, R.P., Reddy, S.: A car-like mobile manipulator with an n-link prismatic arm. In: 2021 IEEE Asia-Pacific Conference on Computer Science and Data Engineering (CSDE), pp. 1–6 (2021)
11. Chand, Ravinesh, Chand, Ronal P., Kumar, Sandeep A.: Switch controllers of an n-link revolute manipulator with a prismatic end-effector for landmark navigation. PeerJ Comput. Sci. **8**(e885), 1–27 (2022)
12. Kourist, R., Sabín, J.: Compartmentalization in Biocatalysis, pp. 89–112 (2021)
13. B. Yousra, Ahmed, E.A.: Collection and transport of waste types by compartmentalised vehicles. In: 2020 5th International Conference on Logistics Operations Management (GOL), pp. 1–5 (2020)
14. Mathews, N., Christensen, A., O'Grady, R., Mondada, F., Dorigo, M.: Mergeable nervous systems for robots. Nat. Commun. **8**, 12 (2017)
15. Vanualailai, J., Sharma, B., ichi Nakagiri, S.: An asymptotically stable collision-avoidance system. Int. J. Non-Linear Mech. **43**(9), 925–932 (2008)
16. Sharma, B., Vanualailai, J., Singh, S.: Lyapunov-based nonlinear controllers for obstacle avoidance with a planar n-link doubly nonholonomic manipulator. Rob. Autonomous Syst. **60**(12), 1484–1497 (2012)
17. Amrita, D., Vanualailai, J., Sandeep. A.K., Bibhya, S.: A cohesive and well-spaced swarm with application to unmanned aerial vehicles. In: Proceedings of the 2017 International Conference on Unmanned Aircraft Systems, Miami, FL, USA, pp. 698–705 (2017)

Chapter 39
Motion Planning and Navigation of a Dual-Arm Mobile Manipulator in an Obstacle-Ridden Workspace

Prithvi Narayan, Yuyu Huang, and Ogunmokun Olufeni

Abstract This paper presents a design of velocity controllers for a 2-link dual-arm mobile manipulator which is required to move from its starting configuration to a final position while targeting to avoid multiple fixed circular obstacles of random sizes and positions, and observing all mechanical singularities which are associated with the system. With the help of Lyapunov-based control scheme (LbCS), non-linear time-invariant continuous velocity-based control laws are formulated which enable the center of the car-like mobile structure to converge to a predetermined target position and the links attain a final orientation. The method also guarantees stability associated with the system proving the use of direct method of Lyapunov. The computer simulations illustrate the effectiveness of the proposed technique.

Keywords Dual-arm mobile manipulator · Motion planning · Obstacles · Lyapunov-based control scheme · Stability

39.1 Introduction

The work related to motion planning and navigation of autonomous mobile robots are a continuous research targeting to find the optimal and collision-free path of motion in different sorts of environments. This would imply that the reliability and accuracy of the path taken are evident [1, 2]. The main targeting factor for this research field is to provide an algorithm, navigation system, or functional controllers in order to navigate a mobile robot to the predetermined target without any assistance of the human operator [3].

Planning paths for autonomous robots have become an interesting field of research nowadays since mobile robots are mostly used in applications such as industrial fields and similarly in academic and research fields. Robotic mechanical systems such as car-like robot, human-like robot, flying robot, industrial robot, and mobile

---

P. Narayan (✉) · Y. Huang · O. Olufeni
The University of the South Pacific, Suva, Fiji
e-mail: prithvinarayan14@gmail.com

© The Author(s), under exclusive license to Springer Nature Singapore Pte Ltd. 2024
P. K. Jha et al. (eds.), *Proceedings of Congress on Control, Robotics, and Mechatronics*, Smart Innovation, Systems and Technologies 364,
https://doi.org/10.1007/978-981-99-5180-2_39

manipulators are appearing in various real-word applications such as transportation, unmanned factories, and intelligent cars. Mobile manipulator system had been used for multiple purposes and plays a vital role in the improvement of human life in the modern era [2, 4].

A challenging problem in robotics is the find-path problem which aims to find the shortest, collision-free path for robots from the initial position toward a predefined ending location [5]. The path of robot and its motion planning algorithms can be categorized into three major approaches: classical approach, heuristic approach, and machine learning. Cell decomposition [6], artificial potential field (APF) [2, 7, 8], sampling-based methods, and sub-goal network are examples of classical algorithms. Neutral network, fuzzy logic, and nature-inspired algorithms are common heuristic-based approaches [9, 10]. Evaluating the associated parameters of information and gathering knowledge to become intelligent is mainly done by using machine learning. Machine learning is based on gathering experience or data and converts the raw data into useful information by mining data. It is mostly used in computer science nowadays and also used by many industries for automating tasks and doing complex data analysis. The classical methods are designed and functioning well in static known environment. As heuristic approach is based on problems in [11] and mostly inspired by nature of multiple interacting objectives as explained in [12] which are more popular in robot navigation field compared to classical methods. On the other hand, the machine learning is a subfield of artificial intelligence (AI) as machines incorporate data and "learn" for themselves. It is currently the most promising tool in the AI kit for businesses as it is also affected by the accuracy of an algorithm. Each approach has its advantages and drawbacks, and the major drawback of classical method is high time-consuming and trapping in local minima which makes them inefficient in practice [8], whereas heuristic approach can overcome the local minima problem. More discussions on advantages and drawbacks can be found in [13].

One of the popular methods in robotic motion and path planning which is artificial potential field of classical approach will be used in this research. Artificial potential field-based methods can be applicable for both the static and dynamic environments as well as for the known or unknown environments. It is based on the second method of Lyapunov, which is currently a powerful mathematical technique for the study of the qualitative behavior of natural or man-made systems that could be modeled, in an approximate way, by differential equations [14]. In this paper, we will design the velocity-based, time-invariant, continuous nonlinear controllers so that the mobile manipulator can move from a starting configuration to a final position targeting to avoid multiple fixed circular obstacles of random sizes and positions, and observing all mechanical singularities associated with the system. For the purpose of this research, we have considered a 2D mobile manipulator with attached two articulated 2-link arms of revolting joints attached to a car-like mobile platform in a two-dimensional plane. To solve this path and motion planning problem, the Lyapunov-based control scheme (LbCS) [14, 15] has been utilized. The main contributions of this research are to develop the velocity controller for the two arms (4 links) mobile manipulator. The proposed LbCS can effectively avoid circular obstacles in order to move from the initial position to its target. Comparing it to the similar works in the literature, Sharma et.al [15] have worked on the motion control of one-arm (having n-links) mobile manipulator. In this research, we extend the work of [15] to a dual-arm mobile manipulator.

This paper is organized as follows. In Sect. 39.2, the schematic representation of the mobile manipulator model is discussed and the kinematic equations governing the motion of the car-like mobile structure is given. In Sect. 39.3, the attraction and relating avoidance functions required for the robot to converge to a designated target, avoid fixed obstacles along its route and satisfy all the system singularities. In Sect. 39.4, the attraction and avoidance functions are combined to formulate a Lyapunov function from which the velocity controllers are extracted. Stability analysis is categorized in Sect. 39.5 with Sect. 39.6 demonstrating the simulation results, and finally Sect. 39.7 gives the concluding comments for work which can be targeted in the future.

39.2 The Mobile Manipulator Model

Let us consider a 2D mobile manipulator that has two articulated 2-link arms with revolute joints attached to a car-like mobile platform in the z_1z_2-plane as shown in Fig. 39.1.

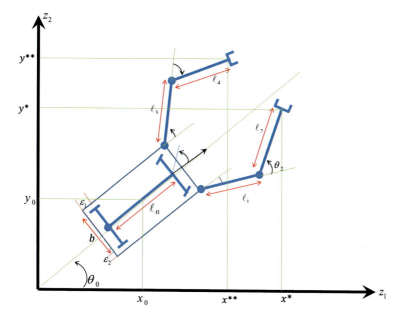

Fig. 39.1 Schematic representation of a two-dimensional mobile manipulator

With the aid of Fig. 39.1, the following assumptions are made:

1. The coordinates of the center of the car-like mobile platform are (x_0, y_0) and the orientation with respect to the z_1-axis is θ_0;
2. ϕ is the steering angle of the mobile platform;
3. The ith Link has a length of ℓ_i and angular position $\theta_i(t)$ as shown;
4. The coordinate of the grippers (end-effectors) is (x^*, y^*) and (x^{**}, y^{**}).

Remark The coordinate of the end-effectors can be expressed as:

$$x^* = x_0 + \frac{\ell_0 + 2\epsilon_1}{2}\cos\theta_0 + \frac{b + 2\epsilon_2}{2}\sin\theta_0 + \ell_1\cos(\theta_0 + \theta_1) + \ell_2\cos(\theta_0 + \theta_1 + \theta_2);$$

$$y^* = x_0 + \frac{\ell_0 + 2\epsilon_1}{2}\sin\theta_0 - \frac{b + 2\epsilon_2}{2}\cos\theta_0 + \ell_1\sin(\theta_0 + \theta_1) + \ell_2\sin(\theta_0 + \theta_1 + \theta_2);$$

$$x^{**} = x_0 + \frac{\ell_0 + 2\epsilon_1}{2}\cos\theta_0 - \frac{b + 2\epsilon_2}{2}\sin\theta_0 + \ell_3\cos(\theta_0 + \theta_3) + \ell_4\cos(\theta_0 + \theta_3 + \theta_4);$$

$$y^{**} = x_0 + \frac{\ell_0 + 2\epsilon_1}{2}\sin\theta_0 + \frac{b + 2\epsilon_2}{2}\cos\theta_0 + \ell_3\sin(\theta_0 + \theta_3) + \ell_4\sin(\theta_0 + \theta_3 + \theta_4).$$

If ℓ_0 is the distance between the two axles of the mobile platform and b is the length of each axle, then the kinematic model of the system mobile manipulator is given by

$$\left. \begin{array}{l} \dot{x}_0 = v\cos\theta_0 - \frac{\ell_0}{2}\omega_0\sin\theta_0, \\ \dot{y}_0 = v\sin\theta_0 + \frac{\ell_0}{2}\omega_0\cos\theta_0, \\ \dot{\theta}_0 = \frac{v}{\ell_0}\tan\phi := \omega_0, \\ \dot{\theta}_i = \omega_i \quad \text{for } i = 1, 2, 3, 4, \end{array} \right\} \quad (39.1)$$

where v and ω_0 are the translational and rotational velocities of the platform, while ω_i are the rotational velocities of the link.

Our objectives are to utilize LbCS to design the controllers $v(t)$, $\omega_0(t)$, and $\omega_i(t)$ (for $i = 1, 2, 3, 4$) so that the dual-arm mobile manipulator can move from an initial configuration to a final configuration while avoiding collisions with fixed circular obstacles of random sizes and positions, and simultaneously observing all mechanical singularities associated with the system. That is, the center of the car-like mobile platform (x_0, y_0) should converge to a predetermined target position and the links should attain the desired final orientation. The LbCS formulates a basis of motion planning and control of robots. In this paper, we will show how this method could be easily applied to design appropriate velocity-based controllers that will guide the mobile robot to its respective targets avoiding all the fixed obstacles, taking into account all the kinematic constraints and the singularities of the arm.

39.3 The Attraction and Avoidance Functions

In this section, suitable attraction and avoidance functions will be constructed which will inherently be part of a Lyapunov function from which the velocity control laws will be extracted according to LbCS.

39.3.1 Target Attraction Function

For the mobile manipulator, we have a designed target center at (τ_1, τ_2) with its radius $r(t)$. Let $\mathbf{x} = (x_0, y_0, \theta_0, \theta_1, \theta_2, \theta_3, \theta_4)$, then we have the target function:

$$V(\mathbf{x}) = \frac{1}{2}\left[(x_0 - \tau_1)^2 + (y_0 - \tau_2)^2 + \sum_{i=1}^{4}(\theta_i - a_i)^2\right],$$

where a_i (for $i = 1, 2, 3, 4$) are the desired final orientations of the four links, respectively, in radian.

39.3.2 Fixed Obstacles Avoidance Function

Let there be $q > 0$ fixed obstacles within the boundaries of the workspace. We assume that the kth obstacle is circular with center (Ox_k, Oy_k) with radius ro_k. For safety reasons, we have enclosed the mobile platform and each of the four links in circular protective regions with centers (x_i, y_i) and radius r_i. Note that (x_i, y_i) (for $i = 1, 2, 3, 4$) is expressed as

$$x_1 = x^* - \frac{\ell_1}{2}\cos(\theta_0 + \theta_1) - \ell_2\cos(\theta_0 + \theta_1 + \theta_2), \quad x_2 = x^* - \frac{\ell_2}{2}\cos(\theta_0 + \theta_1 + \theta_2),$$

$$x_3 = x^{**} - \frac{\ell_3}{2}\cos(\theta_0 + \theta_3) - \ell_4\cos(\theta_0 + \theta_3 + \theta_4), \quad x_4 = x^{**} - \frac{\ell_4}{2}\cos(\theta_0 + \theta_3 + \theta_4),$$

$$y_1 = y^* - \frac{\ell_1}{2}\sin(\theta_0 + \theta_1) - \ell_2\sin(\theta_0 + \theta_1 + \theta_2), \quad y_2 = y^* - \frac{\ell_2}{2}\sin(\theta_0 + \theta_1 + \theta_2),$$

$$y_3 = y^{**} - \frac{\ell_3}{2}\sin(\theta_0 + \theta_3) - \ell_4\sin(\theta_0 + \theta_3 + \theta_4), \quad y_4 = y^{**} - \frac{\ell_4}{2}\sin(\theta_0 + \theta_3 + \theta_4),$$

while the radius r_i is given as

$$r_0 = \frac{1}{2}\sqrt{(\ell_0 + 2\varepsilon_1)^2 + (b + 2\varepsilon_2)^2}, \quad r_i = \ell_i/2 \ (i = 1, 2, 3, 4).$$

Then we have the following obstacles avoidance function:

$$W_{ik}(\mathbf{x}) = \frac{1}{2}\left[(x_i - Ox_k)^2 + (y_i - Oy_k)^2 - (r_i + ro_k)^2\right]$$

for $k = 1, 2, \ldots, q$ and $i = 0, 1, 2, 3, 4$.

39.3.3 System Singularities

A singular configuration arises when the angles of link 2 and link 4 is 0, π or $-\pi$ which means that the links will be folded onto each other or fully stretched [8, 15]. This implies that $0 < |\theta_2| < \pi$ and $0 < |\theta_4| < \pi$. The avoidance function of these singular configurations is:

$$S_1(\mathbf{x}) = |\theta_2|, \qquad S_2(\mathbf{x}) = \pi - |\theta_2|,$$
$$S_3(\mathbf{x}) = |\theta_4|, \qquad S_4(\mathbf{x}) = \pi - |\theta_4|.$$

Another singular configuration is observed for the angles of link 1 and link 3. The θ_1 and θ_3 are bounded since link 1 can freely rotate within $(-3\pi/2, \pi/2)$ while link 3 can freely rotate within $(-\pi/2, 3\pi/2)$ so that these links do not collide with the platform. This gives rise to the following avoidance function:

$$S_5(\mathbf{x}) = \theta_1 + 3\pi/2, \qquad S_6(\mathbf{x}) = \pi/2 - \theta_1,$$
$$S_7(\mathbf{x}) = \theta_3 + \pi/2, \qquad S_8(\mathbf{x}) = 3\pi/2 - \theta_3.$$

39.4 The Lyapunov Function and Controller Extraction

According to LbCS, we combine all the attractive and avoidance functions to form a Lyapunov function. We define a tentative Lyapunov function, which is also known as total potentials as

$$L(\mathbf{x}) = V(\mathbf{x})\left(1 + \sum_{i=0}^{4}\sum_{k=1}^{q}\frac{\alpha_{ik}}{W_{ik}(\mathbf{x})} + \sum_{j=1}^{8}\frac{\beta_j}{S_j(\mathbf{x})}\right),$$

where $\alpha_{ik} > 0$ and $\beta_j > 0$ are constants known as control parameters.

We now apply the LbCS to extract the control laws $v(t)$, $\omega_0(t)$ and $\omega_i(t)$. The time derivative of $L(\mathbf{x})$ is

$$\dot{L}(\mathbf{x}) = \frac{\partial L}{\partial x_0}\dot{x}_0 + \frac{\partial L}{\partial y_0}\dot{y}_0 + \frac{\partial L}{\partial \theta_0}\dot{\theta}_0 + \sum_{i=1}^{4}\frac{\partial L}{\partial \theta_i}\dot{\theta}_i$$

$$= \frac{\partial L}{\partial x_0}\left(v\cos\theta_0 - \frac{\ell_0}{2}\omega_0\sin\theta_0\right) + \frac{\partial L}{\partial y_0}\left(v\sin\theta_0 + \frac{\ell_0}{2}\omega_0\cos\theta_0\right)$$

$$+ \frac{\partial L}{\partial \theta_0}\omega_0 + \sum_{i=1}^{4}\frac{\partial L}{\partial \theta_i}\omega_i$$

$$= \left[\frac{\partial L}{\partial x_0}\cos\theta_0 + \frac{\partial L}{\partial y_0}\sin\theta_0\right]v + \left[-\frac{\partial L}{\partial x_0}\frac{\ell_0}{2}\sin\theta_0 + \frac{\partial L}{\partial y_0}\frac{\ell_0}{2}\cos\theta_0 + \frac{\partial L}{\partial \theta_0}\right]\omega_0$$

$$+ \sum_{i=1}^{4}\frac{\partial L}{\partial \theta_i}\omega_i$$

Putting $\dot{L}(\mathbf{x}) \leq 0$, we obtain the controllers

$$\left.\begin{array}{rl} v &= -\dfrac{1}{\delta_1}\left(\dfrac{\partial L}{\partial x_0}\cos\theta_0 + \dfrac{\partial L}{\partial y_0}\sin\theta_0\right) \\ \omega_0 &= -\dfrac{1}{\delta_2}\left(-\dfrac{\partial L}{\partial x_0}\dfrac{\ell_0}{2}\sin\theta_0 + \dfrac{\partial L}{\partial y_0}\dfrac{\ell_0}{2}\cos\theta_0 + \dfrac{\partial L}{\partial \theta_0}\right) \\ \omega_i &= -\dfrac{1}{\delta_{2+i}}\dfrac{\partial L}{\partial \theta_i} \end{array}\right\} \quad (39.2)$$

where $\delta_i > 0$ (for $i = 1, 2, \ldots, 6$) are convergence parameters.

39.5 Stability Analysis

Let θ_0^* be the orientation of the mobile platform at the target. Then the point $\mathbf{e} = (\tau_1, \tau_2, \theta_0^*, a_1, a_2, a_3.a_3)$ is an equilibrium point of system (1). Theorem 1 discusses the stability issues pertaining to the equilibrium point $\mathbf{e}$.

Theorem 1 *The equilibrium point* $\mathbf{e}$ *of system (1) is stable provided the controllers* v, ω_0 *and* ω_i *(for* $i = 1, 2, 3, 4$*) are defined as in (2).*

Proof We use the direct method of Lyapunov to prove the $\mathbf{e}$ is a stable equilibrium point of system (1). The Lyapunov function $L(\mathbf{x})$ defined in Sect. 39.4 is positive, continuous, has continuous first partial derivatives and bounded over the domain $D = \{\mathbf{x} \in \mathbb{R}^7 : W_{ik}(\mathbf{x}) > 0, i = 0, 1, \ldots 4, k = 1, 2, \ldots q$ and $S_j(\mathbf{x}) > 0$ $j = 1, 2, \ldots, 8\}$. Moreover, $L(\mathbf{e}) = 0$ and $L(\mathbf{x}) > 0$ for all $\mathbf{x} \neq \mathbf{e}$.

With the form of the controllers in (2), the $\dot{L}(\mathbf{x})$ simplifies to

$$\dot{L}(\mathbf{x}) = -\delta_1 v^2 - \delta_1 \omega_0^2 - \sum_{i=1}^{4}\delta_{2+i}\omega_i^2.$$

It is evident that, in the domain D, $\dot{L}(\mathbf{x}) \leq 0$ and $\dot{L}(\mathbf{e}) = 0$. Therefore, the equilibrium point $\mathbf{e}$ is a stable.

Fig. 39.2 Trajectory of the dual-arm mobile manipulator from initial position (5, 5) to target at (55, 55)

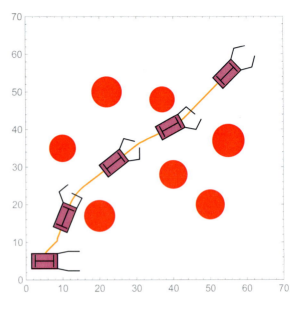

Fig. 39.3 Link angles (θ_1 in red, θ_2 in blue, θ_3 in black, θ_4 in brown) along the trajectory of the dual-arm mobile manipulator

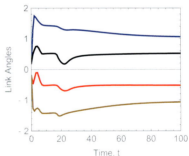

39.6 Simulation Results

In this section, we illustrate the simulation results of our dual-arm mobile manipulator as it navigates in a workspace containing fixed circular obstacles of different size with different position. We also observe the behavior of the Lyapunov function and the designed controllers with time taken to reach its target.

We consider a simple setup as shown in Fig. 39.2 for which the mobile manipulator moves from its initial position (5, 5) to its target at (55, 55) while avoiding collisions with circular obstacles of random positions and sizes. The initial configuration of the links is $(\theta_1(0), \theta_2(0), \theta_3(0), \theta_4(0)) = (-0.2, 0.2, 0.2, -0.2)$, and the desired final configuration is $(a_1, a_2, a_3, a_4) = (-\pi/6, \pi/3, \pi/6, -\pi/3)$. The convergence of the link angles is shown in Fig. 39.3.

Figure 39.4 demonstrates the behavior of Lyapunov function and its time derivative along the system trajectory. The figure provides critical information as to when there is displacement

Fig. 39.4 Evolution of the Lyapnov function and its time derivative ($L(\mathbf{x})$ in red, $\dot{L}(\mathbf{x})$ in blue) along the trajectory of the dual-arm mobile manipulator

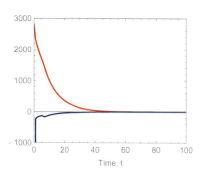

Fig. 39.5 Link angles (θ_1 in red, θ_2 in blue, θ_3 in black, θ_4 in brown) along the Trajectory of the dual-arm mobile manipulator

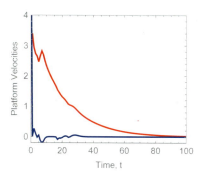

Fig. 39.6 Link angles (θ_1 in red, θ_2 in blue, θ_3 in black, θ_4 in brown) along the Trajectory of the dual-arm mobile manipulator

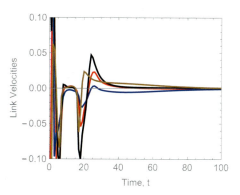

change for time derivative function and also numerically verifies that the energy function is decreasing over time as the robot reaches its target.

The graphs of the nonlinear velocity controllers are shown in Figs. 39.5 and 39.6. For the mobile base, its transitional(blue) v and rotational(red) ω_0 velocities are shown in Fig. 39.5 which explains the non-holonomic constraint of the system as it is dependent on the velocities of the mobile base. It can also be differentiated between the two types of velocity controllers as to ω_0 is changing over time more than v. The rotational velocities of the arm links ω_1, ω_2, ω_3, ω_4 are demonstrated in Fig. 39.6 which explains the movement of the arms over time while avoiding the obstacles and safely attaining the desired final configuration.

39.7 Concluding Remarks

In this paper, a set of time-invariant continuous velocity controllers are presented that successfully can navigate a mobile manipulator with two articulated 2-link arms from its initial configuration to its desired target while avoiding circular obstacles of random size and of random position. The proposed Lyapunov function is a functional equation as it guarantees stability of the system while obtaining a collision-free trajectory and satisfying the holonomic, non-holonomic, kinematic and dynamic constraints associated with the system.

Future work in this area will involve n links for two arms mobile robot, the avoidance function for 4 links car-like robot can be generalize for further computations as for n links. Also, having multiple arms where the protective regions for arms can be replaced by other strategies instead of circular regions. Deriving acceleration-based control laws that include the dynamics (mass and inertia) of the robot, including obstacles of other types such as lines and ellipse will be an interesting extension of this research.

References

1. Kumar, S.A., Chand, R., Chand, R.P., Sharma, B. Linear manipulator: motion control of an n-link robotic arm mounted on a mobile slider. Heliyon, **9**(1) (2023)
2. Sharma, B., Singh, S., Vanualailai, J., Prasad, A.: Globally rigid formation of n-link doubly nonholonomic mobile manipulators. Rob. Autonomous Syst. **105**, 69–84 (2018)
3. Montiel, O., Orozco-Rosas, U., Sepulveda, R.: Path planning for mobile robots using Bacterial Potential Field for avoiding static and dynamic obstacles. Expert Syst. Appl. **42**(12), 5177–5191 (2015)
4. Chand, R., Chand, R.P., Kumar, S.A. Switch controllers of an n-link revolute manipulator with a prismatic end-effector for landmark navigation. PeerJ Comput. Sci. **8** (2022)
5. Karur, K., Sharma, N., Dharmatti, C., Siegel, J.E.: A survey of path planning algorithms for mobile robots. Vehicles **3**(3), 448–468 (2021)
6. Kloetzer, M., Mahulea, C., Gonzalez, R.: Optimizing cell decomposition path planning for mobile robots using different metrics. In: 2015 19th International Conference on System Theory, Control and Computing (ICSTCC), pp. 565–570 (2015)
7. Lee, M.C., Park, M.G.: Artificial potential field based path planning for mobile robots using a virtual obstacle concept. In: Proceedings 2003 IEEE/ASME International Conference on Advanced Intelligent Mechatronics (AIM 2003), **2**, 735–740 (2003)
8. Sharma, B., Vanualailai, J., Prasad, A.: Formation control of a swarm of mobile manipulators. Rocky Mountain J. Math. **41**(3), 909–940 (2011)
9. Chaudhary, K., Prasad, A., Chand, V., Sharma, B.: ACO-Kinematic: a hybrid first off the starting block. PeerJ Comput. Sci. **8** (2022)
10. Nand, R., Chaudhary, K., Sharma, B.: Stepping ahead based hybridization of meta-heuristic model for solving Global Optimization Problems. In: 2020 IEEE Congress on Evolutionary Computation (CEC), Glasgow, UK, pp. 1–8 (2020)
11. Martens, D., Baesens, B., Fawcett, T.: Editorial survey: swarm intelligence for data mining. Mach. Learn. **82**(1), 1–42 (2011)
12. Yang, X.-S.: Nature-inspired mateheuristic algorithms: success and new challenges. J. Comput. Eng. Inf. Tech. **01**(01) (2012)
13. Mac, T.T., Copot, C., Tran, D.T., Keyser, R.D.: Heuristic approaches in robot path planning: a survey. Rob. Autonom. Syst. **86**, 13–28 (2016)

14. Sharma, B., Raj, J., Vanualailai, J.: Navigation of carlike robots in an extended dynamic environment with swarm avoidance. Int. J. Robust Nonlinear Control. **28**, 678–698 (2017). https://doi.org/10.1002/rnc.3895
15. Sharma, B., Vanualailai, J., Singh, S.: Lyapunov-based nonlinear controllers for obstacle avoidance with a planar n-Link doubly nonholonomic manipulator. Robot. Autonomous Syst. **60**(12), 1484–1497 (2012)

Chapter 40
A Real-Time Fall Detection System Using Sensor Fusion

Moape Kaloumaira, Geffory Scott, Asesela Sivo, Mansour Assaf, Shiu Kumar, Rahul Ranjeev Kumar, and Bibhya Sharma

Abstract There can be serious and harmful effects on people following a fall event due to its severity. This paper presents a real-time fall detection using sensor fusion to improve the overall accuracy of the system when subjected to continuous operation. The system combines data from accelerometer and an ultrasonic sensor to detect falls in real time. The developed mobile application accurately identifies fall scenarios and sends SMS notifications to emergency contacts. The proposed system's ultrasonic sensor module has wireless communication capabilities, while the accelerometer readings are acquired from the smartphone. Appropriate feature, threshold values, and program flow have been chosen, such that fall detection is accurate, and the system is operational even if one of the sensors malfunctions. The proposed system has been validated experimentally and the accuracy, sensitivity, and specificity are 87.5%, 89.47%, and 94.74%, respectively.

Keywords Fall detection · Smartphone · Android · Sensor fusion

40.1 Introduction

As compared to only 10% in 2000, the United Nations Population Division estimates this number will rise to 21% in 2050 [1]. The growing number of older people poses increasing challenges for the Ministry of Health, especially in developing countries that have fewer health services and higher spending. A fall is defined as the unexpected contact of the human body with the ground and is a common occurrence in the elderly as they age. With population growth and technological advances, great minds have worked consistently and meticulously to develop solutions to minimize the problem of falls among the elderly. In order to effectively eliminate the problem

M. Kaloumaira · G. Scott · A. Sivo · M. Assaf · R. R. Kumar (✉) · B. Sharma
The University of the South Pacific, Laucala Campus, Suva, Fiji
e-mail: rahul.kumar@usp.ac.fj

S. Kumar
Fiji National University, Suva, Fiji

© The Author(s), under exclusive license to Springer Nature Singapore Pte Ltd. 2024
P. K. Jha et al. (eds.), *Proceedings of Congress on Control, Robotics, and Mechatronics*, Smart Innovation, Systems and Technologies 364,
https://doi.org/10.1007/978-981-99-5180-2_40

of falling, various research techniques and technologies have been introduced, such as: The benefit of using these solutions is that it reduces the risk of serious injury in the elderly from unexpected falls. The main disadvantage identified is that these solutions require time and money, which means that developing countries cannot afford the facilities offered to older people [2].

This paper proposes the development of a low cost real-time fall detection system by utilizing the idea of sensor fusion. In particular, the proposed system uses the accelerometer and ultrasonic sensor readings to indicate fall scenarios in real time. The processing of these sensor signals are governed by the developed mobile application that is able to accurately detect the fall scenario and automatically send SMS notification to the emergency contact. It should be noted that only the wireless communication module has been developed for the ultrasonic sensor, while the accelerometer sensor readings are acquired from the mobile phone itself, which runs the application. The proposed system is designed in such a way that the fall detection is classified accurately by validating across two sensor readings. This idea of sensor fusion has been implemented to solely counter inaccuracies during continuous detection. Worth mentioning is that the system has additional backup routines that can switch to only work with one sensor to detect fall in cases where one of the sensors malfunctions.

40.2 Background

Falls are a major problem in the modern world as they are considered to be the most serious and deadly activity for people aged 60–65. Falls in the elderly occur for a variety of reasons and result in different outcomes, and awareness of these causes and outcomes helps R&D engineers of fall prevention systems to come up with effective solutions to the problem of falls in the elderly [3].

The purpose of fall detector applications is to prevent falls and alert individuals when a fall is likely to occur. This particular fall detection system uses wearable electronic sensors that must be worn by the user on their clothing and consists of two parts: accelerometers attached to the body and a built-in accelerometer in a smartphone [1]. Wearable devices that use smartphone accelerometers have an advantage because most smartphones have sensors such as cameras, microphones, GPS, and digital tools that can help reduce falls in elderly people. The objective of the proposed system is to address the issue of falls during daily activities like walking, sitting, or standing. By utilizing a smartphone system, the movements of your body are recorded, and the resulting data can be analyzed to enhance the system's performance by identifying variances in movement. The human body produces a substantial amount of data when in motion, and acceleration is deemed to be the most significant type of data for detecting falls due to the laws of physics. This is because the acceleration of a body is closely linked to the force acting on it, and during a fall, the force on the body changes correspondingly, making acceleration data an appropriate choice for fall detection [4]. Using acceleration-based fall detection systems can help reduce falls among the elderly and can be easily adopted by a large number of users.

The main goal of the system described in [5] is to detect falls and alert monitoring personnel to reduce the risk of injury. According to the researchers of [5], fall detection can be categorized into three types: camera vision, wearable devices, and ambience sensors. In this particular system, ultrasonic sensor arrays are utilized to detect falls. The proposed solution classifies fall scenarios and positions by analyzing changes in distances recorded by the ultrasonic sensors. Two arrays of ultrasonic sensors, placed at different locations in the room, are used to detect the position of the human body during a fall. The sensors measure the movement and positioning of the body, and the different distance calculations of the recorded pulses are used for fall detection and gesture classification. The human body activity can be divided into five positions which are standing, sitting, lying, running, and jumping. The data received from the ultrasonic sensors has been used to determine these activities. Various falling positions can result in different types of injuries, with forward falls typically causing leg and knee injuries, backward falls leading to severe head injuries, and side falls causing broken bones, ligament damage, and internal complications. The proposed approach incorporates an ultrasonic sensor with a frequency of approximately 40 kHz, employing an 8-pulse waveform. The sensor transmits a Tx pulse signal to the object, then receives the Rx signal that is reflected back to the sensor, enabling the distance to be measured by calculating the time between the reflector targets and the sensor [5]. To install the system on the ceiling and side panels, an ESP8266-07 Wi-Fi module, which is a small and power-efficient module, is utilized. This module also supports a broad range of ESP8266 clients, APs, and + APs [5]. The microcontroller plays a critical role in the system, as it controls both the sensor and the Wi-Fi module, with Arduino microcontrollers serving as the sending and receiving systems. The proposed system comprises two modules, namely the hardware and software modules. The hardware module stores and processes information from the sensors, while the software module determines case conditions and provides monitoring capabilities.

Moreover, as proposed in [6], the fall detection using Arduino-based system is designed to monitor acceleration values using an MPU 6050 accelerometer component. By comparing the total acceleration values with a threshold value, the system can detect whether the user is standing up or lying down on the ground. For the person standing, the system constantly monitors the acceleration, while when the person is lying on the ground, the buzzer emits an acoustic signal to ensure the initialization of the GSM module and to notify the health centers and family members. If a person loses consciousness during a fall, the microcontroller provides location data from the GPS module and sends it to the appropriate services with an alarm message [6].

In [7], a fall detection system that uses accelerometers and gyroscopes is designed to detect two types of human activities: static and dynamic postures. Static postures include a person sitting, standing, or lying down, while dynamic postures involve movement between static postures. The system includes two tri-axial accelerometers that are placed on the person's chest and thigh, respectively, in order to recognize their static postures. The system has been shown to have 91% sensitivity and 92% specificity. However, the system is not without its limitations, as it struggles to detect a person jumping on a bed or falling from a wall while in a seated position.

In addition, the method presented in [8] for fall detection is based on kinematics, and it aims to monitor the different positions of an individual during a fall. The system is classified into two categories, namely threshold-based methods (TBM) and machine learning methods (MLM). TBM is less expensive to implement compared to MLM as it utilizes a threshold to distinguish a fall from activities of daily living (ADL), while MLM requires a dataset containing fall and ADL samples. However, the proposed MLM method requires more computations and is more expensive to implement than TBM.

Likewise, two different fall detection methods are discussed in [9] and [10]. The first approach [9] utilizes a tri-axial accelerometer and gyroscope placed at the upper trunk of the human body. The system is activated when the acceleration surpasses a certain threshold and the gyroscope checks the orientation of the subject. False alarms are eliminated by a push button, and the system response is fast and effective. The second approach [10] uses two or more acoustic sensors to detect a fall event based on measured height and loudness. Two additional microphones are used to minimize noise and distance which is kept at 4 m apart. The system achieved 70% accuracy with no false alarms, which can be improved up to 100% with a penalty of 5 false alarms every hour.

With respect to computer vision-based strategies, three different fall detection systems that use different techniques to detect falls have been highlighted in [11, 12], and [13]. A novel approach in fall detection systems is the use of human shape variations for fall detection. This system relies on detecting differences in the human form, utilizing a combination of the best-fit ellipse around the body, histograms of the silhouette projection, and changes in the head's temporal position. These cues enable the detection of various behaviors [11]. On the other hand, [12] proposes an automated monitoring system with face recognition to detect falls in a specific area. It uses webcams to collect data on the person's movement, position, and distance from the camera to determine if a fall has occurred. As for [13], the system uses an omni-directional camera with a separate PC server that captures images in 360 degrees to solve the problem of blind spots. The system has a sensitivity of 78% without personal information and up to 90% with personal information but requires users to provide personal information such as height and BMI, which increases implementation costs.

Sensor fusion-based strategies have proven to be very effective in fall detection. One such approach is the combination of accelerometer, gyroscopes, and acoustic sensors to detect heartbeats [14]. Typically, the device is positioned at the center of the body, and wireless communication is established using the ZigBee protocol. In the event that the battery level is low, the system employs the ZigBee connection to notify the user to charge the battery, and five different algorithms have been developed, the result of which varies between 92 and 97%. By combining the system with accelerometers and gyroscopes, the detection rate increases to 97% [14].

Moreover, in [15], a fall detection system using mobile devices is designed to detect falls and monitor heart rate by mounting a mobile device on the user's waist, which alerts caregivers in case of an emergency. The detection of falls and heart rate in a system is achieved through the use of tri-axis accelerometers and three-channel

ECG circuits, respectively. In addition to a simple threshold algorithm, the system employs several supplementary techniques to enhance the accuracy of detection [15]. The response time of the system is faster and more efficient, allowing for timely alerts to be sent to caregivers.

To sum up, falls are a serious concern among the elderly and can lead to severe injuries and even death. Wearables, assistive technologies highlighted in [16], and sensor-based fall detection systems have been developed to prevent falls and alert individuals or authorities when a fall occurs. These systems utilize different technologies such as accelerometers, ultrasonic sensors, and GPS modules to detect falls and can be easily adopted by a large number of users. While these systems have shown promising results, they still have limitations and require further improvements to provide accurate and reliable fall detection. This study intends to address some of the shortcomings by using a sensor fusion approach and enhance the overall system accuracy. The next sections will focus on the developed prototype and its performance analysis.

40.3 Design and Implementation

The proposed fall detection system was designed by assembling software and hardware components into a finished prototype, as depicted in Fig. 40.1. The following sections outline the steps taken to implement the software, hardware, system integration, and final prototype. To aid in understanding human body gestures and various fall postures, Figs. 40.2 and 40.3 are provided.

Fig. 40.1 Design prototype

Fig. 40.2 Human body gestures [5]

Fig. 40.3 Human fall postures [5]

40.3.1 Software Component (Mobile Application or Mobile App)

The fall detection app was developed using the MIT APP INVENTOR 2 software and incorporates sensor fusion techniques (Fig. 40.4). The app integrates data from both the smartphone's in-built accelerometer and a separate ultrasonic sensor module to perform real-time fall detection analysis. Two different approaches were programmed in the app using sensor fusion techniques—one using accelerometer readings and the other using ultrasonic sensor readings. The app uses the combined outputs of both approaches to accurately detect fall situations. The developed system has shown high accuracy in detecting falls, and it can help improve the safety and well-being of at-risk individuals.

One of the approach is basically using the accelerometer readings and finding the magnitude of the in-built accelerometer sensor readings. In fall detection, the norm of acceleration typically refers to the magnitude of the acceleration vector obtained from an accelerometer sensor. In addition, it represents the overall acceleration magnitude experienced by the user, regardless of its direction. The magnitude has been calculated by acquiring the built-in accelerometer readings from the smartphone, that is, the A_x, A_y, and A_z, which are the accelerations in the x, y, and z plane, respectively. Since the acceleration is to be measured in a 3D space, the magnitude of the total acceleration is calculated as follows:

Fig. 40.4 App build in MIT APP INVENTOR 2 software

Fig. 40.5 Tri-axial orientation of a smartphone

$$|A| = \sqrt{A_x^2 + A_y^2 + A_z^2} \qquad (40.1)$$

whereby the x, y, and z are acceleration values in three different directions.

The norm of acceleration is often used as a feature to detect falls. A fall typically results in a sudden increase in the norm of acceleration, which can be detected using appropriate threshold values. Figure 40.5 illustrates the tri-axial orientation through which the accelerometer sensor takes its readings on a smartphone.

Moreover, it should be noted that the $|A|$ for each time step was calculated, and for the purpose of measuring the fall, difference between the $|A|$ is calculated between initial $|A_i|$ and the preceding $|A_p|$. Thereafter, a threshold was set to discriminate whether a person is standing or falling. The threshold setting is explained in Sect. 40.4. Figure 40.6 shows the proposed design algorithm that has been programmed and deployed.

40.3.2 Hardware Component

To create a prototype, an Arduino Mega 2560 processor, an HC-SR04 ultrasonic sensor, and an HC-05 ZS-040 Bluetooth module for Arduino were employed. Figure 40.7 shows the assembled component.

To prototype the proposed system using an HC-SR04 ultrasonic sensor, HC-05 ZS-040 Bluetooth module, and an Arduino Mega 2560, the user must gather necessary materials and connect the ultrasonic sensor and Bluetooth module to the Arduino Mega. The Arduino Mega must then be connected to a computer, and the code must be written to read the distance measured by the HC-SR04 ultrasonic sensor and send it to the HC-05 ZS-040 Bluetooth module. The code is uploaded to the Arduino Mega and paired with a Bluetooth-enabled device, such as a phone. A serial terminal app can then be used to connect to the Bluetooth module and display the distance measured by the HC-SR04 ultrasonic sensor.

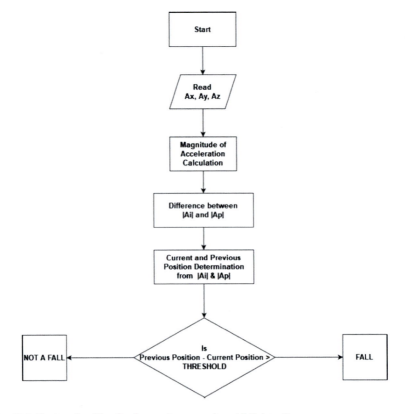

Fig. 40.6 Design algorithm for the accelerometer-based fall detection

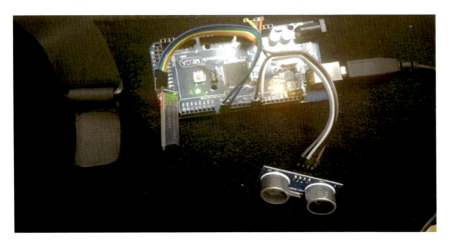

Fig. 40.7 Ultrasonic sensor Arduino-based module

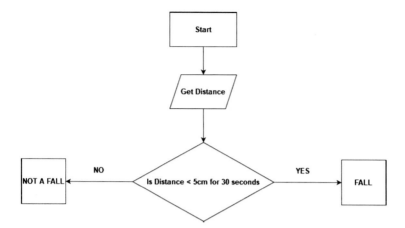

Fig. 40.8 Ultrasonic sensor Arduino-based fall-detecting mode

The Arduino software program was utilized to calibrate and set up the ultrasonic sensor. When the distance detected by the sensor is 5 cm or less, the Arduino sends a high signal '1' to the smartphone via the Arduino Bluetooth module. Conversely, if the distance is greater than 5 cm, a low signal '0' is transmitted to the phone. Figure 40.8 illustrates the ultrasonic sensor's fall-detecting mode which was programmed as per the algorithm given below.

40.3.3 System Integration

The input from both the sensors provide inputs to the app via interfaces; whenever any of the sensors detects a fall scenario, it relays it to the app. The app then sends a SMS to the emergency contact and informs them that the user has fallen after double checking using both the fall detection approaches in a specified interval of time. Thereafter, it proceeds to sending SMS to the emergency contacts as per Fig. 40.10. The whole system can take in both the inputs or either of the sensor readings in case the other one fails. Figure 40.9 presents the developed system while Fig. 40.10 presents the overall algorithm for the fall detection system.

40.3.4 Mobile Application Setup

The user initiates the application and can designate an emergency contact in the event that the "SETTINGS" button becomes inaccessible. With-in the "SETTINGS" section, the user can also specify the amount of time the app should wait before

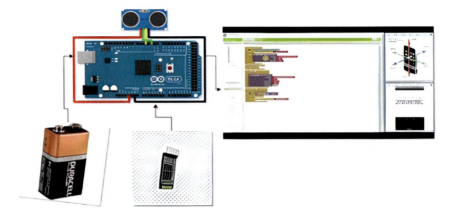

Fig. 40.9 Schematic of the developed system

sending an SMS to the designated emergency contact with details of their fall. Furthermore, by clicking on the "About" button, the user may access information about the developer of the application. The diagram shown in Fig. 40.11 presents the main screens of the fall detection system's user interface.

40.4 Implementation Results and Analysis

Activities of daily living (ADL) and fall events were simulated and evaluated alongside one another. The ultrasonic sensor and SVM values were captured and stored, while the accelerometer was utilized to record simulated ADL events like sitting, standing, walking, and resting. Simulated fall event was made, and its $|A|$ was calculated. Figure 40.12a–e shows the accelerometer readings for the scenarios or states specified in Table 40.1 for the simulated ADLs. Based on five instances for each of the following states, the average values of $|A|$ are given in Table 40.1 whereby the threshold setting for fall detection is set to 10.

Displayed in the table above is the average of various states taken from five separate cases. A threshold of 10 is utilized to accommodate the user's varying activities of daily living (ADLs). Based on the above average values of $|A|$, , only two classes (fall or no-fall) will be evaluated.

40 A Real-Time Fall Detection System Using Sensor Fusion

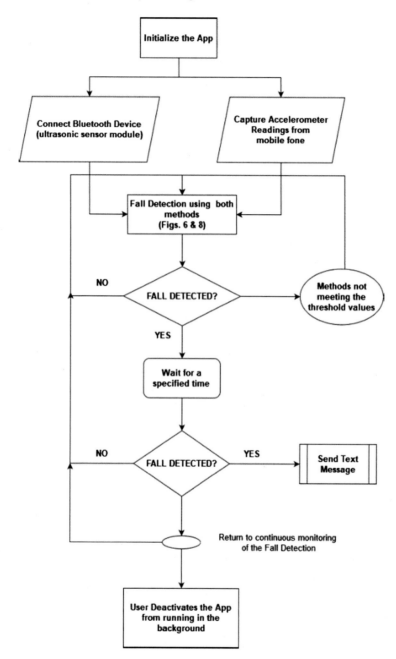

Fig. 40.10 Overall algorithm for the proposed system

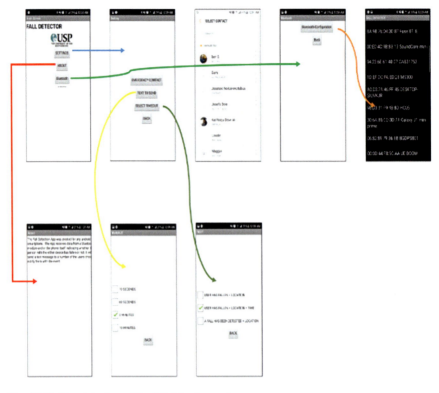

Fig. 40.11 User interface of the fall detection system

Fig. 40.12 Accelerometer readings for five states specified in Table 40.1

Table 40.1 States and its average $|A|$ values

State	Average value
Sitting	10
Standing	10
Walking	8
Rest	10
Fall	12

Fig. 40.13 Ultrasonic sensor outputs

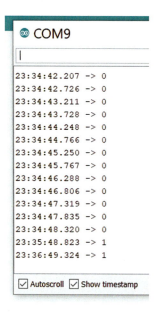

40.4.1 Ultrasonic Sensor Performance and Overall System Evaluation

The ultrasonic sensor module reads the values and sends out a high signal '1' if the distance between the module and any object is equal to less than 5 cm (fall), and low signal '0' if the distance is otherwise (no-fall). The ultrasonic sensor output readings are shown in Fig. 40.13.

To test the developed system that uses the sensor fusion strategy mentioned earlier, 38 different scenarios were created to determine the likelihood of a fall and no-fall. Table 40.2 presents the probability of true and false positives. Using these metrics (outlined in Table 40.2), the system's accuracy, sensitivity, and specificity were calculated using the classical approach for binary classification.

Following the above values of TP, TN, FP, and FN, the accuracy, sensitivity, and specificity are **87.50%, 89.47%,** and **94.74%**, respectively.

Table 40.2 True/false positive and true/false negative rates

Metric	Analysis	State	Accuracy (%)
True positive (TP)	17/20	True states of fall	85
False positive (FP)	1/20	False states of fall	5
False negative (FN)	2/20	False states of not a fall	10
True negative (TN)	18/20	Trues states of not a fall	90

Worth mentioning is that the fall detection system is not perfect and is prone to errors. It uses an ultrasonic sensor module to detect falls, but as there is only one sensor, it can only detect falls in one particular axis. This limitation is addressed by using the phone's built-in accelerometer sensor to detect falls based on the threshold value of the three-dimensional accelerometer values of the phone. However, the accelerometer-based system may also be prone to errors as it cannot distinguish between a fall and a non-fall event, such as when a user lies down abruptly to rest. The system is relatively basic and requires further development to reach perfection. Nevertheless, despite its limitations, the system is still reliable and sensitive enough to detect falls.

40.5 Conclusion

This paper has presented a real-time fall detection system that utilizes the data from two different sensors for detecting a fall. At the heart of the proposed system is the mobile application developed for Android OS that dictates the decisions for detecting a fall. The system utilizes the in-built accelerometer from the mobile phone to acquire data and calculate the magnitude of the total acceleration ($|A|$). On the other hand, the external module has one ultrasonic sensor, which measures the distance of the module from the ground. Both the sensors have already been calibrated and tested to deduce appropriate values of the threshold. For the accelerometer system, the threshold values for $|A|$ were set to 10 to ensure that any values greater than 10 would be classified as "fall." As for the ultrasonic sensor, distance between ground and the module should be above 5 cm. This means that if the calculated distance is less than 5 cm, then it denotes a "fall" scenario. Additionally, it should be noted that the program flowchart has been designed in such a way that the proposed system is functional even if one of the sensors does not work.

Future work will include integration of the GPS location within the application to locate the user easily and accommodate for calling ambulance for quicker action or if the specified emergency contact is out of reach.

References

1. Dang, T.T., Truong, H., Dang, T.K.: Automatic fall detection using smartphone acceleration sensor. Int. J. Adv. Comput. Sci. Appl. 123–129 (2016)
2. Habib, M.A., Mohktar, M., Kamaruzzaman, S.B., Kheng, L.S., Pin, T.M., Ibrahim, F.: Smartphone-based solutions for fall detection and prevention: challenges and open issues. www.mdpi.com/journal/sensors. pp. 7182–7208 (2014)
3. El-Bendary, N., Tan, Q., Pivot, F.C., Lam, A.: Fall detection and prevention for the elderly: a review of trends and challenges. Int. J. Smart Sens. Intell. Syst. **6**(3), 1231–1234 (2013)
4. Dang, T.T., Truong, H., Dang, T.K.: Automatic fall detection using smartphone acceleration sensor. Int. J. Adv. Comput. Sci. Appl. **7**(12), 123–129 (2016)

5. Nadee, C., Chamnongthai, K.: Multi-Sensor system for automatic fall detection. Proc. APSIPA Ann. Summit Conf. **2015**, 930–933 (2015)
6. Vetsandonphong, N.: Arduino Based Fall Detection and Alert System. School Engineering in Malaysia, Malaysia (2016)
7. Li, Q., Hanson, M., Stankovic, J., Barth, A., Lach, J.: Accurate, fast fall detection using gyroscopes and accelerometer-derived posture information. IEEE Comp. Soc. 138–143 (2009)
8. Igual, R., Medrano, C., Plaza, I.: Challenges, issues and trends in fall detection systems. Bio-Med. Eng. 1–20 (2013)
9. Tong, L., Chen, W., Song, Q., Ge, Y.: A research on automatic human fall detection method based on wearable internal force information acquisition system. Int. Conf. Rob. Biometr. 1–5 (2019)
10. Popescu, M., Li, Y., Skubic, M., Rantz, M.: An acoustic fall detector system that uses sound height information to reduce the false alarm rate. In: 30th Annual International IEEE EMBS Conference, pp. 4628–4631 (2008)
11. Foroughi, H., Rezvanian, A., Paziraee, A.: Robust falll detection using human shape and multi-class support vector machine. In: Sixth Indian Conference on Computer Vision, Graphics and Image Processing, pp. 413–420 (2008)
12. Khawandi, S., Daya, B., Chauvet, P.: Implementation of a monitoring system for fall detection in elderly healthcare. Proced. Comp. Sci. **3**, 216–220 (2011)
13. Miaou, S.G., Sung, P.H., Huang, C.Y.: A customized human fall detection system using omni-camera images and personal information. In: Proceedings of the 1st Distributed Diagnosis and Human Healthcare Conference, pp. 39–42 (2006)
14. Dinh, A., Teng, D., Chen, L., Shi, Y., McCrosky, C., Basran, J., Del BelloHass, V.: Implementation of a physical activity monitoring system for the elderly people with built-in vital sign and fall detection. In: Sixth International Conference on Information Technology, pp. 1226–1231 (2019)
15. Nguyen, T.-T., Cho, M.-C., Lee, T.-S.: Automatic fall detection using wearable biomedical signal measurement terminal. In: 31st Annual international conference of the IEEE EMBS, pp. 5203–5206 (2009)
16. Kumar, S.A., Vanualailai, J., Prasad, A.: Assistive technology: autonomous wheelchair in obstacle-ridden environment. PeerJ Comput. Sci. (PeerJ). **7**, e725 (2021). https://doi.org/10.7717/peerj-cs.725

Chapter 41
A Vision-Based Feature Extraction Techniques for Recognizing Human Gait: A Review

Babita D. Sonare and Deepika Saxena

Abstract Gait recognition has become more popular and significant in the recent years due to security concerns since it can be carried out remotely without authorization. This article discusses the vision-based model and model-free feature extraction methods for identifying human gaits. Both methods are distinctive in and of themselves. The structural parts of the human body are dealt with via model-based approaches, including joint locations, joint angles, stride length/cadence, and 2D stick figures. Model-free techniques, including gait energy image, absolute frame difference image, gait history image, etc., give spatiotemporal information on gait silhouettes. Subject identification is made based on the high rate of recognition after approach-wise features are provided to classifiers. The important characteristics will then stand out.

Keywords Biometrics · Gait recognition · Individual identification · Model-based features · Model-free features

41.1 Introduction

Biometric recognition refers to identifying a person by biological or behavioral characteristics. The face, iris, fingerprints, palm, gait, and other biological factors of humans aid in identifying a person. Medical psychiatry claims that a person's gait, or how they walk, is unique and comprises 24 diverse components. Gait is the rhythmic pattern of actions that causes motion.

B. D. Sonare (✉) · D. Saxena
Department of Computer Science and Engineering, Poornima University, Jaipur, Rajasthan, India
e-mail: sonare.babita@gmail.com

D. Saxena
e-mail: deepika.saxena@poornima.edu.in

© The Author(s), under exclusive license to Springer Nature Singapore Pte Ltd. 2024
P. K. Jha et al. (eds.), *Proceedings of Congress on Control, Robotics, and Mechatronics*, Smart Innovation, Systems and Technologies 364,
https://doi.org/10.1007/978-981-99-5180-2_41

Recently, in video surveillance applications such as recognition and access control, human gait recognition is gaining the interest of researchers due to its effectiveness. Gait can be captured remotely and does not require the subject's consent. The advantage of gait is that it is discrete and unobtrusive. The ability to recognize human gait may be impacted by outside elements such as view angle, walking speed, carrying status, clothing, and shoes. Machine learning and deep learning classification approaches are helpful in making it resilient.

The human gait recognition system is divided based on vision-based input, namely model-based and model-free or appearance based [1]. The human body can be modeled by using a stick pattern according to bones, joint placements, joint angles, and limbs [2]. These can be produced by elliptically depicting each body part and then determining which body portion is essential for identifying the gait. As shown in Fig. 41.1, former features can be produced using structural elements of the human body, like stride length and steps per minute, that is, cadence. But it requires high computational cost. Later, elements are based on how people look or appear. These can be produced using the video's extracted frames. Figure 41.2 represents the individual from a binary frame known as a silhouette.

The gait recognition system, as shown in Fig. 41.3, consists of gathering the data, preprocessing the frame by subtracting the background, feature extraction, and then classifying the subject. The output of the system is a person's identity.

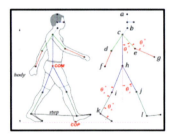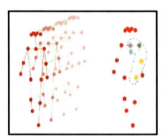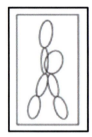

Fig. 41.1 Model-based features (Joint angles/position, stick pattern, body part represented as an ellipse)

Fig. 41.2 Model-free/appearance-based silhouettes (normal walking, wearing a coat, carrying a bag)

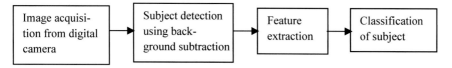

Fig. 41.3 General gait recognition system

The raw gait data can be obtained using the camera, sensors, or accelerometer. Gait images/frames of people's walking gaits can be used to identify them because various people have varied walking styles. When taking videos, a camera is often positioned at a distance of several meters from the subject. Subjects from the frame can be detected using background subtraction [3]. For background subtraction, frame difference, Gaussian mixture model, kernel density estimation, and codebook generation methods are effective. Feature extraction can be done using a model-based or model-free based approach. Model-free extracted features are high-dimensional vectors, which require dimensionality reduction. Features such as joint position, angle, stride height, or some statistical features can be extracted from the model-based approach [2]. Features such as gait energy images, accumulative frame difference images, gait moment images, and gait active images can be extracted from model-free approaches. The extracted features can be given to classification algorithms such as SVM, NB, KNN, RF, MLP, or CNN, giving the subject id. It is crucial to carefully select the classification approach depending on each circumstance to enhance the identification rate. System performance can be checked with performance measure accuracy and error rate.

This review study focuses on feature extraction methods, namely model-based and model-free methods for identifying humans based on gait. We proposed a system based on both approaches, both features will be extracted separately and fed to the classifier and accuracy will be measured where the approach with the higher accuracy would be taken into account, regardless of the gait silhouettes' quality or resolution. Significant characteristics would provide good accuracy. The other portions of the paper are divided as follows: Sect. 2 contains a literature review; Sect. 3 includes a comparative analysis; Sect. 4 contains a suggested system; and Sects. 5 and 6 have the conclusion and references.

41.2 Literature Review

Most of the literature on the human gait recognition system is based on the approach used for extracting features. A review of related work is organized as feature extraction based on model-based and model-free approaches.

41.2.1 Model-Based Gait Recognition System

The human body is intended to be modeled using model-based feature representation, and features are taken from this model. Appearance and carrying status are not covariate factors in model-based feature representation. But it depends on the level of the video. It requires good-quality images to find joint positions, angles, height, etc., features.

The automated system for gender classification using joint angles and points as a gait signature and classified using a support vector machine described in [4], giving 96% accuracy. Three angles between the two front and rear legs can be utilized as features in the model-based approach mentioned in [5], and after dimension reduction techniques like PCA, the distance time warping method is employed to identify humans 92.3% of the time accurately. A novel model built on moving feature extraction analysis demonstrated in [6]. The system uses the Fourier series to extract the gait signature automatically, and the Fourier components of the motion of the leg are used for classification, with a 100% CCR.

The model-based feature vector supplied to the artificial neural network for gait classification in [7], and achieved an accuracy of 89%, and experimented on the CASIA B dataset. The multiple sclerosis and stroke gait class, six joint data were collected and converted into joint angles using an inverse kinematic solution [8] and later fed to machine learning algorithms, namely KNN, SVM, ELM, and MLP. ELM has performed well. Image moments [9] were extracted as features from silhouettes and fed to the nearest neighbor classifier for identifying humans. Demonstrated on CASIA B dataset achieving 71.42% accuracy.

41.2.2 Model-Free Gait Recognition System

In model-free approaches, features such as gait energy images, accumulative frame difference images, gait moment images, and gait active images can be extracted. It is essential to apply dimensionality reduction techniques to reduce the size of feature vectors. Dimensionality reduction techniques such as PCA, LDA, MDA, DCT, and I-vectors are effective.

Gait Energy Image (GEI) [9]: GEI is the mean of all frames during a single gait cycle. The gait energy image (GEI) $G(x, y)$ is defined as $Bt(x, y)$ at time t in a sequence of length N as shown in Eq. 41.1.

$$G(x, y) = 1/N \sum_{t=1}^{N} Bt(x, y) \qquad (41.1)$$

Gait Motion Image (GMI) [10]: Let $D(x, y, t)$ be a binary image sequence representing regions of motion of let $I(x, y, t)$ be an image sequence; several applications, image differencing is sufficient to get D. Finally, the MEI E τ (x, y, t) is represented as shown in Eq. 41.2.

$$E\tau(x, y, t) = \bigcup_{i=0}^{\tau} D(x, y, t-1) \qquad (41.2)$$

τ is important in determining how long a movement will last.

Motion History Images (MHI) [10]: A motion history image illustrates how the image is moving, as opposed to where it is moving. The pixel intensity in an MHI I depends on the time-based event of motion at that particular location. We use a straightforward replacement and decay operator for the results displayed in Eq. 41.3.

$$I\tau(p, q, t) = \begin{cases} \tau & \text{if } I(p, q, t) = 1 \\ \max(0, I\tau(p, q, t-1) - 1) & \text{otherwise} \end{cases} \qquad (41.3)$$

Accumulative frame difference energy image (AFDEI) [11]: By integrating the forward and backward frame differences, the frame difference energy image is created, shown in Eq. 41.6.

$$Fa(p, q, t) = \begin{cases} 0 & \text{if } D(p, q, t) \leq D(p, q, t-1) \\ D(p, q, t) - D(p, q, t-1) & \text{otherwise} \end{cases} \qquad (41.4)$$

where $Fa(p, q, t)$ is the forward frame difference image.

$$Fb(p, q, t) = \begin{cases} 0 & \text{if } D(p, q, t) \leq D(p, q, t-1) \\ D(p, q, t-1) - D(p, q, t) & \text{otherwise} \end{cases} \qquad (41.5)$$

Consider $Fb(p, q, t)$ is the backward frame difference image.

$$F(p, q, t) = Fa(p, q, t) + Fb(p, q, t) \qquad (41.6)$$

where $F(p, q, t)$ is the frame difference image.

The spatiotemporal picture is called the gait energy image [12], and to address the problems in the training template, the synthetic template of real GEI by simulating the distortion in real GEI has been introduced. The proposed recognition approach highly performs well in comparison with published methods. Gait recognition methods based on average silhouette and contour using classifier ensembles [13] have been proposed.

Independent component analysis was used to extract the fusion features of the motion silhouette image and the GEI [14], which were then fed into a multiclass extreme learning machine classifier. A new head-torso-thigh (HTI) [15] easy-to-compute and efficient gait representation is proposed, applying the nearest neighbor classifier to identify low-contrast infrared images. The problem of feature selection for human identification based on gait energy image [16] is addressed, by providing the supervised (Wrapper cross-validation) and unsupervised approach (PCA and MDA), experimented on CASIA B dataset with improved recognition performance. GEI (dynamic and static information) and AFDEI (reflects the temporal characteristics) moment invariants [11] were combined as the gait feature. The nearest neighbor classifier was used since it produced better recognition results than a single feature.

The classic appearance-based gait recognition based on GEI [17] proposed. The TUM GAID dataset is used for extensive experiments. The automatic human identification system using gait sequences, which calculates average silhouettes [18] presented and reduces their dimensionality using principal component analysis. The system then supplies the essential features to a Euclidian distance classifier, giving the system the best recognition rate of 90% on the CASIA B Dataset. MPGR-CF model for human recognition, wherein dynamic features, i.e., optical flow and spatiotemporal feature, i.e., GEI, are extracted and presented [19] and trained using SVM and HMM classifier, respectively, and recognition results are fused at the decision level. Demonstrated on dataset CASIA B and OU-ISIR, generate the high correction recognition rate.

The gait recognition performance in cooperative and antagonistic contexts, the effect of training data size, and the amount of Carrying Status (CS) label recognition difficulty by using GEI as a feature and MSVM classifier presented in [20]. Features based on a higher order of statistical moments on horizontal, vertical and grid structures of silhouettes on OU-ISIR and CASIA B datasets provide shape invariance [21] proposed. The system's robustness is measured by using different classifiers such as decision tree (C4.5), Naïve Bayes, KNN, and random forest. The author has proved the proposed technique, which performs well with statistical hypothesis testing such as the t-test, F-test, and ICC test.

The periodicity of the gait cycle to find gait moment images used in [22]. Moment deviation images are calculated by subtracting the original silhouettes from the feature vector created by the gait moment image and using the nearest neighbor technique to aid in human recognition; this method produces a respectable recognition rate.

41.3 Comparative Analysis

The entire human body is intended to be modeled by model-based feature representation techniques. Joint locations, joint angles, stride length/cadence, four distances (left–right foot, head-foot, head-pelvis, foot-pelvis), 2D stick figure, lower-leg pendulum, and thigh pendulum are only a few examples of model-based feature representation strategies. A few places on human bodies are also included, along with their distances and angles. Model-based methods typically have an accurate classification rate of more than 80% and an error rate of about 10%.

Human silhouettes' entire motion or shape is intended to be processed via model-free feature representation. Silhouettes are used for feature representation

Table 41.1 Table captions should be placed above the tables

References	Gait features	Dataset used	Accuracy (%)
Yoo et al. [4]	Joint angles and points	Not mentioned	96
Benbakreti et al. [5]	Three angles between the two front and rear legs	CASIA B	92.3
Cunado et al. [6]	Fourier components of the motion of the leg	Not mentioned	100
Kong et al. [7]	Model-based feature vector	CASIA B	89
Patil et al. [8]	Six joint angles	Clinical data	–
Dixit et al. [9]	Image moments	CASIA B	71.42
Ismail et al. [18]	Gait energy image	CASIA B	90
Lu et al. [23]	Texture and color characteristics of the body's head, top, leg	CASIA Gait	98.46
Sharif et al. [24]	Multi-level feature extraction	CASIA A	98.6
		CASIA B	93.5
		CASIA C	97.3
Bakchy et al. [25]	Gait energy image	CASIA B	58

without models. Segmentation mistakes, shadows, moving objects in the background, compression artifacts at the edge of the human silhouette, and the threshold for classifying background and foreground are all signs of trouble with silhouette recognition. Techniques for model-free feature representation include GEI, GHI, FDEI, AEI, and GFI. GEI/GHI can depict both dynamic and stationary components. FDEI is a GEI variant used to fix frames that have missing pieces. AEI depicts relationships between frames. Dimensionality reduction is necessary before further processing with model-free feature extraction approaches. Table 41.1 gives the comparative analysis of study of selected papers till 2020 with recognition rate as accuracy in percentage.

41.4 Proposed System

The feature extraction methods currently in use for identifying human gaits are reviewed in this work. The suggested system shown in Fig. 41.4 would be built on a high recognition rate achieved from approached features. It will aid in improving the rate of human identification.

Gathering a gait video sequence with a camera is the first step in determining a person's gait. Depending on the angle, there are various ways to install the camera. Video can be used to extract gait frames. Model-based features can be recovered after the subject has been identified from the frame. Various backdrop removal techniques can be used to create the binary silhouette from the observed gait frame. This phase entails normalizing the silhouettes to a particular size to obtain silhouettes using segmentation, morphological operator erosion, and dilation. Silhouettes

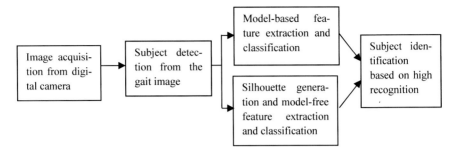

Fig. 41.4 Proposed human gait recognition system

should not be hidden because doing so degrades feature extraction, which increases identification accuracy rates. The reviewed model-free feature extraction techniques can be applied to silhouettes, and feature vectors can be generated. Dimensionality reduction techniques can be applied to the feature vector. The feature vector of model-based and model-free can be supplied to the classification techniques such as support vector machine, decision tree, random forest, and multilayer perceptron. The subject identification can be made based on the high recognition rate that is accuracy of the individual featured approach. Here, it would be easy to identify the key characteristics/features.

41.5 Conclusion

This review paper aims to conduct a literature study on the approaches for finding features used to identify human gaits. It has been discovered that understanding model-based and model-free approaches will aid in improving performance in terms of recognition rate. The model-based approaches provide structural data and cutting-edge technology. Model-free approaches provide spatiotemporal data that is easy to produce and beneficial for recognition. We can investigate which key features are most significant based on the high human recognition rate represented in the proposed system.

References

1. Shirke, S, Pawar, S.S., Shah, K.: Literature review: model free human gait recognition. In: Proceeding—2014 4th International Conference Communication Systems and Network Technologies CSNT 2014, pp. 891–895 (2014). https://doi.org/10.1109/CSNT.2014.252
2. Yam, C.-Y., Nixon, M.S.: Gait recognition, model-based. Encycl. Biometrics, 633–639 (2009). https://doi.org/10.1007/978-0-387-73003-5_37.
3. Desai, H.M., Gandhi, V.: A survey: background subtraction techniques. Int. J. Sci. Eng. Res. **5**(12), 1365–1367 (2014)

4. Yoo, J.H., Hwang, D., Nixon, M.S.: Gender classification in human gait using support vector machine. Lect. Notes Comput. Sci. **3708**:138–145 (2005). https://doi.org/10.1007/11558484_18
5. Benbakreti, S., Benbakreti, S., Benyettou, M.: Recognizing human gait in video sequences. In: Proceeding 2012 International Conference of the Chilean Computer System ICMCS 2012, pp. 323–327 (2012). https://doi.org/10.1109/ICMCS.2012.6320280.
6. Cunado, D., Nixon, M.S., Carter, J.N.: Automatic extraction and description of human gait models for recognition purposes. Comput. Vis. Image Underst. **90**(1), 1–41 (2003). https://doi.org/10.1016/S1077-3142(03)00008-0
7. Kong, W., Saad, M.H., Hannan, M.A., Hussain, A.: Human gait state classification using artificial neural network. In: IEEE SSCI 2014–2014 IEEE Symposium Series on Computational Intelligence—CIMSIVP 2014 2014 IEEE Symposium on Computational Intelligence Multimedia, Signal Vis Process Proc pp. 0–4 (2015). https://doi.org/10.1109/CIMSIVP.2014.7013287
8. Patil, P., Kumar, K.S., Gaud, N., Semwal, V.B.: Clinical human gait classification: extreme learning machine approach. In: 1st International Conference on Advances in Science, Engineering and Robotics Technology 2019, ICASERT 2019, vol. 2019, no. Icasert, pp. 1–6 (2019). https://doi.org/10.1109/ICASERT.2019.8934463
9. Dixit, M.: GAIT signature for identification systems using image moments. Int. J. Eng. Res. Technol. **10**(07), 604–606 (2021)
10. Bobick, F., Davis, J.W.: The recognition of human movement using temporal templates. IEEE Trans. Pattern Anal. Mach. Intell. **23**(3), 257–267 (2001). https://doi.org/10.1109/34.910878
11. Luo, J., Zi, C., Zhang, J., Liu, Y.: Gait recognition using GEI and AFDEI. Guangdian Gongcheng/Opto-Electronic Eng. **44**(4), 400–404 (2017). https://doi.org/10.3969/j.issn.1003-501X.2017.04.003
12. Han, J., Bhanu, B.: Individual recognition using gait energy image. IEEE Trans. Pattern Anal. Mach. Intell. **28**(2), 316–322 (2006). https://doi.org/10.1109/TPAMI.2006.38
13. Romero-Moreno, M., Martínez-Trinidad, J.F., Carrasco-Ochoa, J.A.: Gait recognition based on silhouette, contour and classifier ensembles. In: Lecture notes in computer science (including subseries lecture notes in artificial intelligence and lecture notes in bioinformatics), vol. 5197 LNCS, pp. 527–534 (2008). https://doi.org/10.1007/978-3-540-85920-8_64
14. Nizami, F., Hong, S., Lee, H., Ahn, S., Toh, K.-A., Kim, E.: Multi-view gait recognition fusion methodology. In: 2008 3rd IEEE conference on industrial electronics and applications, ICIEA 2008, 2008, pp. 2101–2105. https://doi.org/10.1109/ICIEA.2008.4582890
15. Tan, D., Huang, K., Yu, S., Tan, T.: Efficient night gait recognition based on template matching. Proc. Int. Conf. Pattern Recognit. **3**, 1000–1003 (2006). https://doi.org/10.1109/ICPR.2006.478
16. Khalid Bashir, S.G., Xiang, T.: Feature selection on gait energy image for human identification, pp. 985–988 (2008)
17. Lenac, D.S., Ramakić, A., Pinčić, D.: Extending appearance based gait recognition with depth data. Appl. Sci. **9**(24) (2019). https://doi.org/10.3390/app9245529
18. Ismail, S.N.S.N., Ahmad, M.I., Anwar, S.A., Isa, M.N.M., Ngadiran, R.: Gait feature extraction and recognition in biometric system. J. Telecommun. Electron. Comput. Eng. **8**(4), 127–132 (2016)
19. Wang, X., Feng, S.: Multi-perspective gait recognition based on classifier fusion. IET Image Process. **13**(11), 1885–1891 (2019). https://doi.org/10.1049/iet-ipr.2018.6566
20. Uddin M.Z., et al.: The OU-ISIR large population gait database with real-life carried object and its performance evaluation. IPSJ Trans. Comput. Vis. Appl. **10**(1) (2018). https://doi.org/10.1186/s41074-018-0041-z
21. Nandy, R.C., Chakraborty, P.: Cloth invariant gait recognition using pooled segmented statistical features. Neurocomputing **191**, 117–140 (2016). https://doi.org/10.1016/j.neucom.2016.01.002
22. Jian, M., Dong, J., Zhang, Y. (2007) Recognizing humans based on gait moment image. In: Proceedings - SNPD 2007: Eighth ACIS International Conference on Software Engineering, Artificial Intelligence, Networking, and Parallel/Distributed Computing., vol. 1, pp. 713–718. https://doi.org/10.1109/SNPD.2007.307

23. Lu, Y., Boukharouba, K., Boonært, J., Fleury, A., Lecœuche, S.: Application of an incremental SVM algorithm for on-line human recognition from video surveillance using texture and color features. Neurocomputing **126**, 132–140 (2014). https://doi.org/10.1016/j.neucom.2012.08.071
24. Sharif, M., Attique, M., Tahir, M.Z., Yasmim, M., Saba, T., Tanik, U.J.: A machine learning method with threshold based parallel feature fusion and feature selection for automated gait recognition. J. Organ. End User Comput. **32**(2), 67–92 (2020). https://doi.org/10.4018/JOEUC.2020040104
25. Bakchy, S.C., Mondal, M.N.I., Ali, M.M., Hoque Sathi, A., Ray, K.C., Jannatul Ferdous, M.: Limbs and muscle movement detection using gait analysis (2018). https://doi.org/10.1109/IC4ME2.2018.8465598

Chapter 42
Right Ventricle Volumetric Measurement Techniques for Cardiac MR Images

Anjali Abhijit Yadav and Sanjay R. Ganorkar

Abstract Cardiac disease diagnosis is very important domain in biomedical and technological innovations. The right ventricle (RV) volumetric analysis is useful for finding anatomical and functional defects and blood loading capacity of heart at right side. The volumetric measurements of RV such as blood volume at systole (ESV) and diastole phases (EDV) are significant for further decisions in cardiac disease diagnosis. For experimentation and to find better one, we have been performed three techniques A. motion-based clustering B. intensity-based clustering C. deep learning-based architecture. These approaches are used for measuring volumetric parameters such as end systole (ESV), end diastole (EDV) and ejection fraction (EF). Then, results are compared with ground truths provided by clinicians. The minimum percentage error occurred in measurements of deep learning techniques as compare to others. The clustering approaches having limitations of data scarcity, which can be eliminated using DL techniques and shows better performance.

Keywords End systole (ESV) · End diastole (EDV) · Ejection fraction (EF)

42.1 Introduction

Cardiovascular diseases are the leading cause of deaths and group of disorders of the heart and blood vessels such as coronary heart disease. CMRI is the standard modality for the non-invasive assessment of high quality, accurate functional and anatomical images in any orientation. CMRI provides accurate information of

---

A. A. Yadav (✉) · S. R. Ganorkar
Research Scholar, SPPU, Sinhgad College of Engineering, Sinhgad Institute of Technology and Science, Pune, India
e-mail: yadavanjali2k18@gmail.com

S. R. Ganorkar
e-mail: srganorkar.scoe@sinhgad.edu

S. R. Ganorkar
SPPU, Sinhgad College of Engineering, Pune, India

© The Author(s), under exclusive license to Springer Nature Singapore Pte Ltd. 2024
P. K. Jha et al. (eds.), *Proceedings of Congress on Control, Robotics, and Mechatronics*, Smart Innovation, Systems and Technologies 364,
https://doi.org/10.1007/978-981-99-5180-2_42

morphology, muscle perfusion, tissue variability and blood flow using adequate protocols. The cardiac contractile function can be quantified and analyzed through ventricle volumes, masses and ejection fraction (EF) by segmenting the left (LV) and right (RV) ventricles from cine MR images [1, 2]. The left and right ventricle volumetric analysis is useful in the process of patient management, disease diagnosis, risk evaluation and therapy decisions. The right ventricle function has been recognized in heart failure, RV myocardial infarction, congenital heart diseases and pulmonary hypertension. Only a few researches on the right ventricle (RV) have been published. We are working on RV analysis based on evaluation of blood volumes [3, 4]. RV contouring is used to transform an image's representation into a more meaningful step. Contouring results are further going to calculate volumetric measurements such as end systole volume (ESV), end diastole volume (EDV), ejection fraction (EF), stroke volume and ventricle mass of right ventricle. Still facing following difficulties:

a. Fast speed pumping of the cardiac motion b. Circulatory system c. Imaging noisy interference d. Diverse and eccentric structure f. Fragile wall g. Inadequate borders [1–6].

42.2 Previous Work

Caroline Petitjean suggested that evaluating several RV segmentation algorithms on common data. Seven automated and semi-automated methods have been considered, along them three atlas-based methods, two prior-based methods and two prior-free, image-driven methods that make use of cardiac motion. The obtained contours were compared against a manual tracing by an expert cardiac radiologist, taken as a reference, using Dice metric and Haussdorff distance. Best results show that an average 80% Dice accuracy and a 1 cm Haussdorff distance can be expected from automated algorithms [1]. The author presents a review of automated and semi-automated methods performing contouring of cardiac cine MRI sequence short-axis images. Medical background and specific segmentation difficulties related to these images are presented in this paper. The author suggests that for RV analysis, complex segmentation methods and prior knowledge-based methods need to be developed [3]. Medical image segmentation has shown enormous success using convolutional neural networks (CNNs), even when constructed on a shape prior. CNNs are prone to create anatomically error free segmentations. These segmentation results are closer to the inter-expert variability. The methodology in this work is developed for creating cardiac image segmentation to evaluate cardiac frames using a trained CNN. It finds anatomically improbable outcomes with constrained variational autoencoder, but distorts the results toward the nearest anatomically accurate heart shape [4]. The short and long axis view of RV is provided in this paper. The intensity inhomogeneity often occurs in real-world images. It is the considerable challenge in image segmentation. The most widely used image segmentation algorithms are region-based and typically rely on the homogeneity of the image intensities in the regions of interest. It often fails to provide accurate segmentation results due to the intensity inhomogeneity [5]. The volumetric functionality of RV is essential for LV volume measurements and pulmonary circulations. The expert clinicians can take decisions for optimization of RV preload and afterload to enhance RV contractility based on volume detection. In congenital heart problems, RV volumetric analysis is important to find preload, afterload of blood volume [6, 7]. The manual method is facing intra and inter-expert variability. For measuring the volume of heart chambers, scientists suggest a semi-automatic VR approach. It is rapid, effective and less sensitive to inter-observer variability then the semi-automated type. This technique is easier to conduct extra tests in contexts with complicated anatomical forms. A 3D phantom of the observed cavity was produced by multiplying the voxel area of each slice by the set of slice thickness. The method of this paper represents volume analysis using voxel area summation and claims that it is a practical and reliable method. We have also preferred the same approach in our work for volumetric analysis [8]. For small datasets in biomedical images contouring of any organ is challenging using deep learning. The UNet architecture is a very powerful deep architecture for such cases. In this technique, end-to-end decisions are made in the decoding levels and RV characteristics are extracted in the encoding layers using a UNet-shaped network topology. Many residual blocks are cascaded to extract RV characteristics in the encoding stages. Convolutional layers are used in the decoding levels to provide the RV predictions accurately [9, 10]. During CMR assessment, RV borders are manually located in clinical routine. This automatic segmentation of RV using CNN with stacked autoencoder suffers from leakage and shrinkage of contours. The fuzzy borders and presence of trabeculations of RV shape can be minimized using deep learning techniques. In the future scope of this paper, the author have suggested that they wish to improve accuracy of results and minimization of computational time. Also results of this technique need to be applied on more no. of patients for authentication purpose and real-time implementation. Also a full heart segmentation method for ventricles and atriums is necessary to invent to make a combined tool for the same [11–13]. A rapid RCNN method for RV wall motion detection is suggested by the author in this paper. A combination of CNN and MLP is designed and studied for heart contouring and received mean localization error as 6.07 mm [14]. A survey paper on various cardiac segmentation methods for various cardiac modalities is presented in this paper. The paper has covered CT, MRI and ultrasound images of RV LV ventricles, atriums, etc. The segmentation algorithm evaluation is implemented based on evaluation parameters such as Jaccard Metric, Dice coefficient and Haussdorff distance. The paper shows that the CNN-based methods gives promising results for 2D and 3D segmentation. Also the study shows the challenges of deep learning methods for real-time implementation are (1) Label shortage, (2) Model uniformity depends on scanners and pathologies (3) Model interchangeability [15]. Q. Zheng and all have described a rugged cardiac segmentation algorithm based on deep learning. For experimentation, they have used three kinds of datasets such as ACDC, Sunnybrook and RVSC. This spatial architecture segments RV, LV named as LVRV Net, which is designed to extract ROI and constructed in 3D form for epi and endo borders. Maximum accuracy received is 0.93 for LVC and 0.88 for RVC, but worked better for distance measure. Finally, this paper claims that apical frames results need to be improved [16]. The main drawback of UNet architecture is loss of information in the reconstruction process. Very fine level segmentation is necessary for retinal images which is implemented with multiscale architecture combined in UNet with lossfree downsampling process [17].

As a result, the earlier study demonstrates how several methods were used to contour the RV in cardiac images. Our goal is to develop a method for measuring right ventricular volumes that will save clinicians time and efforts. And also provide findings that are as accurate as those of pathology specialists.

42.3 Methodology

Here, we are presenting three types of techniques for right ventricle contouring and further volumetric measurements. These results further evaluated clinicians for finding various cardiac disease parameters such as stroke volume.

42.3.1 Motion-Based Clustering Approach

Figure 42.1 shows motion-based clustering model which is primarily utilized to cope with illumination variations and image noise created in cardiac MRI images while capturing. It will be utilized to solve the issues of ill-defined boundaries of the right ventricle. Here, the object detection is relating to motion rather than shape. To estimate motion intensity, an optical flow method is used. The very famous Lukas Kanade (LK) optical flow method is implemented in this model for motion detection. K-means clustering is further added to extract ROI. Energy minimizing in level set formation is employed for getting particular region of interest [5, 6, 18].

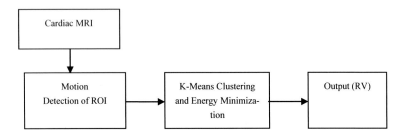

Fig. 42.1 Motion-based clustering model [5]

42.3.2 Fuzzy C Means-Based Clustering Approach

The fuzzy c means-based clustering approach basically forms the clusters based on image intensities. The input cardiac frames are transformed to grayscale images and then scaled to 256 × 256 during preprocessing. Wiener filtering is excellent for diminishing motionally blurred noise. This filter includes the degradation function and statistical noise attributes into the restoration procedure, so we added the Weiner filter to predict uncorrupted frames and decrease mean square error. As intensity clustering can't distinguish ROI due to the same intensity level available in RV as compared to background. Therefore, our model is divided with two steps. In the first stage, we implemented fuzzy c means clustering to divide an image into high level. It employs fuzzy c partitioning, in which every data point may belong to more than one cluster based on its membership value, which corresponds to its cluster value. It creates a fuzzy matrix (U, X) of dimension p x c, where p represents the number of data points and c represents the number of clusters. U represents a membership function that determines whether a data point belongs to that function or not and accepts only values ranging from 0 to 1, also known as a crisp set [19]. Markov Random Field (MRF) is the second block added here for a fine level of contouring. FCM and class label data are treated as a mixture of two random processes in this block. The overall labels will be calculated using iterations. When it reaches to maximum iterations, the result remains unchanged and gives final contouring of the RV endocardium border [20, 21].

- **Algorithm of Fuzzy C Means-Based Clustering Approach**

Step 1: Input cardiac MR images available in .PGM format.
Step 2: Convert into .png format.
Step 3: Resize images into 256 × 256 Gy scale images.
Step 4: Apply Weiner filtering to remove blurring noise.
Step 5: Fuzzy C Means clustering with energy minimization process applied.
Step 6: Using Markov random model apply probabilistic model used to separate ROI.

42.3.3 Deep Learning-Based Approach

We have selected the deep learning-based 'U' form architecture in our third technique [22]. The architecture is split into two manifestations. The first is contraction and another is expansion of images [18]. The series of up convolutions and concatenation with high-resolution features from the contracting path take place. The depth of the up and downsampling steps is set to six. In convolution procedures, the kernel size is set at 3 × 3. The 'he normal' kernel initializer was used to set initial weights and biases. Each phase in downsampling represents four actions, such as adding a convolutional layer followed by a drop-out to limit over-fitting in deep learning, then

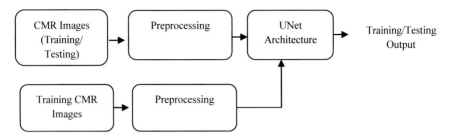

Fig. 42.2 Process diagram of deep learning technique [9, 10]

adding another convolutional layer followed by a max-pooling layer to down sample the frame. The importance of this chosen max-pooling layer is to lower the number of parameters for further feature extraction. It employs a max-pooling layer (2, 2) with the activation function 'elu', Padding = 'same' and de-convolution to improve the contouring of RV. In the first upsampling step, the original convolutional layer output is concatenated with an upsampling layer to obtain the transpose of convolution with '2'-Stride, 'same'-Padding and Dropout with 0.2. 'Sigmoid' is selected as the activation function of the output convolution layer [22]. Thus, the procedure identifies features at the pixel level from higher resolution to lower resolution for impenetrable categorization, and the output map has two classes at the end, one is region of interest and the rest is of image (Fig. 42.2).

42.4 Result and Discussions

The results of our selected three techniques are discussed along with the dataset in this section. At the end of the section, comparative volumetric analysis with respect to each technique is given.

42.4.1 Dataset

The MRI data available for 18 patients for the left ventricular chamber is selected. The heart database is provided by A2SI LABs. It contains short-axis cardiac MR images. PGM format. The images have already been converted from DICOMM to PGM format. This dataset provides images format with extension of PGM format and modified for supporting 3D. Pat??/expert1/ and Pat??/expert2/: contains the manual segmentation of the endocardial and epicardial border at end systolic and end diastolic time given for left ventricle. An interior voxel is set to 255. Each image has a header of that the image has 100 lines, and 100 columns indicates that 255 is the maximum value of the image voxel. 14 is the number of 2D + t sequences, 1.77 mm is pixel spacing (along x and y axis) of the 2D images with 6 thickness of the 2D images [23].

The system used for our experimentation is Intel CPU, CORE i5-1135G7@2.40 GHz, windows 10, 64 bit operating system and 4 GB RAM. The volumetric measurements of the right ventricle are end systole volume and end diastole volume (ESV and EDV) in mL/m^2. We have selected endocardial borders for volume measurements. Volumes are computed as the sum of all areas multiplied by the space between slices. The ESV is used in analysis of right ventricle functionality and calculation of ejection fraction and stroke volume. ESV is the sum of all pixels from basal slice to apex slice, selected as systolic frames multiplied by the value of the space between slices.

$$ESV = \sum \text{Systolic RV Area} * \text{Space between slices} \qquad (42.1)$$

The significance of EDV is to estimate preload of heart, ejection fraction and stroke volume. EDV is calculated as the sum of pixels of all RV regions of diastolic frames multiplied by the value of the space between slices.

$$EDV = \sum \text{Diastolic RV Area} * \text{Space between slices} \qquad (42.2)$$

EF is evaluated by clinical expert's to decide how well the heart is pumping blood. It is the ratio of difference between end diastole volume and end systole volume to end diastole volume (%).

$$EF(\%) = \frac{(EDV - ESV)}{EDV} * 100 \qquad (42.3)$$

where

EF—Ejection fraction
ESV—End systole volume
EDV—End diastole volume.

42.4.2 Local Motion-Based Clustering Approach

The results of local motion intensity clustering are shown in Fig. 42.3. Figure 42.3a shows input CMR image. Figure 42.3b represents motion detection output using Lukas Kanade algorithm in the form of pixel intensity of the right ventricle. This intensity is further referred to as the k-factor for K-means clustering. The clustered output is given in Fig. 42.3c. The localized right ventricle endocardium borders are marked with red borders in Fig. 42.3d, but the borders of RV contouring fail at the time of blood pooling.

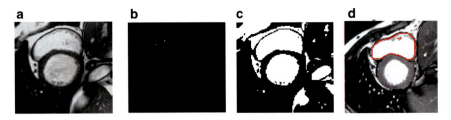

Fig. 42.3 **a** Cardiac short-axis MRI image, **b** motion detection of RV, **c** K-means clustering output, **d** segmented RV output

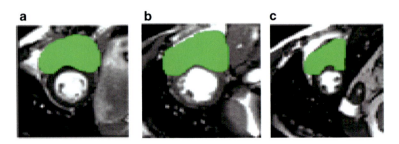

Fig. 42.4 Right ventricle contouring output using fuzzy c means clustering

42.4.3 Fuzzy C Means-Based Clustering Approach

Figure 42.4 shows results of our second approach. It shows a green color portion which is nothing but right ventricle contouring output using fuzzy c means clustering for a few frames of different patients.

42.4.4 Deep Learning-Based Approach

Figure 42.5 shows results of our third approach. Figure 42.5a shows input short-axis CMR images. Figure 42.5b presents ground truth image at endocardium borders and Fig. 42.5c shows predicted output using deep learning techniques.

Table 42.1 gives a list of hyperparameters used in deep learning UNet architecture. The activation function selected is ReLU with ADAM: optimizer, mean_squared_ error:loss function. 25:epochs with 16:batch size given 0.94 training accuracy and 0.95 validation accuracy. Total 412 training images are taken for training and 38 for testing without augmentation.

Table 42.2 presents the comparative results of selected volumetric parameters such as ESV, EDV and EF for selected three techniques. After comparison with ground truth results, the percentage error has been given in the table. The ground truths images for further analysis are provided by clinical experts. The motion-based

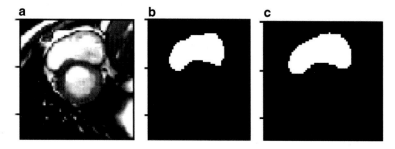

Fig. 42.5 a Cardiac short-axis MRI image, **b** mask image, **c** segmented RV output

Table 42.1 Hyperparameter selection of deep learning model

Type of parameter	Hyperparameter
Activation function	ReLU
Optimizer	ADAM
Loss function	mean_squared_error
Epoch number	25
Batch size	16
Loss	0.04
Training-accuracy	0.94
val_loss	0.04
val_accuracy	0.95

clustering approach has an average % error of 30.14, 10.94 and 21.66 for ESV, EDV and EF, respectively. The fuzzy c means-based clustering and deep learning-based approaches have achieved ESV, EDV and EF 14.15, 3.38, 13.68, 1.55, 1.01, 0.71, respectively. The results show that DL approach is better for future designing of fully automatic cardiac volumetric measurement systems.

Table 42.2 Comparative results: ESV, EDV and EF with ground truth images

	Avg. error in predicted ESV (mL) (%)	Avg. error in predicted EDV (mL) (%)	Avg. error in predicted EF (%)
Motion-based clustering approach	30.14	10.95	21.66
Fuzzy c means-based clustering approach	14.15	13.68	1.01
Deep learning-based approach	3.38	1.55	0.71

42.5 Conclusion

This paper presents three types of techniques used for right ventricle contouring from short-axis cardiac MR images. The right ventricle contouring is further useful for cardiac parameters measurements such as ESV, EDV and EF in the diagnosis process of cardiac diseases. (1) A motion-based clustering strategy (2) Intensity-based clustering approach and (3) Deep learning-based approaches are explored for the study to get state of art technique, which gives promising results as closer as expert clinicians. For volumetric characteristics, the performance of selected technologies is evaluated. The ejection fraction calculations show 0.71% mean error using DL technique, 1% using fuzzy c means clustering and 21.67% with motion intensity clustering. Deep learning approaches outperform the other two in terms of accuracy. It gives the smallest mean difference (%) as compared to others. The fuzzy c means clustering approach with MRF model also gives comparative promising results compared to motion-based clustering, still highly reliable on database. After rigorous testing on real-time CMR, capturing fully automatic volumetric measuring systems can be constructed using deep learning approaches. It makes the CMR process and diagnosis is lesser prone to manual interventions.

Acknowledgements We would like to thank Professor Dr. Santosh Konde and team, Department of Radiology, Smt. Kashibai Navale Medical College and General Hospital, Narhe, Pune, Maharashtra for preparing dataset ground truths and their valuable suggestions.

References

1. Petitjean, et al. [10] Right ventricle segmentation from cardiac MRI: a collation study. Med. Image Anal. **19**, 187–202 (2015)
2. Bartelds, B., Douwes, J.M., Berger, R.M.F.: The right ventricle in congenital heart diseases. In: Gaine, S.P., Naeije, R., Peacock, A.J. (eds.) The Right Heart. Springer, Cham (2021). https://doi.org/10.1007/978-3-030-78255-9_13
3. Yadav, A., Rohokale, V., Shah, S.: Healing of cardiovascular diseases through mind-heart interaction. In: IEEE, 2018 Global Wireless Summit (GWS), pp. 97–101 (2018)
4. Painchaud, N., et al.: [5] Cardiac segmentation with strong anatomical guarantees. IEEE Trans. Med. Imaging 0278-0062 (2020)
5. Guo, et al.: [6] Local motion intensity clustering (LMIC) model for segmentation of right ventricle in cardiac MRI images. IEEE J. Biomed. Health Inform. (2018). https://doi.org/10.1109/JBHI.2018.2821709
6. Li, et al.: [5] A level set method for image segmentation in the presence of intensity inhomogeneities with application to MRI. IEEE Trans. Image Process. **20**(7), (2011)
7. Dini, et al.: [13] Right Ventricular Failure in Left Heart Disease: From Pathophysiology to Clinical Manifestations and Prognosis. Springer Nature (2022). https://doi.org/10.1007/s10741-022-10282-2
8. Yogev, et al.: [12] Proof of concept: comparative accuracy of semiautomated VR modeling for volumetric analysis of the heart ventricles. Heliyon **8**, e11250 (2022). https://doi.org/10.1016/j.heliyon.2022.e11250

9. Liu, Z., Feng, Y., Yang, X.: Right ventricle segmentation of cine MRI using residual U-net convolutional networks. IEEE (2019). https://doi.org/10.1109/PDCAT.2019.00072
10. Ronneberger, O., Fischer, P., Brox, T.: U-Net: Convolutional Networks for Biomedical Image Segmentation. arXiv:1505.04597v1 [cs.CV] 18 May (2015)
11. Peng, et al.: [5] A review of heart chamber segmentation for structural and functional analysis using cardiac magnetic resonance imaging. Magn. Reason. Mater. Phy. **29**, 155–195 (2016)
12. Avendi, M., Kheradvar, A., Jafarkhani, H.: Fully automatic segmentation of heart chambers in cardiac MRI using deep learning. J. Cardiovasc. Magn. Reason. **18**(S1), 351 (2016)
13. Avendi, M., Kheradvar, A., Jafarkhani, H.: Automatic segmentation of the right ventricle from cardiac MRI using a learning-based approach. In: Magnetic Resonance in Medicine. Wiley, New York, NY, USA (2017). https://doi.org/10.1002/mrm.26631
14. Kermani, et al.: [4] NF-RCNN: heart localization and right ventricle wall motion abnormality detection in cardiac MRI. Phys. Med. **70** (2020)
15. Chen, C., et al.: [6] Deep learning for cardiac image segmentation: a review. Front. Cardiovasc. Med. **7** (2020), Article 25
16. Zheng, Q., et al.: [3] 3D consistent & robust segmentation of cardiac images by deep learning with spatial propagation. IEEE Trans. Med. Imaging, Inst. Electr. Electron. Eng. (2018). https://hal.inria.fr/hal-01753086
17. Yin, P., et al.: [4] Deep guidance network for biomedical image segmentation. IEEE Access **8** (2020). https://doi.org/10.1109/ACCESS.2020.3002835
18. Bhan, A., Goyal, Ray, V.: Fast fully automatic multiframe segmentation of left ventricle in cardiac MRI images using local adaptive k-means clustering and connected component labeling. In: International Conference on Signal Processing and Integrated Networks IEEE, pp 114–119 (2015)
19. Song, J., Yuan, L.: Brain tissue segmentation via non-local fuzzy c means clustering combined with Markov random field. Math. Biosci. Eng., MBE **19**(2), 1891–1908 (2021). https://doi.org/10.3934/mbe.2022089
20. Bai, et al.: [4] Intuitionistic center-free FCM clustering for MR brain image segmentation. IEEE J. Biomed. Health Inform. 23(5) (2019)
21. Setyawan, et al.: [5] MRI image segmentation using morphological enhancement and noise removal based on fuzzy C-means. In: IEEE Conference, Proceedings of 2018 5th International Conference on Information Technology, Computer and Electrical Engineering (ICITACEE) (2018)
22. Avendi, M.R., Kheradvar, A., Jafarkhani, H.: A combined deep-learning and deformable-model approach to fully automatic segmentation of the left ventricle in cardiac MRI. MedIA Image Anal. **30**, 108–119 (2016)
23. https://www.laurentnajman.org/heart/H_data.html

Chapter 43
Statistical Evaluation of Classification Models for Various Data Repositories

V. Lokeswara Reddy, B. Yamini, P. Nagendra Kumar, M. Srinivasa Prasad, and Y. Jahnavi

Abstract Exploitation of massive amount of multidimensional data of numerous diversities from heterogeneous sources is emerging. Segments can be expanded by the effective use of the data and simultaneously the warehoused data furnishes significant eventualities for strengthening cardinal decision making. Cutting edge intelligence and efficacious approaches and methods are essential to produce a model by means of various types of data. Machine learning is described as constructing a model for handling unfathomable accumulated or streaming data which is difficult to be handled by conventional data processing techniques. The prerequisite for machine learning is also prompted based on preprocessing, analysis, demonstration and prediction on diverse categories of data. The experimentation has been performed on Mushrooms dataset and Census dataset by carrying out machine learning algorithms. It has been proved that the ensemble algorithm shows better performance on the intended datasets in terms of various performance measures.

Keywords Machine learning · Classification · Ensemble learning · Empirical evaluation

---

V. Lokeswara Reddy
Department of Computer Science and Engineering, K.S.R.M College of Engineering (Autonomous), Kadapa, Y. S. R (Dt), Andhra Pradesh, India

B. Yamini
Department of Computer Science and Engineering, SRM Institute of Science and Technology, Kattankulathur, Chennai, Tamil Nadu, India

P. Nagendra Kumar
Department of Computer Science and Engineering, Geethanjali Institute of Science and Technology, Nellore, Andhra Pradesh, India

M. Srinivasa Prasad
Department of Library Science, Dr V S Krishna Govt Degree and PG College (Autonomous), Visakhapatnam, Andhra Pradesh, India

Y. Jahnavi (✉)
Department of Computer Science, Dr V S Krishna Govt Degree and PG College (Autonomous), Visakhapatnam, Andhra Pradesh, India
e-mail: yjahnavi.2011@gmail.com

© The Author(s), under exclusive license to Springer Nature Singapore Pte Ltd. 2024
P. K. Jha et al. (eds.), *Proceedings of Congress on Control, Robotics, and Mechatronics*, Smart Innovation, Systems and Technologies 364,
https://doi.org/10.1007/978-981-99-5180-2_43

43.1 Introduction and Preliminaries

There exist numerous kinds of machine learning such as supervised learning, unsupervised learning and reinforcement learning. Supervised learning is one of the predominant fundamental categories of machine learning, wherein the machine learning algorithm is trained on labeled data. The algorithms which function using unlabeled data are unsupervised machine learning algorithms. Reinforcement learning outright considers inspiration from the perceptions in day-to-day operations and movements. Favorable outcomes are promoted or 'reinforced' and non-favorable outcomes are downgraded [1].

The basic rationale of Information Retrieval System (IRS) research area is for retrieving relevant information, where the information may be composed of text, images, audio, video, etc. Natural language processing techniques are useful in IRS to improve the accuracy [2]. The purpose of information extraction is exclusively different. Text mining is the extraction of interesting and useful patterns in textual data. Natural language processing techniques such as morphological, syntactic, semantic, pragmatic analysis support the approaches of text mining such as text classification, text clustering, summarization, question-answering and anaphora resolution. There exists different text classification and clustering algorithms such as Naïve Bayes, support vector machines, partitioning methods, and hierarchical methods. Natural language processing techniques are useful to remove indistinctness and ambiguity hidden in text documents [3–8].

Here, the aim is to study various classification algorithms in machine learning applied on different kinds of datasets. Correlation of machine learning with various disciplines has been represented in Fig. 43.1. The datasets that are used are census income and birds' datasets. Accumulation tools and conception algorithms for predictive framework and analytics are used. In this paper, it has been investigated that the algorithms have been experimented on various datasets such as census income which is taken from UCI machine learning warehouse and birds dataset taken from Ponce research group. We can regulate this model to implement at its optimal level while getting related and hence the ensemble algorithm works effectively with high precision.

43.2 Literature Work and Methodologies

Present survey which emphases on exploitation of enormous volume of multidimensional data of various diversities from different sources is presented. The prerequisite for machine learning is also instigated by an allocation based on analysis, demonstration and prediction on deliveries, transportation, etc. Sectors can be expanded by the efficacious use of the data, and at the same time, the warehoused data allows significant eventualities for inflating strategic decision making. State-of-the-art intelligence and efficacious techniques and approaches are essential to generate a standard

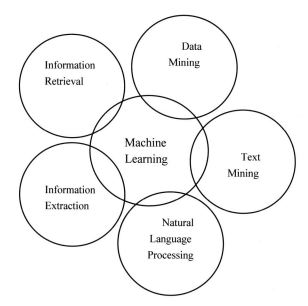

Fig. 43.1 Correlation of machine learning with various disciplines

through the various types of data. Machine learning is represented as evaluating and developing a model for processing composite warehoused or streaming data which is difficult to be operated by traditional computational techniques [9].

Classification is also termed as supervised learning that comprises two steps. First one is model formation and the next is model usage. In the first step, the collection of tuples exploited considering prototype formation is termed the training set. The constructed epitome is delineated as the model that is based on either rules or decision trees or in any other form. In the second phase, the model constructed from the first step is used for classifying the future or unseen data. The investigated label of the test sample is contrasted with the categorized outcome from the contemplated model.

Text mining is differing to Information Retrieval in the perception that it is not founded on specific criteria where it will identify several hidden and unidentified patterns which we do not encounter by exploring the corpus [10–14]. There exist various algorithms for performing classification, extracting salient features, opinion mining, processing of scalable web log data using map reduce framework, etc. [15–22].

The pseudocode for Bagging algorithm [23, 24] of machine learning has been represented as follows.

Algorithm: Bagging

1. **Input:** Tr is the Training Sample of Dataset D;
2. L is the Learning Classifier;
3. C is the count of bootstrap samples
4. **Output:** Bagging ensemble classifiers $E^*(x)$
5. **Procedure:**

6. for $i = 1$ to C do
7. B_i is a Bootstrap sample from D
8. Create Classifier $M_i = L(B_i)$
9. end for
10. Predict the class label for the given class sample
11. $E^*(x) = \sum_{i=1}^{C} M_i(x) = y$

Empirical estimation of each classifier is performed using various evaluation measures. The experimentation needs to be carried out by using machine learning techniques for model building. Effective classification techniques are essential for the success of accurate prediction [25–27].

1. Precision: It is the fraction of the retrieved documents that are relevant. It can be represented as

$$\text{precision} = \frac{|\{\text{relavant documents}\} \cap \{\text{retrieved documents}\}|}{|\{\text{retrieved documents}\}|}$$

It is also called as reproducibility or repeatability which measures the consistency of the results. The value of precision varies from 0 to 1, where the value nearby 1 denotes less false positive prediction. It can also be defined as $\frac{TP}{TP+FP}$, where TP is true positive and FP is false positive.

2. Recall: It is the fraction of the documents that are relevant to the query which are successfully retrieved.

$$\text{recall} = \frac{|\{\text{relevant documents}\} \cap \{\text{retrieved documents}\}|}{|\{\text{relevant documents}\}|}$$

It is also known as sensitivity. The value of recall varies from 0 to 1, where the value nearby 1 denotes less false negative prediction. It can also be defined as $\frac{TP}{TP+FN}$, where TP is true positive and FN is false negative.

3. F measure: It is the harmonic mean of precision and recall values. The value of F measure varies from 0 to 1, where the value nearby 1 denotes the best precision and recall. It can also be defined as $2 * \frac{P*R}{P+R} = \frac{TP}{TP + \frac{1}{2}(FP+FN)}$, where TP is true positive, FN is false negative, P is precision and R is recall.

4. Specificity: It is the ratio of negative results as a percentage of the number of samples which are negative. The range is varying from 0 to 1. Value close to 1 specifies less negative prediction. It is defined as $\frac{TN}{TN+FP}$, where TN is true negative, FP is false positive.

5. Accuracy: It is the most commonly used statistical measure. The range is varies from 0 to 1. Value close to 1 indicates better prediction. It is defined as $\frac{TP+TN}{TP+TN+FP+FN}$, where TP is true positive, TN is true negative, FP is false positive and FN is false negative.

In the experimentation, the algorithms have been evaluated using various primary evaluation measures that have been represented in the results and discussion section [28].

43.2.1 Results and Discussion

The datasets such as Census income dataset, Mushrooms dataset have been considered from UCI machine learning repository for evaluation of various algorithms such as J48, Naïve Bayes, multilayer perceptron, ZeroR and ensemble algorithm. The results of various classification algorithms are represented as follows.

43.2.2 Dataset 1 (Census Income Dataset)

The considered sample Census income dataset has 14 attributes and 48,842 instances that have only 2 classes. This dataset has taken from an UCI machine learning repository.

True positive rate, false positive rate, root mean square error and mean absolute error are calculated for J48, Naive Bayesian, multilayer perceptron, ZeroR and ensemble algorithm, which are represented in Table 43.1.

True positive rate of J48, Naive Bayesian, multilayer perceptron, ZeroR and ensemble algorithm is 0.862, 0.834, 0.827, 0.759 and 0.898, respectively. It shows that the ensemble algorithm is able to extract a greater number of correct instances.

False positive rate of J48, Naive Bayesian, multilayer perceptron, ZeroR and ensemble algorithm is 0.275, 0.382, 0.335, 0.759 and 0.238, respectively. It shows that the ensemble algorithm is better in terms of false positive rate, compared with the other algorithms, i.e., 0.238 only.

Mean absolute error of J48, Naive Bayesian, multilayer perceptron, ZeroR and the ensemble algorithm is 0.1942, 0.1735, 0.1836, 0.3656 and 0.1938, respectively. It shows that J48 and the ensemble algorithm are better in terms of mean absolute error, compared with ZeroR. But for this dataset, Naive Bayesian and multilayer perceptron perform better by exhibiting low mean absolute error, i.e., 0.1735 and 0.1836, respectively.

Root mean squared error of J48, Naive Bayesian, multilayer perceptron, ZeroR and the ensemble algorithm is 0.3196, 0.3723, 0.3749, 0.4276 and 0.3194, respectively.

Table 43.1 Results of the classification algorithms on census income dataset

Census income	True positive rate	False positive rate	Mean absolute error	Root mean squared error
J48	0.862	0.275	0.1942	0.3196
NB	0.834	0.382	0.1735	0.3723
MLP	0.827	0.335	0.1836	0.3749
ZERO R	0.759	0.759	0.3656	0.4276
Ensemble algorithm	0.898	0.238	0.1938	0.3194

Fig. 43.2 Comparison of root mean squared error of classifiers on census income dataset

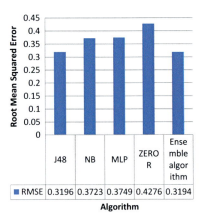

It shows that the ensemble algorithm is better in terms of root mean squared error, compared with the other algorithms, i.e., 0.3194 only.

For the considered dataset, the ensemble algorithm shows better performance than other algorithms in terms of the root mean square error. The root mean square error values of various algorithms are pictorially represented in Fig. 43.2 on census income dataset.

43.2.3 Dataset 2 (Mushrooms Dataset)

The considered sample Mushrooms dataset has 23 attributes and 8124 instances that have only 2 classes. This dataset has taken from an UCI machine learning repository.

True positive rate, false positive rate, root mean square error and mean absolute error are calculated for J48, Naive Bayesian, multilayer perceptron, ZeroR and the ensemble algorithm on Mushrooms dataset, which are represented in Table 43.2.

Table 43.2 Results of the classification algorithms on Mushrooms dataset

Mushrooms dataset	True positive rate	False positive rate	Mean absolute error	Root mean squared error
J48	1	0	0.0987	0.0233
NB	0.645	0.047	0.1026	0.3005
MLP	1	0	0.106	0.2488
ZEROR	0.387	0.387	0.2131	0.3264
Ensemble algorithm	1	0	0.0987	0.0233

Fig. 43.3 Comparison of root mean squared error of classifiers on Mushroom dataset

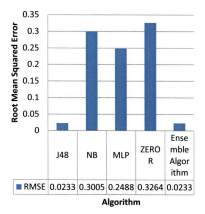

True positive rate of J48, Naive Bayesian, multilayer perceptron, ZeroR and the ensemble algorithm is 1, 0.645, 1, 0.387 and 1, respectively. It shows that the ensemble algorithm is able to extract more number of correct instances.

False positive rate of J48, Naive Bayesian, multilayer perceptron, ZeroR and the ensemble algorithm is 0, 0.047, 0, 0.387 and 0, respectively. It shows that the ensemble algorithm is better in terms of false positive rate, compared with the other algorithms, i.e., 0 only.

Mean absolute error of J48, Naive Bayesian, multilayer perceptron, ZeroR and the ensemble algorithm is 0.0987, 0.1026, 0.106, 0.2131 and 0.0987, respectively. It shows that J48 and the ensemble algorithm are better in terms of mean absolute error, compared with Naive Bayesian, multilayer perceptron, ZeroR, i.e., 0.0987.

Root mean squared error of J48, Naive Bayesian, multilayer perceptron, ZeroR and the ensemble algorithm is 0.0233, 0.3005, 0.2488, 0.3264 and 0.0233, respectively. It presents that the ensemble algorithm demonstrates the same performance as J48 in terms of root mean squared error, i.e., 0.3194 only.

For the considered dataset, the ensemble algorithm shows better performance than other algorithms in terms of the root mean square error. The root mean square error values of various algorithms are pictorially represented in Fig. 43.3 on Mushrooms dataset.

The experimentation has shown that the ensemble algorithm outperforms other existing algorithms on diverse considered datasets.

43.3 Conclusion

This paper focuses on the diverse machine learning algorithms applied on Mushrooms dataset and Census datasets. Tree-based classifier, Bayesian classifier, neural network-based classifier, rule-based classifier and an ensemble classifier are practiced on diverse datasets that are estimated by using several performance measures.

The ensemble algorithm illustrates better results than the remaining algorithms in terms of true positive rate, false positive rate, mean absolute error and root mean squared error.

References

1. Flah, M., et al.: Machine learning algorithms in civil structural health monitoring: a systematic review. Arch. Comput. Methods Eng. **28**(4), 2621–2643 (2021). ML
2. Gu, Y., et al.: Domain-specific language model pretraining for biomedical natural language processing. ACM Trans. Comput. Healthc. (HEALTH) **3**(1), 1–23 (2021). NLP
3. Cai, M.: Natural language processing for urban research: a systematic review. Heliyon **7**(3), e06322 (2021). NLP
4. Jahnavi, Y., Radhika, Y.: A cogitate study on text mining. Int. J. Eng. Adv. Technol. (IJEAT) ISSN (2012): 2249-8958
5. Jahnavi, Y., Radhika, Y.: Hot topic extraction based on frequency, position, scattering and topical weight for time sliced news documents. In: 2013 15th International Conference on Advanced Computing Technologies (ICACT). IEEE, 2013
6. Jahnavi, Y.: Statistical data mining technique for salient feature extraction. Int. J. Intell. Syst. Technol. Appl. **18**(4), 353–376 (2019)
7. Jahnavi, Y.: Analysis of weather data using various regression algorithms. Int. J. Data Sci. **4**(2), 117–141 (2019)
8. Jahnavi, Y., Radhika, Y.: FPST: a new term weighting algorithm for long running and short lived events'. Int. J. Data Anal. Techn. Strategies **7**(4), 366–383 (2015)
9. Jahnavi, et al.: A new algorithm for time series prediction using machine learning models. Evol. Intel. (2022). https://doi.org/10.1007/s12065-022-00710-5
10. Lin, J.: A proposed conceptual framework for a representational approach to information retrieval. ACM SIGIR Forum. 55(2), New York, NY, USA: ACM, 2022. IR
11. Liu, P., et al.: Pre-train, prompt, and predict: a systematic survey of prompting methods in natural language processing. arXiv preprint arXiv:2107.13586 (2021). NLP
12. Liu, X., et al.: Neural feedback facilitates rough-to-fine information retrieval. Neural Netw. (2022). IR
13. Mallick, J., et al.: Proposing receiver operating characteristic-based sensitivity analysis with introducing swarm optimized ensemble learning algorithms for groundwater potentiality modelling in Asir region, Saudi Arabia. Geocarto Int. 1–28 (2021)
14. Pfister, S.M. et al.: A summary of the inaugural WHO classification of pediatric tumors: transitioning from the optical into the molecular era. Cancer Discov. 12(2), 331–355 (2022). Classification
15. Yeturu, J.: A New Term Weighting Algorithm for Identifying Salient Events. LAP LAMBERT Academic Publishing (2018)
16. Tiwari, V., et al.: Applications of the Internet of Things in healthcare: a review. Turk. J. Comput. Math. Educ. **12**(12), 2883–2890 (2021)
17. Haripriya, et al.: Using social media to promote E-commerce business. Int J Recent Res Aspects **5**(1), 211–214 (2018)
18. Vijaya, U., et al.: Community-based health service for Lexis Gap in Online Health Seekers
19. Bhargav, et al.: An extensive study for the development of web pages. Indian J Public Health Res. Dev. 10(5) (2019)
20. Srivani, et al.: An approach for opinion mining by Acumening the data through exerting the insights
21. Jahnavi, et al.: A novel processing of scalable web log data using map reduce framework. In: Proceedings of CVR 2022, Computer Vision and Robotics, ISBN: 978-981-19-7891-3

22. Yeturu, J., et al.: A new algorithm for time series prediction using machine learning models. Evol. Intell. 1–12 (2022). https://doi.org/10.1007/s12065-022-00710-5
23. Jiang, N., Yang, H.: Stabilized scalar auxiliary variable ensemble algorithms for parameterized flow problems. SIAM J. Sci. Comput. **43**(4), A2869–A2896 (2021)
24. Lee, K., et al.: Sunrise: a simple unified framework for ensemble learning in deep reinforcement learning. In: International Conference on Machine Learning. PMLR, 2021
25. Sukanya, et al.: Country location classification on Tweets. Indian J Public Health Res. Dev. 10(5) (2019)
26. Thakkar, et al.: Clairvoyant: AdaBoost with cost-enabled cost-sensitive classifier for customer Churn prediction. Comput. Intell. Neurosci. **2022** (2022)
27. Wang, et al.: Evaluation of constraint in photovoltaic cells using ensemble multi-strategy shuffled frog leading algorithms. Energy Convers. Manag. **244**, 114484 (2021)
28. Alhijawi, et al.: Survey on the objectives of recommender system: measures, solutions, evaluation methodology, and new perspectives. ACM Comput. Surv. (CSUR) (2022). evaluation measures

Chapter 44
Hierarchical Clustering-Based Synthetic Minority Data Generation for Handling Imbalanced Dataset

Abhisar Sharma, Anuradha Purohit, and Himani Mishra

Abstract Predictive modeling is a new area of data science and machine learning that is gaining popularity. It provides sustained business growth, accurate future predictions, and trend estimations. Predictive modeling is the process of creating, processing, and validating a model that may be used to make future predictions using known results. Predictive modeling depends on the complete and precise datasets, however some of the datasets are imbalanced in nature that leads to data misclassification. Models trained on an imbalanced dataset with a small number of minority class instances, despite their high accuracy, would perform poorly during training. In this paper, an approach for synthetic minority class data generation using agglomerative hierarchical clustering and ward's linkage criteria is proposed. Experimentation is carried out using five real-world datasets, namely Abalone, Page Blocks, Pima, Vehicle, and Yeast available at KEEL Data Repository. Testing of the experimentation is done using the SVM classifier with radial-bias kernel function. Data visualization is performed for understanding statistical properties, correlations, and distribution of class instances in the feature space. The classifier model is evaluated for before and after synthetic data generation using f-measure, recall, precision, and accuracy.

Keywords Imbalanced dataset · SMOTE · Hierarchical clustering · SVM classification

44.1 Introduction

Predictive modeling is an emerging and diversified field of data science and machine learning. It is leveraging industry growth and creating sustainable and futuristic predictions [1]. Predictive modeling is essentially a work of classification, to classify things into groups and make predictions for future growth and decision-making. It is

---

A. Sharma (✉) · A. Purohit · H. Mishra
Department of Computer Engineering, S.G.S.I.T.S., Indore, India
e-mail: abhisars09@gmail.com

a center of attraction for business analytics and business modeling [2]. Classification tasks are carried out with various techniques, using various models such as SVM model, neural network, decision trees, Naïve Bayes, logistic regression, etc. A dataset containing a fair number of data instances for all classes is considered best, as in such datasets classifiers get equal chances for fair learning for all classes. But if we consider a condition in which datasets lack an equal proportion of class data, then with such datasets classification tasks will be hindered.

Classification with imbalanced datasets is challenging and a never-ending learning process. Fair classification challenges exacerbate the difficulty of training classifiers with imbalanced datasets. Data, like ever-evolving technology is critical to predict industry development and future growth [3–6]. If a dataset is used that contains the majority of one class data, the classification model will fail to predict another type of class data since the findings will be incorrect. Forecasting sectors get affected by machine learning technologies without a balanced dataset. The entire system will be jeopardized when using an imbalanced dataset [7, 8]. An imbalanced dataset puts the entire system at danger. This is due to the fact that models based on imbalanced datasets are unable to comprehend the properties of the classes with few number of data instances and consider them as outliers or noise.

Handling such unbalanced datasets is critical, not just for classification work but also for other machine learning and data science applications. Various strategies for dealing with imbalanced datasets have been developed over time. Data-level techniques, algorithmic approaches, and hybrid approaches are the three major types of techniques discussed [1–4, 9].

In this paper, the data-level method is explored, and experiments are conducted to learn more about it. A data-level strategy focuses on increasing or decreasing the proportion of data instances that belong to the majority or minority class while working with imbalanced datasets. Hierarchical clustering-based synthetic minority data generation technique for imbalanced datasets is presented in this paper.

In this technique, agglomerative hierarchical clustering will be used for identification of minority class feature space and sites for generation of synthetic data. Interpolation method is used for generating minority class synthetic data using ward's linkage as linkage criteria. Testing of the proposed method will be done using an SVM classifier with a radial-bias kernel function. Data visualization for understanding characteristics of datasets is also performed. Experimentation is carried out on five real-world datasets which are given in Table 44.1 [10]. Classification report for SVM classification is generated for result analysis.

The rest of the paper is organized as follows. Section 44.2 provides background study and related work, which demonstrates various ways for dealing with imbalanced datasets. For addressing imbalanced datasets, Sect. 44.3 presents the proposed approach and algorithm for synthetic data generation of minority class. Section 44.4 offers information about the experiment, including the dataset that was employed. Section 44.5 is the results and discussion section, which contains a tabular comparison of the various results obtained. The conclusion of Sect. 44.6 provides a quick review of the proposed effort and the outcomes acquired.

Table 44.1 Datasets used

S. No.	Dataset name	No. of instances	No. of attributes	Classes/no. of data points in each class
1	Abalone	4176	7	Positive 22.98% Negative 77.01%
2	Page blocks	5472	9	Positive 10.21% Negative 89.79%
3	Pima	768	8	Positive 34.84% Negative 76.47%
4	Vehicle	846	18	Positive 23.53% Negative 76.47%
5	Yeast	1484	8	Positive 28.90% Negative 71.10%

44.2 Background Study and Related Work

This section begins with a description of imbalanced data handling techniques, followed by clustering and SVM classification, and concludes with a summary of evaluation criteria.

44.2.1 Imbalanced Data Handling Techniques

A dataset that has more instances of one class than instances of another is referred to as an imbalanced dataset. Majority class instances are those class instances in imbalanced datasets that are comparatively more in number and can influence the prediction in their favor. Minority class instances are those that are low in number and are classified as outliers or noise by a classifier and have a sparse distribution over the dataset. There are various methods of synthetic data generation, viz., data-level approaches, algorithmic approaches, hybrid approaches, and kernel-based approaches [3, 11, 12].

Undersampling, oversampling, synthetic minority data generations (SMOTE) are data-level approaches whereas ensemble methods and cost-sensitive approaches are categorized under hybrid approaches [3]. All these methods proposed as solutions to the imbalanced dataset challenges. Algorithmic approaches, kernel-based approaches are out of course for this work, hence they are not discussed here. Other than these methods a brief study of other strategies are stated as follows.

44.2.1.1 Data-Level Approaches

Data-level approaches pay specific interest on the data points present in the dataset. These methods manipulate existing data for generation of new data or handling of the preexisting data. There are various types of data-level approaches which are discussed as follows.

- **Undersampling**: Undersampling is a technique for balancing a dataset by removing the bulk of class instances. Because the majority of class instances are eliminated at random, it's also known as random undersampling, [13, 14] show how to delete instances of the majority class. The disadvantage of this strategy is that because undersampling is done at random, potentially useful data instances can be lost.
- **Oversampling**: Oversampling is a strategy for balancing a dataset by oversampling minority data instances. For balancing the dataset, this technique creates clones of existing data instances. The conditions for performing the task of oversampling are well illustrated in Tomek links [15]. It's likely that the model will overfit for some values as replicas of the minority class instances are constructed, resulting in a biased forecast.
- **Synthetic Minority Oversampling Technique (SMOTE)**: On the basis of the distribution of minority classes present in the dataset, the SMOTE technique is used to generate synthetic data for minority classes [4, 5]. This strategy outperforms oversampling and undersampling (according to the survey), because neither majority data instances nor minority class replicas are eliminated, resulting in no loss of potentially valuable examples and a low risk of model overfitting.

44.2.1.2 Hybrid Approaches

Hybrid approaches are those comprising both data-level approaches and algorithmic approaches. They are termed as ensemble techniques and cost-sensitive techniques.

- **Ensemble Technique**: An ensemble technique is one in which several different classifiers are combined to form a single classifier for better learning. They're sometimes referred to as hybrid techniques because they combine data-level and algorithmic methodologies [16, 17]. The bagging or boosting methods are used in the ensemble methodology. In bagging, a dataset is divided into several little datasets of the same size, and the learning procedure is repeated for all of the datasets in order to learn each and every class feature. Weak classifiers are iteratively taught in the boosting approach, as it is claimed that via recurrent training, a weak learner can be transformed into a strong learner [2]. This strategy focuses mostly on struggling students.
- **Cost-sensitive Techniques**: Techniques that are cost-sensitive are applied in a variety of ways. Cost is assigned to multiple instances in this technique, and after each successful learning iteration, the assigned cost is modified for better classifier learning [3].

44.2.2 Clustering

Clustering is an unsupervised learning algorithm. In clustering data samples are categorized under various labels based on similar features. There are many types of clustering algorithms available such as k-means clustering algorithm, hierarchical clustering algorithm, DBSCAN clustering algorithm, BIRCH clustering algorithm, etc. [8, 18]. The proposed work is restricted to hierarchical clustering and thus, it is discussed here in detail.

Hierarchical clustering or hierarchical cluster analysis groups similar objects in a single group. This group is called cluster. This algorithm ends when all the clusters are grouped in a single cluster [3]. Every cluster is distinct from another cluster but instances of clusters have similar properties. Hierarchical clustering is of two types; agglomerative and divisive as shown in Fig. 44.1.

- **Hierarchical Agglomerative Clustering (HAC)**: Hierarchical clustering is a bottom-up approach and is also called AGNES. It treats every single datapoint as a singleton cluster. It groups all the clusters based on their similarity. A similarity matrix is created using norm1 or norm2 distances. Pairs of clusters are successively merged until a big cluster is created containing all clusters. End result is a tree-based representation of objects named dendrograms. Figure 44.2 shows dendrograms generation for agglomerative hierarchical clustering.

Fig. 44.1 Types of hierarchical clustering

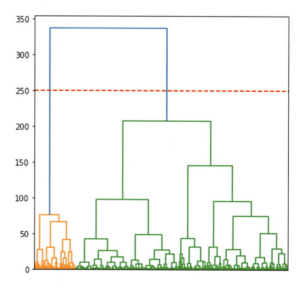

Fig. 44.2 Dendrogram generation

- **Divisive Hierarchical Clustering**: Divisive hierarchical clustering is a top-down approach. It is just the reverse of agglomerative clustering. This technique is also called DIANA. It splits a bigger cluster recursively for creating smaller clusters in a hierarchy. In this approach two clusters are chosen, their sum of squared errors (SSE) is calculated and clusters having larger SSE is splitted.

44.2.3 SVM Classification

Classification is a technique for distinguishing classes under a label. Various classification models are used for this purpose, viz., rule-based classifier, nearest neighbor classifier, decision tree-based classifier, bayesian classifier, artificial neural network classifier, and support vector machine classifier [1]. In the proposed work SVM classifier is to be used. A discussion is presented here on SVM classifiers.

Support vector machine is a supervised learning algorithm which is used for classification in machine learning. Support vector machines work on hyperplanes for segregating n-dimensional feature space into different class labels. The hyperplane here is the decision boundary which differentiates classes of different categories. There are basically two types of SVM; linear SVM and nonlinear SVM [3].

- Linear SVM: If a dataset contains only two classes, then those classes are separable by a single straight line and thus are called as linearly separable and for this linear SVM is used.
- Nonlinear SVM: If a dataset contains more than two classes and it is not possible to separate such datasets with a straight line, then such datasets are termed as nonlinearly separable datasets and nonlinear SVM is used for classification purpose.
- Decision Boundary: Decision-boundary is that point in space where we divide the space into different categories. These straight decision boundaries are termed as hyperplanes. If there are two features then, the hyperplane will be a straight line. If there are three features, then the hyperplane will be of two dimensions. Decision boundaries are traced between maximum distances between the two samples.

The data points or vectors that are the closest to the hyperplane affecting the position of the hyperplane are termed as support vectors. The distance between the vectors and the hyperplane is called as **margin** and the goal of SVM is to maximize this margin. The **hyperplane** with maximum margin is called the **optimal hyperplane**.

44.2.4 Evaluation Criteria

Performance of classification model or classifier is evaluated using a confusion matrix on the set of test data. Confusion matrix is an n cross n square matrix having four distinct values of true positive, true negative, false positive, and false negative which are used to calculate precision, recall, f1-score, and accuracy values [3, 4].

- Precision (P): Precision determines the fraction of records that actually turns out to be positive in the group the classifier has declared as positive. The higher the precision is, the lower the number of false positive errors committed by the classifier.

$$P = \frac{TP}{TP + FP} \quad (44.1)$$

- Recall (R): Recall measures the fraction of positive examples correctly predicted by the classifier. Classifiers with large recall have very few positive examples misclassified as the negative class. The value of recall is equivalent to TP rate.

$$R = \frac{FN}{FN + TP} \quad (44.2)$$

- F1-score (F1): F1-score is summarization of recall and precision values. It is used to see a balance between recall and precision for imbalanced datasets.

$$F1 = 2 * \left(\frac{P * R}{P + R}\right) \quad (44.3)$$

- Accuracy: Accuracy is defined as the number of instances correctly classified. Accuracy can be very high for a model but it is also possible that a model with high accuracy fails in correct classification, especially with imbalanced datasets.

$$Accuracy = \frac{TP + TN}{TP + TN + FP + FN} \quad (44.4)$$

44.3 Proposed Approach

An approach for synthetic data generation for minority class using agglomerative hierarchical clustering and ward's linkage criteria is discussed in this section. Figure 44.3 shows the flow diagram of the proposed approach. The strategy for the synthetic data generation for minority class is to identify minority class data using HAC and after identification of minority class data, again applying HAC on minority class data for the identification of data space for synthetic data generation. Synthetic data will be generated by interpolating the nearest points occupied by HAC. Ward's linkage criterion is used for HAC. Ward's linkage uses a sum of squared errors for the linking of two data points which is based on the Euclidean distance. The classifier which is used is SVM classifier as the datasets used are small and contain only

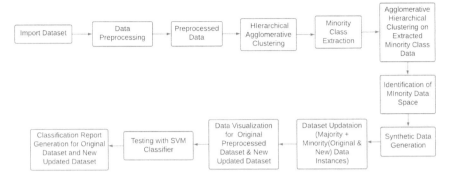

Fig. 44.3 Flow diagram of proposed approach

	Diameter	Height	Shell weight	Class
count	4084.000000	4084.000000	4084.000000	4084.000000
mean	0.406308	0.138530	0.233142	0.223800
std	0.095848	0.037001	0.129482	0.416841
min	0.115000	0.015000	0.005000	0.000000
25%	0.350000	0.115000	0.130000	0.000000
50%	0.420000	0.140000	0.230000	0.000000
75%	0.480000	0.165000	0.320500	0.000000
max	0.605000	0.250000	0.655000	1.000000

(a)

	Diameter	Height	Shell weight	Class
count	4162.000000	4162.000000	4162.000000	4162.000000
mean	0.408245	0.137492	0.232569	0.238347
std	0.097085	0.037921	0.128339	0.426124
min	0.043500	0.015000	0.005000	0.000000
25%	0.350000	0.110000	0.130625	0.000000
50%	0.425000	0.140000	0.225000	0.000000
75%	0.480000	0.165000	0.320000	0.000000
max	0.649500	0.324037	0.655000	1.000000

(b)

Fig. 44.4 a, b Statistical properties of Abalone dataset before and after synthetic data generation

binary class data. The SVM classifier also used Euclidean distance, thus using ward's linkage criteria was most suitable. All the steps are elaborated further.

(i) **Import Dataset**

Dataset will be studied and checked for missing values, average count, mean, standard deviation, minimum value, and maximum value. Imbalance in the dataset will be checked and the percentage of instances for binary classes will be recorded.

(ii) **Preprocessing Data**

In this step, data will be preprocessed using following techniques.

(a) Label Encoding: This is used for converting categorical data into numeric data.
(b) Normalization: Data normalization is applied for getting attribute values in the same scale.

(iii) **Applying Agglomerative Hierarchical Clustering**

Agglomerative hierarchical clustering will be applied on preprocessed datasets using Euclidean distances for generating similarity matrices. For generation of dendrogram, ward's linkage criteria is considered.

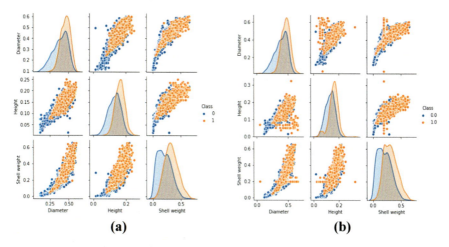

Fig. 44.5 a, b Pairplot after synthetic data generation

(iv) **Extracting Minority Class**

Minority class data will be extracted from clustering applied in the previous step.

(v) **Identification of Minority Data Space**

Agglomerative clustering will be applied again on the extracted minority class data for identification of minority data space, where synthetic data will be generated. This is carried out using the same criteria used in the previous clustering.

(vi) **Generation of Synthetic Data**

Synthetic data for minority classes will be generated using an interpolation method between values of smaller clusters and the process will continue in the same manner as dendrograms are generated.

(vii) **Data Visualization**

Data visualization is done on preprocessed original dataset and new updated dataset after synthetic data generation for understanding patterns of data instances. Knowing the interestingness of data, identification of crucial features of data is carried out using data visualization and change in feature characteristics is also analyzed.

(a) Statistical Properties: Statistical properties show mean, minimum, maximum, I quartile, II quartile, III quartile, and IV quartile values for each feature of the dataset.
(b) Correlation: Correlation shows relation among each feature of dataset and relation of each feature with class. Pearson's coefficient and Spearman's coefficient methods can be used for finding correlation.
(c) Pairplots: Pairplots depict the distribution pattern of data over class and features.

Fig. 44.6 a Heatmap before synthetic data generation. b Heatmap after synthetic data generation

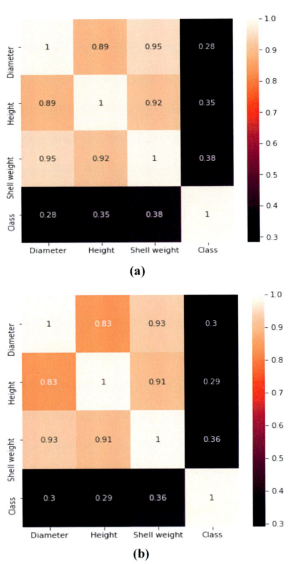

(viii) **Splitting Dataset into Train and Test Set**

The preprocessed original dataset and new updated dataset both will be split into training and test sets. This is done for the training of the classifier and testing of the classifier over its training. Splitting of the dataset is done in the ratio of 80:20. The training dataset will have 80% of data instances and the test dataset will contain 20%.

(ix) Testing with SVM Classifier and Generating Classification Report

SVM classifier will be used for classification of original and new updated dataset. SVM classifiers will be trained with the training dataset and tested with the test dataset with radial-bias kernel function. After training and testing of the SVM classifier is completed, a classification report will be generated for understanding the performance of the classifier on the dataset. The classification report will contain accuracy, precision, recall, and f1-score. Comparison for classification report for the original and updated dataset will be done and results will be deduced according to observation.

Algorithm used for experimentation for the proposed approach is presented below in Algorithm 1.

Algorithm 1: Generation of Synthetic Data
Input: Imbalanced dataset

Output: Dataset with synthetically generated data

Procedure:

(i) Original dataset will be imported for checking imbalance present in binary classes and studying statistical properties (average count, standard deviation, min, max, and quartile values) of data.

(ii) Original dataset will be preprocessed next. Label encoding will be applied for transforming categorical data into numeric. Z-score normalization will be used for normalizing.

(iii) Hierarchical agglomerative clustering (HAC) will be applied on preprocessed dataset using ward's linkage criteria for cluster creation and minority class data instances will be extracted by cutting the dendrogram at a certain height (where clusters $= 2$).

(iv) To get the sequence of clustering of minority data instances, HAC will be applied again on minority class using Euclidean distance as similarity measure and ward's linkage criteria for dendrogram generation, Euclidean distance will be used for similarity measure and ward's linkage criteria will be considered for the generation of dendrograms.

(v) Synthetic data for minority classes will be generated within clusters using the interpolation method.

(vi) The original dataset will be updated with new synthetically generated data. Steps (iv) and (v) will again be applied to the newly updated dataset until correlation matrices do not change to a large extent.

(vii) Data visualization (correlation heatmap, box plot graphs, kernel density graphs, and histograms) for understanding the relation of data attributes will be carried out.

(viii) Original dataset and updated dataset will be splitted into training and test dataset in the ratio of 80:20.

(ix) SVM classifier will be used for learning over the training data instances of both original and updated datasets. Radial-bias kernel function will be used. A classification report will be generated for SVM classification.
(x) Based on classification reports generated for the original dataset and updated dataset results will be concluded.

44.4 Experimentation Details

The datasets and experimental setup are presented. The proposed approach is carried out on five datasets available at KEEL Data Repository. Datasets along with no. of attributes, no. of instances, and class percentage is tabulated in Table 44.1 [10]. Entire experiment is carried out in the Jupyter Notebook of Anaconda Navigator. Python programming is used to carry out desired tasks. "Positive Class" is the class of interest for the experimentation. Synthetic data is generated for this class till statistical properties of the dataset do not change with greater margin.

44.5 Results and Discussion

The outcomes of the experiment are discussed in this section. Table 44.2 displays the percentage difference between before and after synthetic data generation in class instances. Table 44.3 gives a comparison of minority class precision (P), recall (R), and f1-score (F1). Precision (P), recall (R), and F1-score (F1) values for majority class instances before and after synthetic data generation, as well as accuracy before and after synthetic data generation, are given in Tables 44.4 and 44.5. Table 44.6 [3] gives a comparison of the F1-score of SMOTE for single neural network, k-nearest neighbor, and SVM (used in the paper) classifiers.

From Figs. 44.4, 44.5 and 44.6, statistical features and data visualization graphs for the abalone dataset are shown before and after synthetic data production.

Table 44.2 Difference of percentage in class instances before and after synthetic data generation

S. No.	Dataset name	Before synthetic data generation (counts in percentage)		After synthetic data generation (counts in percentage)	
		Negative class	Positive class	Negative class	Positive class
1	Abalone	77.01	22.98	76.16	23.83
2	Page blocks	89.78	10.21	77.04	22.95
3	Pima	65.10	34.89	63.06	36.93
4	Vehicle	76.47	23.52	74.47	25.52
5	Yeast	71.09	28.90	62.66	37.33

Table 44.3 Comparison of precision, recall, and f1-score for minority class before and after synthetic data generation

S. No.	Dataset name	Before synthetic data generation			After synthetic data generation		
		P	R	F1	P	R	F1
1	Abalone	0.88	0.12	**0.22**	0.68	0.20	**0.31**
2	Page blocks	0.95	0.29	**0.45**	0.99	0.80	**0.88**
3	Pima	0.43	0.30	**0.35**	0.64	0.50	**0.56**
4	Vehicle	0.91	0.21	**0.34**	0.81	0.40	**0.53**
5	Yeast	0.77	0.28	**0.41**	0.79	0.42	**0.55**

Bold highlights a significance of f1-score

Table 44.4 Comparison of precision, recall and f1-score for majority class before and after synthetic data generation

S. No.	Dataset name	Before synthetic data generation			After synthetic data generation		
		P	R	F1	P	R	F1
1	Abalone	0.81	1.00	**0.89**	0.81	0.97	**0.88**
2	Page blocks	0.95	1.00	**0.98**	0.95	1.00	**0.97**
3	Pima	0.72	0.82	**0.76**	0.77	0.86	**0.81**
4	Vehicle	0.75	0.99	**0.86**	0.83	0.97	**0.89**
5	Yeast	0.76	0.96	**0.85**	0.70	0.92	**0.80**

Bold highlights a significance of f1-score

Table 44.5 Comparison of accuracy before and after synthetic data generation

S. No.	Dataset name	Accuracy before synthetic data generation	Accuracy after synthetic data generation
1	Abalone	0.81	0.80
2	Page blocks	0.95	0.95
3	Pima	0.65	0.73
4	Vehicle	0.76	0.82
5	Yeast	0.76	0.72

SVM classifier with radial-bias kernel function is used for classification. Before the generation of synthetic data, the used classifier was unable to learn the properties of minority class data. After increasing the number of minority class instances by generating synthetic data, classifiers became able to learn the characteristics of minority data. Statistical properties are also retained throughout the synthetic data generation. Statistical properties and data visualization graphs are generated for all the datasets used. Following results can be deduced from the results obtained.

Table 44.6 Comparison of SMOTE technique on various classifiers

S. No.	Dataset name	Single neural network classifier	k-Nearest neighbor classifier	SVM classifier			
				Minority class		Majority class	
				Before synthetic data generation	After synthetic data generation	Before synthetic data generation	After synthetic data generation
1	Abalone	0.44	0.50	0.22	0.31	0.89	0.88
2	Page blocks	0.97	0.98	0.45	0.88	0.98	0.97
3	Pima	0.66	0.64	0.35	0.56	0.76	0.81
4	Vehicle	0.92	0.86	0.34	0.53	0.86	0.89
5	Yeast	0.68	0.66	0.41	0.55	0.85	0.80

- The proposed approach worked well with minority class instances as f1-score for minority class is increased.
- Learning of classifiers depends on the distribution of data in feature space as shown by various graphs.

44.6 Conclusion

In this paper, an approach for synthetic data generation for minority class using agglomerative hierarchical clustering and ward's linkage as linkage criteria has been proposed. Experiments were carried out on five real-world datasets, namely Abalone, Page Blocks, Pima, Vehicle, and Yeast which are available at KEEL Data Repository. Data visualization is performed for understanding statistical properties of the datasets used. Pair-plot visualization graph is plotted to understand the data distribution. For understanding correlation, correlation matrix using Spearman's method is generated as heatmap.

Testing of the proposed approach after experimentation is carried out using an SVM classifier with radial-bias kernel function. The work is carried out on Jupyter Notebook using Python. Classification report is generated for analyzing results. The results show that f1-score of the minority class increased from 34.4 to 56.6% for the various datasets used, a 21.2% increase overall. The recall rate for minority class has risen from 24 to 46.4%, a 22.4% increase overall. Precision has declined slightly from 78.8 to 78.2%, a 0.6% decrease overall. The total accuracy of the SVM model increased by 1.4%, to 80.4% from 79% according to the experiments. These results show that the proposed approach by significantly increasing the number of minority class data instances has worked well with learning of classifiers for understanding characteristics of minority class data.

Densely dispersed data instances make learning of the model challenging, and locating the minority dataspace for synthetic data synthesis becomes problematic. Both classes overlap in a dense distribution, causing problems with model learning and minority dataspace identification. The distribution of data instances in feature space could be considered while developing unique solutions for coping with imbalanced datasets.

Acknowledgements I am heartily thankful to my guide Dr. Anuradha Purohit, Associate Professor, S.G.S.I.T.S. Indore and co-guide Ms. Himani Mishra, Assistant Professor, S.G.S.I.T.S. Indore, for their guidance, support, and subject expertise because of which this work has been completed.

References

1. Ortiz, M.P., Gutierrez, P.A., Tino, P., Martinez, C.H.: Oversampling the minority class in the feature space. IEEE Trans. Neural Netw. Learn. Syst. **27**(9), 1947–1961 (2016). https://doi.org/10.1109/TNNLS.2015.2461436
2. Ahmed, S., Mahbub, A., Rayhan, F., Jani, M.R., Shatabda, S., Farid, D.M.: Hybrid methods for class imbalance learning employing bagging with sampling techniques. In: 2nd IEEE International Conference on Computational Systems and Information Technology for Sustainable Solutions 2017, 30 August 2018. https://doi.org/10.1109/CSITSS.2017.8447799
3. Sharma, A., Purohit, A., Mishra, H.: A survey on imbalanced data handling techniques for classification. Int. J. Emerg. Trend. Eng. Res. **9**(10), 1341–1347 (2021). https://doi.org/10.30534/ijeter/2021/089102021
4. Galar, M., Fernandez, A., Barrenechea, E., Bustince, H., Herrera, F.: A review on ensembles for the class imbalance problem: Bagging-, Boosting-, and hybrid-based approaches. IEEE Trans. Syst. Man. Cybern. C Appl. Rev. **42**(4), 463–484 (2012). https://doi.org/10.1109/TSMCC.2011.2161285
5. Barua, S., Islam, M.M., Yao, X., Murase, K.: MWMOTE—majority weighted minority oversampling technique for imbalanced data set learning. IEEE Trans. Knowl. Data Eng. **26**(2), 405–425 (2014). https://doi.org/10.1109/TKDE.2012.232
6. He, H., Garcia, E.A.: Learning from Imbalanced Data. IEEE Trans. Knowl. Data Eng. **21**(9) (2009) https://doi.org/10.1109/TKDE.2008.239
7. Sun, Y., Wong, A.K.C., Kamel, Md.S.: Classification of imbalanced data: a review. Int. J. Pattern Recogn. Artif. Intell. **23**(4), 687–719 (2009). https://doi.org/10.1142/S0218001409007326
8. Lim, P., Goh, C.K., Tan, K.C.: Evolutionary cluster-based synthetic oversampling ensemble (ECO-ensemble) for imbalance learning. IEEE Trans. Cybern. **47**(9), 2850–2861 (2017). https://doi.org/10.1109/TCYB.2016.2579658
9. Kang, Q., Zhou, L.S.M.C., Wang, X., Wu, Q.D., Wei, Z.: A distance-based weighted undersampling scheme for support vector machines and its application to imbalanced classification. IEEE Trans. Neural Netw. Learn. Syst. **29**(9), 4152–4165 (2018). https://doi.org/10.1109/TNNLS.2017.2755595
10. URLs for datasets used: Abalone: https://archive.ics.uci.edu/ml/datasets/abalone; Page Blocks: https://sci2s.ugr.es/keel/dataset.php?cod=147; Pima: https://sci2s.ugr.es/keel/dataset.php?cod=155; Vehicle: https://sci2s.ugr.es/keel/dataset.php?cod=149; Yeast: https://sci2s.ugr.es/keel/dataset.php?cod=153
11. Rout, N., Mishra, D., Mallick, M.K.: Handling imbalance data: a survey. In: International Proceedings on Advances in Soft Computing, Intelligent Systems and Applications, Advances

in Intelligent Systems and Computing, p. 628. Springer Nature Singapore Pte Ltd. 2018. https://doi.org/10.1007/978-981-10-5272-9_39
12. Batista, G.E.A.P.A., Prati, R.C., Monard, M.C.: A study of the behavior of several methods for balancing machine learning training data. ACM SIGKDD Expl. Newsl. **6**(1), 20–29 (2004). https://doi.org/10.1145/1007730.1007735
13. Yang, K., Yu, Z., Wen, X., Cao, W., Chen, C.L.P., Wong, H.-S., You, J.: Hybrid classifier ensemble for imbalanced data. IEEE Trans. Neural Netw. Learn. Syst. **31**(4), 1387–1400 (2020). https://doi.org/10.1109/TNNLS.2019.2920246
14. Bhowan, U., Johnston, M., Zhang, M., Yao, X.: Evolving diverse ensembles using genetic programming for classification with unbalanced data. IEEE Trans. Evol. Comput. **17**(3) (2013). https://doi.org/10.1109/TEVC.2012.2199119
15. Barua, S., Islam, Md.M., Murase, K.: A novel synthetic minority oversampling technique for imbalanced data set learning. In: International Conference on Neural Information Processing, vol. 7063 (2011). https://doi.org/10.1007/978-3-642-24958-7_85
16. Devi, D., Biswas, S.K., Purkayastha, B.: A boosting based adaptive oversampling technique for treatment of class imbalance. In: 2019 International Conference on Computer Communication and Informatics, 2 Sept 2019. https://doi.org/10.1109/ICCCI.2019.8821947
17. Feng, W., Dauphin, G., Huang, W., Quan, Y., Bao, W., Wu, M., Li, Q.: Dynamic synthetic minority over-sampling technique-based rotation forest for the classification of imbalanced hyperspectral data. IEEE J. Select. Top. Appl. Earth Observ. Rem. Sens. **12**(7) (2019). https://doi.org/10.1109/JSTARS.2019.2922297
18. Zang, Y.P., Zhang, L.N., Wang, Y.C.: Cluster-based majority under-sampling approaches for class imbalance learning. In: 2010 2nd IEEE International Conference on Information and Financial Engineering. https://doi.org/10.1109/ICIFE.2010.5609385

Chapter 45
Regenerative Braking in an EV Using Buck Boost Converter and Hill Climb Algorithm

Vandana Kumari Prajapati, Arya Jha, C. R. Amrutha Varshini, and P. V. Manitha

Abstract Mankind cannot thrive without automobiles due to the greater productivity they provide. Traditionally, most of the energy used to power automobiles comes from fossil fuels. These vehicles are being replaced by electric vehicles over time. Typically, a car's braking system uses hydraulic braking technology. However, because it generates more heat when braking, this conventional braking technique wastes a lot of energy. Therefore, the development of regenerative braking introduced in electric vehicles has eliminated these drawbacks, in addition to assisting in energy conservation and increasing the vehicle's efficiency. When operating in regenerative mode, the motor transforms kinetic energy into electrical energy to recharge the batteries or capacitors. The hill climb algorithm is used to analyze the regenerative braking of EVs, and MATLAB software is used to create and model electrical circuits for the required configuration for regenerative braking. The parameters of lead acid battery as well as parameters of DC machine have been analyzed. The performance of the regenerative braking model with hill climb algorithm is analyzed against the regenerative braking without any algorithm which is considered as the basic model.

Keywords Battery · SoC · Regenerative braking · Motoring · Generator · Hill climbing algorithm

V. K. Prajapati · A. Jha · C. R. A. Varshini · P. V. Manitha (✉)
Department of Electrical and Electronics Engineering, Amrita School of Engineering Bengaluru, Amrita Vishwa Vidyapeetham India, Bengaluru, India
e-mail: pv_manitha@blr.amrita.edu

V. K. Prajapati
e-mail: bl.en.u4eee20051@bl.students.amrita.edu

A. Jha
e-mail: bl.en.u4eee20002@bl.students.amrita.edu

C. R. A. Varshini
e-mail: bl.en.u4eee20007@bl.students.amrita.edu

© The Author(s), under exclusive license to Springer Nature Singapore Pte Ltd. 2024
P. K. Jha et al. (eds.), *Proceedings of Congress on Control, Robotics, and Mechatronics*, Smart Innovation, Systems and Technologies 364,
https://doi.org/10.1007/978-981-99-5180-2_45

45.1 Introduction

The increasing interest towards electric vehicles (EV) is due to the fact that EVs are considered as a solution to the environmental problems, traditional fossil fuel operated vehicles generate. Bi-directional DC-DC converters have recently been widely available for usage in a variety of applications such as electric vehicles, where electrical energy is transferred between the motor and the battery side [1]. Batteries are the only source of power for zero emission vehicles. Large-scale batteries with a high energy density, a small size, and characteristics for vehicle reliability have been operated on the ground [2, 3].

State of charge (SoC) is the proportion of the battery's overall capacity that represents the amount of electrical energy that is currently stored in a battery. The battery's state of charge (SoC), determines the amount of energy it can store before recharge is required [4–6]. The battery can only hold a specific amount of charge at once, and any more energy produced by regenerative braking will be dissipated once it reaches its maximum capacity.

On the other hand, regenerative braking can be quite efficient in recharging the battery if it is not fully charged. This is especially true when commuting in stop-and-go scenarios, where the vehicle continually accelerates and brakes. In these circumstances, it's possible that the battery will never get a chance to fully charge through the vehicle's charging system, although regenerative braking can help to keep the battery at a safe state of charge (SoC).

The hill climbing algorithm is used to extend the driving range of electric vehicles as well as improve performance by utilizing EV and the overall cost will be less competitive than conventional vehicles. With the help of the Buck Boost converter, the DC motor can be operated in motoring as well as regenerative braking mode depending on buck mode or boost mode of operation of the converter. Regenerative braking has been found to enhance driving range up to 15%. Conventional vehicle braking is not an option. For the EVs, a bidirectional DC-DC converter is employed for operating it in motoring and regenerative braking mode [7, 8].

45.2 Implementation of Regenerative Braking

45.2.1 System Description

The basic model of regenerative braking without implementing any algorithm is shown in Fig. 45.1.

Fig. 45.1 Block diagram without HC algorithm implementation [3]

In Fig. 45.1, the system representation without any algorithm has been illustrated where the major components taken under consideration are DC machine, DC-DC converter and battery (240 V). Here both modes of operation such as motoring and regenerative through generator are demonstrated. The forward direction implementation of the system makes the machine behave as motor and reverse direction implementation refers to the regenerative type through generator mode of operation.

45.2.2 Regenerative Braking

Regenerative braking is a type of braking system that converts kinetic energy—the motion of a moving object—into electrical energy, which can then be used to power the moving vehicle or stored in a battery [9]. The system operates by turning the electric motor into a generator when the brake pedal is pressed and is used in electric or hybrid automobiles [10, 11]. The generator action which transforms the vehicle's kinetic energy into electrical energy, limits the motor's rotation. The battery then accumulates this energy for later use. The technology can also increase the vehicle's overall efficiency and reduce the wear and tear on the mechanical brakes [10, 11].

Figure 45.2 illustrates the representation of the system incorporating hill climbing algorithm where the forward and reverse functioning has been shown.

The proposed layout of the system consists of battery and DC machine. The system performs the task of Buck Boost operation with and without HC algorithm implementation in it. The components with their respective parameters and values are given in Table 45.1.

a. **Buck Boost Mode of Operation in Regenerative Braking Without HC Algorithm**

The converter that is implied for this system is the Buck Boost converter. A Buck Boost converter is required to change the voltage level at the output. The basic components of the regenerative braking system include a DC/DC Buck Boost converter, 240 V battery, and DC machine. The scopes are connected to the battery to capture voltage, current, and state of charge waveforms. The 240 V battery powers the motor during forward operation, providing enough electricity to carry the machine

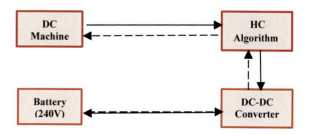

Fig. 45.2 Block diagram with HC algorithm implementation

Table 45.1 System parameters

S. No.	Components	Parameters	Values
1	Battery-lead acid	Nominal voltage	240 V
		Rated capacity	15.5 Ah
		Initial SoC	50%
2	DC machine	Rated power	5 HP
		Rated voltage	240 V
		Rated speed	1750 RPM
		Rated field voltage	300 V
		Field type	Wound
		Mechanical input	Torque = 20 Nm
3	DC voltage source	Field voltage	240 V
4	Inductor (L)	Inductance (H)	5e−3
5	Capacitor	Capacitance (F)	150e−6
6	PWM generator	Switching frequency	20,000 Hz

forward which in turn makes the vehicle travel forward. This mode, which is used for motoring, (BUCK mode) operates in normal condition.

In the proposed layout of regenerative braking, the Buck Boost converter has been designed and simulated to drive the DC motor of the electric vehicle. The MOSFET switch of a one-quadrant converter is fired by an output pulse from the PWM generator block (buck or boost). The generator's duty cycle is determined by the input D. A way to generate low-frequency output signals (voltage) from high-frequency pulses is made possible with pulse width modulation (PWM). The key benefit of PWM is the extremely low-power loss in the switching devices. There is no current flowing through a switch when it is off, and there is absolutely no voltage drop across the switch when power is being transmitted to the load under ideal condition.

In the system, the SoC is monitored and based on that the duty cycle corresponds to buck or boost mode which is given as input to PWM generator which generates the pulses at that duty cycle. For motoring action converter is operated in buck mode and for regenerative braking converter is operated in boost mode by the controller. Figure 45.3 represents the simulation circuit with hill climbing algorithm.

b. **Buck Boost Mode of Operation in Regenerative Braking Using Hill Climbing Algorithm**

The hill climbing algorithm is an optimization method for locating a function's local maximum. It starts with an arbitrary answer and iteratively increases the value until no more such steps are possible. The approach is easy to use and computationally quick, but it may become trapped in a local maximum and fail to locate the function's global maximum [12]. The hill climbing (HC) algorithm, a real-time optimization technique, may effectively control and manage the electric motor's power output based on the vehicle's current driving conditions, which can be advantageous in electric vehicles.

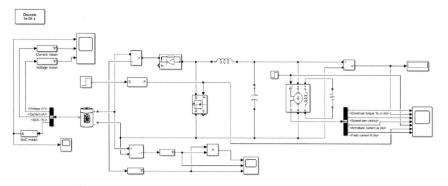

Fig. 45.3 Simulation circuit without HC algorithm block

By doing this, there may be increase in the vehicle's energy efficiency and range while making sure the motor runs safely and optimally. Additionally, HC is a cost-effective option for electric car producers because it can be deployed with relatively simple hardware and software. The HC algorithm is a quick, effective, and reliable optimization algorithm that has a wide range of applications including engineering, finance, and artificial intelligence. HC can withstand noisy or erratic input data, making it appropriate for applications where such data is present.

In the context of state of charge (SoC) prediction for batteries, the algorithm can be used to optimize a model that predicts the remaining battery capacity based on various input parameters such as current, temperature, and voltage [13]. The algorithm starts with an initial guess for the model parameters and then iteratively modifies them to find the set of parameters that results in the most accurate predictions. The algorithm stops when it reaches a local maximum or minimum, which is a set of parameters that cannot be improved upon by making small changes. This set of parameters can then be used to make accurate predictions of the battery's remaining capacity [14]. The ideal operating point for the Buck Boost converter may be found using the hill climbing method. In order to do this, the algorithm is used to find the converter's maximum power point (MPP), or the point at which it can transfer the greatest energy from the regenerative braking system to the battery [15, 16]. Figure 45.4 represents the MATLAB Simulink circuit of the system with HC algorithm.

45.3 Results

45.3.1 Battery in Charging Mode (Generator Operation)

Figure 45.5 shows the voltage, current, and SoC characteristics of a battery taking into considerations of its charging state during the generator operation. It is inferred from the graph as the voltage of the battery is increasing the SoC of the battery is

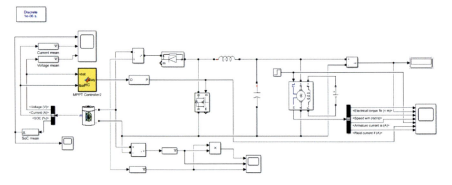

Fig. 45.4 Represents the MATLAB Simulink circuit of the system with HC algorithm

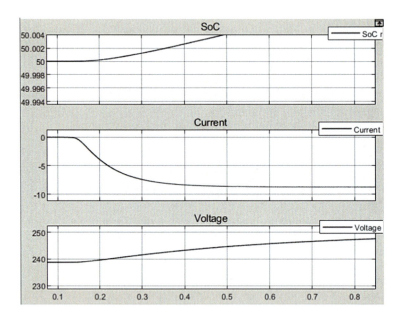

Fig. 45.5 Charging mode of battery

also increasing, as well as the current decreases which shares the scenario of boost mode.

45.3.2 Battery in Discharging Mode (Motoring Operation)

Figures 45.6 and 45.7 depict the voltage, current, and SoC characteristics of a battery while it is discharging during motoring operation. According to the graph, when the voltage of the battery decreases, so does the SoC of the battery, while the current rises, revealing the buck mode scenario.

In Figs. 45.6 and 45.7, y-axis and x-axis depict the measured entities (as SoC in percentage, current in ampere, and voltage in volts) and time in seconds, respectively.

In Fig. 45.8, motor and generator action of the DC machine with respect to its changes in parameters such as electrical torque, speed and armature current has been illustrated leading to showcasing of difference between both operations. Also, in

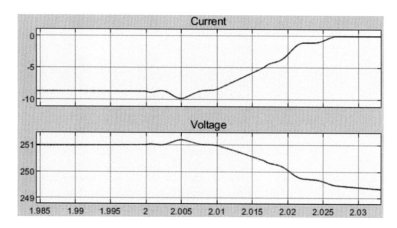

Fig. 45.6 Voltage and current during discharging mode of battery

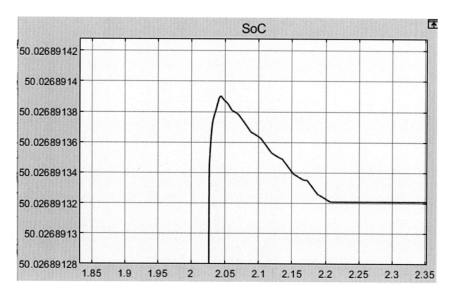

Fig. 45.7 State-of-charge during discharging mode of battery

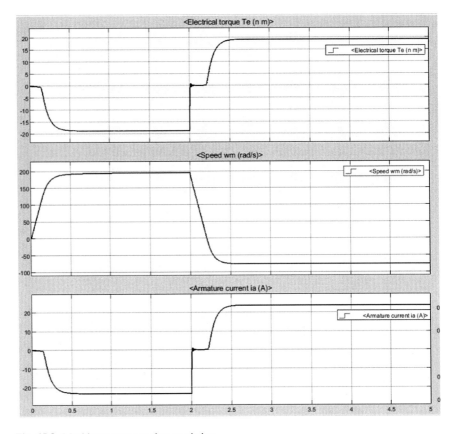

Fig. 45.8 Machine parameter characteristics

y-axis parameters are entitled as stated above and x-axis refers to time interval (in secs).

The result in Fig. 45.8 illustrates that the generating action of the machine occurs between 0- and 2-time units, while its motoring action occurs after 2-time units. And in both modes of operation, the torque is proportional to the armature current but have an inverse relationship to speed. However, the direction of the torque is negative in the generating mode compared with the motoring mode.

45.3.3 SoC Comparison

Figures 45.9 and 45.10 indicate that the state of charge of the battery varies significantly when operated with and without the hill climbing algorithm. In both of the figures, y-axis and x-axis suggest SoC and time (in secs), respectively.

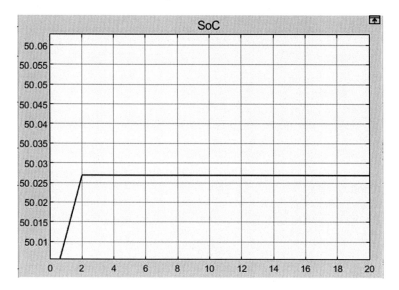

Fig. 45.9 State-of-charging without hill climbing algorithm

Fig. 45.10 State-of-charging with hill climbing algorithm

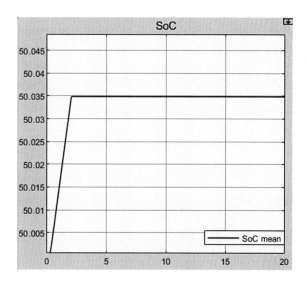

Table 45.2 displays the measured SoC in percentage during both the scenarios, with and without the hill climbing algorithm.

When compared with SoC measured without the hill climbing algorithm, SoC obtained using the algorithm shows improved accuracy and precision in measuring the battery's actual state of charge. This can lead to more effective battery management and longer battery life, as well as more reliable performance of the overall

Table 45.2 SoC comparison with and without HC algorithm

Parameter	With HC algorithm	Without HC algorithm
SoC	50.035%	50.0275%

system that the battery powers. On the other hand, without the hill climbing algorithm, the SoC level may be less, leading to suboptimal battery management and potentially shorter battery life. This can result in unreliable performance of the overall system and decreased overall efficiency.

Simulation is carried out to analyze the efficiency of battery with respect to SoC estimation. Buck boost converter is connected for the bidirectional power flow between battery and DC machine. This exercise is repeated with and without hill climbing algorithm and the results are analyzed. Primary aim of this paper is to encounter the change in SoC level and charging and discharging characteristics of the battery.

Figure 45.5 depicts the working of DC machine as generator of the torque is negative (seen in Fig. 45.8). Here, while the DC machine is working as a generator, the 240 V battery with initial SoC level as 50% is getting charged and is acting like a load; moreover, an incremental change in the SoC has been witnessed indicating the charging of the battery.

Figure 45.6 provides the insight of working of the DC machine as motor where the torque is positive (seen in Fig. 45.8). Also, in this arrangement the battery is acting like a source and powering the DC machine. The SoC of the battery is showing a decremental change indicating discharging of the battery.

In the basic model, hill climbing algorithm has been invoked due to which an improved level of charging of the battery is seen which enables to conclude that by using HC algorithm the highest peaks have been taken when compared with the previous stage value resulting in increased SoC percentage in the same interval of time.

45.3.4 Limitations

The total performance of the regenerative braking system may be impacted by the effectiveness of the Buck Boost converter. The energy recovered during regenerative braking may be dissipated in the conversion process if the converter is not sufficiently efficient, which would lower the overall energy savings. One of the claim which can be considered opposing against the proposed system is the fact that use of Buck Boost converter with hill climbing algorithm may increase the cost of the system. But the trade-off need to be done to achieve better efficiency of the battery against this increase in price.

The capacity and state of charge (SoC) of the battery determine how much energy can be stored during regenerative braking. Moreover, the battery may sustain damage if it is unable to withstand the high electric current during regenerative braking.

45.3.5 Conclusion

The use of the hill climbing algorithm can lead to significant improvements in the accuracy and precision of battery SoC, while its absence may result in suboptimal performance.

It is crucial to properly choose and design the Buck Boost converter to achieve high efficiency and little energy losses during the conversion process in order to enhance the effectiveness of regenerative braking. The hill climbing algorithm can also be improved through modification in order to obtain the optimum solution during regenerative braking. Regenerative braking is a technology that has a lot of potential for reducing energy use and boosting the range of electric vehicles.

In conclusion, regenerative braking is an approach that can decrease power consumption in electric vehicles. In this paper, it uses a Buck Boost converter and a hill climbing algorithm. However, there are some facts that must be taken into account, such as the converter's performance, the battery's restrictions, the driving environment, and maintenance costs.

References

1. Yu, W., Lai, J.-S.: Ultra high efficiency bidirectional dc-dc converter with multi-frequency pulse width modulation. In: 2008 Twenty-Third Annual IEEE Applied Power Electronics Conference and Exposition, Austin, TX, pp. 1079–1084. https://doi.org/10.1109/APEC.2008.4522856
2. Varshini, C.R.A., Tiwari, A., Jha, A., Annamalai, K.R., Deepa, K., Sailaja, V.: Survey on fast charging with SoC estimation. In: 2022 3rd International Conference on Smart Electronics and Communication (ICOSEC), pp. 59–64, Trichy, India (2022). https://doi.org/10.1109/ICOSEC54921.2022.9952016
3. Baumann, M., Simon, B., Dura, H., Weil, M.: The contribution of electric vehicles to the changes of airborne emissions. In: 2012 IEEE International Energy Conference and Exhibition (ENERGYCON), pp. 1049–1054, Florence, Italy (2012). https://doi.org/10.1109/EnergyCon.2012.6347724
4. Madhav Sai, N., Lekshmi, S., Manitha, P.V.: Modelling of an electric vehicle. J. Phys. Conf. Ser. **2312**(1), 012055
5. Muralidharan, A., Sreelekshmi, R.S., Nair, M.G.: Cell modelling for battery management system in electric vehicles. In: 2020 Third International Conference on Smart Systems and Inventive Technology (ICSSIT), pp. 558–564, Tirunelveli, India (2020). https://doi.org/10.1109/ICSSIT48917.2020.9214253
6. Nair, T.M., Nair, M.G., et al.: Energy management for hybrid energy storage in electric vehicles using neural network. In: 2021 Second International Conference on Electronics and Sustainable Communication Systems (ICESC), Coimbatore, pp. 407–411, India (2021). https://doi.org/10.1109/ICESC51422.2021.9532878
7. Karthikeyan, P., Siva Chidambaranathan, V.: Bidirectional buck–boost converter-fed DC drive. In: Artificial Intelligence and Evolutionary Computationsin Engineering Systems, pp. 1195–1203. Springer, New Delhi (2016)
8. Likhith, S., Bhargavi, S., Manitha, P.V., Lekshmi, S.: DC-DC converter for EV charger with controlling unit. Int. Conf. Adv. Technol. (ICONAT) **2022**, 1–4 (2022). https://doi.org/10.1109/ICONAT53423.2022.9726034
9. Mohanty, A., Manitha, P.V., Lekshmi, S.: Simulation of electric vehicles using permanent magnet DC motor. In: 2022 4th International Conference on Smart Systems and Inventive

Technology (ICSSIT), pp. 1020–1025 (2022). https://doi.org/10.1109/ICSSIT53264.2022.9716549
10. Gao, Y., Ehsani, M.: Electronic Braking System of EV and HEV-Integration of Regenerative Braking, Automatic Braking Force Control And ABS
11. Caratti, A., Catacchio, G., Gambino, C., Kar, N.: Development of a Predictive Model for Regenerative Braking System, pp. 1–6 (2013). https://doi.org/10.1109/ITEC.2013.6573497
12. Patri, S.D., Vali, S.H., Rafi, V.: Battery charging using a hill climbing MPPT algorithm and heuristic algorithm (MPPT). In: 2021 International Conference on Recent Trends on Electronics, Information, Communication & Technology (RTEICT), pp. 91–96, Bangalore, India (2021). https://doi.org/10.1109/RTEICT52294.2021.9573849
13. Kim, M.-J., Chae, S.-H., Moon, Y.-K.: Adaptive battery state-of-charge estimation method for electric vehicle battery management system, in 2020 International SoC Design Conference (ISOCC), Yeosu, Korea (South), pp. 288–289 (2020). https://doi.org/10.1109/ISOCC50952.2020.9332950
14. Duraisamy, T., Deepa, K.: Evaluation and comparative study of cell balancing methods for lithium-ion batteries used in electric vehicles. Int. J. Renew. Energy Dev. **10**, 471–479 (2021). https://doi.org/10.14710/ijred.2021.34484
15. Albiol-Tendillo, L., Vidal-Idiarte, E., Maixé-Altés, J., Bosque-Moncusí, J.M., Valderrama-Blaví, H.: Design and control of a bidirectional DC/DC converter for an electric vehicle. In: 2012 15th International Power Electronics and Motion Control Conference (EPE/PEMC), pp. LS4d.2-1–LS4d.2-5, Novi Sad, Serbia (2012). https://doi.org/10.1109/EPEPEMC.2012.6397462
16. Bahari, M.I., Tarassodi, P., Naeini, Y.M., Khalilabad, A.K., Shirazi, P.: Modeling and simulation of hill climbing MPPT algorithm for photovoltaic application. In: 2016 International Symposium on Power Electronics, Electrical Drives, Automation and Motion (SPEEDAM), pp. 1041–1044, Capri, Italy (2016). https://doi.org/10.1109/SPEEDAM.2016.7525990

Chapter 46
An Enhanced Classification Model for Depression Detection Based on Machine Learning with Feature Selection Technique

Praveen Kumar Mannepalli, Pravin Kulurkar, Vaishali Jangade, Ayesha Khan, and Pardeep Singh

Abstract Facebook, Twitter, and Instagram are just a few examples of how social media have changed our lives. People are more linked than ever before, leading to the development of a distinct online identity. Recent studies have revealed that an increased number of hours spent on social media platforms is connected with an increased likelihood of developing depression. Depression is characterized by pervasive sadness and a general absence of interest in most activities. Severe depression, often known as major depressive disorder, is a serious mental illness that can have far-reaching effects. The purpose of this study is to analyze depression, utilizing a variety of socio-demographic and psychological data to determine if a person is depressed or not. Different operations have been performed, including data collection, preprocessing, feature selection, classification, and evaluation. This research is evaluated on the depression detection dataset. Data is processed in the data preprocessing step by checking null and missing values and performing data encoding using a label encoder. Further, the recursive feature elimination technique has extracted the most important features from the dataset in the feature selection. On the other hand, machine learning-based SVM and DT techniques are used for classification. The performance of these models is measured using different performance metrics. After applying these methods, the proposed decision tree model obtains the highest 98% accuracy, which is better than the other models.

Keywords Depression detection · Support vector machine · Machine learning · Recursive feature elimination · Decision tree

---

P. K. Mannepalli (✉) · P. Kulurkar · V. Jangade · A. Khan
G. H. Raisoni Institute of Engineering and Technology (Autonomous), Nagpur, India
e-mail: praveen.hawassa@gmail.com

P. Singh
Department of CSE, Graphic Era Hill University, Dehradun, India
e-mail: pardeepsingh@gehu.ac.in

© The Author(s), under exclusive license to Springer Nature Singapore Pte Ltd. 2024
P. K. Jha et al. (eds.), *Proceedings of Congress on Control, Robotics, and Mechatronics*, Smart Innovation, Systems and Technologies 364,
https://doi.org/10.1007/978-981-99-5180-2_46

46.1 Introduction

A depression illness that might involve emotions of grief, loss, or rage is known as depression. It makes it difficult for a person to carry out their everyday tasks. There are many different ways that people might show their frustration. People react differently, some on social media (SM) and others in their personal life. People communicate with their pals and exchange information using SM platforms. This results in an enormous volume of new data being produced every day. These pieces of information can be collected in the form of photographs, videos, or text representing the individual's mental state [1].

Depression means "a state of mind that expresses mood disorders such as depressed, unhappy, bored, loss of appetite, lack of concentration, anxiety, etc." [2]. Researchers have recently discovered that a confluence of variables other than a chemical imbalance in the brain causes depression. These factors include improper mood regulation by the brain, genetic predisposition, stressful life events, medicine, and psychological difficulties. Symptoms of depression and its physical manifestations may result from a combination of causes. There is evidence that a chemical imbalance contributes to depression. However, this imbalance is not the result of a single molecule but of multiple chemicals interacting. This level of complexity allows us to make sense that two persons can experience the same sickness symptoms yet have vastly different physical and social circumstances [3].

Researchers can pinpoint the gene responsible for a person's susceptibility to low moods and the gene's response to pharmacological therapy. The purpose of these studies is to develop more effective, individualized treatments for the symptoms of depression. The effect of brain plays a crucial part in the onset of depression-related symptoms. The connections between nerve cells, the proliferation of nerve cells, and the activity of nerve cells all contribute to minor fluctuations in mood. Despite the inaccuracy of understanding neurological links, it is possible to determine a person's mental state using the right understanding of neurological equilibrium. Globally, more than 300 million individuals suffer from depression, and WHO estimates that 1 billion people have some form of mental illness [4]. ML is widely recognized as a powerful method for sifting through the huge amounts of data in the healthcare industry. Predicting the likelihood of MD and executing likely treatment results is an area where ML approaches are used. ML [5, 6] computer programs that learn and get better at their ML is a branch of artificial intelligence (AI) that allows computers to acquire new skills and hone existing ones without human involvement or supervision. ML techniques construct a mathematical model using sample data to forecast or make judgments without being explicitly taught. This data is referred to as training data. There are many uses for ML techniques, such as email filtering and computer vision (CV), where it would be impractical or impossible to create conventional techniques to carry out the necessary tasks. Techniques for supervised ML can use labels to transfer knowledge from one dataset to another. To forecast future output values, a learning algorithm first analyzes a known training dataset and generates an inferred function [7].

Reviewing the existing literature, the proposed research has concluded that no comprehensive examinations of Facebook exist [8, 9]. Therefore, the suggested study constructed a database of Bengali Facebook comments, posts, and status changes. Several supervised learning methods have been used to estimate problems and their solutions. The suggested study has attempted to extrapolate depression rates from existing data. In this case, we apply and compare the outcomes from state-of-the-art techniques such as decision tree (DT) and support vector machine (SVM). The suggested research study analyzes depression and non-depression data with high accuracy. The comparison findings show that different methods provide identical outcomes and results for all classifiers used in the data.

The main contributions of this paper are listed as follows:

- Initially, a literature review was conducted on several emotion detection approaches for detecting depression.
- Secondly, the experiments are carried out on the depression detection dataset.
- Thirdly, suggest ML techniques like a decision tree and support vector machine to utilize all factors and maintain robustness.
- In the fourth step, the model's performance is measured by different performance metrics like accuracy, precision, recall, f1-score, confusion matrix, and ROC curve.
- Lastly, our research demonstrates the significance of depression identification for detecting mental disorders.

This paper's remaining sections are structured as follows. Section 46.2 reviews the relevant literature, while Sect. 46.3 details the dataset and study approach employed. Section 46.4 contains a description of the experiment and its results. Conclusions and summaries are included at the end of Sect. 46.5.

46.2 Related Work

Numerous types of research have been conducted throughout the years on using ML to increase the investigation of MD. In [10], in this study, they evaluate and contrast several ML approaches and classifiers—specifically, decision tree (DT), logistic regression (LR), support vector machine (SVM), Naive Bayes (NB) classifier, and the K-nearest neighbor (KNN) classifier to determine the mental health of a population of interest. High school kids, college students, and working adults are all included in this identifying procedure. In addition, the article provides a case study of how a specific Twit's expressibility may be determined using the Twitter scraping program Twins.

Aggarwal and Goyal [11] use a combination of data about the player and the game and a measure of the player's self-esteem to make predictions about whether or not the player is experiencing mental health issues like anxiety or sadness. Using a dataset of online gamers, the tenfold cross-validation approach was employed to compare four distinct ML classifiers. DT classifier has the highest accuracy of the four algorithms tested across the board. The DT achieves a precision of 100% for the GAD questionnaire and 84.71% for the SWL questionnaire.

Liu et al. [12] provide a multimodal brain features-based, data-driven, impartial, and rigorous ML strategy to capture the complexities of brain structure representations. Graph convolutional neural networks (GCN), SVM, and regularized LR were implemented. Our models' cross-validated classification accuracy is 97% (LR), 93% (SVM), and 89.58% (GCN), which is very encouraging.

Parikh et al. [13] suggest a method utilizing data analysis and ML to identify depressed individuals. The same online site has two different types of depression exams. The maximal accuracy for SVM on the PHQ-9 dataset is 99.74%. Given the more casual tone of tweets, the accuracy of the Twitter dataset suffers slightly in comparison. The proposed approach greatly boosts accuracy by incorporating the SVM method, which provides maximum accuracy with properly preprocessed data.

José Solenir et al. [14], using a CNN, context-free word embeddings, and early and late fusion techniques, you suggest a technique for early identification of depression in SM. The findings suggest that the proposed technique successfully identifies users experiencing depression, with a precision of 0.76 and equal to or better than many baselines in terms of efficacy (F1(0.71)). Emoticon semantic mapping also improved recall and accuracy (46.3 and 32.1%, respectively).

Nowadays, the repercussions of depression [15] are astonishing. The suicidal propensity, and several other exhaustion and melancholy, have nearly engulfed the whole planet. An early warning system like this one can help you avoid these problems. A person's everyday regular actions are represented by sensorimotor sensor readings, which may be used to detect brief shifts in behavior. The integration of MS data with socio-demographic information can greatly aid depression identification. This paper used ML methods such as RF, AdaBoost, and ANN to combine sensor readings from motors and demographic data, resulting in reliability and an F1-score of 98 percent in both situations. The coefficients of correlation computed by Cohen and Matthew are both 0.96.

In this investigation, Tlachac and Rundensteiner [16] looked at the detection capabilities of two different melancholy testing methods: one that was accessible secretly and one that was available publicly. Incorporating crowd-sourced texts and tweets into the merged datasets of MA and EMU is the first time done. The strategy included using CFE, FS, and ML techniques, among other things. The 245 features include phrase group rates, part of utterance tag frequency range, emotions, and loudness, all contained in their database. Regression analysis is the right model for text information recorded for two weeks. In terms of F1, AUC, and recall, this model averages 0.806 points on a scale from 1 to 10.

46.2.1 Research Gap

As part of this study, numerous articles also look at how to increase the suggested system's accuracy in recognizing tweets that contain depression-related terms and phrases and classifying them appropriately. People who are mentally sick but afraid to seek treatment or unsure of how to self-diagnose will benefit from the success of this project, since it awakens the administration and psychologists of the need to identify and treat mental health problems. Nevertheless, the study relied on self-reported instances, as with most previous studies attempting to forecast depression, and to date, efforts trying to detect patients who are as yet unaware of their depression condition are still uncommon. The symptoms of depression also vary widely from person to person. A symptom-based diagnosis of depression makes it extremely challenging, if not difficult, to objectively assess psychological events instead of physiological ones. Depression may be reliably diagnosed using machine learning algorithms, according to several research. In contrast, past studies have mostly dealt with small groups of people. Because of this, they work with datasets with a limited number of features or datasets with few examples of each characteristic. Finally, they choose a random prediction technique without knowing how other algorithms fared on the dataset.

46.3 Research Methodology

This section provides the research methodology for depression detection. Also, the section discusses the problem statement and the solution to overcome depression problems.

46.3.1 Problem Statement

Depression may be a mental illness. It's the biggest global health issue. Depression increases the chance of early death. Seventy percent of people would not see a doctor if they were depressed. Social media is being used to express feelings and monitor mental health. Preparing social network data, particularly Facebook comments can reveal depression-related information. ML algorithms are increasingly used to find industry-specific data patterns. Psychological sciences have seldom employed ML. Machine learning addressed the aforesaid problems.

Few machine learning researches have predicted depression. To solve this gap, a state-of-the-art solution has been presented by researchers. This study seeks to detect depressed people and their causes of using ML.

46.3.2 Proposed Methodology

The depression identification technique consists of several steps, including collecting datasets, feature selection (FS), data preprocessing techniques for choosing the

Fig. 46.1 Proposed flowchart

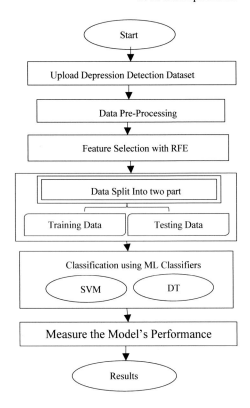

needed characteristics for recognizing symptoms of depression, dataset splitting, and classification systems in machine learning used to divide raw data into distinct classes. In this part, we'll go through each of these stages and the many strategies that may be employed to put them into action.

This research methodology steps in Fig. 46.1 and below briefly show subpoints.

46.3.2.1 Data Collection and Preprocessing

In this research, we have used a depression detection dataset from the GitHub repository. This dataset needs to be preprocessed, so applied data preprocessing. Check null values, fill in missing values and label the data with the help of the data preprocessing procedure. Also, encode the dataset with the label encoder.

Data Encoding is carried out on the gathered datasets for training and testing. In general, ML techniques perform better when presented with numerical data. While encoding data, use Scikit-Label learn Encoder to convert the categorizations found in the training and testing datasets into a numeric representation.

46.3.2.2 Feature Selection Using RFE

FS was handled through recursive feature elimination (RFE) in this particular development. FS describes methods through which a dataset's most useful features (columns) are chosen. ML techniques may be made more economical (lower space or time complexity) and successful by restricting the number of features they must consider. Misleading input characteristics can lower the predicted performance of some ML systems [17].

Recursive feature elimination (RFE) or RFE feature selection, the procedure minimizes the complexity of a model by selecting the most important characteristics and eliminating the less important ones. RFE helps to make models more effective by reducing the number of features they use.

46.3.2.3 Data Splitting

Data splitting is the essential step for implementation in which data is divided into two or three ways. In most cases, data is splitted into two parts, i.e., training and testing. As a result, we have 80% of training and 20% of testing datasets.

46.3.2.4 Classification Using the ML Model

Classification refers to organizing data by grouping like items into distinct categories. Business norms, arbitrary boundaries, or a mathematical formula can all be used to establish these groups. There can be a correlation between the attributes of the thing being classed and a previously established class assignment, which can inform the classification procedure. SVM and DT are two models that may be used to provide depression forecasts.

Support Vector Machine (SVM)

Another widely used state-of-the-art machine learning technique is support Vector machine (SVM) [18]. Support vector machines are a supervised learning model used in machine learning for classification and regression analysis [19, 20]. In addition to linear classification, SVMs may easily conduct nonlinear classification by implicitly mapping their inputs into high-dimensional feature spaces via the kernel method. It essentially creates distinctions between social classes [21]. The margins are drawn to maximize the distance between the margin and the classes, hence decreasing classification error [22].

Decision Tree (DT)

Step 1: S suggests beginning with the tree's root containing the original data.

Step 2: Find the best attribute in the dataset using the attribute selection measure (ASM).

Step 3: Split the S into groups containing values for the top characteristics.

Step 4: Generate the decision tree node containing the best attribute [23].

Step 5: Generate extra decision trees in a recursive manner utilizing the selections of the information produced in step 3. Repeat the same process until the networks can no longer be classified, at which point the terminal node gets designated as a leaf node [24].

46.3.3 Proposed Flowchart

In this proposed flowchart, we are representing the process of methodology. In this flowchart, firstly, we input a machine learning-based model of depression detection dataset, then preprocess data by performing data preprocessing, then with the help of the RFE model, we select a feature and split the dataset and then the process of implementing SVM and DT classifier. In all this process, we get an output with a high classification outcome and then terminate the model.

46.4 Results Analysis

The simulation and experimental results of the suggested model are presented in this section. This investigation uses the Python Simulation tool to carry out its implementation. All experiment incorporates the Python programming language and the Jupyter Notebook platform with several Python libraries. A data visualization graph and an examination of the performance matrix in terms of precision, accuracy, f1-score, recall, sensitivity, ROC, and specificity demonstrate the ML models' success for the depression dataset.

46.4.1 Dataset Description

In this research, we have collected the depression detection dataset,[1] which is collected from the GitHub repository. Traditional methods of depression detection rely on in-depth clinical interviews, during which the psychologist carefully examines the subject's responses to a battery of questions to form an opinion about the person's mental health for the analysis of a machine learning-based model of depression detection. Symptoms of depression are a general lack of motivation and interest in formerly pleasurable activities and feelings of emptiness, melancholy, worry, and sleep disruption. Added to these characteristics are a lack of motivation, an inability

---

[1] https://github.com/ranju12345/Depression-Anxiety-Facebook-pageComments-Text.

46 An Enhanced Classification Model for Depression Detection Based ...

Table 46.1 Data labels

Category	Label
Depression	1
Not depression	2

to focus, feelings of guilt or worthlessness, suicidal ideation, and psychotic symptoms. Table 46.1 gives the data labels, 1 denotes the depression, and 2 denotes the not depression.

Figure 46.2 shows the count plot of the input dataset label. The depressive person approx. 2500 count value and non-depressed person approx. 4800 count values.

Table 46.2 gives the performance of the proposed SVM and DT classifier with the help of five performance matrices. The proposed SVM classifier gets 94% accuracy and 98.32% accuracy by the decision tree. Similarly, precision, f1-score, specificity, and sensitivity parameters show their performance in terms of percentage. We can see that the DT gets the highest performance. Correspondingly, Figs. 46.3 and 46.4 show the bar graph of the simulation results; in the *x*-axis, the performance parameters, the *y*-axis, and the %, respectively.

In Fig. 46.5 of the ROC curve, we found the decision tree classifier's true or false positive rate. The blue line and the FP rate by the orange line show the TP rate. The *x*-axis represents the epochs 0.965% FPR of SVC, while the DT model gets 74.1%, which is the highest accuracy where it is stable. And in the *y*-axis, as we can see

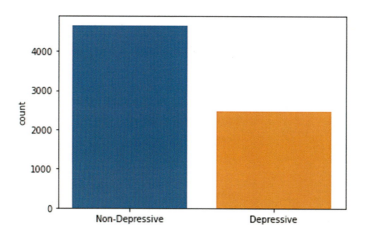

Fig. 46.2 Bar graph of the dataset

Table 46.2 Performance of proposed machine learning classifier

Models	Acc	Precision	F1-score	Specificity	Sensitivity
SVM	94.02	89.00	88.75	88.57	94.24
DT	9832	72.17	71.91	70	79.59

Fig. 46.3 Bar graph of parameter performance of SVM classifier

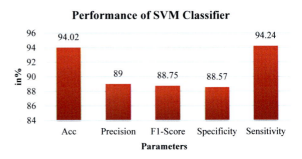

Fig. 46.4 Bar graph of parameter performance of SVM classifier

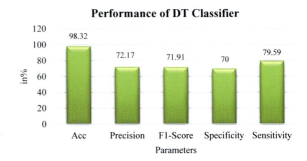

in this ROC curve, the true positive rate performed well compared with the false positive rate. And the ROC curve achieved the highest accuracy value. Table 46.3 gives comparison between the base and proposed ML classifiers.

Figure 46.6 bars compare the base classifier's accuracy to the suggested classifier. The base bagging classifier gets 83% accuracy, and 80% accuracy obtained by the gradient boosting classifier, while proposed SVM and DT obtain 98 and 94% accuracy. Our suggested model outperforms the baseline classifiers, as expected.

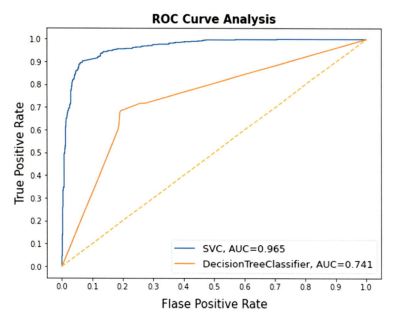

Fig. 46.5 ROC curve

Table 46.3 Comparison between the base and proposed ML classifiers

Parameter	Proposed classifiers		Base classifiers	
	SVM	DT	Bagging	Gradient boosting
ACC (%)	94	98	83	80

Fig. 46.6 Accuracy comparison between the base and proposed classifier

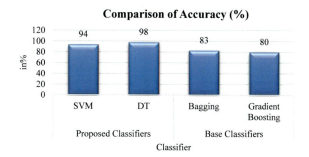

46.5 Conclusion

In this research, identifying depression status by Facebook behavior is defined as a binary classification issue. Several feature datasets and ML methods are tested. Data cleaning, feature selection, labeling, and classification are just a few preprocessing tasks that must be completed before any analysis can begin. By combining the strengths of the SVM and DT classifiers, we can transform a highly nonlinear classification issue into a linearly separable one. We present a working version of the suggested technique. Utilizing various psycholinguistic aspects, we have determined whether or not our suggested strategy is effective. This study demonstrates that our suggested strategy may greatly enhance accuracy and classification performance. In addition, the data demonstrates that across several tests, DT provides the greatest accuracy (98.32%) compared with other ML techniques when trying to identify depression. This research may be seen as a first step in developing a comprehensive social media-based system for assessing users' mental health, evaluating their potential for mental health problems and providing appropriate treatment options. In the future, this research might be expanded by examining more ML models that are unlikely to overfit the utilized data and by discovering a more reliable technique to quantify the influence of the features.

Future studies will use a different method for extracting paraphrases from various affective variables. The usefulness and efficiency of our methods will be tested using other datasets we want to collect. We concur with the growing corpus of research arguing for more targeted investigations of depression.

References

1. Manish, R.J., Tripathi, M.: Multimodal depression detection using machine learning. In: Artificial Intelligence, Machine Learning, and Mental Health in Pandemics, pp. 53–72 (2022)
2. Angskun, J., et al.: Big data analytics on social networks for real-time depression detection. J. Big Data (2022)
3. Aleem, S., ul Huda, N., Amin, R., Alshehri, A.: Machine learning algorithms for depression: diagnosis, insights, and research directions. Electronics 11(7), 1111 (2022)
4. Ramalingam, D., Sharma, V., Zar, P.: Study of depression analysis using machine learning techniques. Int. J. Innov. Technol. Explor. Eng. (2019). https://doi.org/10.35940/ijitee.h7163.0881019
5. Jagtap, N., Shukla, H., Shinde, V., Desai, S., Kulkarni, V.: Use of Ensemble Machine Learning to Detect Depression in Social Media Posts (2021). https://doi.org/10.1109/ICESC51422.2021.9532838
6. Chiong, R., Budhi, G.S., Dhakal, S., Chiong, F.: A textual-based featuring approach for depression detection using machine learning classifiers and social media texts. Comput. Biol. Med. (2021). https://doi.org/10.1016/j.compbiomed.2021.104499
7. Divakara, B.C., Lavanya, M.R., et al.: Detection of early stage depression. 11(6) (2022)
8. Ashok, M., Nagaraju, R., Chandra Sekhar Reddy, P., Prashanthi, P.: A method for detection for anxiety and depression of human brain using machine learning and artificial intelligence. Int. J. Adv. Sci. Technol. (2020)

9. Islam, M.R., Kabir, M.A., Ahmed, A., Kamal, A.R.M., Wang, H., Ulhaq, A.: Depression detection from social network data using machine learning techniques. Heal. Inf. Sci. Syst. (2018). https://doi.org/10.1007/s13755-018-0046-0
10. Vaishali Narayanrao, P., Lalitha Surya Kumari, P.: Analysis of Machine Learning Algorithms for Predicting Depression (2020). https://doi.org/10.1109/ICCSEA49143.2020.9132963
11. Aggarwal, R., Goyal, A.: Anxiety and depression detection using machine learning. In: 2022 International Conference on Machine Learning, Big Data, Cloud and Parallel Computing (COM-IT-CON), vol. 1, pp. 141–149 (2022). https://doi.org/10.1109/COM-IT-CON54601.2022.9850532
12. Liu, A., et al.: Machine Learning Aided Prediction of Family History of Depression (2017). https://doi.org/10.1109/NYSDS.2017.8085046
13. Sumitra Motade, F.M., Parikh, K., Hassan, A.: Machine Learning-Based Approach for Depression Detection Using PHQ-9 and Twitter Dataset (2022)
14. José Solenir, R.T., Figuerêdo, L., Maia, A.L.L.M.: Early depression detection in social media based on deep learning and underlying emotions. Online Soc. Netw. Media **31**, 100225 (2022)
15. Raihan, M., Bairagi, A.K., Rahman, S.: A Machine Learning Based Study to Predict Depression with Monitoring Actigraph Watch Data (2021). https://doi.org/10.1109/icccnt51525.2021.9579614
16. Tlachac, M.L., Rundensteiner, E.: Screening for depression with retrospectively harvested private versus public text. IEEE J. Biomed. Heal. Inform. (2020). https://doi.org/10.1109/JBHI.2020.2983035
17. Brownlee, J.: Recursive Feature Elimination (RFE) for Feature Selection in Python (2020)
18. Mahesh, B.: Machine learning algorithms—a review machine learning algorithms—a review view project self flowing generator view project Batta Mahesh independent researcher machine learning algorithms—a review. Int. J. Sci. Res. (2018)
19. Wu, X., et al.: Top 10 algorithms in data mining. Knowl. Inf. Syst. (2008). https://doi.org/10.1007/s10115-007-0114-2
20. Patel, M.J., Khalaf, A., Aizenstein, H.J.: Studying depression using imaging and machine learning methods. NeuroImage: Clin. (2016). https://doi.org/10.1016/j.nicl.2015.11.003
21. Andrew, A.M.: An introduction to support vector machines and other kernel-based learning methods. Kybernetes (2001). https://doi.org/10.1108/k.2001.30.1.103.6
22. Jamil, Z., Inkpen, D., Buddhitha, P., White, K.: Monitoring Tweets for Depression to Detect At-risk Users (2017). https://doi.org/10.18653/v1/w17-3104
23. Han, J., Kamber, M., Pei, J.: Data Mining: Concepts and Techniques (2012)
24. Podgorelec, V., Zorman, M.: Decision tree learning. In: Encyclopedia of Complexity and Systems Science (2015)

Chapter 47
Design & Analysis of Grey Wolf Optimization Algorithm Based Optimal Tuning of PID Structured TCSC Controller

Geetanjali Meghwal, Shruti Bhadviya, and Abhishek Sharma

Abstract Many problems related to instability are faced by power system which crates huge oscillations and make a system unstable. Such problems can easily overcome by a good damping controller. To find out the solutions of stability problems many methods including analytical methods or the numerical-based have been applied in many ways to obtain extreme values. And now these methods are developed into the more advanced form of themselves known as optimization techniques. To solve complex engineering design problems and real application, GWO is one of the more suitable optimization techniques. Applications of GWO algorithm are investigated in this paper to tune the parameters of proportional integral derivative-structured TCSC-based controller to damp out the power system oscillations and improves settling time subjected to various loading conditions. Dynamic performance of proposed controller is analyzed for SMIB power system using MATLAB/ Simulink. An obtained simulation result shows the performance of GWO-tuned PID-structured Thyristor Controlled Series Capacitor-based controller and compare with previously published non-dominated sorting genetic algorithm-II (NSGA-II) and cuckoo search algorithm (CSA) for the same power system.

Keywords Grey wolf optimization algorithm · NSGA-II · CSA · TCSC · Power system stability · SMIB system

G. Meghwal (✉) · S. Bhadviya · A. Sharma
Geetanjali Institute of Technical Studies, Dabok, India
e-mail: genius1998.sy@gmail.com

S. Bhadviya · A. Sharma
Engineering, Geetanjali Institute of Technical Studies, Udaipur, India

47.1 Introduction

Power system is very complex and power transfer from one circuit to another circuit very complex. So it is required to power transfer without disturbance. So this is necessary to different technique applied to secure power transfer [1]. To maintain power system stability and security within margin, it's required to take a closer look and analysis of power system parameters [2]. To improve stability flexible AC transmission system (FACTS) devices are initiated, series compensation is demonstrated based on FACTS devices are highly suitable for controlling power flow. From FACTS family TCSC is one of the important controllers which increases the capacity of transmittable power by regulating its line reactance periodically. It can be also utilized to reduce power system oscillations, resonance mitigation, load flow control, reducing net losses and unsymmetrical components, limiting short circuit currents and improves stability [3]. Using GA [4], PSO [5], DE [6], CSA [7] parameters of TCSC-based controllers are optimized. A new meta heuristic algorithm named as GWO technique is used to meet the given goal which inspired using grey wolves (Canis Lupus) which mimic the leadership hierarchy and the hunting idea of grey wolves in nature. Leadership strategy is simulated using four types of operators named as α, β, Δ and ω are employed along with three main steps which is hunting, searching for prey and attacking prey, respectively [9]. A total of 29 well-known test functions are benchmarked in this algorithm which is used to develop by Mirjalili et al. [8]. This is a latest nature-inspired meta heuristic algorithm. In this paper, GWO algorithm is used for optimizing the TCSC parameters. To get the optimal parameters of Proportional Integral Derivative structured TCSC based controller towards various loading conditions and disturbances with MATLAB/Simulink environment GWO has successfully applied. Simulation results represent the merits of the proposed TCSC-based PID controller to enhance stability and comparative analysis also performed for with and without GWO-tuned TCSC controller.

47.2 Literature Review

There are many author reviews and analysis of power system that use many techniques. There are some authors who give their different contributions related to power system. Different authors solve the power system issue in different manner. Panda et al. [12, 13, 15–18, 20] presented a new nonlinear control scheme for TCSC for transient stability analysis of the SMIB power system.

Zero dynamic design approach is used to formulate nonlinear control strategy of the TCSC controller. Tales et al. [19] presented Thyristor Controlled Series Capacitor (TCSC), which enables an increase in power transfer and control of a system. TCSC widely used to reduce synchronous resonance (SSR) as a dynamic device. Ganthia et al. [14] showed an interval fuzzy logic control technique 2 that is used to design a stable control for a TCSC to get better petite signal stability of the power system. For a single machine infinite bus power system (SMIB), Sethi et al. [7] analyzed FACTS devices of two distinct kin-TCSC (varies transmission line reactance with enhanced stability) and SVC (betters voltage regulation). Both the devices have a good and fast response. Rupashree et al. defined a CSA optimal tuning of proportional integral derivative-structured TCSC controller. Sidhartha Panda et al. explained the stability to improve by using NSGA-II.

These proposed controllers along with TCSC, enhance the power system stability. Performance analysis is done using MATLAB/Simulink with respect to power angle, speed deviation and variation of TCSC reactance for fault. Our contribution and research gaps the same power system applied Grey Wolf optimizer (GWO) technique and enhances power system stability of the system by reducing settling time.

47.3 TCSC and PID Structure of TCSC-Based Controller

TCSC is having three components, i.e. bypass inductor (L), capacitor bank (C) and bidirectional thyristor. TCSC reactance is changed by adjusting its thyristor firing angle. To modulate the reactance offered by TCSC, the structure of the TCSC-based damping controller is shown in Fig. 47.1 where α is firing angle, K_p is proportional gain, K_i is integral gain and K_d is derivative gain of proportional integral derivative (PID) controller [1]. Description of TCSC-based controller is presented in [1]. The projected controller input signal is speed deviation ($\Delta\omega$) error and the output signal is reactance $X_{TCSC}(\alpha)$ [12].

PID controller parameters (K_p, K_i, K_d) are optimized using speed error signal so that TCSC reactance modulate effectively to cancel some portions of reactance of line and improves damping of power system oscillations [8].

Formula for effective reactance is given by:

$$X_{Eff} = X - X_{TCSC}(a) \qquad (47.1)$$

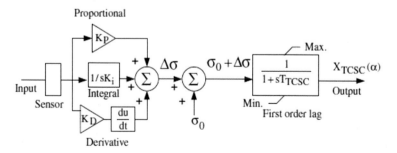

Fig. 47.1 Structure of TCSC-based PID controller [7]

47.4 Modelling of Power System with TCSC

Model of TCSC-based SMIB system is shown in Fig. 47.2. Here synchronous generator is used to supply power to an infinite bus through double loop transmission line and TCSC. Here V_t represents the terminal voltage of generator and E_b denotes voltage of infinite bus. For receiving end transformer reactance, reactance of transmission line per circuit and equivalent Thevenin's impedance is denoted by X_T, X_L and X_{TH} respectively [23].

Model [12] represented synchronous generator, i.e. its field circuit, equivalent snubber winding on quadrature axis, electromechanical oscillation equation and internal voltage equation of generator. System data is assigned in [2]. Use of state equations [7], in MATLAB/Simulink environment modelling of power system is explained in [16]. Different Eqs. (47.2–47.8) of modelling of the system is shown below [22]:

$$\frac{dE'_q}{dt} = -\frac{E'_q}{T'_{d0}} + \frac{(X_d - X'_d)t'_d}{T'_{d0}} + \frac{E_{fd}}{T'_{d0}} \quad (47.2)$$

$$\frac{dE'_d}{dt} = -\frac{E'_d}{T'_{q0}} + \frac{(X_q - X'_q)i'_q}{T'_{q0}} \quad (47.3)$$

$$V_q = E_q{'} + x_d{'}i_d{'} - R_q i_q{'} \quad (47.4)$$

$$V_d = E_d{'} - x_q{'}i_q{'} - R_d i_d{'} \quad (47.5)$$

$$E_q{'} = E_{fd} + (x_d - x_d{'})i_{d0}{'} \quad (47.6)$$

$$E_d{'} = -(x_q - x_q{'})i_{q0}{'} \quad (47.7)$$

$$\frac{d\delta}{dt} = \omega_B(\delta_m - \delta_{m0}) = (\omega_m - \omega_\delta) \quad (47.8)$$

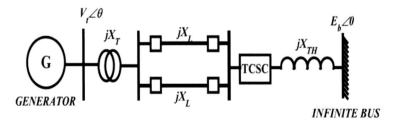

Fig. 47.2 TCSC with SMIB power system [7]

47.5 Objective Function

When the rotor of generator machine faces any disturbance in its motion, it introduced oscillations in power system. To minimize such oscillations and improve stability, TCSC-based controller is designed. In this paper, objective function J is chosen as an integral time absolute error (ITAE) of speed deviation $\Delta\omega$.

Objective function J is defined as follows:
where,
$\Delta\omega(t)$ = Absolute value of speed deviation which follows a disturbance,
t_{sim} = Range of simulation time.

To improve system response objective function, parameters are minimized using parameters of TCSC-based controller that are K_P, K_I and K_D.

$$J = \int_0^{t\,sim} t|\Delta\omega(t)|dt \tag{47.9}$$

47.6 Grey Wolf Optimizer (GWO)

To solve complex engineering problems and real application, Grey Wolf optimizer (GWO) results are proven such that the proposed algorithm is more suitable for challenging tasks with unknown search space. GWO results are found to be more competitive and challenging than optimization techniques like PSO, GSA, differential evolution, evolutionary computation (ES) and expectation propagation (EP). So we choose GWO technique for TCSC-based structured PID controller.

In GWO technique, wolves live pack basically belongs to Canidae family having a leader Alpha (α) who indicates their strict social dominance hierarchy and most decisions for the group are also made by him. Therefore, the leader's decision is followed by the other members of the group. Some common decisions are where to hunt, when to wake up, where to sleep, etc. α is not the strongest member of the pack, but he manages all of the activity of the pack, showing that discipline and organization in the pack take precedence over strength [8].

A group having a pack of subordinates who helps α for making decisions are known as advisors represented by β. They are in the next line to become α if the present leader passes away or becomes old and also maintain the discipline and other activity of the pack. Commands to other wolves are also given by β and feedback is also delivered to α [9].

The lowest category in wolves is Omega (ω), they follow all other dominant wolves. Role of scapegoat played with them and are assigned last to eat in the pack. Although, omega wolves have less significance in a group, but it may cause for internal fighting if pack loses them. It happened due to the absence of all frustrations and violence of all wolves by Omegas. Omega plays a role of babysitter and also maintains the dominance structure in the group as well as is satisfactory between them [10].

If wolf doesn't relate to any rank, then it comes into the category of Delta wolves. They are carrying two ranks in the pack which are above omega wolf and under delta wolves. This category of wolves assigned works that are hunting, caretaking, sentinels and scouts. Scouts are responsible for danger alert and keeping watch on pack territory, same as sentinels play the duty of a guard and protection to pack. Elder wolves are those who once may be α or β in their life time and kept more experience. α and β are assisted by hunters for hunting and arrange the food for the pack. Caretaker wolves provide care to weak, ill and wounded wolves in pack [11].

The main phases of hunting have following features:

- Stalk the prey, then chase them and finally approach them.
- Encircling, pursuing and harassing prey until it stops moving.
- Last is attack on prey.

GWO Mathematical Model and Algorithm: The various steps follow GWO algorithm.

47.6.1 Social Hierarchy

First best solution is α, second one best solution is β and the third best solution is δ. Apart from these, all the candidate solutions fall under ω [11].

47.6.2 Prey Encircling

Prey is encircled by grey wolves while hunting. To reflect it mathematically,

$$\vec{D} = \left| \vec{C} \cdot \vec{X}_p(t) - \vec{X}(t) \right| \quad (47.10)$$

$$\vec{X}(t+1) = \vec{X}_p(t) - \vec{A} \cdot \vec{D} \quad (47.11)$$

t represents iteration of current.

From Fig. 47.3, it's clear that as the position of prey (X^*, Y^*) varies, a grey wolf will also be varying its position to (X, Y). Adjustment in $\vec{A}$ and $\vec{C}$ is done for different places around to meet best agent with respect to present location. Like, to reach (X^*-X, Y^*),
$\vec{A} = (1, 0)$ and $\vec{C} = (1, 1)$ is set.

In the Fig. 47.5 wolves can reach any position between points via random vectors $r1$ and $r2$. Hence, using Eq. (1.10) and (1.11) wolves can change their position around prey in the given space [10].

Fig. 47.3 Social hierarchy of grey wolf

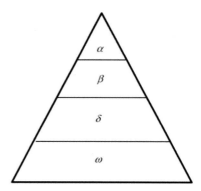

47.6.3 Hunting

A guides the hunting while β and δ might take part in it sometimes. Grey wolves can identify prey location and encircle them. Although in a search space, location of prey is unknown, i.e. optimum. To simulate hunting by grey wolves, Alpha candidates are assumed as the best candidate while β and δ have better ideas regarding prey possible location. Hence, the first three solutions are saved and the reset search agents including ω are obliged to change their position as per the best search agent. It is given as follows [11],

$$\vec{D}_\alpha = |\vec{C}_1.\vec{X}_\alpha - \vec{X}|, \ \vec{D}_{\alpha\beta} = |\vec{C}_2.\vec{X}_\beta - \vec{X}|, \ \vec{D}_\delta = |\vec{C}_3.\vec{X}_\delta - \vec{X}| \quad (47.12)$$

$$\vec{X}_1 = \vec{X}_\alpha(t) - \vec{A}_1.(\vec{D}_\alpha), \ \vec{X}_2 = \vec{X}_\beta(t) - \vec{A}_2.(\vec{D}_\beta), \ \vec{X}_3 = \vec{X}_\delta(t) - \vec{A}_3.(\vec{D}_\delta) \quad (47.13)$$

$$\vec{X}(t+1) = \frac{\vec{X}_1 + \vec{X}_2 + \vec{X}_3}{3} \quad (47.14)$$

Figure 47.4 shows in 2D search space that alpha, beta and delta estimate prey location, while others change their position randomly around prey.

47.6.4 Attacking

At this stage the grey wolf begins to attack, i.e. after the hunt. It is the exploitation process in which the value of $\vec{a}$ is decreased to obtain a mathematical realization.

Hence, the variation range for an $\vec{A}$ reduces where it has a random value fallen in $[-2a, 2a]$ and during the iteration $\vec{a}$ changes from 2 to 0.

Fig. 47.4 2D position vector with possible next location

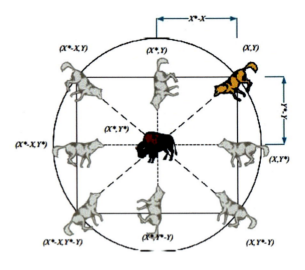

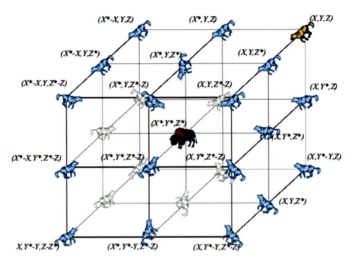

Fig. 47.5 3D position vector with possible next location

If, $\vec{A}$ has the random value in $[-1,1]$, then search agent can have its search position anywhere between its existing position and prey's position, as depicted in the Fig. 47.6, where, $|A| < 1$ shows that wolves are attacking prey [21].

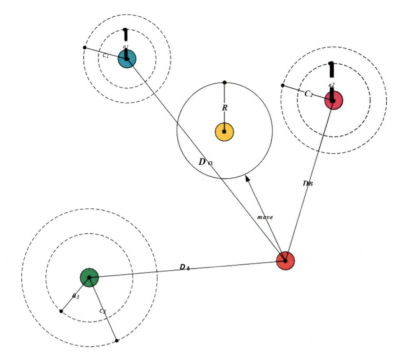

Fig. 47.6 GWO position updating

47.6.5 Prey Searching

Searching depends upon the alpha, beta and delta positions. At, first they diverge and then they converge for attack. For mathematical realization of divergence, $\vec{A}$ is used to have random values > 1 or < -1 to depict the action of diverge of search agent. It emphasizes exploration allowing GWO to search globally with $|A| > 1$, forgetting the fitter prey. Exploration favours by vector $\vec{C}$. . It is clear from the equation, it has random values in [0, 2]. Figure 47.7 shows a flow chart of GWO algorithm [11].

47.7 Result Analysis and Discussion

In this paper, the system used Grey Wolf optimizer (GWO)-tuned proportional integral derivative-structured TCSC controller that is applied SMIB system. The system is tested on various types of disturbance as nominal, light and heavy loading conditions. When various disturbances are applied system shows the high oscillatory response, but the system is tuned with GWO algorithm PID structure TCSC controller when system shows the superior result. The oscillation is damped out very fast and

Fig. 47.7 Flow chart of GWO algorithm [11]

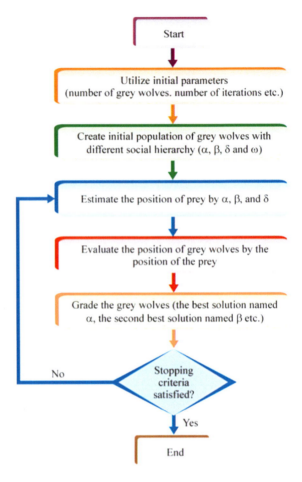

improves stability of the system. The simulation result compares with three algorithms and is tested with various types of fault applied on the system at nominal, light and heavy loading conditions. The various comparison tables are presented in tabulation form in the system and it shows the superiority of the system. Table 47.1 gives values of different loading conditions. Table 47.2 gives condition of different faults and Table 47.3 gives TCSC-based tunes PID controller parameters with GWO, CSA and NSGA-II algorithm. Figure 47.8 shows a best cost v/s iteration graph of GWO algorithm. Table 47.4 gives GWO parameters.

47 Design & Analysis of Grey Wolf Optimization Algorithm Based …

Table 47.1 Loading condition considered

Type of loading conditions	P (pu)	Q (pu)
Nominal loading	0.8	0.017
Light loading	0.5	0.006
Heavy loading	1.2	0.038

Table 47.2 Conditions of different faults

Types of fault	Conditions
ftype-1	10% step increase in references voltage setting
ftype-2	Symmetrical three-phase faults
ftype-3	10% step decrease in mechanical torque input
ftype-4	Permanent line outage disturbance

Table 47.3 TCSC-based controller parameters

Optimized algorithm	Parameters of TCSC-based controller		
	K_P	K_I	K_D
GWO [7]	500	2.177×10^{-4}	8.1260
NSGA-II [7]	29.5477	2.4866	0.1533
CSA	48.7238	4.1930	0.0010

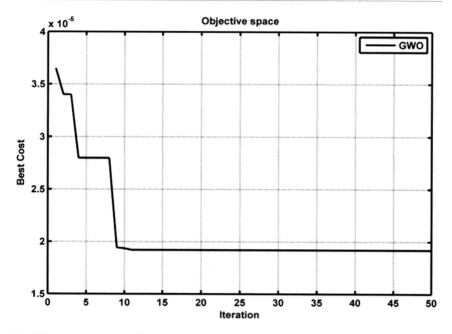

Fig. 47.8 Best cost v/s iteration graph

Table 47.4 GWO parameters

Parameters	Value
Dimensions	3
Max iteration	50
Search agent	30
Lower limit	1.0000×10^{-4}
Upper limit	500

47.7.1 Case-1: Nominal Loading at Fault Type-2 Conditions

In this condition, we apply nominal loading at fault type-2 conditions. Figures 47.9, 47.10 and 47.11 show various responses as with GWO, CSA and NSGA-II tune PID-structured TCSC-based controller. From above simulation result, we conclude that GWO algorithm defines superior response than CSA and NSGA-II algorithm in the same system. GWO, CSA and NSGA-II tuned system shows blue, black and red lines. The various graphs define and prove that GWO algorithms superior to CSA and NSGA-II algorithm enhance power system stability. Table 47.5 shows settling time for various algorithms at nominal conditions.

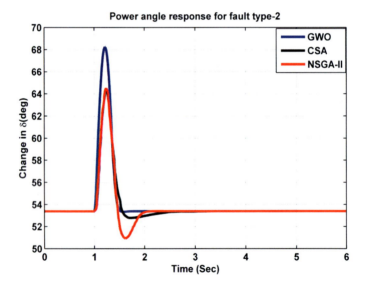

Fig. 47.9 Response of power angle for ftype-2

47 Design & Analysis of Grey Wolf Optimization Algorithm Based … 615

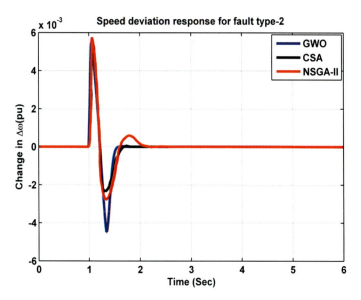

Fig. 47.10 Response of speed for fault ftype-2

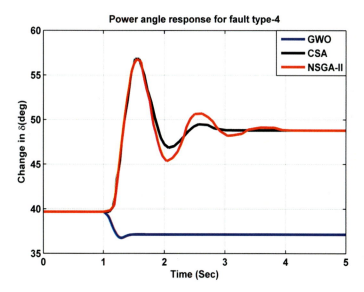

Fig. 47.11 Response of power angle for fault ftype-4

Table 47.5 Case-1 Comparison of settling time for ftype-2

Types of deviation	GWO-tuned system (Sec.)	CSA-tuned system (Sec.)	NSGA-II-tuned system (Sec.)
Power angle deviation	1.4811	2.2662	1.9493
Speed deviation	1.5291	1.6370	2.0220

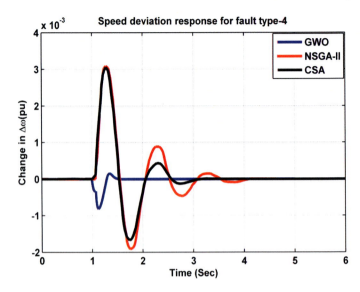

Fig. 47.12 Speed deviation response for ftype-4

Table 47.6 Case-2 Comparison of settling time for ftype-4

Types of deviation	GWO-tuned system (Sec.)	CSA-tuned system (Sec.)	NSGA-II-tuned system (Sec.)
Power angle deviation	1.4268	2.8661	3.8661
Speed deviation	1.4881	2.9605	3.9281

47.7.2 Case-2: Light Loading at Fault Type-4 Conditions

In this condition, we apply light loading at fault type-4 conditions. Figures 47.11, 47.12 show various responses as with GWO, CSA and NSGA-II tuned PID-structured TCSC-based controller. From above simulation result, we conclude that GWO-tuned algorithm response is superior to other algorithms. The comparisons of various graphs define and prove that GWO algorithms are superior to CSA and NSGA-II algorithm and improves power system stability. Table 47.6 gives settling time for various algorithms at light conditions.

47.7.3 Case-3: Heavy Loading at Fault Type-3 Conditions

In this condition, we apply heavy loading at fault type-4 conditions. Figures 47.13 and 47.14 show various responses as with GWO, CSA and NSGA-II tuned PID-structured TCSC-based controller. GWO-tuned TCSC controller gives better response than

47 Design & Analysis of Grey Wolf Optimization Algorithm Based …

other algorithms as CSA and NSGA-II. So GWO-tuned TCSC controller is an improved dynamic performance of the system. GWO, CSA and NSGA-II-tuned system shows blue, black and red lines. Table 47.7 gives settling time for various algorithms at heavy conditions.

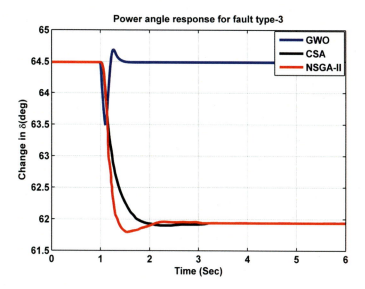

Fig. 47.13 Response of power angle deviation for ftype-3

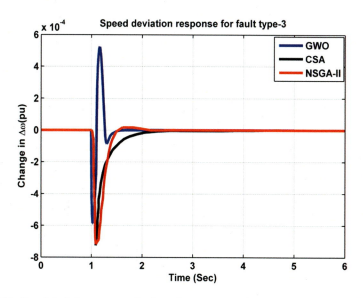

Fig. 47.14 Speed deviation response for ftype-3

Table 47.7 Case-3 Comparison of settling time for ftype-3

Types of deviation	GWO-tuned system (Sec.)	CSA-tuned system (Sec.)	NSGA-II-tuned system (Sec.)
Power angle deviation	1.4521	1.7705	1.9556
Speed deviation	1.4722	2.0663	1.8529

47.8 Conclusion

Proposed PID structure TCSC-based controller parameters are successfully optimized for SMIB power system using new evolutionary nature-inspired Grey Wolf Optimization algorithm. Proposed controller test for different disturbance and loading conditions to evaluate its performance using MATLAB/Simulink based platform. The comparative result analysis shows the power angle and speed deviation response for different faults with GWO, CSA and NSGA-II PID-tuned TCSC controller. From result analysis, it has been observed that GWO provides fast response towards various disturbances, hence it is proved that the optimal solution for PID controller obtained with GWO algorithm was found superior to the previously used algorithm.

References

1. Wang, H., Wenjuan, D.: Analysis and Damping Control of Power System Low-Frequency Oscillations. Springer, New York (2016)
2. Zhang, X.-P., Rehtanz, C., Pal, B.: In: Flexible AC Transmission Systems: Modelling and Control. Springer Science & Business Media (2012)
3. Vittal, V., McCalley, J.D., Anderson, P.M., Fouad, A.A.: In: Power System Control and Stability. Wiley (2019)
4. Mirjalili, S., Mirjalili, S.: Genetic algorithm. In: Evolutionary Algorithms and Neural Networks: Theory and Applications, pp. 43–55. (2019)
5. Marini, F., Walczak, B.: Particle swarm optimization (PSO). a tutorial. Chemom. Intell. Lab. Syst. **149**, 153–165 (2015)
6. Karaboğa, D., Ökdem, S.: A simple and global optimization algorithm for engineering problems: differential evolution algorithm. Turk. J. Electr. Eng. Comput. Sci. **12**(1), 53–60 (2004)
7. Sethi, R., Panda, S., Sahoo, B.P.: Cuckoo search algorithm based optimal tuning of PID structured TCSC controller. In: Computational Intelligence in Data Mining-Volume 1: Proceedings of the International Conference on CIDM, 20–21 December 2014, pp. 251–263. Springer, India (2015)
8. Panda, S.: Multi-objective PID controller tuning for a FACTS-based damping stabilizer using non-dominated sorting genetic algorithm-II. Int. J. Electr. Power Energy Syst. **33**(7), 1296–1308 (2011)
9. Mirjalili, S., Mirjalili, S.M., Lewis, A.: Grey wolf optimizer. Adv Eng Softw **69**, 46–61 (2014)
10. Faris, H., Aljarah, I., Al-Betar, M.A., Mirjalili, S.: Grey wolf optimizer: a review of recent variants and applications. Neural Comput. Appl. **30**, 413–435 (2018)

11. Al-Tashi, Q., Rais, H.M., Abdulkadir, S.J., Mirjalili, S., Alhussian, H.: A review of grey wolf optimizer-based feature selection methods for classification. Evolut. Mach. Learn. Techniques: Algorithms and Appl. 273–286 (2020)
12. Panda, S., Padhy, N.P.: MATLAB/SIMULINK based model of single-machine infinite-bus with TCSC for stability studies and tuning employing GA. Int. J. Energy Power Eng. **1**(3), 560–569 (2007)
13. Panda, S., Padhy, N.P.: Power system with PSS and FACTS controller: modelling, simulation and simultaneous tuning employing genetic algorithm. Int. J. Energy and Power Eng. **1**(3), 493–502 (2007)
14. Ganthia, B.P., Rana, P.K., Patra, T., Pradhan, R., Sahu, R.: Design and analysis of gravitational search algorithm based TCSC controller in power system. Mater. Today: Proc. **5**(1), 841–847 (2018)
15. Kar, P., Panda, P.C., Swain, S.C., Kumar, A: Dynamic stability performance improvement of SMIB power system using TCSC and SVC. In: 2015 IEEE Power, Communication and Information Technology Conference (PCITC), pp. 517–521. IEEE (2015)
16. Haldera, A., Palb, N., Mondalc, D.: Transient stability analysis of a multi-machine power system with TCSC controller—a zero dynamic design approach. Electr Power Energy Syst **97**, 51–71 (2018)
17. Kumar, M.D., Sujatha, P.: Design and development of self tuning controller for TCSC to damp inter harmonic oscillation. Energy Proc **117**, 802–809 (2017)
18. Mishra, D.K., Mohanty, A., Ray, P.: MATLAB/SIMULINK based FA for optimizing TCSC controller in a power system. ICACCS-2017, January (2017)
19. Taleb, M., Salem, A., Ayman, A., Azma, M.A.: Optimal allocation of TCSC using adaptive cuckoo search algorithm, IEEE (2016)
20. Deb, K., Samir, A., Amrit, P., Tanaka, M.: A fast elitist non-dominated sorting genetic algorithm for multi-objective optimization: NSGA-II. In: Parallel Problem Solving from Nature PPSN VI: 6th International Conference Paris, France, September 18–20, 2000 Proceedings, vol. 6, pp. 849–858. Springer, Berlin, Heidelberg (2000)
21. Mahato, D.P.: Grey wolf optimizer for load balancing in cloud computing. In: Research Advances in Network Technologies, pp. 205–221. CRC Press (2023)
22. Abedini, M.: A novel controller algorithm to improve stability of power system based on hybrid of fuzzy controller and gray wolf optimization by coordinating PSS and TCSC with considering uncertainty (2023)
23. Makhmudov, T.: Influence of TCSC control systems on oscillations damping. In: AIP Conference Proceedings, vol. 2552(1), pp. 040009. AIP Publishing LLC (2023)

Chapter 48
Design of an Adaptive Neural Controller Applied to Pressure Control in Industrial Processes

Lucas Vera, Adela Benítez, Enrique Fernández Mareco, and Diego Pinto Roa

Abstract In the present work, an adaptive neural controller is designed and applied to pressure control in industrial processes, implementing artificial neural networks and adding an algorithm based on adaptive interaction theory to them, with which it is possible to obtain an intelligent controller. The controller based on neural networks has adaptive properties through the new Brandt-Lin algorithm; this will allow controlling the process without a training phase and prior knowledge of the plant. Therefore, the intelligent controller can adapt online to changes in industrial processes. The challenge of this work is the implementation of this intelligent controller in a real plant (Festo MPS PA Compact Workstation), whose study will be carried out in a training kit that will simulate the industrial process to be controlled. Finally, the work will demonstrate the supremacy of the proposed controller compared with a classic PID controller, being far superior in all the simulations carried out in the control of the real plant.

Keywords Neural networks · Adaptive interaction · Adaptive control · Neural control

48.1 Introduction

Control strategies were constantly being developed, resulting in a continuous advance, going from classical control techniques in later years to modern control. At the end of the twentieth century, the artificial intelligence research line applied to control systems emerged, thus giving birth to Intelligent Control and Advanced

L. Vera · A. Benítez · E. Fernández Mareco (✉) · D. Pinto Roa
Universidad Nacional de Asunción, San Lorenzo, Paraguay
e-mail: efernandezmareco@pol.una.py

D. Pinto Roa
e-mail: dpinto@pol.una.py

Control Systems, obtaining successful results that consolidated intelligent control techniques.

The development of intelligent controllers has integrated areas such as control, communication theory, computer science, artificial intelligence, operational research, and neuroscience. Intelligent controllers have shown significant improvements in control system quality compared with conventional methods and offer advantages such as decreased resource consumption, operating times, waste, costs, and increased yields.

The main dynamic features of intelligent control systems are adaptability, flexibility, autonomy, robustness, feedback, and cooperation. These features enable the system to modify its configuration, adapt to changing conditions, make agile decisions, recover from failures, abstract information for better execution, and coordinate between agents to achieve a specific goal. The development of intelligent controllers has allowed the integration of different areas and the use of advanced techniques, resulting in improved control systems quality and advantages such as decreased resource consumption, operating times, waste, costs, and increased yields [1].

These intelligent control techniques are needed (artificial neural networks, fuzzy logic, genetic algorithms, among others) because of the progressive demand for better solutions, more efficient and automatic control compared with human intervention, and the management of complex processes, among other benefits [2].

This work presents the design and implementation of an intelligent controller based on artificial neural networks for a pressure control system for industrial processes. Using the Brandt-Lin algorithm, the proposed algorithm is based on the adaptive interaction theory, thus comparing its efficiency against a classic controller.

48.2 Intelligent Control in Industrial Processes

According to Kozak [3], we can divide the stages of developing control engineering methods into four main groups: conventional control strategies, advanced control I, advanced control II, and advanced control III.

Conventional control was used for several decades in the industrial field. The proportional–integral–derivative (PID) controller is best known for its popularity as it has a wide range of operating conditions, functional simplicity, and ease of implementation.

Among the conventional control techniques, we can find manual control, feedback control (FB), cascade control (CC), prior control (FFW), ratio control (RC), and control of composite structures (FB + FFW + CC). Then, we find the advanced control strategies, already known as classics, because they have been implemented for over 40 years in the industry. These controllers perform in only some plants, but they provide cost-effective solutions. The advanced control I strategies would be: adaptive and self-adjusting control, the gain scheme method, multivariable control methods (state space models and transfer functions), multivariable control methods (decoupling and decentralized control), pinning processes (SISO, MIMO), and nonlinear control methods (I/O linearization). The control strategies belonging to advanced control II would be optimal control methods (LQ and LQG), robust control methods (H2, Hinf, IMC), model predictive control (MPC-DMC, MPC-GPC), decentralized control (time domain, frequency domain), algebraic control methods (polynomial synthesis), robust QFT control methods.

Finally, the control strategies belonging to advanced control III would be hybrid predictive control, fuzzy control (PID, MPC, FPGA), neural network control (optimal, MPC, FPGA), discrete event control (hybrid with Petri nets), nonlinear hybrids, soft computing, and expert control methods. Advanced controls would be the most current control strategies, a recent trend in using artificial neural networks and hybrid models in their structures.

On the other hand, according to Ponce [4], the field of artificial intelligence is made up of several areas of study. The most common and important are: search for solution, expert systems, natural language processing, pattern recognition, robotics, machine learning, logic uncertainty and fuzzy logic.

More recently, techniques like neural networks, fuzzy logic, and genetic algorithm have been successfully applied with some success to problems of control and identification of nonlinear systems for several application domains. Moreover, in some works, both neural networks and advanced methods are used to get better performances [5].

Recently, much success has been achieved using neural networks (NN) to control nonlinear dynamic systems. It has been shown that NN has emerged as a successful tool in dynamic control systems [6].

It is well known that NN approximation-based control relies on universal approximation property in a compact set in order to approximate unknown nonlinearities in the plant dynamics. The widely used structures of neural network-based control systems are similar to those employed in adaptive control, where a neural network is used to estimate the unknown nonlinear system, the network weights need to be updated using the network's output error, and the adaptive control law is synthesized based on the output of networks. Therefore the major difficulty is that the system to be controlled is nonlinear with its diversity and complexity as well as its lack of universal system models. It has been proved that the neural network is a complete mapping. Therefore, an adaptive predictive control algorithm is developed to solve the problems of tracking control of the systems [7].

48.3 Design Proposal

48.3.1 Design Description

A control system is robust if it is insensitive to differences observed between the plant and the model used for controller synthesis. Every modeling procedure, either for linear or nonlinear processes, suffers from the so-called model mismatch. The model

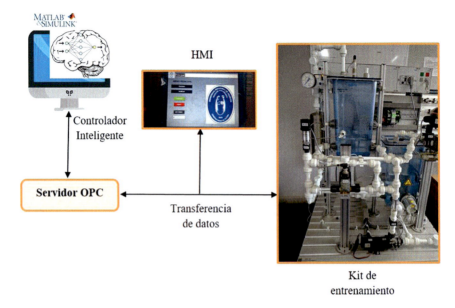

Fig. 48.1 Control system diagram. *Source* Own authorship

of the system is not a faithful replica of plant dynamics. On this basis, the uncertainty can be seen as a measure of unmodeled dynamics, noise, and disturbances affecting the plant [8]. Therefore this work involves designing a controller based on artificial neural networks and adaptive interaction theory using the Matlab/Simulink tool. In this context, the controller acquires adaptability characteristics, thus achieving an intelligent control methodology.

The developed artificial neural network will fulfill the controller's role in the control system. This intelligent controller will try to keep the pressure stable in the real plant, the training kit Festo MPS PA Compact Workstation [9], through the communication established with the Matlab software and the plant to be controlled. This communication is done through a PLC SIEMENS S7 314C-2 and an OPC server, Kepserverex, allowing the artificial neural network to control the centrifugal pump to modify the water flow, thus maintaining the pressure in the system.

Figure 48.1 presents a diagram of the developed control system. An HMI KTP700 is included so the operator can set the desired pressure values and view the pressure and centrifugal pump behavior.

48.3.2 Engineering Design

The project's design is based on four sections for its development: design, simulation, implementation, and results.

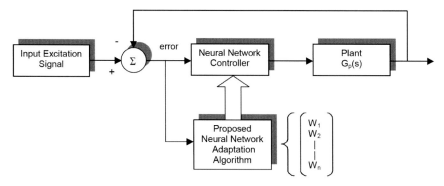

Fig. 48.2 Control system based on an artificial neural network. *Source* Saikalis et al. [15]

The control system is designed based on a closed-loop control. This system comprises two main blocks, the plant to be controlled and the artificial neural network controller (ANN).

An ANN is part of the supervised learning algorithms and can be used to model highly nonlinear systems. The network is trained iteratively, modifying the strengths of the connections so that the inputs are assigned to the correct response [10–13].

The connection weights of the ANN are adjusted by an adaptation algorithm known as Brand-Lin [14], causing the neural network to provide the optimal control signal for the system.

Figure 48.2 shows the block diagram proposed for the intelligent controller's solution, where the adaptation algorithm proposal can be seen working with the controller based on neural networks.

Figure 48.3 shows that the Simulink block diagrams consider the structure of a neural network with the inputs, the synaptic weights, the summation, and the activation function.

A general adaptation algorithm developed under adaptive interaction theory is used to adapt the synaptic weights of the driver seen in [15]. It is applied to a simple neural network structure to extract an interpretation more in line with the tool in Matlab.

The meanings of the labels placed inside Fig. 48.3 are:

$x1, x2$	the input variables to the neural network.
$w1, w2, w3, w4, w5, w6$	the weights of the corresponding neurons.
Σ	addition operation.
$a1, a2, a3$	results of sum operation.
f	activation function.
$u3, u4, u5$	outputs of the corresponding neurons.

The activation function used in this simple neural network structure is the sigmoid function.

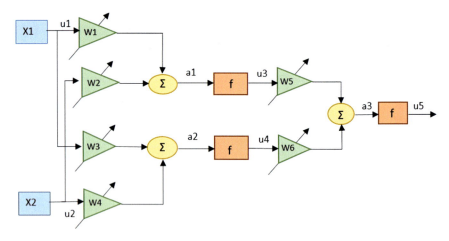

Fig. 48.3 Simple structure of a neural network. *Source* Saikalis et al. [15]

$$\sigma(x) = \frac{1}{1+e^{-x}} \qquad (48.1)$$

The neural network and the adaptation algorithm are described mathematically, as shown below. Equation (48.2) is the previous value of neuron "n" before being evaluated in the activation function corresponding to each neuron.

$$a_n = \Sigma_s w_s * u_k \qquad (48.2)$$

where

n belongs to the number of neurons,
s pertains to the number of connections per neuron,
k is the number of entries in the weights established in the structure.

In Eq. (48.3), we have the output value of the neuron "n" activation function.

$$u_n = \sigma(a_n) \qquad (48.3)$$

Now, if we establish that:

$$\phi_n = \frac{1}{2} * \frac{d}{dt} \Sigma_n w_s^2 = \Sigma_n w_s * w_s' \qquad (48.4)$$

And by applying the law of general adaptation already shown in Saikalis [15]: we obtain the following:

$$\alpha_c' = F_{\text{post } c}' * [x_{\text{post } c}] * \left(\frac{y_{\text{pre } c}}{y_{\text{post } c}}\right) * \Sigma_{S f} O_{\text{post } c} \alpha_s * \alpha_s'$$

48 Design of an Adaptive Neural Controller Applied to Pressure Control …

$$- \gamma F'_{post\ c} * \left[x_{post\ c} \right] * y_{pre\ c} * \frac{\delta E}{\delta y_{post\ c}} \quad (48.5)$$

We can get:

$$w'_s = u_{pre} \left(\phi_{post} * \sigma \left(-a_{post} \right) + \gamma * f_{post} \right) \quad (48.6)$$

where "pre" means the signal before that neuron and "post" is the signal after it.

f is the direct feedback signal and
γ is the learning coefficient.

Equations 48.4 and 48.5 describe the Brandt-Lin algorithm, allowing us to adapt the weights in neural networks. This algorithm is similar to the reverse propagation algorithm but without error propagation.

This artificial neural network has as input the error signal of the system and as output the control signal that will seek to minimize the error in the system. The sigmoid function is the activation function used in this simple neural network structure. The type of structure we will use is the neural network with two hidden neurons, but it is possible to use another, more advanced neural network to optimize performance.

Then the controller configuration will be as follows in Fig. 48.4.

The network's internal structure has three neurons, each having two connection weights self-adjusted without human intervention by the adaptation algorithm by calculating the increase of each weight in each iteration, modifying its value.

The variable k at the controller's output in Fig. 48.4 is simply a constant. This variable will amplify or reduce the output values of the control effort signal "ec"

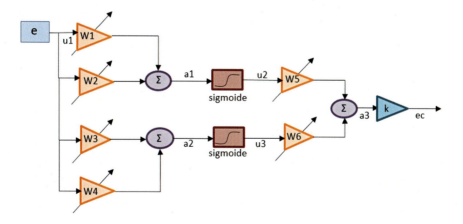

Fig. 48.4 Configuring the artificial neural network driver. *Source* Own authorship

according to the operating range required by the pump working in a range of 0–10 V DC; the value of k is defined according to studies carried out in the simulation consequent to this study.

Through mathematical expressions, we can establish the relationships between the inputs and outputs of neurons, obtaining the following equations:

$$u_1 = e \tag{48.7}$$

$$a_1 = u_1 * w1 + u_1 * w2 \tag{48.8}$$

$$a_2 = u_1 * w3 + u_1 * w4 \tag{48.9}$$

$$u_2 = \sigma(a_1) \tag{48.10}$$

$$u_3 = \sigma(a_2) \tag{48.11}$$

$$a_3 = u_2 * w5 + u_3 * w6 \tag{48.12}$$

$$ec = a_3 * k \tag{48.13}$$

We have also defined the performance index:

$$E = e^2 = (u - y)^2 = u^2 - 2 * u * y + y^2 \tag{48.14}$$

Therefore, we will have:

$$\frac{dE}{dy} = -2 * u + 2 * y = -2 * (u - y) = -2 * e \tag{48.15}$$

If we apply the Brandt-Lin algorithm of Eqs. (11) and (12), we can have the following:

$$w_1' = e * \phi_3 * \sigma(-a_1) \tag{48.16}$$

$$w_2' = e * \phi_3 * \sigma(-a_1) \tag{48.17}$$

$$w_3' = e * \phi_4 * \sigma(-a_2) \tag{48.18}$$

$$w_4' = e * \phi_4 * \sigma(-a_2) \tag{48.19}$$

$$\phi_3 = w_5 * w_5' \quad (48.20)$$

$$\phi_4 = w_6 * w_6' \quad (48.21)$$

$$w_5' = \gamma * u_2 * e \quad (48.22)$$

$$w_6' = \gamma * u_3 * e \quad (48.23)$$

The constant γ is called the learning rate, and its value is setup for studies. It allows fast learning and next to the optimal performance according to the control system that has been designed. Once all the equations that would enable us to calculate the increments that each weight will have in the neural network have been defined, including the learning rate, we will go on to design the intelligent controller under the Matlab/Simulink development software.

Figure 48.5 presents the design of the intelligent controller made in Simulink/Matlab. The connection weights of each neuron, the output amplification constant "k," and the gamma learning coefficient "γ" referring to the adaptation algorithm are also highlighted. As can be seen, the artificial neural network has the "error signal" as input and the control signal as "output."

Figure 48.6 shows the neural network within a block linked to the plant's transfer function identified with Matlab. With this, the simulations presented in the results are carried out. The structure of the neural network controller module is detailed in Fig. 48.5.

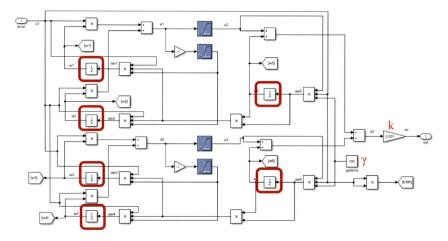

Fig. 48.5 Structure of the adaptive neural network controller. *Source* Own authorship

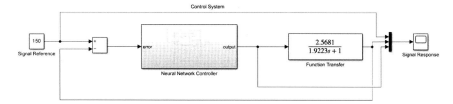

Fig. 48.6 Control system with neural controller. *Source* Own authorship

48.4 Experiment

48.4.1 Numerical Simulation

For the simulation of the intelligent controller, it is necessary to identify the plant by calculating its transfer function using the Matlab "System Identification" tool, which obtains a 67% fidelity transfer function with the dead time of the system compensated.

To evaluate the efficiency of the intelligent controller, we will carry out simulations controlling a plant, thus comparing the results obtained with two conventional PID controllers. Figure 48.7 shows the identified plant and proposed controller. In the first PID controller, PID1, the parameters are tuned using Matlab. For the second PID, PID2, the parameters will be changed by the operator's experience to perform the control simulation, as given in Table 48.1.

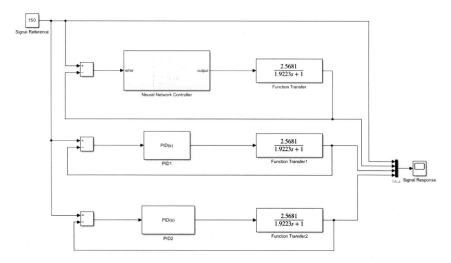

Fig. 48.7 Design of control structures for simulation considering the three controllers. *Source* Own authorship

48 Design of an Adaptive Neural Controller Applied to Pressure Control ...

Table 48.1 Parameters of conventional PID controllers

PID1 PID fitted by Matlab	PID2 PID adjusted by operator experience
$K_p = 0.51$	$K_p = 0.05$
$1/T_i = 1.12$	$1/T_i = 0.5$
$T_d = 0$	$T_d = 0.2$

Figure 48.8 presents a comparison of the system responses. The graph shows that the PID1 controller gives better results by having a shorter reaction and settling time than the other controllers but with a more significant overshoot.

Next, we show Table 48.2, where we have the types of controllers with their respective establishment times.

This result does not mean that the PID controller is superior to the intelligent controller in the system.

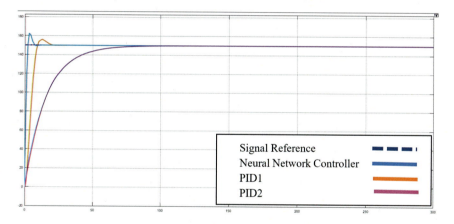

Fig. 48.8 Simulation results with all three controllers. *Source* Own authorship

Table 48.2 Settling time for each controller

Controller	Establishment time (seg.)
PID1	11
Intelligent controller	26
PID2	98

Fig. 48.9 Training Kit FESTO. *Source* CTA, Advanced Technology Center, San Lorenzo, Paraguay

48.4.2 Real Implementation

The same tests will be applied to verify the results obtained with the Matlab simulations, but implementing the controllers in a control system with the real plant is shown in Fig. 48.9.

The Festo MPS PA Compact Workstation training kit is the plant to be controlled. It has three main components for the operation of the system:

- A pressure sensor with an operating range of 0–9 bar.
- The PLC S7 314C-2PN/DP with PROFIBUS/PROFINET communication.
- The centrifugal pump operates under a 0–24 V voltage with a maximum flow of 10 l/min.

To implement the intelligent controller is necessary to establish communication between the Matlab software and the Festo MPS training kit through a communication architecture. This architecture consists of two essential elements, the programmable logic controller (PLC) and the OPC server (Kepserverex).

The PLC will take the data from the plant pressure sensor and send it through the Kepserverex software, creating a shared IP address communicating with the Matlab

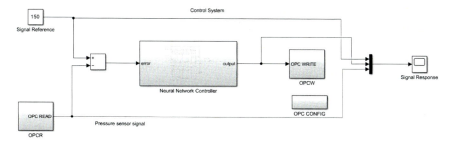

Fig. 48.10 Integrated control system with blocks OPC. *Source* Own authorship

software. Then Simulink will calculate the error referring to the setpoint to enter it in the intelligent controller to determine the control signal to correct the error.

This signal will be sent again through the Kepserverex software to the PLC to convert it into a voltage value that modifies the operation of the centrifugal pump and changes the water flow, thus correcting the pressure error in the system. Figures 48.10 and 48.11 show this scheme.

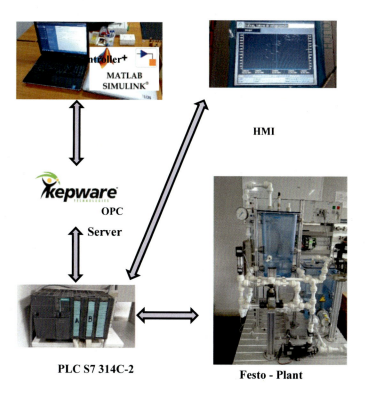

Fig. 48.11 Control system structure diagram. *Source* Own authorship

Simulink has a tool to establish OPC communication with Kepserverex through a block for data reception (OPC READ) and another block for data transmission (OPC WRITE), and the main block to establish communication with Kepware (OPC Config Real-Time).

Once the communication with the intelligent controller is set up, the control system can evaluate the controllers' efficiency and effectiveness.

The results are shown in Figs. 48.12, 48.13, 48.14, 48.15, 48.16, 48.17, 48.18 and 48.19. These show the different tests carried out, seeking to maintain a constant setpoint value, varying the setpoint with other reference signals, and adding a disturbance to the system.

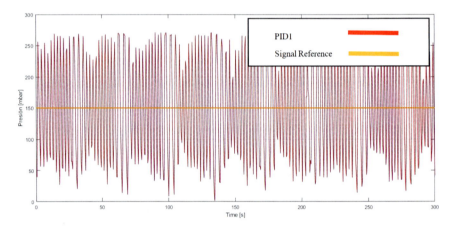

Fig. 48.12 System response implementing the PID1 controller. *Source* Own authorship

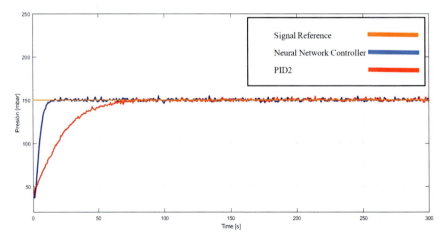

Fig. 48.13 Comparing system response with the intelligent and PID2 controllers. *Source* Own authorship

48 Design of an Adaptive Neural Controller Applied to Pressure Control ...

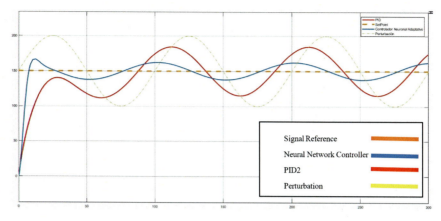

Fig. 48.14 Controller response to a disturbance. *Source* Own authorship

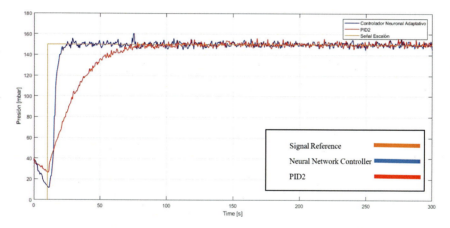

Fig. 48.15 System response to a step input signa. *Source* Own authorship

The PID controller adjusted by Matlab presents instability in the system control since the transfer function identification could have been more efficient by providing only 67% fidelity for adjusting its parameters (see Fig. 48.12). Therefore, this controller was discarded for the other tests.

The adaptive neural controller presents a better response than the PID controller adjusted by operator experience, as it has a shorter response time and settling time, as shown in Fig. 48.13.

More tests were conducted to demonstrate its effectiveness, adding a disturbance to the system, as shown in Figs. 48.14, 48.15, 48.16, 48.17, 48.18 and 48.19. The test signals introduced to the real plant are the disturbance signal, unit step signal, ramp type signal, square wave signal, sinusoidal signal, and sawtooth signal in Figs. 48.14, 48.15, 48.16, 48.17, 48.18 and 48.19, respectively.

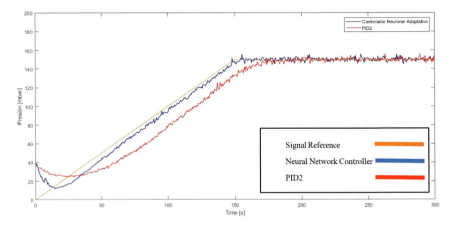

Fig. 48.16 System response to a ramp-type input signal. *Source* Own authorship

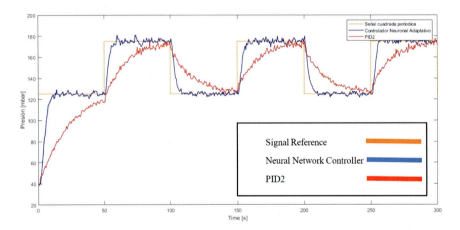

Fig. 48.17 System response to a periodic square wave-type input signal. *Source* Own authorship

The intelligent controller presents a better approximation to the desired signal for the system in all cases compared with the PID2 controller.

Finally, we can calculate each controller's average errors through all the tests carried out with both controllers. These results are summarized in Table 48.3.

Therefore, the intelligent controller proved to have greater efficiency and effectiveness in controlling the real plant, surpassing the classic PID2 controller with an average error of 1.28% against 5.13%.

On the other hand, regarding the absolute average error, it is seen that the neural network controller has a value of 4.77% against the 9.59% of the traditional controller. It is also seen that for the mean square error, the value of the intelligent controller is 304.9%, and that of the PID2 is 530.1%; the same occurs with the root of the mean

48 Design of an Adaptive Neural Controller Applied to Pressure Control ...

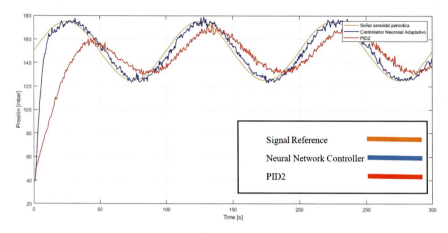

Fig. 48.18 System response to a sinusoidal input signal. *Source* Own authorship

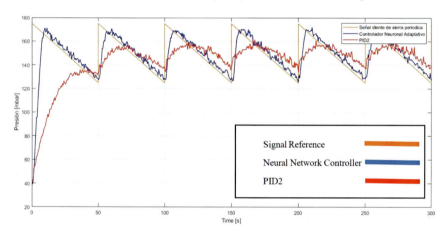

Fig. 48.19 System response to a sawtooth input signal. *Source* Own authorship

Table 48.3 Performance evaluation through error comparison

Error	Intelligent controller (%)	Controller PID 2 (%)
Mean error	1.28	5.13
Mean absolute error	4.77	9.59
Mean square error	304.9	530.1
Root mean square error	12.03	17.59

square error that has a value for the controller by neural networks of 12.03% against the PID2 of 17.59%.

Comparing both controllers, it is observed that the controller's behavior by neural networks tends to have fewer errors, especially in the variations of the process variable, which gives precision, accuracy, and reliability to the controlled system.

48.5 Conclusions

The initial purposes of the work in question are designing and implementing a conventional controller and a controller based on artificial neural networks and comparative analysis through the simulations. The method of an intelligent controller based on artificial neural networks and the implementation of the neural controller designed in a Festo Didactic—MPS® PA Compact Workstation pressure control plant allowed the initial purposes to be achieved.

The intelligent control method (adaptive neural controller) is superior and more efficient than the conventional control (PID controller), taking into account specific performance parameters of the samples extracted from the pressure control system under the plant conditions with the initial state, with disturbances, maximum and minimum pressure pressure-demand different samples of reference signals, among other controller conditions, observing a good response behavior of the system in a steady state and a fast compensation in a short form.

Through the availability of this type of pressure control system, a rapid response of the system to sudden changes in pressure is obtained, which allows us to conclude that it is a robust and highly efficient controller, summarizing lower cost and higher quality of control.

All these techniques and even the proposed controller can be developed in other programming languages and other development platforms; therefore, their investigation is recommended. Also, this controller can be implemented for embedded systems.

References

1. Bello, I.M.C., Rojas, K.A.R., Arévalo, L.E.B.: Intelligent controllers: a review of the implications of design business organizations under intelligent controller's mechanisms. Int. J. Appl. Eng. Res. **13**(18), 13744–13753 (2018)
2. Santos, M.: Un enfoque aplicado del control inteligente. Revista Iberoamericana de Automática e Informática Industrial RIAI **8**(4), 283–296 (2011)
3. Kozák, Š: State-of-the-art in control engineering. J. Electr. Syst. Inf. Technol. **1**(1), 1–9 (2014)
4. Ponce, P.: Artificial Intelligence with Engineering Applications. Alfaomega, Mexico DF (2010)
5. Nasr, M.B., Chtourou, M.: Neural network control of nonlinear dynamic systems using a hybrid algorithm. Appl. Soft Comput. **24**, 423–431 (2014)

6. Chow, T.W., Fang, Y.: A recurrent neural-network-based real-time learning control strategy applying to nonlinear systems with unknown dynamics. IEEE Trans. Industr. Electron. **45**(1), 151–161 (1998)
7. Niu, L., Zhou, S., Xie, H.: Neural network-based adaptive dynamic structure control for a class of uncertain nonlinear systems in strict-feedback form. In: 2015 8th International Symposium on Computational Intelligence and Design (ISCID), vol. 1, pp. 521–524 (2015)
8. Patan, K.: Two-stage neural network modelling for robust model predictive control. ISA Trans. **72**, 56–65 (2018)
9. FESTO Process automation, http://www.festo.com/didactic/de/ProcessAutomation. Last accessed 10 Jan 2020
10. Sendoya-Losada, D.F., Vargas-Duque, D.C., Ávila-Plazas, I.J.: Implementation of a neural control system based on PI control for a non-linear process. In: 2018 IEEE 1st Colombian Conference on Applications in Computational Intelligence (ColCACI), pp. 1–6 (2018)
11. Yoo, S.J., Park, J.B., Choi, Y.H.: Adaptive neural control for a class of strict-feedback nonlinear systems with state time delays. IEEE Trans. Neural Netw. **20**(7), 1209–1215 (2009)
12. Tiliouine, H.: A modified neural network controller configuration. IFAC Proc. Volumes **42**(13), 103–108 (2009)
13. Srivignesh, N., Sowmya, P., Ramkumar, K., Balasubramanian, G.: Design of neural based PID controller for nonlinear process. Procedia Eng. **38**, 3283–3291 (2012)
14. Brandt, R.D., Lin, F.: Adaptive interaction and its application to neural networks. Inf. Sci. **121**(3–4), 201–215 (1999)
15. Saikalis, G., Lin, F.: Adaptive neural network control by adaptive interaction. In: Proceedings of the 2001 American Control Conference (2001)

Chapter 49
A Comparative Analysis of Real-Time Sign Language Recognition Methods for Training Surgical Robots

Jaya Rubi, R. J. Hemalatha, I. Infant Francis Geo, T. Marutha Santhosh, and A. Josephin Arockia Dhivya

Abstract This project proposes a real-time robot that can interact with humans based on the gestures fed to it as input. The proposed proposal aims to develop a constructive design of a robot that has computer vision and is trained to read human gestures. There is a need for intelligent robots in the healthcare industry. The impact of this project will be on sophisticated healthcare systems, especially the surgical system. The implementation is achieved by training the robot using deep CNN and making the robot perform certain functions like moving the arm upwards and downwards as well as opening and closing the robot grippers. It is also important to mention that a comparative analysis has been made with the existing system and advanced technology called MediaPipe framework for the acquisition of input signals. The comparative analysis will give us a clear picture of the usage of different types of classifiers for training robotic models. With the impact of this project, it would be easy for the physicians to pick and place the medical equipment in a correct manner and provide assistance to the surgeon during surgery. This device can also be very useful in robot-assisted surgeries as it can be further developed to perform actions like drilling and making incisions.

Keywords Hand gesture · Complex background · Robotic surgery · Assistive robot image recognition · Image segmentation · Feature extraction

---

J. Rubi (✉) · R. J. Hemalatha · I. Infant Francis Geo · T. Marutha Santhosh ·
A. Josephin Arockia Dhivya
Department of Biomedical Engineering, Vels Institute of Science, Technology and Advanced Studies, Pallavaram, Chennai, India
e-mail: jayarubiap@gmail.com

R. J. Hemalatha
e-mail: hemalatharj@velsuniv.ac.in

A. Josephin Arockia Dhivya
e-mail: a.dhivya.se@velsuniv.ac.in

© The Author(s), under exclusive license to Springer Nature Singapore Pte Ltd. 2024
P. K. Jha et al. (eds.), *Proceedings of Congress on Control, Robotics, and Mechatronics*, Smart Innovation, Systems and Technologies 364,
https://doi.org/10.1007/978-981-99-5180-2_49

49.1 Introduction

Interaction with computers is now inevitable in the modern world. Researchers are motivated to create a robust human–computer interaction (HCI) by getting rid of unnecessary equipment like the keyboard and mouse. The human hand is thought to be one of the simplest ways for a person to interact with a computer. Its 27 degrees of freedom provide it extra flexibility while making gestures [1]. In the previous years, it was discovered that the majority of researchers used the hand as the object to be detected when gesturing, meaning that 21% of research is related to the hand [2]. It is not always viable to have a basic background in gesture recognition applications that run in real time. A quick, reliable gesture detection system is typically hindered by difficulty localizing the hand against a complicated and cluttered background. Pre-processing and segmentation, feature extraction, and classification are the three primary stages of hand gesture recognition [3]. The majority of earlier studies used a binary image of a hand to extract features. It can be turned into any language once it has been transformed using computer vision. There are many different types of research being done to develop an effective and accurate system, and the majority of these efforts are based on pattern recognition. However, the system employing a single feature is frequently insufficient, thus the hybrid approach is developed to address this issue. However, in a real-time system, we require quicker approaches to problem-solving [4]. Nowadays, we use parallel implementation to increase the processing speed of our computers. The motivation for developing this project is to improve the idea and accessibility of robotics in health care [5]. The advancement in 3D printing and advanced prosthetic technologies is also paving a way for further advancements like robotics in prosthetics [6, 7].

49.2 Literature Review

49.2.1 The Literature Survey of Some Existing Systems Is Done

1. Zhoa et al. "Real-time sign language recognition based on video stream": This proposal example demonstrates how sign language is used by deaf-dumb people around the world to communicate, making the design of a sign language recognition system extremely important and helpful for helping hearing people understand them. When it comes to real-world uses, RGB cameras rather than RGB-D cameras are used to collect video streams. A 3D-CNN approach coupled with optical flow processing to increase recognition accuracy is suggested. An optimized dense optical flow is used to filter the gathered RGB video stream before it is fed into a 3D-CNN to extract feature vectors [8].

2. Panda et al. "Hand Gesture Recognition using Flex Sensor and Machine Learning Algorithms": In this research, we present a method for flex sensors and Arduino UNO-based hand gesture identification. Traditional machine learning methods are used to assess the data collected from the sensors corresponding to various hand motions. Additionally, we provided a proposal for an adversarial learning strategy that outperforms these conventional learning techniques [9].
3. A technique for real-time static hand gesture identification was put out by Jayshree et al. A Sobel edge detector was utilized to retrieve the hand region. Utilizing the centroid and edge area, a feature vector was created. For matching motions in the final stage, the smallest Euclidean distance was applied. The system's overall recognition rate was 90% [10].
4. Signs can be recognized automatically according to Subhash et al. This feature fusion involved the scaling-invariant feature transform (SIFT), shape contexts, and HOG. Multi-SVM was employed during the recognition phase. HOG outperforms the other two in terms of these properties. By combining these features, the system's overall accuracy was 92.6% [11].
5. Recognition of hand gestures against complicated backgrounds was suggested by Pramod et al. They produced a database of 10 gestures on their own. The hand region was detected and identified from a complicated background using a Bayesian model. They blend high-level (based on shape and texture) and low-level (based on color) elements SVM which was used for classification [12].
6. Ishak's "Design and implementation of robot-assisted surgery based on the internet of things": This paper presents a controllable robotic arm via the use of IoT. Accelerometer and gyroscope are used to capture the gestures and postures of the smartphone. The signals will be captured by the android application and sent to a Raspberry Pi to control the robotic arm. Python script is employed in a Raspberry Pi to develop a program that will be able to control the robotic arm and to receive commands from the smartphone [4].

49.3 Methodology

The survey of the existing work gives a clear picture of issues related to acquiring the data as input. As we see, multiple papers have referred to acquiring the data but the processing of data gradually decreases the shape and texture of the image acquired [13]. Background noises are part of the complicated system that is sign language. Convolutional neural networks are applied to the image in an effective method to improve classification accuracy and for practical use [14]. The image was provided as input, and the fundamental processes were involved. The hand object was discovered when the camera was first turned on. The program was used to extract the feature. Later, classification and forecasting were carried out. The majority of the object detection issues use an image dataset and bounding box mapping to train the model. It costs money to label the bounding box for each image [15]. Along with that, we also put out a region of interest prediction that makes use of skin segmentation.

We cropped the image from the segmented, delimited region, and provided it to the classifier for prediction. The comparative analysis for the MediaPipe framework was also done for the same gestures. To achieve precise results, picture processing is carried out in a fully calculative and sequential manner. A crucial stage is for the camera to capture the image. It was observed that two important elements that might have an impact on the outcome were the lighting and the camera's readability [13]. In order to comprehend the convergences, classifications, and predictions, the entire picture processing process has been broken down into its component parts [16]. Within a split second of the camera is started, the picture is being captured. The data was gathered to set the camera's frame-capturing function in motion. Hue, saturation, and value (HSV) pictures were converted to BGR. This format enables us to clearly show the image [17]. The skin segmentation procedure comes next, which aids in the pre-processing step even more. The system is supplied with the trained model, which enables comparison of the input frames and gesture prediction; in addition to being a Google open-source framework, MediaPipe provides a framework for use in a machine learning pipeline. The MediaPipe framework is advantageous for cross-platform programming because it was built by manually tracking statistical data. Figure 49.1 depicts the process in which the acquisition of an image is initialized and compared with the existing dataset. The MediaPipe framework supports many audio and video formats, making it multimodal. The MediaPipe framework is used by the developer to design and analyze systems using graphs, as well as to develop systems with the aim of making appliances. The pipeline configuration is where the steps involved in the MediaPipe using system are managed [18].

49.4 Results and Discussion

Real-time sign language recognition for robotic surgery was demonstrated in this research. A program was developed using Python to recognize signs in texts using the convolutional neurulation approach (Kera's implementation). Coding for dataset capturing has been completed and put into action. The gestures produced HSV as a result [19].

Figure 49.2 gives a brief idea about the gestures that were given as input to train the model. The epoch of the gestures was varied based on different values, and the accuracy of the same was also improved. The graphical output of the same can be given below. Initially, the data was collected and placed in folders called training folder and test folder. Around 350 images of each gesture was taken to train the model. The comparison of epoch with accuracy and loss has been shown in Fig. 49.3.

Further, the same process was carried out using a MediaPipe framework and the results were obtained. This framework was much advanced and it was observed that it is not necessary to train it in order to obtain the output. The MediaPipe framework was able to identify the gestures and deliver the results within a fraction of a second. As we see, Fig. 49.4 depicts the identification of gestures and the frames per second in which it captures the image. The framework delivers the output in 18–19 frames per

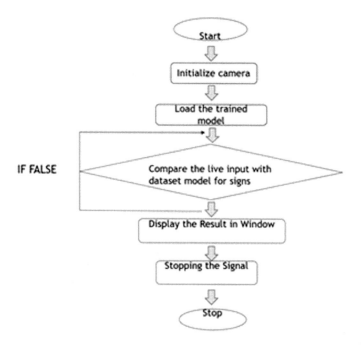

Fig. 49.1 Comparison of input images with dataset model

second. Another advantage of using a MediaPipe framework is the background. In the former method, the lighting conditions and white background played a major role, whereas in the MediaPipe framework, the background doesn't affect the quality of output. While comparing the results, it is very clear that the addition of a MediaPipe framework improves the quality of output. The framework is able to capture and detect signs much faster than the older methods.

As a result, we assign a function for each gesture and the results obtained were 92% accurate. The results can be seen below. Thus, the proposed work concludes that this gesture recognition can be further fed to the robotic system in order to train the robots to assist the doctors in the surgical environment [20].

The last two decades have seen significant advancements in the study of hand gesture recognition, which cleared the way for organic human–computer interaction systems [5]. Numerous problems remain unresolved, including those relating to the accurate identification of the gesturing phase, sensitivity to variations in speed, form, and size, and issues arising from background noises [6]. Research into hand gesture recognition is still very busy in order to occlusion. This paper provides a survey of current reports on vision- and sensor-based hand gesture recognition systems that are built on machine learning techniques [7]. Systems for recognizing hand gestures that rely on machine learning algorithms suffer from overfitting with small sample sizes and need appropriate signal pre-conditioning for effective identification. Furthermore, the classification job for the learning algorithm performs poorly with

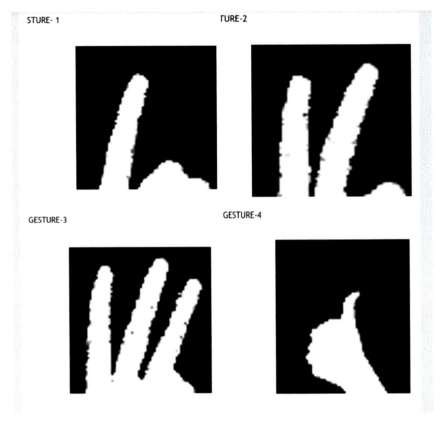

Fig. 49.2 Multiple gestures obtained to train the dataset

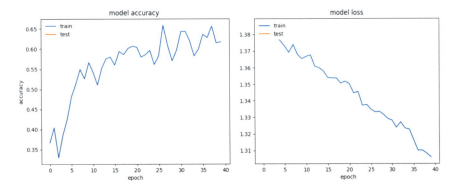

Fig. 49.3 Comparison of epoch with accuracy and loss

Fig. 49.4 Gestures identified through MediaPipe framework

untrained or unseen data. Due to its impact on recognition accuracy, many studies on gesture recognition focus on pre-processing, extracting features, and choosing the best machine learning algorithms based on the amount of data collected. The techniques for acquiring gestures, the feature extraction process, the classification of hand gestures, the applications recently put forth in the fields of sign language, robotics, and other fields, and the difficulties presented by the environment have been discussed in this paper. The future work includes the development of a robotic arm and training the arm to carry out multiple functions in the robotic environment. The multifunctional robotic arm will be very useful as the future idea is to develop an arm that doesn't need any human intervention. As experienced during the surge of COVID-19, the exclusion of manpower from the surgical environment and improvising the robotic-based surgeries will be very useful to avoid infections caused by humans [21]. The research on hand gesture recognition and improvising the classifiers for improving the gestures will be carried on to build an efficient and effective human–machine interaction machinery.

References

1. Sheenu, G.J., Vig, R.: A multi-class hand gesture recognition in complex background using sequential minimal optimization. In: 2015 International Conference on Signal Processing, Computing and Control (ISPCC), pp. 92–96 (2015). https://doi.org/10.1109/ISPCC.2015.7375004
2. Karam, M.: A Framework for Research and Design of Gesture-Based Human Computer Interactions, PhD. Thesis. University of Southampton (2006)
3. Rautaray, S.S., Agrawal, A.: Vision based hand gesture recognition for human computer interaction: a survey. Artif. Intell. Rev. **43**(1), 1–54 (2012)
4. Ishak, M.K., Mun Kit, N.G.: Design and implementation of robot assisted surgery based on internet of things (IOT). In: International Conference on Advanced Computing and Applications (ACOMP) (2017)
5. Rubi, J., Kanna, K.R., et al.: Bringing intelligence to medical devices through artificial intelligence. In: Recent Advancements in Smart Remote Patient Monitoring, Wearable Devices, and Diagnostics Systems, pp. 154–168 (2023). https://doi.org/10.4018/978-1-6684-6434-2.ch007

6. Rubi, J., Dhivya, A.J.A.: Wearable health monitoring systems using IoMT. In: The Internet of Medical Things (IoMT), pp. 225–246 (2022). https://doi.org/10.1002/9781119769200.ch12
7. Rubi, J., Hemalatha, R.J., Dhivya, J.A., Thamizhvani, T.R.: 3D printed eco friendly smart prosthetic arm with rotating wrist. Indian J. Publ. Health Res. Dev. **10**(5), 825 (2019). https://doi.org/10.5958/0976-5506.2019.01117.3
8. K. Zhao, K. Zhang, Y. Zhai, D. Wang, J. Su, Real-time sign language recognition based on video stream. In: 2020 39th Chinese Control Conference (CCC), pp. 7469–7474 (2020). https://doi.org/10.23919/CCC50068.2020.9188508
9. Panda, A.K., Chakravarty, R., Moulik, S.: Hand gesture recognition using flex sensor and machine learning algorithms. IEEE-EMBS Conf. Biomed. Eng. Sci. (IECBES) **2021**, 449–453 (2020). https://doi.org/10.1109/IECBES48179.2021.9398789
10. Pansare, J.R., Gawande, S.H., Ingle, M.: Real-time static hand gesture recognition for American Sign Language (ASL) in complex background. J. Sign. Image Process. **3**, 364–367 (2012)
11. Agrawal, S.C., Jalal, A.S., Bhatnagar, C.: Recognition of Indian sign language using feature fusion. In: IEEE Proceedings of 4th International Conference on Intelligent Human Computer Interaction, pp. 1–5 (2012)
12. Pisharady, P.K., Vadakkepat, P., Loh, A.P.: Attention based detection and recognition of hand postures against complex backgrounds. Int. J. Comput. Vis. **101**, 403–419 (2012)
13. Zhao, X., Zhu, Z., Liu, M., Zhao, C., Zhao, Y., Pan, J., Wang, Z., Wu, C.: A smart robotic walker with intelligent close-proximity interaction capabilities for elderly mobility safety. Front. Neurobot. **14** (2020)
14. Sagitov, A., Gavrilova, L., Tsoy, T., Li, H.: Design of simple one arm- surgical robot for minimally invasive surgery. In: 12th International Conference on Development in E-System Engineering (DeSE) (2019)
15. Miao, Y., Jiang, Y., Muhammad, G.: Telesurgery robot based on 5G tactile internet. Mob. Netw. Appl. **23**, 1645–1654 (2018)
16. de Smet, M.D., Gerrit, J.L, Faridpooya, K., Mura, M.: Robotic-assisted surgery in ophthalmology. Curr. Opin. Ophthalmol. **29**(3), 248–253 (2018)
17. Jason, D., Wright, M.D.: Robotic assisted surgery (Balancing evidence and implementation). JAMA **318**, 1545 (2017)
18. Jin, J., Chung, W.: Obstacle avoidance of two-wheel differential robots considering the uncertainty of robot motion on the basis of encodes odometry information. Sensors **19**(2), 289 (2019)
19. Hamza, K.K., Zhang, X., Mu, X., Odekhe, R., Bala Alhassan, A.: Modeling and stimulation of transporting elderly posture of multifunctional elderly-assistance and walking-assistant robot. In: 8th International Conference on CYBER Technology in Automation Control and Intelligent Systems (CYBER) (2018)
20. Ahmed, S.F., Ali, A., Kamran Joyo, M., Rehan Fahad, M., Siddiqui, A., Bhatti, J.A., Liaquat, A., Dezfouli, M.M.S.: Mobility assistance robot for disabled persons using electromyography sensor. In: IEEE International Conference on Innovative Research and Development (ICIRD) (2018)
21. Dhivya, A.J.A., Rubi, J., Hemalatha, R.J., Thamizhvani, T.R.: Drone—an assistive device for aquacare monitoring. In: Peng, S.L., Hsieh, S.Y., Gopalakrishnan, S., Duraisamy, B. (eds.) Intelligent Computing and Innovation on Data Science. Lecture Notes in Networks and Systems, vol. 248. Springer, Singapore (2021). https://doi.org/10.1007/978-981-16-3153-5_26

Chapter 50
Design and Development of Rough Terrain Vehicle Using Rocker-Bogie Mechanism

Vankayala Sri Naveen, Veerapalli Kushin, Kudimi Lohith Kousthubam, Kudimi Lokesh Nandakam, R. S. Nakandhrakumar, and Ramkumar Venkatasamy

Abstract In the scientific community, there is a lot of interest in studying Mars with tiny meanderers that can travel great distances and carry a few scientific instruments. In order to find instruments against outcrops or free shakes, scan a region for an example of interest, and gather rocks and soil tests for return to Earth, such meanderers would travel to locations that were separated by a few kilometers. Within the mission's constraints of mass, power, volume, and cost, our research objective is to develop innovations that make such situations possible. For data on the planet's climatic history, fixed-landers will provide excellent, logical information about the air and dirt. As we are executing the damper suspension to diminish the vibrations brought about by the meanderer when it is moving or moving all over the world, the wanderer can convey payload more than 10kg upon its back, we carried out the pick and spot arm to pick the examples for the lab research, we executed the rocker-bogie system as the wanderer can move all over the planet in any landscape.

V. Sri Naveen (✉) · V. Kushin · K. L. Kousthubam · K. L. Nandakam · R. S. Nakandhrakumar · R. Venkatasamy
Department of Mechatronics Engineering, Centre for Automation and Robotics (ANRO), Hindustan Institute of Technology and Science, Chennai, Tamil Nadu 603103, India
e-mail: 21130038@student.hindustanuniv.ac.in

V. Kushin
e-mail: 21130043@student.hindustanuniv.ac.in

K. L. Kousthubam
e-mail: 21130048@student.hindustanuniv.ac.in

K. L. Nandakam
e-mail: 21130049@student.hindustanuniv.ac.in

R. S. Nakandhrakumar
e-mail: rsnkumar@hindustanununiv.ac.in

R. Venkatasamy
e-mail: ramkv@hindustanuniv.ac.in

Keywords Rocker-bogie · Manipulator · Mars

50.1 Introduction

Over the past investigation and advancements, the rocker-bogie suspension system is widely utilized for different working comes and models in light of its prevalent vehicle security since it will back off the harsh territories essentially. This sort of component will oppose the mechanical disappointments that return on account of brutal or lopsided surfaces. The Harsh lot vehicle planned abuse rocker-bogie component is typically delayed in speed in light of the fact that the fundamental target of abuse of this system is to accomplish high security [1]. The vehicle is provided with contrasting kinds of electrical sensors, and gadgets what's more, microcontrollers all together that the authorized individual will get significant time data of line regions and may act subsequently. The primary benefit of this vehicle is that we will actually want to downsize the life hazard of our officers and increment the power of armed force all together that the weaponry of our nation is vigorous and furthermore the military can take the productive endowments of science and cutting edge innovation. In this day and age, there is partner degree expanding would like versatile robots that square measure ready to work in partner degree unstructured or then again unforgiving air with a very lopsided lot. These robots square measure utilized for those errands that people can't do and that don't appear to be protected. Among these portable frameworks, it is the rocker-bogie mechanical framework to counter repulsive force effect, and NASA and the response impetus research center have consolidated and fostered an instrument known as the rocker-bogie suspension style has become a tried quality application remarkable for its prevalent vehicle steadiness [2]. The term rocker' depicts the shaking component of the connections and joints of the mechanical framework by determination of changed differential. The body plays a critical job to deal with the commonplace pitch point of every rocker by allowing every rocker to move according to the case (see Fig. 50.1). Wheel connected to each completion. Intruders were to stacking as tracks of armed force tanks as idlers appropriating the heap. Intruders were very utilized on the trailers of semi-big rigs as each time, the trucks were constrained to convey the part of the heavier burden [3] (see Fig. 50.2). There are 2 critical advantages to the ongoing component. The essential endowments region unit that the wheels strain on the base is going to be equilibrated. The subsequent benefit is that while rising over difficult, lopsided land parcels, every one of the six wheels can ostensibly remain associated with the surface what's, underneath load, driving the vehicle over the land parcel. Like rocks which are presumably up to doubly the wheel breadth in size while holding each of the six wheels on the least. Like any mechanical apparatus, the lean balance is limited through the pinnacle of the center of gravity (see Fig. 50.2). During this undertaking, we will more often than not could join a camera to structure it as extra supportive and affordable. The rocker-bogie uncommon (see Fig. 50.3) was intended to be utilized at steady speeds [4]. Fit for hybrid impediments that rectangular degree request of

Fig. 50.1 Damper suspension had been used in the robot to avoid the vibrations caused by the robot while moving around the terrains

a wheel. Nonetheless, it is far from conquering an enormous hybrid, the movement of the vehicle productively stopped though the front wheel climbs the impediment. Once in activity at low speed [5] (more than 10 cm/s), unique shocks are exceptionally (see Fig. 50.4) a decent arrangement diminished. For several planetary missions, wanderers can be constrained to perform at human-stage speeds (~ 1 m/2 d). Shocks resulting from the impact of the front wheel contrary to partner deterrent could harm the payload or the vehicle. This depicts a way through which of utilizing a rocker-bogie vehicle just so it'll adequately step over most boundaries as opposed to affecting and developing over them [6]. The vast majority of the advantages of this framework (see Fig. 50.5) are consistently achieved with practically no mechanical alteration to current styles—exclusively a change sufficient technique. A few mechanical changes in rectangular measure encouraged amassing the most profit and proper development of the common spelling functional speed of future wanderers [7] (see Fig. 50.6).

50.2 Material and Design Analysis Mars Rover

See Figs. 50.2, 50.3, 50.4, 50.5 and 50.6.

50.3 Analysis of Circuit Design

In the circuit design of the robot, we have used six DC motors of 20 V capacity, we are using the Arduino MEGA board to connect the motors, sensors, and soil detectors.

Fig. 50.2 Rocker-Bogie mechanism link had been used in the rover to overcome the obstacles while moving on the un even terrains

Fig. 50.3 In this figure, we can see the diameter and radius of the wheel which are helpful to move the robot in un even terrains

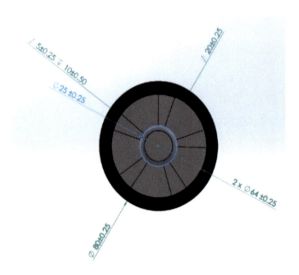

We installed the ultrasonic sensors to avoid obstacles while the robot is in the movement on the terrain, and we installed the moisture sensor to know the conditions of the soil present on the terrain [8] (see Fig. 50.7).

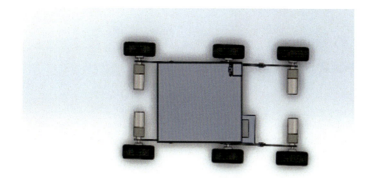

Fig. 50.4 Side view of the assembled robot were we assembled all the parts including circuit design and we included the dimensions of the chassis

Fig. 50.5 In this figure we come to see the manipulator which is used to pick and place the obstacles for the further research work in the laboratory were the obstacles are carried upon the rover

50.4 Mathematical Modeling of DC Motor

Speed Equation

$$\omega = \text{Angular velocity can also be used as } \theta$$
$$\text{as} = \omega$$

$$\omega = V * K \tag{50.1}$$

Fig. 50.6 In this figure, we are able to see the fully assembled rover with the manipulator placed on the rover to pick and place the wanted obstacles

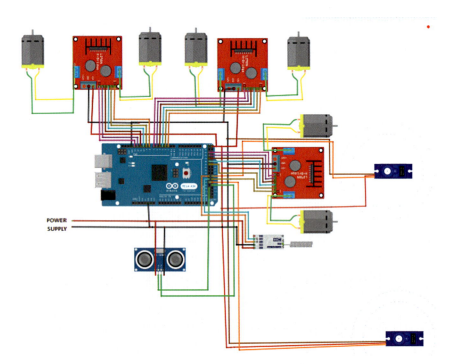

Fig. 50.7 Circuit design of the robot

so,

$$V = 1/Kv * \omega \tag{50.2}$$

As 50.1 and 50.2,

$$\frac{1}{Kv} = Ke$$

Therefore,

$$Ve = Ke.\omega$$
$$(Ve = \text{back emf generated})$$

Torque equation:

As voltage is directly proportional to speed, the current is directly proportional to torque.

$$T = Kt.I$$
$$(T \text{ is the torque and } Kt \text{ is torque constant})$$

As to remove the proportionality symbol, the torque constant Kt is been introduced. These equations show the relationship between voltage, speed and back emf.

The variations in speed, load, and torque are depicted in this graph. As the speed increases, the load decreases. This demonstrates that while a lower load draws less current, the voltage remains the same as the rated voltage. The motor stops turning and draws the most current when the load exceeds the rated load, which results in an increase in current. The stall torque is the name given to this state [9, 10].

The motor has been chosen based on these considerations, tested in this equation to determine the constants, and MATLAB has solved the equation.

50.5 Results of Robot

The output is a sine wave in which the voltage is continuously varied, and the resulting graph shows the ratio of torque (N/M) to speed (RPM). The graph shows that where there is more torque, there is less speed [11]. Where there is more speed, there is less torque. The PID and motor modeling are used to create the complete motor model. To make the AGV run, this model needs to be used with four motors that are synchronized. The Encoders' feedback will be used by the PID algorithm to continuously determine the motor's speed and maintain it. The idea of Master–Slave has been used, with one controller serving as master and the others as slaves [12].

Tests prove when the load is increased the speed of the motor reduces. In order to compensate with that PID controller has been chosen to be used and compute the speed and maintain the speed in the suitable conditions of the motor, i.e., working under the rated conditions of the motor. The output may be uncertain if the load applied is more than that of what this motor can handle. The load value is been used within the range of the motor and been simulated [13]. The V, i.e., voltage is the given input and the out is the torque and the speed of the motor been measured and seen through scope. As the load increases, the speed decreases and that is been given to the P controller to maintain the speed of the motor (see Fig. 50.8).

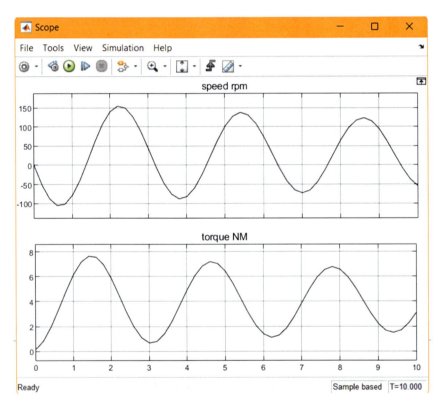

Fig. 50.8 In this picture, we can see that hoe the DC motor speed and torque are calculated with the help of Simulink software

50.6 Final Model

In this final model, we are going to see how the rover is going to be controlled with the help of this block diagram we can understand briefly [14] (see Fig. 50.9).

- The power will be supplied from the power supply unit to the main control unit.
- The main control unit is the head of the block diagram.
- The main control will pass the power supply to the ultrasonic sensors which are placed on the robot to avoid the obstacles.
- The main control unit will pass the power supply to the differential steering control to steer the robot.
- The differential steering control will pass the signal to steering motors which are placed on the robot to steer.
- The main control unit will pass the signal to the motor drivers to run the motor.
- The motor driver is used to drive the wheel motors and actuator to run.
- The main control unit will pass the signal to the pan/tilt control system, robotic arm control system, and as well as for the gripper turret.

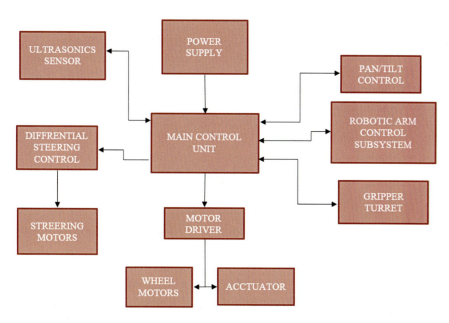

Fig. 50.9 Block diagram of the rover it is also known as main circuit of the rover

50.7 Conclusion and Future Work

The aim of this project was to create a robot that could be used to carry the specimen for test on planet mars so that it would be able to get the specimen where human beings can't go into a particular place or territory. Using robots also reduces human effort and can also minimize the need for manpower age of robots is increasing day by day with the advancements in technology and it is really important for us to keep up with it. As said "Modern problems require solutions [15]", hence using our proposed robot will really help increase our chances of reducing the unwanted materials to be carried out to the laboratory.

The administration administrator robot can stream without all in all a piece of a tune, getting photos, and sending them from a distance, at that factor, the fighters give counsel around the risks and conditions inside the discipline of war. The robot developments depend upon the motors, which can be dependent on the data we convey about the transmitter. RF signals are used to oversee cautions. By utilizing these characters, the coding is performed and the sign is dispatched by means of the source [16]. At the recipient stop, this decoded pennant is given as a pledge to the force of the motors. The mechanical is used for short separation and close by these strains guarantees the prosperity of the domain. This makes the powers see precisely the exact thing happening inside the enveloping district furthermore, set it up because it should. With the help of this proposed progression, there is a couple of help for our well-being controls in the area of intruders. This automated design can similarly be applied in exorbitant top domains where it's far more challenging for human creatures, as a trait of our edges falls into high-height locales. The proposed computerized design can in like manner be applied inside the look for the hurt individuals amid debacles.

References

1. Hirose, S., Ootsukasa, N., Shirasu, T., Kuwahara, H., Yoneda, K.: Central contemplations for the plan of a planetary meanderer. In: Proceedings of 1995 IEEE Global Meeting on Mechanical technology and Mechanization, vol. 2, pp. 1939–1944. IEEE (1995)
2. Xu, Y., Lee, C.: Earthy colored HB. A detachable blend of wheeled wanderer and arm mechanism:(DM)/sup 2. In: Proceedings of IEEE Global Gathering on Mechanical technology and Robotization, vol. 3, pp. 2383–2388. IEEE (1996)
3. Pedersen, L., Kortenkamp, D., Wettergreen, D., Nourbakhsh, I., Korsmeyer, D.: An overview of room mechanical technology. In: ISAIRAS 2003 Blemish 17
4. Fuke, Y., Apostolopoulos, D., Rollins, E., Silberman, J., Whittaker, W.: A model motion idea for a lunar mechanical pioneer. In: Proceedings of the Savvy Vehicles' 95. Conference, pp. 382–387. IEEE (1995)
5. Mill operator, D.P., Lee, T.L.: Fast crossing of unpleasant landscape utilizing a rocker-bogie versatility framework. In: Space 2002 and Advanced mechanics, pp. 428–434 (2002)
6. Yoshida, K., Hamano, H.: Movement elements of a meanderer with slip-based foothold model. In: Proceedings 2002 IEEE Global Gathering on Advanced Mechanics and Computerization (Feline. No. 02CH37292), vol. 3, pp. 3155–3160. IEEE (2002)

7. Lamon, P., Krebs, A., Lauria, M., Siegwart, R., Shooter, S.: Wheel torque control for a rough terrain rover. In: IEEE International Conference on Robotics and Automation, 2004. Proceedings. ICRA'04, vol. 5, pp. 4682–4687. IEEE (2004)
8. Tarokh, M., McDermott, G.J.: Kinematics displaying and examinations of enunciated wanderers. IEEE Exchanges Adv. Mech. **21**(4), 539–553 (2005)
9. Mann, M.P., Shiller, Z.: Dynamic strength of a rocker bogie vehicle: longitudinal movement. In: Proceedings of the 2005 IEEE Global Gathering on Advanced Mechanics and Computerization, pp. 861–866. IEEE (2005)
10. Lindemann, R.: Dynamic testing and recreation of the mars investigation meanderer. In: International Configuration Designing Specialized Meetings and PCs and Data in Designing Gathering, vol. 47438, pp. 99–106 (2005)
11. Lindemann, R.A., Bickler, D.B., Harrington, B.D., Ortiz, G.M., Voothees, C.J.: Mars exploration rover mobility development. IEEE Robot. Autom. Mag. **13**(2), 19–26 (2006)
12. Yu, X., Deng, Z., Tooth, H., Tao, J.: Research on headway control of lunar wanderer with six chamber tapered wheels. In: 2006 IEEE Global Meeting on Advanced mechanics and Biomimetics, pp. 919–923. IEEE (2006)
13. Thianwiboon, M., Sangveraphunsiri, V.: Foothold control for a rocker-bogie robot with wheel-ground contact point assessment. In: Robot Soccer World Cup, pp. 682–690. Springer, Berlin (2005)
14. Deng, Z., Tooth, H., Dong, Y., Tao, J.: Research on wheel-strolling movement control of lunar wanderer with six chamber cone shaped wheels. In: 2007 Worldwide Meeting on Mechatronics and Computerization, pp. 388–392. IEEE (2007)
15. Bai-Chao, C., Rong-Ben, W., Lu, Y., Li-Sheng, J., Falsehood, G.: Plan and reenactment research on another kind of suspension for lunar wanderer. In: 2007 Global Discussion on Computational Knowledge in Advanced mechanics and Mechanization, pp. 173–177. IEEE (2007)
16. Gu, K., Wang, H., Zhao, M.: The dissect of the impact of outer unsettling influence on the movement of a six-wheeled lunar meanderer. In: 2007 Worldwide Meeting on Mechatronics

Chapter 51
Development of Swarm Robotics System Based on AI-Based Algorithms

Aniket Nargundkar, Shreyansh Pathak, Anurodh Acharya, Arya Das, and Deepak Dharrao

Abstract In the recent past, swarm robotics technology has been widely applied in the variegated industrial domains. It essentially incorporates the multi-robot system with robots communicating with each other. In this work, a swarm of 30 robots is considered for warehouse management applications. A particle swarm optimization (PSO)-based swarm robotics system is manually designed and then simulated. The goal of the swarm is to employ the particle swarm optimization technique to efficiently complete loading and unloading duties in a warehouse. The swarm of robots that communicate with one another via radio frequency (RF) communication is subjected to particle swarm optimization. The physical prototype is built with two robot system equipped with sensors such as infra-red (IR), ultrasonic; motor drivers and RF communication unit. The major purpose of this project is to replace conventional approaches for finding the shortest path, such as the A* algorithm and Djikstra algorithm, with particle swarm optimization in order to load and unload items quickly and without the involvement of people. The physical prototype with sensors and RF communication demonstrates the feasibility of the proposed approach and provides a basis for further experimentation and improvement. Further research and experimentation are necessary to address the challenges and limitations of swarm robotics in warehouse management.

A. Nargundkar (✉) · S. Pathak · A. Acharya · A. Das · D. Dharrao
Symbiosis Institute of Technology, Symbiosis International (Deemed University), Lavale, Pune, India
e-mail: aniket.nargundkar@sitpune.edu.in

S. Pathak
e-mail: shreyansh.pathak.btech2019@sitpune.edu.in

A. Acharya
e-mail: anurodh.acharya.btech2019@sitpune.edu.in

A. Das
e-mail: arya.das.btech2019@sitpune.edu.in

D. Dharrao
e-mail: deepak.dharrao@sitpune.edu.in

Keywords Swarm robotics · Particle swarm optimization · RF communication · Obstacle avoidance · Warehouse management system

51.1 Introduction

There is an increased need for change in the way things work in this rapidly changing world of ongoing change and growing human workload. Over the past ten years, industries have risen rapidly, which has raised demand for workers. There is a persistent shortage of labor due to the growing demand for laborers and the requirement for attention to detail. The world is entering into the realm of automated systems [1]. Humans must provide the bare minimum, and it makes the barest mistakes [2]. A big coordinated collection of animals working together toward a common objective is referred to as a swarm. In the considered work, a collection of coordinated robots working toward a goal constitutes the "creatures" in the definition of swarm [3]. Finding a route from point A to point B to military applications is only a few examples of possible objectives. Swarm robots have countless and innumerable uses. Algorithms can be used to instruct the swarm to create or even destroy objects. The particle swarm optimization (PSO) method is one of the primary algorithms used in swarm robotics. Since it is utilized in searching strategies, it is crucial in swarm robotics. Swarm robots also employ additional techniques, such as ant colony optimization (ACO) and glowworm swarm optimization (GSO) [4]. It focuses on biologically inspired robots that mimic and follow the behavior of a wide variety of species and flying systems. The program is based on natural animal behavior patterns and algorithms. To mimic swarm behavior, a development technique called particle swarm optimization is used [5]. Some of the applications are fire and rescue services, a conservation organization, and bridge inspections are key areas of focus. The above-mentioned areas are being investigated, as well as the existing barriers and tools used in these sectors, as well as the need for swarm robots in these sectors [6]. To find the right time to plant and produce crops, many factors such as climate, humidity, and temperature must be taken into account. These conditions and factors are important in increasing crop production [7]. Also, the efficient use of intelligent sensors, communication, and organizational performance of small robots, allowing for information, performance, and information from the environment interactively, is the key to a successful swarm robotics app [1]. All physical contact between swarm bots plays a major role in completing many engagement tasks [8]. Dixit et al. [9] presented an application of AI-based smart robotics system for intelligence warehouse automation.

The particle swarm optimization algorithm will be used in the subsequent paper since it incorporates strategies for path discovery, navigation, and particle localization [10, 11]. Various studies have been performed for path planning and robot movement. Chaudhary et al. [12] designed the hybrid algorithm combining ACO algorithm and kinematic equations of the robot for solving robot navigation problem. The robot navigation problem, particularly the path planning problem that includes fixed obstacles was solved using neural networks by Chaudhary et al. [13]. Tawhid and Ibrahim [14] proposed a new hybrid swarm intelligence optimization algorithm referred to as Monarch Butterfly Optimization (MBO) algorithm with Cuckoo Search (CS) algorithm, named MBOCS and applied for solving nonlinear systems and clustering problems. PSO is being used to address a challenge in warehouses [15, 16]. The issue is with loading and unloading the goods at the proper locations. Since humans are involved, the task is prone to mistakes and moves slowly. The swarm robots are used to address this issue. The swarm bots p to capture some of the ants 'properties and properties of the ants' integration and transfer them to the robot system [17]. The objective of this work is to create a swarm of robots, creating an environment of warehouse with obstacles for simulation purposes, and making line follower robots that avoid obstacles and reach the target point. Robots will follow the path and emulate the loading and unloading of objects in warehouses with the implementation of PSO.

The remainder of the paper is organized as: Sect. 51.2 describes the methodology adopted, hardware, and simulations carried out. In Sect. 51.3, the results are discussed. The conclusions and future directions are mentioned in Sect. 51.4.

51.2 Methodology

This section discusses about the different procedures and processes adopted in our project workflow. PSO simulations were initially done on MATLAB, but then shifted to JAVA to get a better understanding of the results. The objective was to make a swarm of robots that perform line following and obstacle avoidance with supremacy. The objective of JAVA was to perform PSO simulations and find the closest path vector. The model was prepared using certain hardware and the coding was done on JAVA with a function created. Coding was done on Arduino to test the motors and their actual response with sensors and ultrasonic sensors. Global minimums for particles were found using PSO algorithm. All this is shown in Fig. 51.1.

51.2.1 Hardware

After going through the research papers, it was decided to make the model of swarm robots that can do the tasks of loading and unloading efficiently without much trouble. It was decided to make a line follower robot with obstacle avoidance on a path made by us on a chart paper. Modeling was then done through the following steps:

(1) Firstly, Arduino UNO was used to program the functionality of the robot.
(2) Chassis of wood was made as shown in Fig. 51.3.
(3) Motors were interfaced with Arduino.
(4) The IR sensors were used and interfaced with Arduino to make the robot detect lines.

Fig. 51.1 Methodology

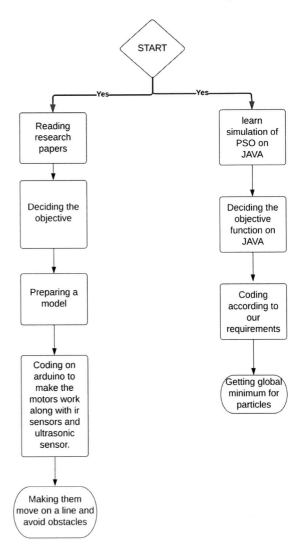

Initially, the Arduino UNO is connected to the IR sensor and motors and to the ultrasonic sensor using Arduino IDE. The model was tested along these lines. The following images depict how we went about coding and making the circuit for the same as shown in Fig. 51.2.

In place of Arduino Nano, Arduino UNO was used. Rest of the connections are the same. Also, 9 V DC batteries were used for power supply. In place of Arduino Motor Driver, we use Arduino UNO and basic l298 N motor driver as shown in Fig. 51.3.

51 Development of Swarm Robotics System Based on AI-Based Algorithms

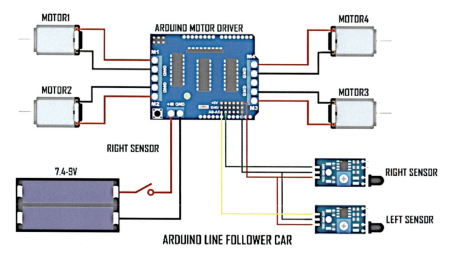

Fig. 51.2 Probability distribution

Fig. 51.3 Model

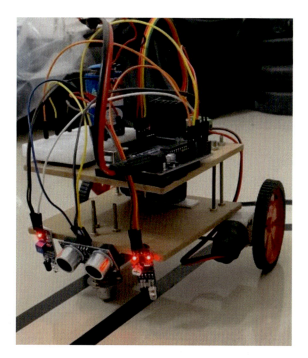

51.2.2 Simulations

The following three factors are involved in the movement of particle in the space:
1. Velocity of the particle
2. Personal best position of the particle
3. Global best position

Position vector of the particle at time t is denoted by x_i and the velocity vector of the particle is denoted by V_t.

The formula for position vector of the particle at time $(t + 1)$ is

$$x_{(t+1)} = V_{i(t+1)} + x_i$$

And the formula for the velocity vector at time $(t + 1)$ is.

$$V_{i(t+1)} = \omega V_{i(t)} + cr_1\left(y_{(t)} - x_{i(t)}\right) + c_2 * r_2\left(z_{(t)} - x_{i(t)}\right)$$

where
ω is the inertia coefficient, the value is set at 0.72984.
c_1, c_2, the values are set at 1.496.
r_1 and r_2 are constant vectors which have values between 0 and 1.

These values are determined based on past experiments and trials and expected to perform well and lead to convergence. Assuming the number of bots the swarm to be 30, the number of inputs being 30 resulting in 30 dimensions. Further, the position, velocity, and the personal best of the particle were determined. We assume the velocity of every particle to be 0 before starting iterating through the particle. Until we reach convergence, we keep iterating through the particles.

51.3 Results and Discussions

The algorithm was run with 30 dimensions, 30 particles, $\omega = 0.729844$, $c1 = c2 = 1.496180$. With this setup, the value is converged to ~ 82.489428976216 as given in Table 51.1. PSO requires a lot of testing, trial, and error methods to reach the suboptimal solution. A swarm of robots was created and it was successfully found out to be following a line and avoiding obstacles while following the PSO.

51 Development of Swarm Robotics System Based on AI-Based Algorithms

Table 51.1 Convergence values for PSO simulations

S. No.	Convergence value
1	102.52845926233425
2	81.19541623817724
3	87.9634888001822
4	70.6895391687537
5	91.38930966131781
6	123.5072632696351
7	75.40357127271108
8	69.37174188500686
9	79.92347551599096
10	116.96912049529931
11	76.74131049711764
12	59.141989638172646
13	61.383036809346976
14	57.437271564438134
15	70.91723807575974
16	85.78899427990058
17	89.33882778528849
18	108.38902964558355
19	54.38386359086408
20	87.32563206844756

Figure 51.4 shows how the particles look in a 2D plane and what their best positions are with respect to the global reference.

Figure 51.5 shows the 3D plot of the particles in space and how they will be visible in space at their best positions.

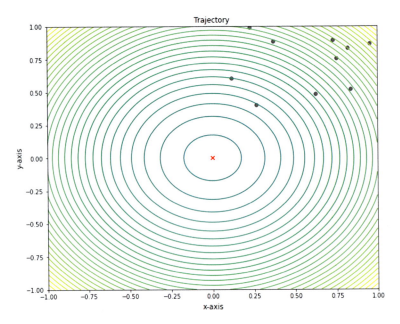

Fig. 51.4 2D depiction of particle swarm optimization

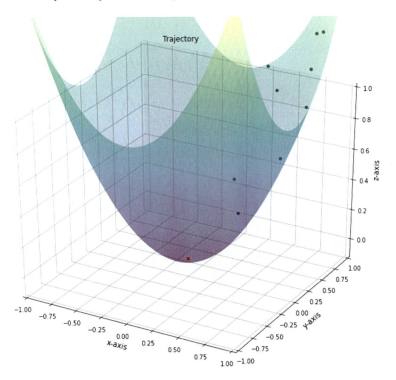

Fig. 51.5 3D depiction of particle swarm

51.4 Conclusions and Future Directions

In this work, PSO algorithm is implemented for swarm robotic system. In the virtual simulation environment, a swarm of 30 robots is considered, and PSO algorithm has been applied for finding the best global minimum position vector for each robot. Further, a physical prototype of swarm robotic system with two robots is built equipped with the sensors and actuators. The application considered is for the warehouse management system. The motive of creating a swarm of robots that save our time and go on to load and unload objects efficiently was achieved using the help of different materials. The robots essentially performed best while following the line and avoiding the object. All the simulations have been performed on Python. The physical prototype with sensors and RF communication demonstrates the feasibility of the proposed approach and provides a basis for further experimentation and improvement. Further research and experimentation are necessary to address the challenges and limitations of swarm robotics in warehouse management. In the near future, authors intend to extend this work for contemporary AI-based algorithms with complex swarm systems.

References

1. Kennedy, J., Eberhart, R.C.: Particle swarm optimization. In: Proceedings of IEEE International Conference on Neural Networks, pp. 1942–1948. IEEE Press, Piscataway, NJ (1995)
2. Dorigo, M., Stützle, T.: Ant Colony Optimization. MIT Press, Cambridge, MA (2004)
3. Di Caro, G., Dorigo, M.: AntNet: Distributed stigmergetic control for communications networks. J Artif. Intell. Res. **9**, 317–365 (1998)
4. Bonabeau, E., Dorigo, M., Theraulaz, G.: Swarm Intelligence: From Natural to Artificial System. Oxford University Press, New York (1999)
5. Deneubourg, J.-L., Aron, S., Goss, S., Pasteels, J.-M.: The self-organizing exploratory pattern of the argentine ant. J Insect Behav. **3**, 159–168 (1990)
6. Di Caro, G., Ducatelle, F., Gambardella, L.M.: AntHocNet: an adaptive nature-inspired algorithm for routing in mobile ad hoc networks. Eur. Trans. Telecommun. **16**(5), 443–455 (2005)
7. Dorigo, M., Maniezzo, V., Colorni, A.: Positive feedback as a search strategy. Technical Report 91–016, Dipartimento di Elettronica, Politecnico di Milano, Milan, Italy (1991). Revised version published as: Dorigo, M., Maniezzo, V., Colorni, A.: Ant system: optimization by a colony of cooperating agents. IEEE Trans. Syst. Man. Cybern. Part B **26**(1), 29–41 (1996)
8. Trianni, V., Dorigo, M.: Self-organisation and communication in groups of simulated and physical robots. J. Biol. Cybern. Arch. **95**(3), 213–231 (2006)
9. Dixit, P., Nargundkar, A., Suyal, P., Patil, R.: Intelligent warehouse automation using robotic system. In: Intelligent Systems and Applications: Select Proceedings of ICISA 2022, pp. 435–443. Springer Nature Singapore, Singapore (2023)
10. Socha, K, Dorigo, M.: Ant colony optimization for continuous domains. Eur. J. Oper. Res. (in press)
11. Adriansyah, A.: Xbee implementation on mini multi-robot system. In: The Proceedings of The 7th ICTS
12. Chaudhary, K., Prasad, A., Chand, V., Sharma, B.: ACO-Kinematic: a hybrid first off the starting block. PeerJ Comput. Sci. **8**, e905 (2022)

13. Chaudhary, K., Lal, G., Prasad, A., Chand, V., Sharma, S., Lal, A.: Obstacle avoidance of a point-mass robot using feedforward neural network. In: 2021 3rd Novel Intelligent and Leading Emerging Sciences Conference (NILES), pp 210–215. IEEE (2021)
14. Tawhid, M.A., Ibrahim, A.M.: An efficient hybrid swarm intelligence optimization algorithm for solving nonlinear systems and clustering problems. Soft Comput. 1–29 (2023)
15. Fitch, R., Lal, R.: Experiments with a ZigBee wireles communication system for self-reconfiguring modular robots. In: IEEE International Conference on Robotics and Automation, Kobe, Japan, May 2009, pp. 1947–1952
16. Benavidez, P., Nagothu, K., Ray, A.K., Shaneyfelt, T., Jamshidi, M.: Multi-domain swarm communication systems. In: International Conference on System of Systems Engineering, SoSE'08, Singapore, 2–4 June 2008, pp. 1–6
17. Lee, J.S., Huang, Y.-C.: ITRI ZB node: a Zigbee/IEEE 802.15.4 platform for wireless sensor networks. In: IEEE International. Conference on Systems, Man, Cybernetics, May 2006, pp. 1462–1467

Chapter 52
Application of Evolutionary Algorithms for Optimizing Wire and Arc Additive Manufacturing Process

Vikas Gulia and Aniket Nargundkar

Abstract Additive manufacturing is a recent trend in production processes owing to its sustainable approach. As a process in itself, additive manufacturing represents a more sustainable means of production as it eliminates the use of excess material and thus unnecessary waste. Wire and ARC additive manufacturing (WAAM) process is an important metal additive manufacturing process. The improvements in surface quality and dimensional accuracy are critical for WAAM. In the current work, two contemporary artificial intelligence (AI)-based algorithms, teaching learning–based optimization (TLBO) and particle swarm optimization (PSO), are applied for optimizing the five process parameters such as wire feeder, pulse voltage, frequency, pulse time, and welding speed with four objectives such as current, voltage, heat input, and width-to-reinforcement ratio which are considered as referred from Youheng et al. (Int. J. Adv. Manuf. Technol. 91:301–313, 2017). The results obtained with TLBO and PSO are comparable and improved by 24%, 32%, and 42% for the minimization of voltage, heat input, and maximization of width-to-reinforcement ratio, respectively. Both algorithms are observed to be robust. The convergence analysis for the algorithms is also discussed.

Keywords Wire and ARC additive manufacturing · Evolutionary algorithms · TLBO algorithm · PSO algorithm · Optimization

---

V. Gulia (✉) · A. Nargundkar
Symbiosis Institute of Technology, Symbiosis International (Deemed University), Lavale, Pune 412115, India
e-mail: vikas.gulia@sitpune.edu.in

A. Nargundkar
e-mail: aniket.nargundkar@sitpune.edu.in

© The Author(s), under exclusive license to Springer Nature Singapore Pte Ltd. 2024
P. K. Jha et al. (eds.), *Proceedings of Congress on Control, Robotics, and Mechatronics*, Smart Innovation, Systems and Technologies 364,
https://doi.org/10.1007/978-981-99-5180-2_52

52.1 Introduction

In recent years, industries are highly dependent on innovation and cutting-edge research owing to customer demands for low prices, and better quality (Abdulhameed et al. [1]). Researchers classified manufacturing processes into five categories as subtractive manufacturing, joining processes, dividing processes, transformative processes, and additive manufacturing (AM). AM is a process of making 3D objects by depositing material layer-on-layer directly from CAD model geometry (Sun et al. [2]). Adekanye et al. [3] categorized AM into various systems of operation such as powder-based bed process, extrusion-based process, sheet lamination process, directed energy deposition process, etc. Extensive research has been carried out for several applications in the field of AM. Rouf et al. [4] focussed on dentistry, orthopaedics, and food and textile industries. Nazir et al. [5] reviewed the design, optimization, and additive manufacturing of cellular structures. Velasco et al. [6] applied AM to design and fabricate bioinspired structures. Apart from the applications, huge work has been done on AM of various materials as well. Chen et al. [7] presented additive manufacturing for piezoelectric materials. Chaudhary et al. [8] surveyed additive manufacturing of polymer-derived ceramics. Madhavadas et al. [9] gave a perspective on metal additive manufacturing for intricately shaped aerospace components. In addition to the above context, Wire and ARC additive manufacturing (WAAM) process is an important metal 3D printing method (Liu et al. [10]). It produces large components (greater than 10 kg) from titanium, steel, aluminium (Kohler et al. [11]), and other metals. This process uses arc welding tools and wire as feedstock (Lin et al. [12]). The notable merits of WAAM are high deposition rates, excellent structural integrity, less material, equipment cost, etc. (Williams et al. [13]).

The typical process parameters which govern WAAM are wire feeder rate, pulse voltage, frequency, pulse time, and welding speed. The wire feeder in WAAM feeds the metal wire into the welding arc for creating a 3D part with layer-by-layer deposition of metal. Consistent and accurate wire feeding controls the precision of the amount of metal deposited with its location. Hence, this is critical for achieving product quality and reliability. Pulse voltage in WAAM refers to the voltage applied for creating an electric arc between the wire and the workpiece. The heat generated during WAAM process is mainly dependent on pulse voltage. Moreover, the droplet transfer mechanism is also observed to be governed by pulse voltage. It is observed from the literature that frequency and pulse time significantly affect the microstructure, surface finish, and mechanical properties of the deposited metal. In addition, pulse time controls the amount of heat input to the workpiece. Welding speed directly contributes to the productivity of the WAAM process. In the light of above, the optimization of such parameters is critical for reducing the cost and improving the product quality as in the case of conventional manufacturing processes. Various approaches have been adopted in the literature to optimize the WAAM process (Srivastava et al. [14] and Singh and Khanna [15]). However, the majority of the research work focuses on topology optimization (support structure optimization) (Karimzadeh and Hamedi

[16]). Recently, machine learning algorithms are also applied for the same. It is evident from the literature that the application of artificial intelligence (AI)-based evolutionary algorithms yields optimized results for conventional as well as advanced manufacturing processes (Gulia and Nargundkar [17], Nargundkar et al. [18] and Pansari et al. [19]). The current work focuses on applying teaching learning–based optimization (TLBO) and particle swarm optimization (PSO) algorithms for optimizing five process parameters viz. wire feeder, pulse voltage, frequency, pulse time, and welding speed. Four objectives such as Avg current, Avg voltage, heat input, and width-to-reinforcement ratio which improves productivity, reliability, and product quality are considered. The problem formulation is referred to by Youheng et al. [20].

The rest of the paper is organized as follows: Sect. 52.2 describes the problem formulation with mathematical functions. The TLBO and PSO algorithms are presented in Sect. 52.3. The solutions obtained for considered objectives with TLBO and PSO along with the comparison and convergence plots are discussed in Sect. 52.4. In the end, conclusions and future directions are mentioned in Sect. 52.5.

Article Highlights

1. TLBO and PSO algorithms are successfully applied for wire and arc additive manufacturing process.
2. Five process parameters such as wire feeder, pulse voltage, frequency, pulse time, and welding speed are optimized.
3. Minimization of voltage, heat input, and maximization of width-to-reinforcement ratio is achieved.
4. The results obtained with TLBO and PSO are comparable and improved by 24%, 32%, and 42% for the minimization of voltage, heat input, and maximization of width-to-reinforcement ratio, respectively.

52.2 Problem Formulation

The objective functions are referred from Youheng et al. [20] viz. Avg. Current (I), Avg. Voltage (U), Heat Input (Q), and Width-to-Reinforcement ratio (W/R) along with five variables such as wire feeder rate (x_1), pulse voltage (x_2), frequency (x_3), pulse time (x_4), and welding speed (x_5). Welding productivity is correlated to the melting rate of the filler material.

52.2.1 Average Current (I)

The mathematical function which is to be minimized is given in Eq. 52.1.

$$\begin{aligned}\text{Min } I = {} & 1444.64286 + 37.78869\,x_1 - 110.29167 x_2 + 1.36935\,x_3 \\ & - 81.875\,x_4 + 0.015\,x_5 - 1.02976\,x_1^2 + 1.8869\,x_2^2 \\ & - 0.00278\,x_3^2 + 3.5\,x_2\,x_4 \end{aligned} \tag{52.1}$$

52.2.2 Avg. Voltage (U)

The objective function for minimization of average voltage is described in Eq. 52.2.

$$\begin{aligned}\text{Min } U = {} & -150.74167 - 8.35\,x_1 + 5.3375\,x_2 + 0.69759\,x_3 + 29.35476\,x_4 \\ & + 0.02604\,x_5 - 0.0004\,x_3^2 - 1.37619\,x_4^2 - 0.00001\,x_5^2 \\ & + 0.02\,x_1\,x_3 + 0.9\,x_1\,x_4 + 0.003\,x_1\,x_5 \\ & - 0.015\,x_2\,x_3 - 0.5\,x_2\,x_4 - 0.065\,x_3\,x_4 - 0.00015\,x_3\,x_5 \end{aligned} \tag{52.2}$$

52.2.3 Heat Input (Q)

Distortion, mechanical properties, and crack resistance are primarily depending on the heat input as it directly affects the heat-affected zone (HAZ). Hence, the objective function of heat input is to be minimized and the mathematical expression is mentioned in Eq. 52.3.

$$\begin{aligned}\text{Min } Q = {} & 2000.81083 - 127.43792\,x_1 - 200.81633\,x_2 + 5.73756\,x_3 \\ & + 295.76633\,x_4 + 1.20401\,x_5 + 4.56062\,x_2^2 - 0.00856\,x_3^2 \\ & - 17.72177\,x_4^2 + 0.00078\,x_5^2 + 0.48615\,x_1\,x_3 + 21.819\,x_1\,x_4 \\ & - 0.0482\,x_2\,x_5 - 0.9583\,x_3\,x_4 - 0.0049\,x_3\,x_5 - 0.19315\,x_4\,x_5 \end{aligned} \tag{52.3}$$

52.2.4 Width-to-Reinforcement Ratio (W/R)

The larger value of width to reinforcement is desirable due to the interaction effect with surface roughness and strength. Hence, the objective function of width-to-reinforcement ratio (W/R) is to be maximized and the mathematical expression is mentioned in Eq. 52.4.

$$\text{Max } W/R = -25.06907 + 1.419\,x_1 - 3.08368\,x_2 + 0.40456\,x_3 \\ - 2.90343\,x_4 + 0.08288\,x_5 + 0.13881\,x_2^2 + 0.00002\,x_5^2 + 0.50819\,x_1\,x_4 \\ - 0.00464\,x_1\,x_5 - 0.01403\,x_2\,x_3 - 0.00262\,x_2\,x_5 \quad (52.4)$$

For all the above-mentioned objective functions, the lower limits and upper limits of design variables are considered as mentioned in Eqs. 52.5–52.9. These are referred to from Youheng et al. [20].

$$7 \leq x_1 \geq 9 \quad (52.5)$$

$$27 \leq x_2 \geq 29 \quad (52.6)$$

$$180 \leq x_3 \geq 220 \quad (52.7)$$

$$1.5 \leq x_4 \geq 2.5 \quad (52.8)$$

$$500 \leq x_5 \geq 700 \quad (52.9)$$

52.3 Methodology

52.3.1 Teaching Learning-Based Optimization Algorithm (TLBO)

Teaching learning–based optimization algorithm (TLBO) is a population-based socio-inspired optimization technique based on the teaching–learning process and proposed by Rao et al. [21, 22]. The algorithm imitates the teaching–learning model in the classroom. In this technique, the population size represents the students in the classroom, and variables are considered as various subjects/ courses allotted to the students. The output of the algorithm is in terms of learners' results which is described as the 'objective function value'. The best solution is represented as a teacher. Figure 52.1 represents flow chart of TLBO algorithm. For a detailed explanation of the algorithm, please refer to Rao et al. [22].

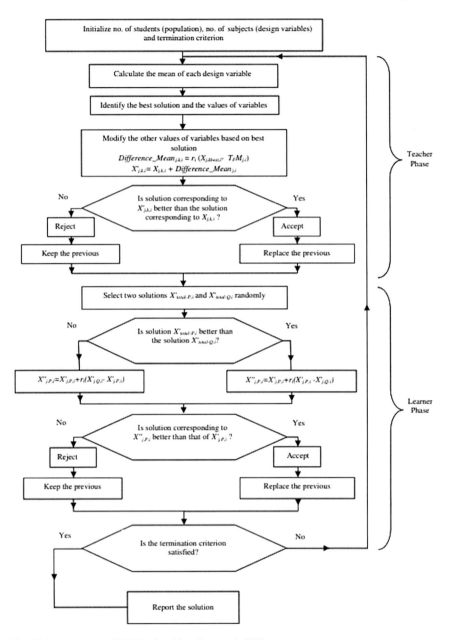

Fig. 52.1 Flow chart of TLBO algorithm (Rao et al. [22])

52.3.2 Particle Swarm Optimization (PSO) Algorithm

Particle swarm optimization (PSO) algorithm is a widely applied contemporary and derivative-free nature-inspired algorithm and was proposed by Kennedy and Eberhart [23]. In this technique, the search behaviour particles in a swarm is mathematically modelled, and the individual best solutions are referred to as 'p-best' (Personal best viz. local optimum), while the best solution in an entire swarm is considered as 'g-best' (Global best viz. Global optimum). Figure 52.2 shows flow chart of TLBO algorithm. For a detailed explanation of the algorithm, please refer to Jain et al. [24].

52.4 Results and Discussion

In this section, the results for four objective functions viz. Avg. Current (I), Avg. Voltage (U), Heat Input (Q), and Width-to-Reinforcement ratio (W/R) using TLBO and PSO algorithms are discussed. The problems are solved for 30 times, and the standard deviation (SD) is also been presented for every problem. The TLBO and PSO algorithms are coded in MATLAB R2017 on the Windows Platform with an Intel Core i3 processor and 4 GB RAM. Table 52.1 presents the control parameters set for the TLBO and PSO algorithms.

The solutions obtained with TLBO and PSO algorithms are presented in Tables 52.2 and 52.3, respectively. Every problem is solved 30 times, and the mean and best solutions are shown. The standard deviation (SD) is also mentioned in Table 52.2. The results are compared with Youheng et al. [20]. The optimum design variables are also indicated along with optimum objective function values. Moreover, the run time in seconds is also mentioned. It is important to note that the SD for TLBO and PSO is practically zero which indicates the robustness of these techniques.

It is observed from Tables 52.1, 52.2, and 52.3 that TLBO and PSO algorithms yield optimized results with adequate robustness. It is evident from Table 52.4 that results obtained with TLBO and PSO are improved by 24%, 32%, and 42% for the minimization of voltage, heat input, and maximization of width-to-reinforcement ratio, respectively. It is important to note that as evident from Table 52.4, the best objective function values are comparable for both TLBO and PSO algorithms. Hence, it can be concluded that both algorithms are capable of exploration and exploitation of feasible space for searching global optimum points.

Figure 52.3a–h presents the convergence plots for the four objective functions with TLBO and PSO algorithms. The jumping of solutions for the PSO algorithm is evident from Fig. 52.3b, d, f, h. This exhibits the exploration capability of the PSO algorithm which empowers it to explore and search the entire feasible space for global optimum without getting trapped into local optima. On the other hand, the quick convergence to the optimized solution is observed in the case of the TLBO algorithm according to Fig. 52.3a, c, e, g. This shows the exploitation capabilities of the TLBO algorithm due to which it can quickly scan, exploits the feasible space, and converges to the optimum value. Since PSO shows better exploration capabilities, the run time for it is higher as compared to the TLBO algorithm.

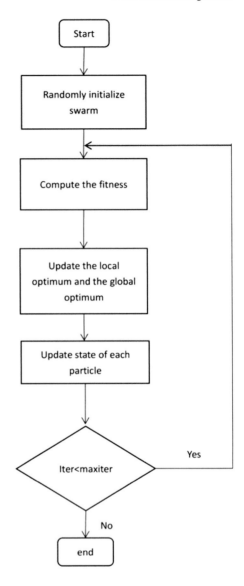

Fig. 52.2 Flow chart of PSO algorithm (Ren et al. [25])

Table 52.1 Control parameters and stopping criteria

Solution methodology	Parameter	Stopping criteria
TLBO	Population size = 100	Objective function value is less than 10^{-16}
	Generations = 20	
PSO	Inertia coefficient = Max 0.9 Min 0.2	
	Acceleration coefficient = 2	
	Generations = 50	

Table 52.2 Results for TLBO algorithm

Solution technique	TLBO			
Variables	Avg. current (I)	Avg. voltage (U)	Heat input (Q)	Width-to-reinforcement (W/R)
x_1	7	9	9	9
x_2	27.83	27	27	29
x_3	180	180	180	180
x_4	1.5	1.5	1.5	2.5
x_5	500	500	700	500
Best value	237.9151	16.1327	232.9109	6.3443
Run time (s)	0.08	0.09	0.16	0.12
x_1	7	9	9	9
x_2	27.81	27	27	29
x_3	180	180	180	180
x_4	1.51	1.5	1.5	2.5
x_5	500	500	700	500
Mean value	237.9149	16.1327	232.9109	6.3443
Run time (s)	0.11	0.12	0.18	0.14
SD (Std. deviation)	0.0023	0.0000	0.0000	0.0000

52.5 Conclusion

In this work, wire and ARC additive manufacturing (WAAM) process parameters are optimized with TLBO and PSO algorithms. The parameters considered are wire feeder rate, pulse voltage, frequency, pulse time, and welding speed. Four objectives defining the manufactured product quality in terms of minimal distortion and improved mechanical properties along with maximizing productivity are considered. The aforementioned objectives are current, voltage, heat input, and width-to-reinforcement ratio. The considered problems are observed to be nonlinear, multimodal complex problems. The results obtained with TLBO and PSO are improved by 24%, 32%, and 42% for the minimization of voltage, heat input, and maximization of width-to-reinforcement ratio, respectively, when compared with the existing results. The robustness of the algorithms is evident from the standard deviation. This demonstrates the successful application of TLBO and PSO algorithms for solving complex advanced manufacturing problems. In the near future, multi-objective and constrained optimization problems for WAAM can be solved.

Table 52.3 Results for PSO algorithm

Solution technique	PSO			
Variables	Avg. current (I)	Avg. voltage (U)	Heat input (Q)	Width-to-reinforcement (W/R)
x_1	7	9	9	9
x_2	27.83	27	27	29
x_3	180	180	180	180
x_4	1.5	1.5	1.5	2.5
x_5	500	500	700	500
Best value	237.9151	16.1327	232.9109	6.3443
Run time (s)	0.20	0.12	0.17	0.16
x_1	7	9	9	9
x_2	27.81	27	27	29
x_3	180	180	180	180
x_4	1.51	1.5	1.5	2.5
x_5	500	500	700	500
Mean value	237.9149	16.1327	232.9109	6.3443
Run time (s)	0.22	0.14	0.19	0.18
SD (Std. deviation)	0.0018	0.0000	0.0000	0.0000

Table 52.4 Comparison of solutions

Objective function/solution technique	TLBO algorithm	PSO algorithm	Results (Youheng et al. [20])
Avg. current (I)	237.9151	237.9151	221.8
Avg. voltage (U)	16.1327	16.1327	21.24
Heat input (Q)	232.9109	232.9109	341.42
Wire (W/R)	6.3443	6.3443	4.46
Avg. run time (s)	0.11	0.16	NA

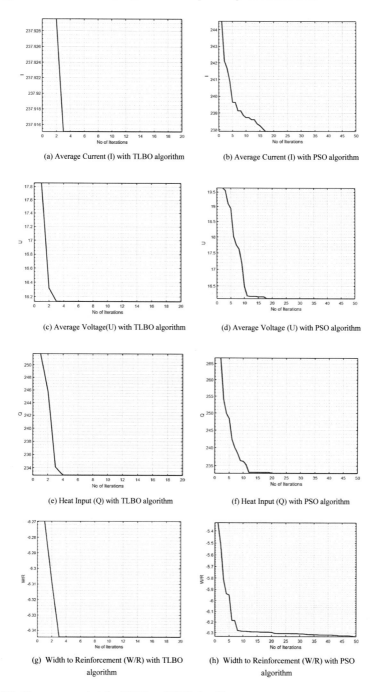

Fig. 52.3 Convergence plots for TLBO and PSO algorithms

References

1. Abdulhameed, O., Al-Ahmari, A., Ameen, W., Mian, S.H.: Additive manufacturing: challenges, trends, and applications. Adv. Mech. Eng. **11**(2), 1687814018822880 (2019)
2. Sun, C., Wang, Y., McMurtrey, M.D., Jerred, N.D., Liou, F., Li, J.: Additive manufacturing for energy: a review. Appl. Energy **282**, 116041 (2021)
3. Adekanye, S.A., Mahamood, R.M., Akinlabi, E.T., Owolabi, M.G.: Additive manufacturing: the future of manufacturing. Addit. Manuf. **709**, 715 (2017)
4. Rouf, S., Malik, A., Singh, N., Raina, A., Naveed, N., Siddiqui, M.I.H., Haq, M.I.U.: Additive manufacturing technologies: industrial and medical applications. Sustain. Oper. Comput. **3**, 258–274 (2022)
5. Nazir, A., Abate, K.M., Kumar, A., Jeng, J.Y.: A state-of-the-art review on types, design, optimization, and additive manufacturing of cellular structures. Int. J. Adv. Manuf. Technol. **104**, 3489–3510 (2019)
6. Velasco-Hogan, A., Xu, J., Meyers, M.A.: Additive manufacturing as a method to design and optimize bioinspired structures. Adv. Mater. **30**(52), 1800940 (2018)
7. Chen, C., Wang, X., Wang, Y., Yang, D., Yao, F., Zhang, W., Hu, D.: Additive manufacturing of piezoelectric materials. Adv. Func. Mater. **30**(52), 2005141 (2020)
8. Chaudhary, R.P., Parameswaran, C., Idrees, M., Rasaki, A.S., Liu, C., Chen, Z., Colombo, P.: Additive manufacturing of polymer-derived ceramics: materials, technologies, properties and potential applications. Prog. Mater. Sci. 100969 (2022)
9. Madhavadas, V., Srivastava, D., Chadha, U., Raj, S.A., Sultan, M.T.H., Shahar, F.S., Shah, A.U.M.: A review on metal additive manufacturing for intricately shaped aerospace components. CIRP J. Manuf. Sci. Technol. **39**, 18–36 (2022)
10. Liu, J., Xu, Y., Ge, Y., Hou, Z., Chen, S.: Wire and arc additive manufacturing of metal components: a review of recent research developments. Int. J. Adv. Manuf. Technol. **111**, 149–198 (2020)
11. Köhler, M., Fiebig, S., Hensel, J., Dilger, K.: Wire and arc additive manufacturing of aluminum components. Metals **9**(5), 608 (2019)
12. Lin, Z., Song, K., Yu, X.: A review on wire and arc additive manufacturing of titanium alloy. J. Manuf. Process. **70**, 24–45 (2021)
13. Williams, S.W., Martina, F., Addison, A.C., Ding, J., Pardal, G., Colegrove, P.: Wire+ arc additive manufacturing. Mater. Sci. Technol. **32**(7), 641–647 (2016)
14. Srivastava, M., Rathee, S., Tiwari, A., Dongre, M.: Wire arc additive manufacturing of metals: A review on processes, materials and their behaviour. Mater. Chem. Phys. 126988 (2022)
15. Singh, S.R., Khanna, P.: Wire arc additive manufacturing (WAAM): A new process to shape engineering materials. Mater. Today: Proc. **44**, 118–128 (2021)
16. Karimzadeh, R., Hamedi, M.: An intelligent algorithm for topology optimization in additive manufacturing. Int. J. Adv. Manuf. Technol. **119**(1–2), 991–1001 (2022)
17. Gulia, V., Nargundkar, A.: Optimization of process parameters of abrasive water jet machining using variations of cohort intelligence (CI). In: Applications of Artificial Intelligence Techniques in Engineering: SIGMA 2018, vol. 2 (pp. 467–474). Springer, Singapore (2019)
18. Nargundkar, A., Gulia, V., Khan, A.: Nano-abrasives assisted abrasive water jet machining of bio-composites–an experimental and optimization approach. J. Adv. Manuf. Syst. (2023)
19. Pansari, S., Mathew, A., Nargundkar, A.: An investigation of burr formation and cutting parameter optimization in micro-drilling of brass C-360 using image processing. In: Proceedings of the 2nd International Conference on Data Engineering and Communication Technology: ICDECT 2017 (pp. 289–302). Springer, Singapore (2019)
20. Youheng, F., Guilan, W., Haiou, Z., Liye, L.: Optimization of surface appearance for wire and arc additive manufacturing of Bainite steel. Int. J. Adv. Manuf. Technol. **91**, 301–313 (2017)
21. Rao, R.V., Kalyankar, V.D.: Parameters optimization of advanced machining processes using TLBO algorithm. EPPM, Singapore **20**(20), 21–31 (2011)
22. Rao, R.V., Savsani, V.J., Vakharia, D.P.: Teaching–learning-based optimization: an optimization method for continuous non-linear large scale problems. Inf. Sci. **183**(1), 1–15 (2012)

23. Eberhart, R., Kennedy, J.: Particle swarm optimization. In: Proceedings of the IEEE International Conference on Neural Networks, vol. 4 (pp. 1942–1948) (1995)
24. Jain, N.K., Nangia, U., Jain, J.: A review of particle swarm optimization. J. Inst. Eng. (India): Ser. B, **99**, 407–411 (2018)
25. Ren, G., Wen, S., Yan, Z., Hu, R., Zeng, Z., Cao, Y.: Power load forecasting based on support vector machine and particle swarm optimization. In: 2016 12th World Congress on Intelligent Control and Automation (WCICA) (pp. 2003–2008). IEEE (2016)

Chapter 53
Healthcare System Based on Body Sensor Network for Patient Emergency Response with Monitoring and Motion Detection

Maaz Ahmed, Diptesh Saha, Aditya Pratap Singh, Gunjan Gond, and S. Divya

Abstract Body sensor network (BSN) is a new technology. BSN care system begins by placing tiny, lightweight sensors on the patient's body that communicate with one another and the body-connected co-ordination node. This system is primarily concerned with measuring and estimating critical parameters such as ECG, temperature, and blood level. This real-time system focuses on a number of parameters, including patient health, motion detection and data transmission, and message transmission to the primary responder and hospital server. We use four types of sensors in this system: temperature sensor, pulse sensor, oxygen sensor, and fall detection sensor, which collect patient information and send it to the microcontroller. From there, the information is transferred to an android smartphone and server via the internet. We propose an IoT-based health system based on body sensors that meet the requirements effectively.

Keywords IoT · BSN · Healthcare

53.1 Introduction

The fact that people live longer than expected should be celebrated as one of the biggest victories in history. As they say, "Getting old is better than the alternative." We will see for the first time, more people of more than 65 years living on this planet than those of less than 5 years, apart from the increase in chronic diseases causing many seniors to have problems taking care of themselves. Therefore, providing a noble quality of life for seniors became a serious social issue at that time, given the contributions of these people to society. Many people want to live in the same place to live progressively and independently for a longer period. One possible way of getting

M. Ahmed (✉) · D. Saha · A. P. Singh · G. Gond · S. Divya
HKBK College of Engineering, Bangalore, India
e-mail: mz.maaz@gmail.com

this is to use connected smart devices that are collectively called "Internet of Things" (IoT) becoming a day-to-day reality at home. The devices on the Internet of Things can communicate with one another and operate as sensors and environment monitors using cloud-based software. They can also interpret information and do necessary tasks, such as managing the temperature, windows and doors, and reminding people.

53.2 Related Work

According to world population record of 2013, number of older population has increased in the developed nations. Two-thirds of total population of old people are living in developed nation. Older population in less developed regions is growing faster. Conclusion shows that up to 2050, nearly 8 in 10 of the world's older population will live in the less developed regions.

Code-Blue is popular healthcare research project based on BSN healthcare which was developed at Harvard sensor network. It is a health monitoring infrastructure made for continuous patient monitoring. This system provides security for all wireless devices, personal computer which may be used to monitor and treat patients. Code-Blue supports filtration and aggregation of events as they flow through network. Code-Blue can operate on multiple devices. This system lacks security also require more data storage [1].

In [2], the author presents a system which uses removable media or also uses mobile phone for storing patient biomedical data which is encrypted form, which can be accesses by authorized medical staff only for further consultation. Here, the information is stored on removable disk or data storage so it can be accessed even device is not working. The system may be implemented in Java. This system is implemented with the currently available and relatively cheap technology. This system is more secure than Code-Blue. This system requires more memory as well as power consumption.

Alarm-net was designed at the University of Virginia. In this, health monitoring of patient in assisted living and home environment can be done. It is the network of MICAZ sensors target gateways. This system consists of IPAQ, PDA'S, and PC's. Here in this, customized dust sensors and integrated temperature, light, pulse and body oxygelation sensors are present. Alarm-net facilitates network and data security for environment. Alarm-net is susceptible to confidentiality attack which leaks resident's location [3].

Union [4] was proposed in the Department of Computing, Imperial College, London. This system includes context awareness aspect. Here the continuous monitoring of patient is done under physiological parameter. This system consists of biosensors which continuously gather, collect, and analyze data. Due to context awareness aspect, it enhances the capturing of any clinical relevant episode.

Medicare [5] was first designed by Chakravorty in 2006. This is the healthcare system which is having full control on human body via network of sensors. Due to this, health provider have control, access on patient body on always on basis. In this systeam, physician can have time access, review, update, and send patient data from wherever they want. Continuous supervising of patient is done here. Medicare is an ongoing project and much work is left. There are many security issues such as secure localization.

Medisn [6] is a network that monitors physiological data from patients in hospitals and during disasters. This system consists of physiological monitors (PM) which collects, encrypt, and sign patient's data before transmitting them to a network of relay points (RPs). Medisn also comes with a backend server.

In [7], S. Pai et al., provide an overview of sensor networks, which are used to monitor physical phenomena and are made up of numerous small, independent wireless devices with limited memory and energy. They can analyze the data they collect to find important information about events, things, and people. In-home patient care, environmental assessment and research, disaster mitigation, energy demand and response, inventory monitoring, surveillance, and law enforcement are just some of the applications for which sensor networks, a new technology, are expected to play a significant role in monitoring individuals, things, and infrastructure.

In [8], K. Lorincz et al., have designed an algorithm "Code-Blue" in its preliminary stages for critical care environment that is rapidly changing.

Security is a top priority for healthcare applications, especially when it comes to protecting patient's privacy in the event that the patient has an embarrassing condition. The security and privacy concerns surrounding WMSN-based healthcare applications are the subject of this paper [9]. Here, the authors discuss about the security of some well-known healthcare projects that use wireless medical sensor networks.

In [10], P. Gope and T. Hwang have proposed a distributed IoT system architecture. They have also presented an anonymous authentication method that has the potential to guarantee some of the notable properties, including sensor anonymity, sensor untraceability, and resistance to replay attacks, cloning attacks, and so forth.

In [11], the authors aim to use technology to improve the quality of dementia care, strengthen internal safety monitoring at care organizations, and improve the professional judgment of caregivers. To give caregivers more flexibility and the ability to respond in real time, an eXtensible Markup Language-based dementia assessment system that combines program code and assessment content is used.

Numerous significant algorithmic issues that arise in mobile and wireless networks have been discussed in [12].

In [13], M. Tentori, J. Favela, and M. D. Rodriguez have illustrated privacy-aware facilities incorporated into Simple Agent Library for Smart Ambient (SALSA). This application enables hospital workers to communicate through contextual messages and provides pertinent information to them based on contextual information like their locations and roles. Utilizing privacy-aware agents, the privacy concerns have been addressed that potential users had regarding this application.

In [14], J. Misic has developed two algorithms. The first algorithm generates the session key by relying on a central trusted security server (CTSS) to confirm that participants are indeed members of the patient's group. Participants in the second algorithm use certificates to verify one another and are largely independent of CTSS.

A flexible cryptographic key management solution is proposed in this paper [15] to facilitate the interoperability of the utilized cryptographic mechanisms in order to comply with HIPAA regulations. Furthermore, instance of agree special cases planned to work with crisis applications and other potential exemptions can likewise be taken care of without any problem.

53.3 Methodology

Using sensors that may be worn on the body, which can monitor a person's health status and give the user as well as the doctor timely insights into various health metrics, is one way to identify patients more quickly without reducing the distance between the patient and a hospital. Utilizing the patient's anamnesis in addition to the measurements provided by this health monitoring system is a crucial step in determining whether or not a visit to the specialist is essential for his or her symptoms because every patient is completely unique. The symptoms of the virus include fever, tiredness, shortness of breath, etc. Here, we frequently discuss the COVID patient's health monitoring system.

Figure 53.1 shows the block diagram of the device. It consists of the embedded model consist of Arduino micro-controller, temperature sensor is connected to controller to measure the patient's body temperature and pulse sensor to measure the pulse rate of the patient for checking patient's condition, blood oxygen level sensor measures the oxygen level in blood, all these data are stored in cloud and monitoring will be done by mobile application. And data stored in cloud can also be accessed by the web page. There will a frame where the motors are fixed for the movement of the two sheets for the open and close for the ventilator. In between the two sheets, the oxygen can is kept, so when the two sheets while closing the oxygen can get compressed and the oxygen is supplied to the patients when the sheets opening the air is again.

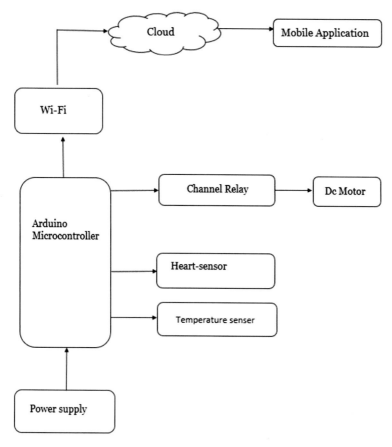

Fig. 53.1 Healthcare system

53.4 Results and Analysis

This tele-medical system primarily focuses on the measurement of critical parameters such as ECG, temperature, blood level, and so on. Body area network and an Android smartphone are used in this system. The system includes additional patient localization and data storage, movement of patient details and data transmission, and alert messages to first responders and hospital servers. Data are transferred to the coordinator node via Bluetooth to an Android-based smartphone in both approaches. We propose creating an automated epileptic patient monitoring system that detects vital statistics such as temperature, saline, and so on. It also detects seizures/epilepsy and sends vital information to doctors via SMS. Detects the patient's state (conscious, death, or sleep) based on their motion, and update alarms as needed. System maintains a complete log of patient activity on a PC at the doctor's office (customized software will be developed for this purpose), which also generates audiovisual alarms for nurses at the patient's end.

53.5 Conclusion

This paper completely meets the primary requirements of dependability, scalability, and range. We discovered an appropriate automated epileptic patient monitoring system that can detect patients' vital statistics such as temperature, saline, and so on, as well as detect when a patient is having seizures/epilepsy. The WBAN and Android smartphone are used in tandem for data collection, storage, and analysis. Security and privacy must be investigated further at all layers of the system, particularly to define trade-offs in terms of performance and comfort of use. This system can be linked to other surveillance systems for enhanced security and remote location management. We used the customized algorithm in this paper. We proposed an IoT-based healthcare system that uses body sensors to efficiently meet requirements.

References

1. Gope, P., Hwang, T.: BSN-care: a secure IOT-based modern healthcare system using body sensor network. IEEE Sens. J. **16**(5), 1368–1376 (2015)
2. Chen, C.-L., Zhan, Y., Xuan, Z.: Medical record encryption storage system based on Internet of Things. Wirel. Commun. Mob. Comput. **2021**, 2109267 (2021)
3. Selavo, L., Wu, Y., Cao, Q., Lin, S., Fang, L., Stankovic, J., He, Z., Wood, A., Virone, G., Doan, T.: Alarm-net: wireless sensor networks for assisted-living and residential monitoring (2006)
4. More, P., Memane, K., Londhe, T., Thanki, H.: Advance IoT-based BSN healthcare system for emergency response of patient with continuous monitoring and motion detection. Int. J. Mod. Trends Sci. Technol. **2**(12) (2016)
5. Chakravorty, R. et al.: Mobicare: a programmable service architecture for mobile medical care. In: Proceedings of UbiCare, pp. 532–536 (2006)
6. Ko, J., Lim, J.H., Chen, Y., Musvaloiu-E, R., Terzis, A., Masson, G.M., Gao, T., Destler, W., Selavo, L., Dutton, R.P.: Medisn: medical emergency detection in sensor networks. ACM Trans. Embed. Comput. Syst. (TECS) **10**(1), 1–29 (2010)
7. S. Pai, M. Meingast, T. Roosta, S. Bermudez, S. Wicker, D. K. Mulligan, and S. Sastry, "Confidentiality in sensor networks: Transactional information," *IEEE Security and Privacy magazine*, 2008.
8. K. Lorincz, D. J. Malan, T. R. Fulford-Jones, A. Nawoj, A. Clavel, V. Shnayder, G. Mainland, M. Welsh, and S. Moulton, "Sensor networks for emergency response: challenges and opportunities," *IEEE pervasive Computing*, vol. 3, no. 4, pp. 16–23, 2004.
9. P. Kumar and H.-J. Lee, "Security issues in healthcare applications using wireless medical sensor networks: A survey," *sensors*, vol. 12, no. 1, pp. 55–91, 2011.
10. Gope, P., Hwang, T.: Untraceable sensor movement in distributed IoT infrastructure. IEEE Sens. J. **15**(9), 5340–5348 (2015)
11. Lin, C., et al.: "A Healthcare Integration System for Disease Assessment and Safety Monitoring of Dementia Patients", Trans. Info. Tech. Biomedicine **12**, 579–586 (2008)
12. Boukerche, Azzedine, ed. *Algorithms and protocols for wireless and mobile ad hoc networks.* John Wiley & Sons, 2008.
13. Tentori, M., Favela, J., Rodriguez, M.D.: Privacy Aware Autonomous Agents for Pervasive Healthcare. IEEE Intelligent Sys **21**, 55–62 (2006)

14. Misic, J.: Enforcing patient privacy in healthcare WSNs using ECC implemented on 802.15.4 Beacon Enabled Clusters," Proc. 6th Annual IEEE Int'l. Conf. Pervasive Comp. Communication., 2008
15. Lee, W.B., Lee, C.D.: A cryptographic key management solution for HIPAA privacy regulations. IEEE Trans. Inf Technol. Biomed. **12**(1), 34–41 (2008)

Printed in the United States
by Baker & Taylor Publisher Services